Problems & Solutions in

NONRELATIVISTIC

QUANTUM

MECHANICS

Problems & Solutions

in

NONRELATIVISTIC

QUANTUM

MECHANICS

Anton Z. Capri

Department of Physics
University of Alberta, Canada

World Scientific
New Jersey • London • Singapore • Hong Kong

Published by

World Scientific Publishing Co. Pte. Ltd.

5 Toh Tuck Link, Singapore 596224

USA office: 27 Warren Street, Suite 401-402, Hackensack, NJ 07601

UK office: 57 Shelton Street, Covent Garden, London WC2H 9HE

Library of Congress Cataloging-in-Publication Data
Capri, Anton Z.
    Problems & solutions in nonrelativistic quantum mechanics / Anton Z. Capri.
        p. cm.
    Includes bibliographical references.
    ISBN-13 978-981-02-4633-4 (alk. paper) -- ISBN-10 9810246331 (alk. paper)
    ISBN-13 978-981-02-4650-1 (pbk.: alk. paper) -- ISBN 9810246501 (pbk.: alk. paper)
    1. Nonrelativistic quantum mechanics -- Problems, exercises, etc. I. Title: Problems and
solutions in nonrelativistic quantum mechanics. II. Title: Nonrelativistic quantum
mechanics. III. Title.
    QC174.24.N64 C374 2002
    530.12'076--dc21   2002029614

British Library Cataloguing-in-Publication Data
A catalogue record for this book is available from the British Library.

To Skaidrite, who knows that physics is simple because "everything equals zero".

Preface

Soon after the first edition of *Nonrelativistic Quantum Mechanics* appeared, I received numerous requests for solutions to the problems in that book. To remedy this situation I started by writing out solutions to the more difficult problems, but as I proceeded with the third edition of *Nonrelativistic Quantum Mechanics* I also revised some of the problems and added quite a few others. Since in constructing these new problems I had to solve them, in the first place, to be sure that they were indeed problems that students could solve, I finally went on to write out solutions to all the problems. However, I did not simply want a compendium of solutions of the Schrödinger equation since with programs such as Maple or Mathematica these solutions are accessible to every student. Instead I wanted to concentrate on problems that teach quantum mechanics. It is with this in mind that I began to collect and solve problems. My idea was to provide a means for students to learn quantum mechanics by "doing it". This is why the book begins with extremely simple problems and progresses to more difficult ones.

Some of the problems extend results that are usually taught in a course on quantum mechanics. But, by having the students obtain the results themselves they are more likely to retain the ideas and at the same time gain confidence in their own abilities.

As usual, I tested most of these problems on my students. Sometimes they came up with very original ways of looking at old problems. I have learned a lot from my students. It is this learning process that led me to occasionally introduce more than one way of solving a problem since the solutions are intended to help students to obtain a better understanding of the techniques involved in tackling problems in quantum mechanics.

The notation and methods used are those explained in *Nonrelativistic Quantum Mechanics* and I frequently refer to chapters from that book. The chapter headings are also the same as in *Nonrelativistic Quantum Mechanics*. Nevertheless, the present book is independent and should serve as a companion to any of the numerous excellent books on quantum mechanics. Throughout the book I have used Gaussian units since these are the units most commonly used in atomic physics. I also tried to arrange the problems according to increasing degree of difficulty. This, was not always possible since it would have meant losing the possibility of arranging them according to topic.

It is a pleasure to thank Professor M. Razavy for his generous help in, not only providing me with some wonderful problems and supplying me with numerous references, but also for his constant moral support.

Of course the students who suffered through the courses in which I subjected them to all sorts of quantum problems also deserve my heartfelt thanks. To their credit, the undergraduates seldom complained. On the other hand, there was many an evening, after I had assigned some more than usually difficult problems in the graduate course on quantum mechanics, that walking down the hall of the fourth floor of the physics building I heard my name muttered with less than flattering epithets. Nevertheless, the graduate students survived and many, after they completed their degree, even thanked me for what they had learned.

It is my hope that these problems and solutions will be of use to future generations of physics students. At any rate they should provide more entertainment than solving cross word puzzles.

A.Z.Capri
Edmonton, Alberta
July, 2002.

Contents

11 Transformation Theory   **196**
 11.1 Fourier Transform of Hermite Functions 196
 11.2 Schrödinger Equation in Momentum Space 197
 11.3 Heisenberg Equation for a Free Particle 198
 11.4 Dirac Picture for Displaced SHO 199
 11.5 Heisenberg Picture for Displaced SHO 202
 11.6 Heisenberg Picture: SHO and Constant Force 203
 11.7 Heisenberg Picture: Constant Force 204
 11.8 Schrödinger Picture: Constant Force 205
 11.9 Dirac Picture: Constant Magnetic Field 206
 11.10 Coherent State 208
 11.11 Coherent State: Overlap of Two States 208
 11.12 Coherent State: Wavefunction 209
 11.13 Squeezing Operator 210
 11.14 No Eigenstates for Creation Operator 211
 11.15 Spin Coherent State: Euler Angles 212
 11.16 Minimum Uncertainty Spin Coherent States 214
 11.17 Spin Coherent States: Complex Variables 216
 11.18 Useful Commutator 219
 11.19 Forced SHO 219
 11.20 3-D Simple Harmonic Oscillator 221
 11.21 Quadrupole Tensor 221
 11.22 Eigenfunction of J^2, L^2, and J_z 223
 11.23 SHO: A Time-independent Operator 225
 11.24 SHO with Time-Dependent Spring 226

12 Non-degenerate Perturbation Theory   **232**
 12.1 Expansion of $1/|\vec{r}_1 - \vec{r}_2|$ 232
 12.2 Second Order Correction to State 234
 12.3 $1/2\,\lambda x^2$ Perturbation of SHO 235
 12.4 $1/4\,\lambda x^4$ Perturbation of SHO 237
 12.5 $1/4\,\lambda x^4$ - Brillouin-Wigner Perturbation 238
 12.6 Two-level System 239
 12.7 Approximate SHO 242
 12.8 Two-dimensional SHO 243
 12.9 Kuhn-Thomas-Reiche Sum Rule 246
 12.10 Electron in Box Perturbed by Electric Field 248
 12.11 Positronium 249
 12.12 Rigid Rotator in Electric Field 250
 12.13 Electric Dipole Moment Sum Rule 251
 12.14 Another Sum Rule 253
 12.15 Gaussian Perturbation of SHO Bosons 254
 12.16 Gaussian Perturbation of SHO Fermions 255
 12.17 Polarizability: Particle in a Box 256
 12.18 Atomic Isotope Effect 258
 12.19 Relativistic Correction to H atom 260

Chapter 1

The Breakdown of Classical Mechanics

1.1 Quantum Number of the Earth

Calculate the principle quantum number for the earth in its orbit about the sun. What is the energy difference between two neighbouring energy levels? Hint: For large n, $E_n \approx E_{classical}$.

Solution

The classical energy of the earth in its orbit about the sun is

$$E = -\frac{GmM}{2R} \tag{1.1.1}$$

where
m is the mass of the earth $= 6 \times 10^{27}$ g ,
M is the mass of the sun $= 2 \times 10^{33}$ g ,
G is the gravitational constant $= 6.67 \times 10^{-8}$ dyn cm^2/g^2 and
R is the sun-earth distance $= 1.5 \times 10^{13}$ cm .
Proceeding as for the hydrogen atom we find that we need only replace e^2 by GmM. Thus, we find

$$E = -\frac{1}{2} \left(\frac{2\pi}{h} GmM \right)^2 m\frac{1}{n^2} . \tag{1.1.2}$$

Solving for n we get

$$n^2 = R \left(\frac{2\pi}{h} \right)^2 GMm^2 = 18 \times 10^{146} . \tag{1.1.3}$$

Therefore,

$$n \approx 4 \times 10^{73} . \tag{1.1.4}$$

1

Also we see that since

$$\Delta E = \frac{1}{2}\left(\frac{2\pi}{h}GmM\right)^2 m\frac{\Delta n}{n^3} = |E|\frac{\Delta n}{n} \qquad (1.1.5)$$

we get $\Delta E = 6.4 \times 10^{-34}$ erg .

1.2   Thermal Wavelength

What is the wavelength associated with gas molecules at a temperature T? Estimate this wavelength for a typical gas at room temperature and compare it to visible light.

Solution

The energy U of a molecule at temperature T is given by the equipartition principle

$$U = \frac{3}{2}k_BT = \frac{hc}{\lambda} . \qquad (1.2.6)$$

Therefore, the wavelength is

$$\lambda = hc/U . \qquad (1.2.7)$$

At room temperature $T \approx 300$ K. So,

$$U = 1.5 \times 1.38 \times 10^{-23} \times 300 \text{ J} = 6.21 \times 10^{-21} \text{ J} . \qquad (1.2.8)$$

$$\lambda = \frac{6.63 \times 10^{-34} \times 3.00 \times 10^8}{6.21 \times 10^{-21}} = 3.20 \times 10^{-5} \text{ m} . \qquad (1.2.9)$$

This is considerably longer than visible light which has a wavelength of about 5×10^{-7} m .

1.3   Photons in a Beam

For a monochromatic beam of electromagnetic radiation of wavelength ($\lambda \approx 5000$ Å) , intensity $I = 1$ watt/m^2, calculate the number of photons passing an area of $A = 1$ cm^2 normal to the beam in one second.

Solution

The energy of a photon is given by

$$E_{\text{photon}} = h\nu = \frac{hc}{\lambda} . \qquad (1.3.10)$$

The total energy of the beam is $E = IA = N E_{\text{photon}}$ where N is the number of photons. Therefore,

$$N = \frac{IA\lambda}{hc} = \frac{1 \times 10^{-4} \times 5000 \times 10^{-10}}{6.63 \times 10^{-34} \times 3.00 \times 10^8} = 2.5 \times 10^{14} \text{ photons} . \qquad (1.3.11)$$

1.4 Hydrogen Atom and de Broglie

Show that if one assumes that the circumference of a stationary state orbit of an electron in a hydrogen atom is an integral multiple of the de Broglie wavelength, one also obtains the correct energy levels.

Solution

The de Broglie wavelength is given by

$$\lambda = h/p .$$

The circumference is $2\pi r$. Thus, we write

$$2\pi r = n\lambda .$$

The energy is given by

$$E = \frac{p^2}{2m} - \frac{e^2}{r} . \tag{1.4.12}$$

We substitute for p and r in terms of λ and introduce the fine structure constant

$$\alpha = \frac{2\pi e^2}{hc} \tag{1.4.13}$$

to get

$$E = \frac{h^2}{2m\lambda^2} - \frac{hc\alpha}{n\lambda} . \tag{1.4.14}$$

We now equate the Coulomb force with the mass times the centripetal acceleration

$$\frac{mv^2}{r} = \frac{2\pi e^2}{n\lambda} . \tag{1.4.15}$$

This yields

$$\frac{h^2}{m\lambda^2} = \frac{2\pi e^2}{n\lambda} . \tag{1.4.16}$$

We can now solve for λ.

$$\frac{1}{\lambda} = \frac{m\alpha c}{h} \frac{1}{n} . \tag{1.4.17}$$

Combining this with our previous result we find

$$E = \frac{m\alpha^2 c^2}{2n^2} - \frac{m\alpha^2 c^2}{n^2} = -\frac{1}{2} \frac{m\alpha^2 c^2}{n^2} . \tag{1.4.18}$$

1.5   Vibrations in NaCl

The shortest possible wavelength of sound in sodium chloride is twice the lattice spacing, about 5.8×10^{-8} cm. The sound velocity is approximately 1.5×10^5 cm/sec.

a) Compute a rough value for the highest sound frequency in the solid.

b) Compute the energy of the corresponding phonons, or quanta of vibrational energy.

c) Roughly what temperature is required to excite these oscillations appreciably?

Solution

The shortest wavelength is given by

$$\lambda_{min} = 5.8 \times 10^{-8} \text{ cm} = 2 \times \text{lattice spacing} \tag{1.5.19}$$

$$v_0 = 1.5 \times 10^5 \text{ cm/s} . \tag{1.5.20}$$

a)

$$\nu = \frac{v_0}{\lambda} . \tag{1.5.21}$$

Therefore,

$$\nu_{max} = \frac{v_0}{\lambda_{min}} = 2.6 \times 10^{12} \text{ Hz} . \tag{1.5.22}$$

b)

$$E_{max} = h\nu_{max} = 1.7 \times 10^{-21} \text{ J} = 0.011 \text{ eV} . \tag{1.5.23}$$

c)

$$u = \frac{3}{2} k_B T . \tag{1.5.24}$$

Assume $u \approx E_{max}$. Therefore,

$$T \approx \frac{2E_{max}}{3k_B} = 83 \text{ K} . \tag{1.5.25}$$

1.6   Crystal Powder

Estimate the effect on the specific heat of reducing a crystal to a fine powder of dimensions of about 10^{-5} cm.

Hint: Study the Debye model [1.1] of specific heat and realize that the size of the crystal now also imposes an upper limit on the wavelength of the sound waves in the crystal.

Solution

The effect of grinding up the crystal into a powder is to limit the maximum wavelength of a standing wave in the crystal to roughly the size of the crystal particles. This changes the integral in the Debye expression [1.1] for the internal energy from

$$U = \frac{12\pi(k_BT)^4}{h^3v_0^3} \int_0^{x_{max}} \frac{x^3\,dx}{e^x - 1} \tag{1.6.26}$$

to

$$
\begin{aligned}
U_{\text{powder}} &= \frac{12\pi(k_BT)^4}{h^3v_0^3} \int_{x_{min}}^{x_{max}} \frac{x^3\,dx}{e^x - 1} \\
&= U - \frac{12\pi(k_BT)^4}{h^3v_0^3} \int_0^{x_{min}} \frac{x^3\,dx}{e^x - 1} \, .
\end{aligned}
\tag{1.6.27}
$$

Here, we have introduced

$$x_{min} = \frac{h}{k_BT}\nu_{min} = \frac{h}{k_BT}\frac{v_0}{\lambda_{max}} \tag{1.6.28}$$

where $\lambda_{max} = 10^{-5}$ cm $\approx$ size of the powder particles. If we now estimate x_{min} at room temperature by using that $v_0 \approx 500$ m/s , we get $x_{min} \approx 8\times10^{-4} << 1$. Therefore, we can approximate the exponential in the last integral by $e^x \approx 1+x$ and get

$$\int_0^{x_{min}} \frac{x^3\,dx}{e^x - 1} \approx \int_0^{x_{min}} x^2\,dx = \frac{1}{3}x_{min}^3 \, . \tag{1.6.29}$$

This means that the internal energy U is reduced by

$$\Delta U = \frac{12\pi(k_BT)^4}{h^3v_0^3} \times \left(\frac{1}{3}\frac{hv_0}{k_B\lambda_{max}}\right)^3 = \frac{4\pi k_BT}{\lambda_{max}^3} \, . \tag{1.6.30}$$

Therefore, the specific heat per unit volume at constant volume

$$\frac{1}{V}\frac{\partial U}{\partial T}$$

is decreased by a constant amount, namely

$$\frac{4\pi k_B}{\lambda_{max}^3} = \frac{4\pi \times 1.38 \times 10^{-23}\,\text{J/K}}{(10^{-7})^3\,\text{m}^3} \approx 0.173\,\text{J/(K m}^3) \, . \tag{1.6.31}$$

1.7 Einstein Coefficients

For a collection of atoms with energies E_n , $n = 1, 2, 3, \ldots$ submerged in a background of radiation at a temperature T, the following transitions may occur:

1) spontaneous from $n \to m$ $E_n > E_m$

2) induced from $n \to m$

3) induced from $m \to n$.

At equilibrium, at a temperature T, the emission and absorption probabilities are given by

$$P_{mn} = K N_n [B_{mn} \rho(\nu) + A_{mn}] \quad \text{emission} \tag{1.7.32}$$

$$P_{nm} = K N_m [B_{nm} \rho(\nu)] \quad \text{absorption .} \tag{1.7.33}$$

Here, $h\nu = E_n - E_m$, K is a proportionality constant, and N_n, N_m are respectively the number of atoms in the states n and m . The coefficients A_{mn}, B_{mn} are known respectively as the "Einstein Coefficients" of spontaneous and induced emission, whereas the coefficient B_{nm} is known as the "Einstein Coefficient" of induced absorption. [1.2] Use these equations together with Planck's radiation law for the radiation density $\rho(\nu)$ at equilibrium to show that
1) the Einstein coefficients of induced absorption and emission are the same, that is that $B_{nm} = B_{mn}$ and that
2) the Einstein coefficients of spontaneous and induced emission are related by

$$A_{nm} = \frac{8\pi}{c^3} h\nu^3 B_{nm} \quad .$$

Solution

In equilibrium, at a temperature T, if the number of atoms in the state n and m is given by N_n and N_m respectively, we have that

$$N_n = N e^{-E_n/k_B T} \quad , \quad N_m = N e^{-E_m/k_B T} \tag{1.7.34}$$

where N is the total number of atoms. Therefore,

$$N_m = N_n e^{(E_n - E_m)/k_B T} = N_n e^{h\nu/k_B T} \quad . \tag{1.7.35}$$

Also at equilibrium the number of transitions from $n \to m$ equals the number of transitions from $m \to n$. Thus, we have

$$N_m P_{nm} = N_n P_{mn} \tag{1.7.36}$$

or

$$N_n e^{h\nu/k_B T} B_{mn} \rho(\nu) = N_n [B_{nm} \rho(\nu) + A_{nm}] \quad . \tag{1.7.37}$$

Thus, solving for the radiation density ρ we get

$$\rho(\nu) = \frac{A_{nm}/B_{mn}}{e^{h\nu/k_B T} - B_{nm}/B_{mn}} \quad . \tag{1.7.38}$$

But, at equilibrium the radiation density is given by Planck's Law

$$\rho(\nu) = \frac{8\pi}{c^3} \frac{h\nu^3}{e^{h\nu/kT} - 1} \tag{1.7.39}$$

Therefore, comparing these two equations we see that we have

$$B_{nm} = B_{mn} \qquad (1.7.40)$$

and

$$A_{nm} = \frac{8\pi}{c^3} h\nu^3 B_{nm} \ . \qquad (1.7.41)$$

Bibliography

[1.1] A.Z. Capri, *Nonrelativistic Quantum Mechanics* 3rd edition, World Scientific Publishing Co. Pte. Ltd., section 1.12, (2002) .

[1.2] F.K. Richtmyer, E.H. Kennard, and J.N. Cooper, *Introduction to Modern Physics*, 6th edition, sec 13.12, McGraw-Hill, New York, (1969).

Chapter 2

Review of Classical Mechanics

2.1 Lagrangian and Hamiltonian for SHO

Find the Lagrangian for a harmonic oscillator. Use the definition of conjugate momentum p to find it and also the Hamiltonian H.

Solution

For a simple harmonic oscillator we have

$$
\begin{aligned}
T &= \frac{1}{2}mv^2 \\
V &= \frac{1}{2}kx^2 \ .
\end{aligned}
\tag{2.1.1}
$$

Therefore,

$$
L = T - V = \frac{1}{2}mv^2 - \frac{1}{2}kx^2 \ .
\tag{2.1.2}
$$

Next we compute the canonical momentum

$$
p = \frac{\partial L}{\partial v} = mv \ .
\tag{2.1.3}
$$

Then, we can substitute p/m for v and get

$$
H = \frac{\partial L}{\partial v}v - L = \frac{p^2}{2m} + \frac{1}{2}kx^2 \ .
\tag{2.1.4}
$$

2.2 Lagrangian and Hamiltonian: Simple Pendulum

Repeat problem 2.1 for the simple pendulum. Interpret the momentum p conjugate to the angle variable θ .

Solution

Here, the kinetic energy is given by

$$T = \frac{1}{2}mv^2 = \frac{1}{2}m\left(\dot{r}^2 + r^2\dot{\theta}^2\right) . \tag{2.2.5}$$

But, $r = l = $ constant. Therefore,

$$T = \frac{1}{2}ml^2\dot{\theta}^2 . \tag{2.2.6}$$

The potential energy is

$$V = mgl(1 - \cos\theta) . \tag{2.2.7}$$

So,

$$L = T - V = \frac{1}{2}ml^2\dot{\theta}^2 - mgl(1 - \cos\theta) . \tag{2.2.8}$$

The conjugate momentum is given by

$$p_\theta = \frac{\partial L}{\partial\dot{\theta}} = ml^2\dot{\theta} . \tag{2.2.9}$$

This provides an equation for $\dot{\theta}$ in terms of p_θ. Then,

$$
\begin{aligned}
H &= \frac{\partial L}{\partial\dot{\theta}}\dot{\theta} - L \\
&= ml^2\dot{\theta} - \frac{1}{2}ml^2\dot{\theta}^2 + mgl(1 - \cos\theta) \\
&= \frac{1}{2}ml^2\dot{\theta}^2 + mgl(1 - \cos\theta) .
\end{aligned}
\tag{2.2.10}
$$

But the Hamiltonian is a function of the generalized coordinate and the conjugate momentum. So we get

$$H = \frac{p_\theta^2}{2ml^2} + mgl(1 - \cos\theta) . \tag{2.2.11}$$

Furthermore, the expression for the conjugate momentum (2.2.9) shows that it is just the *angular momentum*.

2.3 Bohr-Sommerfeld Quantization: SHO

Use Bohr-Sommerfeld quantization to calculate the energy levels of a one-dimensional simple harmonic oscillator.

Solution

We use Bohr-Sommerfeld quantization

$$\oint p\,dq = nh = 2\pi n\hbar \ . \tag{2.3.12}$$

Also, the total energy is

$$E = \frac{p^2}{2m} + \frac{1}{2}kx^2 \ . \tag{2.3.13}$$

Therefore,

$$p = \pm\sqrt{2mE}\sqrt{1 - \frac{kx^2}{2E}} \ . \tag{2.3.14}$$

The motion is bounded by $\pm x_0 = \pm\sqrt{2E/k}$. The two signs in front of p indicate that we have momentum to the right $+$ and then momentum to the left $-$. Therefore,

$$
\begin{aligned}
\oint p\,dx &= \sqrt{2mE}\left[\int_{-x_0}^{x_0}\sqrt{1 - \frac{kx^2}{2E}}\,dx - \int_{x_0}^{-x_0}\sqrt{1 - \frac{kx^2}{2E}}\,dx\right] \\
&= 2\sqrt{2mE}\int_{-x_0}^{x_0}\sqrt{1 - \frac{kx^2}{2E}}\,dx \ .
\end{aligned}
\tag{2.3.15}
$$

We now call $\omega^2 = k/m$ and, after evaluating the integral, set it equal to nh to get

$$2\pi n\hbar = 2\frac{2E}{\omega}\frac{\pi}{2} \ . \tag{2.3.16}$$

So, the quantized energy levels are

$$E_n = n\hbar\omega \ . \tag{2.3.17}$$

2.4   Bohr-Sommerfeld: Particle in a Box

Use Bohr-Sommerfeld quantization to calculate the energy levels of a particle confined to a box of length L. For simplicity assume this is a "one-dimensional box".

Solution

As in the problem above we use Bohr-Sommerfeld quantization

$$\oint p\,dq = nh = 2\pi n\hbar \ . \tag{2.4.18}$$

Again we have that the momentum is related to the energy by

$$p = \pm\sqrt{2mE} \tag{2.4.19}$$

so that we have the particle moving to the right with constant momentum $\sqrt{2mE}$ and then returning to the left with constant momentum $-\sqrt{2mE}$. Hence, we get

$$
\begin{aligned}
\oint p\,dq &= nh \\
&= \sqrt{2mE}\left[\int_0^L dx - \int_L^0 dx\right] \\
&= \sqrt{2mE} \times 2L \ .
\end{aligned}
$$

(2.4.20)

Solving for E we find

$$
E = \frac{h^2}{8mL^2}n^2 \ .
$$

(2.4.21)

2.5 Larmor Frequency

Suppose a gyroscope has a magnetic moment $\vec{\mu}$ proportional to its angular momentum $\vec{L}$ according to

$$
\vec{\mu} = M\vec{L} \ .
$$

The potential energy due to placing the gyroscope in a magnetic field $\vec{B}$ is

$$
V = -\vec{\mu} \cdot \vec{B} \ .
$$

Assume that $\vec{B}$ is constant and derive the equation of motion for $\vec{L}$. Show that the gyroscope precesses with the angular Larmor frequency

$$
\omega_L = MB \ .
$$

Solution

For a gyroscope with angular momentum $\vec{L}$ we have that if the angular velocity is $\vec{\omega}$ and the moment of inertia is I then

$$
\vec{L} = I\vec{\omega} \ .
$$

(2.5.22)

Thus, the kinetic energy is

$$
T = \frac{1}{2}I\vec{\omega}^2 = \frac{1}{2}I\dot{\theta}^2 \ .
$$

(2.5.23)

Also,

$$
V = -\vec{\mu} \cdot \vec{B} = -M\vec{L} \cdot \vec{B} = -MI\vec{\omega} \cdot \vec{B} \ .
$$

(2.5.24)

Then, the Lagrangian is

$$
L = \frac{1}{2}I\dot{\theta}^2 - V = \frac{1}{2}I\dot{\theta}^2 + MI\dot{\theta}B \ .
$$

(2.5.25)

The equation of motion is

$$\frac{d}{dt}\left[I\dot{\theta} + MIB\right] = 0 \ .$$

(2.5.26)

This yields immediately a first integral

$$\dot{\theta} + MB = \text{constant} \ .$$

(2.5.27)

But, for no magnetic field $(B = 0)$ we have the gyroscope at rest and we immediately get $\dot{\theta} = 0$ for $B = 0$. So we see that the integration constant vanishes. Hence,

$$\dot{\theta} = -MB = -\omega_L \ .$$

(2.5.28)

2.6   Applicability of Bohr-Sommerfeld Quantization

The system of quantization proposed by Bohr in 1913 and late gneralized by Sommerfeld is not applicable to all systems. To what general kinds of physical systems is Bohr's procedure applicable? For what kinds of systems is it not applicable.

Solution

The system of quantization proposed by Bohr and Sommerfeld is applicable to systems with repetitive (periodic) motion. It is not applicable to systems with unbounded motion.

2.7   Schrödinger and Hamilton-Jacobi

Consider the time-dependent Schrödinger equation and put

$$\Psi = Ae^{iS/\hbar}$$

where $A = \text{constant}$. Show that in the limit as $\hbar \to 0$ the equation

$$-\frac{\hbar^2}{2m}\frac{\partial^2 \Psi}{\partial x^2} + V\Psi = i\hbar\frac{\partial \Psi}{\partial t}$$

reduces to the Hamilton-Jacobi equation [2.1]

$$\frac{1}{2m}\left(\frac{\partial S}{\partial q}\right)^2 + V(q) = \frac{\partial S}{\partial t} \ .$$

Solution

If we put

$$\Psi = A\,e^{iS/\hbar}$$

in the Schrödinger equation we get

$$-\frac{\partial S}{\partial t}A\,e^{iS/\hbar} = -\frac{\hbar^2}{2m}\left[\frac{i}{\hbar}\frac{\partial^2 S}{\partial x^2} - \frac{1}{\hbar^2}\left(\frac{\partial S}{\partial x}\right)^2\right]A\,e^{iS/\hbar} + VA\,e^{iS/\hbar}\,. \quad (2.7.29)$$

Now let $\hbar \to 0$ then we get

$$\frac{\partial S}{\partial t} + \frac{1}{2m}\left(\frac{\partial S}{\partial x}\right)^2 + V = 0\,. \quad (2.7.30)$$

This is just the Hamilton-Jacobi equation.

2.8 WKB Approximation

In the problem above set $S = W - Et$ and let $W = W_0 + \hbar W_1 + \hbar^2 W_2 + \cdots$ for the case of a one-dimensional Schrödinger equation. Find the equations for W_0 and W_1 and solve the equation for W_0. This is the so-called *Wentzel-Kramers-Brillouin* or WKB approximation.

Solution

Clearly if in the problem above we set $S = W - Et$ (before letting $\hbar \to 0$) we get

$$E = -\frac{i\hbar}{2m}\frac{d^2W}{dx^2} + \frac{1}{2m}\left(\frac{dW}{dx}\right)^2 + V\,. \quad (2.8.31)$$

Letting $W = W_0 + \hbar W_1 + \ldots$ and equating the coefficients of powers of $\hbar$ we get

$$E = \frac{1}{2m}\left(\frac{dW_0}{dx}\right)^2 + V\,. \quad (2.8.32)$$

So,

$$W_0 = \pm\int\sqrt{2m(E - V)}\,dx \quad (2.8.33)$$

and

$$-\frac{i}{2m}\frac{d^2W_0}{dx^2} + \frac{1}{2m}\left(\frac{dW_1}{dx}\right)^2 = 0\,. \quad (2.8.34)$$

So we have an equation for W_1.

2.9   Dumbbell Molecule: Bohr-Sommerfeld

A dumbbell molecule of moment of inertia I is rotating about its centre of mass.
a) Write the langrangian for this system and find the canonical momentum.
b) Use Bohr-Sommerfeld quantization to find the energy levels of this molecule.

Solution

a) If we take the plane in which the molecule is rotating as the plane $z = 0$ then
the Lagrangian is just the kinetic energy

$$L = \frac{1}{2}I\dot{\varphi}^2 \ . \tag{2.9.35}$$

The canonical momentum conjugate to φ is

$$p_\varphi = \frac{\partial L}{\partial \dot{\varphi}} = I\dot{\varphi} \ . \tag{2.9.36}$$

b) Bohr-Sommerfeld quantization states that

$$\oint p_\varphi \, d\varphi = \int_0^{2\pi} p_\varphi \, d\varphi = nh \ . \tag{2.9.37}$$

But,

$$p_\varphi = I\dot{\varphi} = \sqrt{2IE} \tag{2.9.38}$$

where E is the total energy which is conserved and therefore a constant. Thus,

$$\sqrt{2IE} \int_0^{2\pi} d\varphi = 2\pi\sqrt{2IE} = nh \ . \tag{2.9.39}$$

So,

$$E = \frac{1}{2I}n^2 \left(\frac{h}{2\pi}\right)^2 = \frac{n^2\hbar^2}{2I} \ . \tag{2.9.40}$$

Bibliography

[2.1] A.Z. Capri, *Nonrelativistic Quantum Mechanics* 3rd edition, World Scientific Publishing Co. Pte. Ltd., section 2.4, (2002) .

Chapter 3

Elementary Systems

3.1 Commutator Identities

a) Verify the identity

$$[AB, C] = A[B, C] + [A, C]B .$$

b) Using the result above and $[x, p] = i\hbar$ prove that

$$[x^2, p] = 2i\hbar x$$

and

$$[x^n, p] = ni\hbar x^{n-1} .$$

Hence prove that for any function $g(x)$ analytic at the origin

$$[g(x), p] = i\hbar \frac{dg(x)}{dx} .$$

Solution

a) To verify the identity we simply write it out

$$
\begin{aligned}
[AB, C] &= ABC - CAB \\
&= ABC - ACB + ACB - CAB \\
&= A[B, C] + [A, C]B .
\end{aligned}
\tag{3.1.1}
$$

b) By applying this identity we immediately get

$$[x^2, p] = x[x, p] + [x, p]x = xi\hbar + i\hbar x = 2i\hbar x . \tag{3.1.2}$$

If we replace x^2 above by x^n we can provide a proof by induction since

$$[x^n, p] = x^{n-1}[x, p] + [x^{n-1}, p]x \tag{3.1.3}$$

and assuming the result holds true for $n - 1$ we immediately get the result for n. Also, we have already shown the result to be true for $n = 2$.

The second part may be proved in two different ways.

i) Since $g(x)$ is analytic at $x = 0$ we can write a Maclaurin series

$$g(x) = \sum_n c_n x^n \quad , \quad c_n = \frac{1}{n!} \frac{d^n g}{dx^n}\bigg|_{x=0} . \tag{3.1.4}$$

Hence,

$$[g(x), p] = \sum_n c_n [x^n, p] \tag{3.1.5}$$

and using the result above

$$[g(x), p] = \sum_n c_n ni\hbar x^{n-1} = i\hbar \frac{d}{dx} \sum_n c_n x^n = i\hbar \frac{dg(x)}{dx} . \tag{3.1.6}$$

ii) A second method is to use the representation of p as the differential operator

$$p = \frac{\hbar}{i} \frac{d}{dx} \tag{3.1.7}$$

and act with the commutator on some function $f(x)$. Then,

$$[g(x), p]f(x) = g(x)\frac{\hbar}{i}\frac{df}{dx} - \frac{\hbar}{i}\frac{d}{dx}\left(g(x)f(x)\right) = \left(i\hbar\frac{dg}{dx}\right)f(x) . \tag{3.1.8}$$

So, we again find

$$[g(x), p] = i\hbar \frac{dg}{dx} . \tag{3.1.9}$$

3.2   Complex Potential

Assume that the potential V is complex of the form $V = U + iW$. Show that W corresponds to a sink or source of probability.

Hint: Show that

$$\frac{\partial \rho}{\partial t} + \nabla \cdot \mathbf{j} = \frac{2}{\hbar} W\rho .$$

This proves that unless $W = 0$ probability is not conserved.

Solution

We begin with the Schrödinger equation and its complex conjugate.

$$i\hbar \frac{\partial \Psi}{\partial t} = -\frac{\hbar^2}{2m}\nabla^2\Psi + (U + iW)\Psi$$

$$-i\hbar \frac{\partial \Psi^*}{\partial t} = -\frac{\hbar^2}{2m}\nabla^2\Psi^* + (U - iW)\Psi^* . \tag{3.2.10}$$

Next, we multiply the first equation by Ψ^* and the second by Ψ and subtract the two equations to get

$$i\hbar \frac{\partial}{\partial t}(\Psi \Psi^*) = -\frac{\hbar^2}{2m}(\Psi^* \nabla^2 \Psi - \Psi \nabla^2 \Psi^*) + 2iW\Psi^* \Psi . \qquad (3.2.11)$$

After rearranging and using the definitions

$$\begin{aligned} \rho &= \Psi^* \Psi \\ \vec{j} &= \frac{\hbar}{2im}[\Psi^* \nabla \Psi - (\nabla \Psi^*)\Psi] \end{aligned} \qquad (3.2.12)$$

we get the desired result

$$\frac{\partial \rho}{\partial t} + \nabla \cdot \vec{j} = \frac{2}{\hbar}W\rho . \qquad (3.2.13)$$

3.3 Group and Phase Velocity

a) In deep water the phase velocity of water waves of wavelength λ is

$$v = \sqrt{\frac{g\lambda}{2\pi}} .$$

What is the group velocity?

b) The phase velocity of a typical electromagnetic wave in a wave guide has the form

$$v = \frac{c}{\sqrt{1 - (\omega_0/\omega)^2}}$$

where ω_0 is a certain characteristic frequency. What is the group velocity of such waves?

Solution

a) The phase velocity is quite generally given by

$$v_p = \frac{\omega}{k} = \sqrt{\frac{g\lambda}{2\pi}} = \sqrt{g/k} . \qquad (3.3.14)$$

So we see that $\omega = \sqrt{gk}$. Therefore,

$$v_g = \frac{d\omega}{dk} = \frac{1}{2}\sqrt{g/k} = \frac{1}{2}v_p = \frac{\omega}{2k} = \frac{1}{2}\sqrt{\frac{g\lambda}{2\pi}} . \qquad (3.3.15)$$

b) This time we have

$$v_p = \frac{\omega}{k} = \frac{c}{\sqrt{1 - (\omega_0/\omega)^2}} . \qquad (3.3.16)$$

So, $c^2 k^2 = \omega^2 - \omega_0^2$. Differentiating this equation with respect to k we get

$$c^2 k = \omega \frac{d\omega}{dk} \ . \tag{3.3.17}$$

This implies that

$$v_g = \frac{d\omega}{dk} = \frac{c^2 k}{\omega} = \frac{c^2}{v_p} \ . \tag{3.3.18}$$

3.4 Linear Operators

Which of the following operators are linear?
a) $K\psi(x) = \int K(x, y)\psi(y)dy$.
b) K^3 where K is defined above.
c) AB if A and B are linear.
d) B^{-1} if B is linear and B^{-1} is defined by $B^{-1}B = BB^{-1} = 1$.
e) $\exp A = \sum_{n=0}^{\infty} \frac{A^n}{n!}$ if A is linear.
f) $A\psi = \exp(\lambda\psi)$.

Solution

In all cases, to see if the operator is linear, we have to check whether

$$A(\lambda_1 \psi_1 + \lambda_2 \psi_2) = \lambda_1 A\psi_1 + \lambda_2 A\psi_2 \tag{3.4.19}$$

where λ_1 and λ_2 are constants. The results are:
a) Linear since

$$\int K(x, y) [\lambda_1 A\psi_1(y) + \lambda_2 A\psi_2(y)] \, dy$$

$$= \lambda_1 \int K(x, y)\psi_1(y) \, dy + \lambda_2 \int K(x, y)\psi_2(y) \, dy \ . \tag{3.4.20}$$

b) Linear since

$$K^3\psi(x) = \int K(x, y) \, dy \int K(y, z) \, dz \int K(z, w)\psi(w) \, dw \ . \tag{3.4.21}$$

c) Linear since

$$\begin{aligned} AB(\lambda_1 \psi_1 + \lambda_2 \psi_2) &= A(\lambda_1 B\psi_1 + \lambda_2 B\psi_2) \\ &= \lambda_1 AB\psi_1 + \lambda_2 AB\psi_2 \ . \end{aligned} \tag{3.4.22}$$

d) Linear because the identity operator is linear and we have $BB^{-1} = 1$. Thus,

$$\begin{aligned} BB^{-1}(\lambda_1 \psi_1 + \lambda_2 \psi_2) &= \lambda_1 BB^{-1}\psi_1 + \lambda_2 BB^{-1}\psi_2 \\ &= B(\lambda_1 B^{-1}\psi_1 + \lambda_2 B^{-1}\psi_2) \end{aligned} \tag{3.4.23}$$

where we have used the fact that B is linear. We now operate on this equation with B^{-1} from the left to get the desired result

$$B^{-1}(\lambda_1\psi_1 + \lambda_2\psi_2) = \lambda_1 B^{-1}\psi_1 + \lambda_2 B^{-1}\psi_2 \ . \tag{3.4.24}$$

e) We first use the definition of $\exp A$

$$(\exp A)\psi = \sum_{n=0}^{\infty} \frac{A^n}{n!}\psi \ . \tag{3.4.25}$$

But by an argument the same as for part b) above and induction we see that A^n is linear. Furthermore, a sum of linear operators is again linear. Therefore, $\exp A$ is linear. In fact, by the same sort of argument, if A is linear then any function of A say $f(A)$ is also linear.

f) Not linear because

$$\exp[\lambda(\lambda_1\psi_1 + \lambda_2\psi_2)] \neq \lambda_1 \exp[\lambda\psi_1] + \lambda_2 \exp[\lambda\psi_2] \tag{3.4.26}$$

3.5 Probability Density

a) Compute, in closed form, the probability density $\rho(t, x)$ for the wave function.

$$\Psi(t, x) = \int_{-\infty}^{\infty} dk \, A(k) \, \exp -i \left(\frac{\hbar k^2}{2m} t - kx \right)$$

where

$$A(k) = e^{-L^2 k^2 / 2} \ .$$

b) What is the "width" of the probability density at time $t = 0$ and at time t ?

Solution

We first compute the wavefunction $\Psi(t, x)$ explicitly.

$$\Psi(t, x) = \int_{-\infty}^{\infty} dk \, \exp \left[-k^2(L^2/2 + i\hbar t/2m) + ikx \right] \ . \tag{3.5.27}$$

Now, we complete the square in the exponent

$$
\begin{aligned}
&-k^2(L^2/2 + i\hbar t/2m) + ikx \\
= \ &-\frac{1}{2}(L^2 + i\hbar t/m)\left[k^2 - \frac{2ikx}{L^2 + i\hbar t/m} - \frac{x^2}{(L^2 + i\hbar t/m)^2} \right] \\
&-\frac{x^2}{2(L^2 + i\hbar t/m)} \ .
\end{aligned}
\tag{3.5.28}
$$

Also, we define

$$q = k - \frac{ix}{L^2 + i\hbar t/m} \ . \tag{3.5.29}$$

Then,

$$\Psi(t, x) = \exp\left[-\frac{x^2}{2(L^2 + i\hbar t/m)}\right] \int_{-\infty - \frac{ix}{(L^2 + i\hbar t/m)}}^{\infty - \frac{ix}{(L^2 + i\hbar t/m)}} dq$$

$$\times \quad \exp\left[-q^2(L^2/2 + i\hbar t/2m)\right] \quad . \tag{3.5.30}$$

The contour of integration may now be shifted to the real axis since the integrand has no singularities (poles or cuts) in the region between the real axis and a line parallel to the real axis. Also, for real $q \to \infty$, the integrand vanishes. Thus,

$$\Psi(t, x) = \exp\left[-\frac{x^2}{2(L^2 + i\hbar t/m)}\right] \int_{-\infty}^{\infty} dq \, \exp\left[-q^2(L^2/2 + i\hbar t/2m)\right]$$

$$= \sqrt{\frac{2\pi}{(L^2 + i\hbar t/m)}} \exp\left[-\frac{x^2}{2(L^2 + i\hbar t/m)}\right] \quad . \tag{3.5.31}$$

The probability density is $\rho = \Psi^*\Psi$. Therefore,

$$\rho = \frac{2\pi}{\sqrt{L^4 + (\hbar t/m)^2}} \exp\left[-\frac{L^2 x^2}{(L^4 + (\hbar t/m)^2)}\right] \quad . \tag{3.5.32}$$

To see the spreading of the wave packet we find the values of x such that ρ has dropped to roughly $1/e$ of its maximum value. These points are obtained by setting the argument of the exponential equal to -1. Then we find

$$\frac{L^2 x^2}{(L^4 + (\hbar t/m)^2)} = 1 \quad . \tag{3.5.33}$$

So the width is given at any time t by

$$w = L\sqrt{1 + \frac{\hbar^2 t^2}{m^2 L^4}} \quad . \tag{3.5.34}$$

This shows how the wave packet gets wider with time.

3.6 Angular Momentum Operators

Show that if we write

$$\vec{L} = \vec{r} \times \vec{p} = \frac{\hbar}{i} \vec{r} \times \vec{\nabla}$$

then

$$L_x L_y - L_y L_x = i\hbar L_z$$

$$L_y L_z - L_z L_y = i\hbar L_x$$

$$L_z L_x - L_x L_z = i\hbar L_y \quad .$$

Solution

The three components of $\vec{L}$ are explicitly given by

$$
\begin{aligned}
L_x &= yp_z - zp_y \\
L_y &= zp_x - xp_z \\
L_z &= xp_y - yp_x \; .
\end{aligned}
\tag{3.6.35}
$$

Now, use the results of problem 3.1. Then,

$$
\begin{aligned}
[L_x, L_y] &= [yp_z - zp_y, zp_x - xp_z] \\
&= [yp_z, zp_x] + [zp_y, xp_z] \\
&= -i\hbar yp_x + i\hbar xp_y \\
&= i\hbar L_z \; .
\end{aligned}
\tag{3.6.36}
$$

The other pairs of commutators are obtained in exactly the same manner and yield

$$
\begin{aligned}
[L_y, L_z] &= [zp_x - xp_z, xp_y - yp_x] \\
&= [zp_x, xp_y] + [xp_z, yp_x] \\
&= -i\hbar zp_y + i\hbar yp_z \\
&= i\hbar L_x
\end{aligned}
\tag{3.6.37}
$$

as well as

$$
\begin{aligned}
[L_z, L_x] &= [xp_y - yp_x, yp_z - zp_y] \\
&= [xp_y, yp_z] + [yp_x, zp_y] \\
&= -i\hbar xp_z + i\hbar zp_x \\
&= i\hbar L_y \; .
\end{aligned}
\tag{3.6.38}
$$

3.7 Beam of Particles

A beam of free particles is moving along the x-axis with velocity v such that there is one particle in a volume V.
a) What is the corresponding normalized, time-dependent wave function for such a particle?
b) What is the number of particles crossing a unit area, normal to the x-axis, per unit time?

Solution

a) For a free particle moving with velocity v along the x-axis we have

$$
\Psi(t, x) = A \, e^{i(kx - \omega t)}
\tag{3.7.39}
$$

where

$$
k = \frac{mv}{\hbar} \;, \quad \omega = \frac{mv^2}{2\hbar} \; .
\tag{3.7.40}
$$

The particle is confined to the volume V. This means that the normalization integral is over the volume V. So,

$$1 = \int_V |\Psi|^2 \, dV = \int_V |A|^2 \, dV = |A|^2 V \; . \tag{3.7.41}$$

Therefore,

$$A = \frac{1}{\sqrt{V}} \; . \tag{3.7.42}$$

b) The current density is given by

$$
\begin{aligned}
j &= \frac{\hbar}{2im} \left(\Psi^*(t,x) \frac{\partial \Psi(t,x)}{\partial x} - \Psi(t,x) \frac{\partial \Psi^*(t,x)}{\partial x} \right) \\
&= \frac{\hbar}{2im} \frac{1}{V} \left[e^{-i(kx-\omega t)} \frac{\partial}{\partial x} e^{i(kx-\omega t)} - e^{i(kx-\omega t)} \frac{\partial}{\partial x} e^{-i(kx-\omega t)} \right] \\
&= \frac{\hbar k}{mV} = \frac{mv}{mV} = \frac{v}{V} \; . \tag{3.7.43}
\end{aligned}
$$

The number of particles passing through an area S normal to the x-axis is therefore

$$jS = v \frac{S}{V} \; .$$

3.8   Time Evolution of Wave Function

A free particle has the wave packet at time $t = 0$ given by

$$\Psi(0,x) = \int_{-\infty}^{\infty} dk \, \frac{e^{ikx}}{k^2 + a^2} \; .$$

Determine an expression for the wavefunction for a later time t. Do not attempt to evaluate the resulting integral.

Hint: Use the equation

$$\Psi(x,t) = \int_{-\infty}^{\infty} f(k) e^{i[kx - \omega(k)t]} \, dk$$

and remember that the energy for a free particle is given by the equation

$$\hbar \omega = \frac{(\hbar k)^2}{2m} \; .$$

Solution

For a free particle the wavefunction at any time t is of the form

$$\Psi(t,x) = \int_{-\infty}^{\infty} dk \, f(k) e^{i[kx - \omega(k)t]} \tag{3.8.44}$$

where

$$w(k) = \frac{E(k)}{\hbar} = \frac{\hbar k^2}{2m} \ . \qquad\qquad (3.8.45)$$

So we have to find an appropriate $f(k)$ and carry out the integral. To do this
we use the initial condition that

$$\Psi(0, x) = \int_{-\infty}^{\infty} dk \ \frac{e^{ikx}}{k^2 + a^2} \ . \qquad\qquad (3.8.46)$$

This means that

$$f(k) = \frac{1}{k^2 + a^2} \ . \qquad\qquad (3.8.47)$$

Therefore,

$$\Psi(t, x) = \int_{-\infty}^{\infty} dk \ \frac{e^{i[kx - \hbar k^2 t/2m]}}{k^2 + a^2} \ . \qquad\qquad (3.8.48)$$

Fortunately we have been told not to attempt the integral. Therefore, we are
done.

3.9   Operator Hamiltonian

You are given the classical Hamiltonian for the motion of a particle in the form

$$H = \frac{p^2}{2m} e^{-\alpha x}$$

where α is a constant. Find an acceptable Hamiltonian operator for this system.
Notice that the answer is by no means unique.

Solution

The classical Hamiltonian that we are given is

$$H_{\text{classical}} = \frac{p^2}{2m} e^{-\alpha x} \ . \qquad\qquad (3.9.49)$$

To make an operator out of this we must do two things:
a) We must replace p by the operator

$$p_{\text{op}} = \frac{\hbar}{i} \frac{d}{dx} \qquad\qquad (3.9.50)$$

and
b) rearrange the order of the operators p_{op} and x_{op} so that the resultant Hamiltonian is hermitian. The easiest, and by no means unique, way of doing this is
to symmetrize the operators. Therefore we propose

$$H_{\text{op}} = \frac{1}{2m} \frac{1}{2} \left[p_{\text{op}}^2 \, e^{-\alpha x} + e^{-\alpha x} p_{\text{op}}^2 \right] \ . \qquad\qquad (3.9.51)$$

Other possibilities are

$$H_{op} = \frac{1}{2m} p_{op} \, e^{-ax} \, p_{op} \tag{3.9.52}$$

or any convex linear combination of (3.9.51) and (3.9.52).

3.10 Zero of Energy

In classical mechanics, the reference level for the potential energy is arbitrary. What are the effects on the wave function and energy of adding a constant potential in the time-dependent Schrödinger equation?

Solution

Suppose we have the Hamiltonian H and we add a constant V_0 to it to get the Hamiltonian H'. Then we have that

$$i\hbar \frac{\partial \Psi}{\partial t} = H \Psi \tag{3.10.53}$$

as well as

$$i\hbar \frac{\partial \Psi'}{\partial t} = H' \Psi' = (H + V_0) \Psi' \ . \tag{3.10.54}$$

If we try adding a constant phase $-\alpha/\hbar$ to Ψ we have

$$\Psi' = e^{-i\alpha/\hbar} \Psi \ . \tag{3.10.55}$$

Then,

$$i\hbar \frac{\partial \Psi'}{\partial t} = e^{-i\alpha/\hbar} \left[\alpha \Psi + i\hbar \frac{\partial \Psi}{\partial t} \right] \ . \tag{3.10.56}$$

So, if we choose

$$\alpha = V_0$$

we get that

$$\Psi' = e^{-iV_0/\hbar} \Psi \ . \tag{3.10.57}$$

satisfies equation (3.10.54) .

Now, suppose Ψ and Ψ' are stationary states so that we have

$$\Psi(t, x) = e^{-iEt/\hbar} \, \psi(x) \tag{3.10.58}$$

and

$$\Psi'(t, x) = e^{-iE't/\hbar} \, \psi'(x) \ . \tag{3.10.59}$$

Then, we see that if we set

$$E' = E + V_0 \tag{3.10.60}$$

we find that

$$ i\hbar\frac{\partial\Psi'}{\partial t} = \left[V_0 + e^{-iV_0t/\hbar}i\hbar\frac{\partial\Psi}{\partial t} \right] $$
$$ = e^{-iV_0t/\hbar}[H + V_0]\Psi $$
$$ = e^{-iV_0t/\hbar}[E + V_0]\Psi \quad . \tag{3.10.61} $$

So, we have found that

$$ \Psi'(t, x) = e^{-iV_0t/\hbar}\,\Psi(t, x) \tag{3.10.62} $$

and

$$ E' = E + V_0 \quad . \tag{3.10.63} $$

So, we only get the same constant added to the energy and a constant phase shift to the wavefunction.

3.11 Some Commutators

Given the Hamiltonian operator

$$ H = \frac{p^2}{2m} + V(x) $$

where the operators x and p satisfy the commutation relation

$$ [x, p] = i\hbar $$

find the following commutators
a) $[x, H]$
b) $[p, H]$.
Hint: Use the representation

$$ p = \frac{\hbar}{i}\frac{d}{dx} $$

and let the expressions $[x, H]$ and $[x, H]$ operate on an arbitrary function $f(x)$.

Solution

a) Using the hint we have

$$ [x, H]f(x) = \frac{1}{2m}\left[x\{-\hbar^2\frac{d^2}{dx^2}f(x)\} - (-\hbar^2\frac{d^2}{dx^2})\{xf(x)\} \right] $$
$$ = \frac{1}{2m}2\hbar^2\frac{df}{dx} = \frac{i}{m}\hbar\frac{\hbar}{i}\frac{df}{dx} \quad . \tag{3.11.64} $$

So,

$$ [x, H] = i\hbar\frac{p}{m} \quad . \tag{3.11.65} $$

b) Proceeding in the same way

$$[p, H]f(x) \;=\; \frac{\hbar}{i}\frac{d}{dx}\{V(x)f(x)\} - V(x)\frac{\hbar}{i}\frac{df}{dx}$$

$$=\; \frac{\hbar}{i}\frac{dV}{dx}f(x) \;.$$ (3.11.66)

Thus,

$$[p, H] = -i\hbar\frac{dV}{dx} \;.$$ (3.11.67)

3.12 Eigenfunction for a Simple Hamiltonian

A system (in a particular set of units) is described by the Hamiltonian operator

$$H = -\frac{d^2}{dx^2} + x^2$$

a) Show that both

$$\psi_1 = A_1\,e^{-x^2/2}$$

and

$$\psi_2 = A_2 x\,e^{-x^2/2}$$

are eigenfunctions of this Hamiltonian and find the corresponding eigenvalues.
b) Find the constants A_1 and A_2 that correctly normalize these eigenfunctions.
c) Find the expectation value of x for these eigenfunctions.

Solution

a) To verify that the given functions are indeed eigenfunctions we simply act on them with the given Hamiltonian operator. Thus,

$$H\,A_1\,e^{-x^2/2} \;=\; \left[-\frac{d^2}{dx^2} + x^2\right]A_1\,e^{-x^2/2}$$

$$=\; [1 - x^2 + x^2]A_1\,e^{-x^2/2}$$

$$=\; 1\,.\,A_1\,e^{-x^2/2} \;.$$ (3.12.68)

Thus, $A_1\,e^{-x^2/2}$ is indeed an eigenfunction of the given hamiltonain and the corresponding eigenvalue is 1. Similarly,

$$H\,A_2 x\,e^{-x^2/2} \;=\; \left[-\frac{d^2}{dx^2} + x^2\right]A_2 x\,e^{-x^2/2}$$

$$=\; [2 - x^2 + x^2]A_2 x\,e^{-x^2/2}$$

$$=\; 2\,.\,A_2 x\,e^{-x^2/2} \;.$$ (3.12.69)

Thus, $A_2 x\, e^{-x^2/2}$ is also an eigenfunction of the given hamiltonain and the corresponding eigenvalue is 2.

b) Normalization requires that we calculate the integrals

$$\int_{-\infty}^{\infty} |\psi_i|^2\, dx \quad i = 1, 2 \ . \tag{3.12.70}$$

One way to proceed is to first compute the integral

$$I(a) = \int_{-\infty}^{\infty} e^{-ax^2}\, dx \ . \tag{3.12.71}$$

To do this we use a trick

$$
\begin{aligned}
I(a)^2 &= \int_{-\infty}^{\infty} e^{-ax^2}\, dx \int_{-\infty}^{\infty} e^{-ay^2}\, dy \\
&= \int_0^{2\pi} \int_0^{\infty} e^{-ar^2}\, r\, dr \\
&= \frac{\pi}{a} \int_0^{\infty} e^{-z}\, dz \\
&= \frac{\pi}{a} \ .
\end{aligned}
\tag{3.12.72}
$$

Therefore,

$$I(a) = \sqrt{\frac{\pi}{a}} \ . \tag{3.12.73}$$

This gives us immediately the normalization for ψ_1 by setting $a = 1$.

$$A_1 = \frac{1}{\pi^{1/4}} \ . \tag{3.12.74}$$

For A_2 we need the integral

$$\int_{-\infty}^{\infty} e^{-ax^2} x^2\, dx \ .$$

But,

$$\frac{dI(a)}{da} = -\int_{-\infty}^{\infty} e^{-ax^2} x^2\, dx = -\frac{1}{2}\frac{\sqrt{\pi}}{a^{3/2}} \ . \tag{3.12.75}$$

So, again setting $a = 1$ we get

$$A_2 = \left(\frac{4}{\pi}\right)^{1/4} \ . \tag{3.12.76}$$

c) The expectation values of x in the two states are given by

$$(\psi_1, x\psi_1) = \int_{-\infty}^{\infty} e^{-x^2}\, x\, dx = 0 \tag{3.12.77}$$

since the integrand is an odd function. Similarly,

$$(\psi_2, x\psi_2) = \int_{-\infty}^{\infty} e^{-x^2}\, x^3\, dx = 0 \tag{3.12.78}$$

since in this case also the integrand is an odd function.

Bibliography

[3.1] A.Z. Capri, *Nonrelativistic Quantum Mechanics* 3rd edition, World Scientific Publishing Co. Pte. Ltd., section 1.12, (2002) .

Chapter 4

One-Dimensional Problems

4.1 Potential Step

A beam of particles of mass m moving from left to right encounters a sharp potential drop of amount V_0. Let E be the kinetic energy of the incoming particles and show that the fraction of particles reflected at the edge of the potential (located at $x = 0$) is given by [4.1]

$$\left(\frac{\sqrt{E + V_0} - \sqrt{E}}{\sqrt{E + V_0} + \sqrt{E}} \right)^2 .$$

In view of this result, what will happen to a car moving at 10 km/hr if it meets the edge of a 200 m cliff? Is this answer reasonable? If not, why not?

Solution

The Hamiltonian for this problem may be taken to be

$$H = \frac{p^2}{2m} + V(x)$$

where

$$V(x) = \begin{cases} 0 & x < 0 \\ -V_0 & x > 0 \end{cases} .$$

The relevant differential equations now are

$$-\frac{\hbar^2}{2m} \frac{d^2 \psi_1}{dx^2} = E\psi_1 \quad x < 0$$

$$-\frac{\hbar^2}{2m} \frac{d^2 \psi_2}{dx^2} - V_0 \psi_1 = E\psi_2 \quad x > 0 \tag{4.1.1}$$

29

Since the incoming beam is from the left we have only a transmitted beam on the right. Thus, the appropriate solutions are

$$\begin{aligned}
\psi_1(x) &= A\left(e^{ik_1x} + Re^{-ik_1x}\right) \quad x < 0 \\
\psi_2(x) &= Be^{ik_2x} \quad\quad\quad\quad\quad x > 0
\end{aligned} \tag{4.1.2}$$

where

$$k_1^2 = \frac{2mE}{\hbar^2} \quad , \quad k_2^2 = \frac{2m(E - V_0)}{\hbar^2} \quad . \tag{4.1.3}$$

Using the continuity of the wavefunction and its derivative at $x = 0$ we get

$$A(1 + R) = B \quad , \quad ik_1A(1 - R) = ik_2B \ . \tag{4.1.4}$$

Therefore, solving for R we get

$$R = \frac{k_1 - k_2}{k_1 + k_2} \ . \tag{4.1.5}$$

Cancelling a factor of $2m/\hbar^2$ we get

$$|R|^2 = \left|\frac{\sqrt{E + V_0} - \sqrt{E}}{\sqrt{E + V_0} + \sqrt{E}}\right|^2 \ . \tag{4.1.6}$$

Regarding the car, the question is a red herring. The car going over a cliff has nothing to do with the problem we have just solved. For the car going over the cliff the potential is not a step function but varies smoothly from a potential that is 0 for $x < 0$ to a potential which is mgy for $x > 0$. This is a step function in the configuration of matter, not in the potential energy of the car. The car problem involves a potential that acts perpendicular to the car's original direction of motion.

This explains why a probability of $|R|^2 = 0.84\%$ is unreasonable, apart from the fact that we know that cars don't as a rule bounce away from unobstructed cliff edges.

The problem of a car going over a cliff is considerably more complicated and is handled at the end of this chapter (problem 4.15).

4.2   Deep Square Well

A particle moving in one dimension interacts with a potential of the form

$$V(x) = 0 \quad\quad |x| > a$$

$$V(x) = -\frac{v_0}{2a} \quad |x| < a \ .$$

Find the equation determining the energy eigenvalues of this system. Solve it approximately assuming a is very small. What happens in the limit $a \to 0$?

Solution

Here,

$$V(x) = \begin{cases} 0 & \text{for} \quad |x| > a \\ -\frac{v_0}{2a} & \text{for} \quad |x| < a \end{cases} \tag{4.2.7}$$

We define $U = v_0/2a$. The problem now is exactly the same as the problem discussed in [4.2]. The solutions divide into positive and negative parity solutions and read

$$\psi_+(x) = \begin{cases} A\,e^{Kx} & x < -a \\ B\,\cos kx & |x| < a \\ A\,e^{-Kx} & x > a \end{cases}$$

$$\psi_-(x) = \begin{cases} A\,e^{Kx} & x < -a \\ B\,\sin kx & |x| < a \\ -A\,e^{-Kx} & x > a \end{cases} \tag{4.2.8}$$

where

$$K^2 = \frac{2mE}{\hbar^2} \quad , \quad k^2 = \frac{2m(E+U)}{\hbar^2} \quad . \tag{4.2.9}$$

The equations that yield the energy eigenvalues are:
Positive parity

$$k \tan ka = K \tag{4.2.10}$$

Negative parity

$$k \cot ka = -K \quad . \tag{4.2.11}$$

Next, we solve for small a and use the explicit form of U

$$k^2 = \frac{2m(E + v_0/2a)}{\hbar^2} \quad . \tag{4.2.12}$$

Thus, as a gets smaller the term $v_0/2a$ dominates so we can replace k^2 by $2mU/\hbar^2$. This means that $ka \ll 1$ so we can set $\tan ka = ka$ (and $\cot ka = 1/ka$). The even parity solution now yields $k^2 a = K$. This gives that

$$E = -\frac{mv_0}{\hbar^2} \quad . \tag{4.2.13}$$

The odd parity solution requires that $k/ka = -K$. This leads to an energy that diverges as $1/a^2$. Thus, we get only one bound state in the limit as $a \to 0$. The corresponding wavefunction is clearly

$$\psi_+(x) = \begin{cases} A\,e^{Kx} & x < -a \\ A\,e^{-Kx} & x > a \end{cases} = A\,e^{-K|x|} \quad . \tag{4.2.14}$$

Furthermore in this limit we find that $U \to \infty$. The result is a δ-function potential.

4.3 Hydrogenic Wavefunction

The wavefunction of an electron in the ground state of an hydrogen-like atom is

$$\psi(r) = Ae^{-Zr/a} \ , \quad a = \frac{\hbar^2}{me^2}$$

where Z is the charge on the nucleus.

a) Determine the constant A, so that the wavefunction is normalized to unity. Remember that you are in three dimensions and r represents the radial variable in spherical coordinates.

b) At what distance from the origin is the probability of finding the electron a maximum?

c) Determine the average value of: the kinetic energy, the potential energy and the total energy.

This verifies the virial theorem for the Coulomb potential.

Solution

a) To normalize we simply integrate and impose

$$1 = \int |\psi|^2 \, dv \ . \tag{4.3.15}$$

Since the wavefunction is spherically symmetric we have $dv = 4\pi r^2 dr$ and get

$$\begin{aligned}
1 &= 4\pi \int_0^\infty |\psi(r)|^2 r^2 dr \\
&= 4\pi |A|^2 \int_0^\infty e^{-2Zr/a} r^2 dr \\
&= 4\pi |A|^2 \frac{a^3}{4Z^3} \ . \tag{4.3.16}
\end{aligned}$$

So, after choosing the phase of A

$$A = \sqrt{\frac{Z^3}{\pi a^3}} \ . \tag{4.3.17}$$

b) Let the probability of finding the electron between r and $r + dr$ be $P(r) \, dr$. Let the probability of finding it in the volume between v and $v + dv$ be $p \, dv$. Due to the spherical symmetry we again have $dv = 4\pi r^2 \, dr$ and so the two probabilities describe the same thing.

$$P(r) \, dr = 4\pi p(r) \, r^2 \, dr \ . \tag{4.3.18}$$

Hence,

$$P = |\psi|^2 4\pi r^2 = \frac{Z^3}{\pi a^3} e^{-2Zr/a} 4\pi r^2 \ . \tag{4.3.19}$$

We now maximize P.

$$\frac{dP}{dr} = 0 \ . \tag{4.3.20}$$

This yields

$$2r\, e^{-2Zr/a}\left[1 - Zr/a\right] = 0 \ . \tag{4.3.21}$$

So, the answer is $r = a/Z$.

c) We know that the average energy in this state equals the energy in this state since we have an eigenstate of the Hamiltonian. Therefore,

$$\langle H \rangle = -\frac{Z^2 e^2}{2a} \ . \tag{4.3.22}$$

The average potential energy is given by

$$\langle V \rangle = \int |\psi|^2 V(r)\, dv \ . \tag{4.3.23}$$

So,

$$\begin{aligned} \langle V \rangle &= \int_0^\infty e^{-2Zr/a} \frac{-Ze^2}{r} r^2 dr \\ &= \frac{-Ze^2}{a} \ . \end{aligned} \tag{4.3.24}$$

Then, the average kinetic energy $\langle T \rangle$ is given by

$$\langle T \rangle = \langle H \rangle - \langle V \rangle = \frac{Z^2 e^2}{2a} \ . \tag{4.3.25}$$

4.4   Bound State Wavefunction, Current, Momentum

a) Show that the wavefunction for a particle in a bound state may always be chosen to be real.

b) By computing the current density give an explanation of the physical meaning of this result.

c) Use the results obtained to show that for a bound state the expectation value $\langle \vec{p} \rangle$ of the momentum vanishes.

Solution

a) To obtain the wavefunction for a bound state we have to solve a differential equation with all the coefficients real and real boundary conditions. As a consequence, we may always choose the normalization constant so that the solution is real. This is never the case for scattering problems where the boundary conditions describe travelling waves and are of necessity complex.

b) Now, suppose that we have a real wavefunction ψ. The probability current in this case is, as always,

$$\vec{j} = \frac{\hbar}{2im}\,(\psi^*\nabla\psi - \psi\nabla\psi^*) = -\frac{\hbar}{im}\Im\{\psi^*\nabla\psi\} \ . \tag{4.4.26}$$

Since ψ is real, this current vanishes. This means, using the equation of continuity, that the probability density remains constant in time. Thus, a particle in a certain region of space, determined by this wavefunction, remains in that region of space. We have a bound state.

c) Writing out the expression for the current we have

$$\vec{j} = 0 = \frac{\hbar}{2im}\,(\psi^*\nabla\psi - \psi\nabla\psi^*) = \frac{1}{2m}[\psi^*\vec{p}\psi + (\vec{p}\psi^*)\psi] \ . \tag{4.4.27}$$

Therefore,

$$\langle\vec{p}\rangle = \frac{1}{2}\int d^3x\,[\psi^*\vec{p}\psi + (\vec{p}\psi^*)\psi] = 0 \ . \tag{4.4.28}$$

Thus, the average momentum of a particle in a bound state vanishes. On the average the particle remains where it is.

4.5   Time Evolution for Particle in a Box

For a particle in a box with sides at $x = \pm a$ the eigenfunctions and eigenvalues are

$$\psi_{+,n} = A_n\cos[(n+1/2)\frac{\pi x}{a}] \ , \quad E_{+,n} = (n+1/2)^2\frac{\hbar^2\pi^2}{2ma^2}$$

$$\psi_{-,n} = B_n\sin[n\frac{\pi x}{a}] \quad , \quad E_{-,n} = n^2\frac{\hbar^2\pi^2}{2ma^2} \ .$$

If we have a particle in such a box and its wavefunction at time $t = 0$ is given by

$$\Psi(0,x) = \frac{1}{\sqrt{a}}\sin[5\frac{\pi x}{a}]$$

find $\Psi(t,x)$.

Solution

We are given the wavefunction at time $t = 0$ as

$$\Psi(0,x) = \frac{1}{\sqrt{a}}\sin(5\pi x/a) \ . \tag{4.5.29}$$

This is an energy eigenfunction with energy

$$E = 5^2\frac{\hbar^2\pi^2}{2ma^2} \ . \tag{4.5.30}$$

Thus, we have a stationary state and the corresponding time-dependent wave-function is

$$\Psi(t,x) = \frac{1}{\sqrt{a}}\sin(5\pi x/a)\exp(-iEt/\hbar) \tag{4.5.31}$$

or

$$\Psi(t,x) = \frac{1}{\sqrt{a}}\sin(5\pi x/a)\exp(-i\frac{25\hbar\pi^2 t}{2ma^2}) \quad . \tag{4.5.32}$$

4.6 Particle in Box: Energy and Eigenfunctions

Find the energy levels and wavefunctions for a particle in the potential

$$V(x) = \begin{cases} \infty & \text{if } x < 0 , \, x > a \\ -V_0 & \text{if } 0 < x < a \end{cases} \quad .$$

Solution

The Hamiltonian for this problem is

$$H = \frac{p^2}{2m} - V_0 \quad 0 < x < a \quad . \tag{4.6.33}$$

The fact that the potential is ∞ outside the interval $0 < x < a$ means that the wavefunction must vanish at both $x = 0$ and $x = a$. If we call

$$k^2 = \frac{2m(E + V_0)}{\hbar^2} \tag{4.6.34}$$

the Schrödinger equation becomes

$$\frac{d^2\psi}{dx^2} + k^2\psi = 0 \quad 0 < x < a \quad . \tag{4.6.35}$$

The physically acceptable solution is therefore of the form

$$\psi(x) = A\sin(kx) \quad 0 < x < a \tag{4.6.36}$$

and we require that

$$\psi(a) = 0 \tag{4.6.37}$$

so that

$$ka = n\pi \quad n = 1, 2, 3, \ldots \quad . \tag{4.6.38}$$

Thus, the (unnormalized) wavefunctions are

$$\psi_n(x) = \begin{cases} A_n \sin(n\pi x/a) & \text{for } 0 < x < a \\ 0 & \text{for } x < 0 , \, x > a \end{cases} \quad . \tag{4.6.39}$$

The corresponding energies are

$$E_n = \frac{\hbar^2\pi^2 n^2}{a^2} - V_0 \quad n = 1, 2, 3, \ldots \quad . \tag{4.6.40}$$

The normalization is given by

$$A_n = \sqrt{\frac{2}{a}} \quad . \tag{4.6.41}$$

4.7   Particle in a Box

If a particle, in a box with sides at $x = \pm a$, is in a state described by the wavefunction

$$f(x) = A\cos(\pi x/2a) + B\sin(\pi x/a)$$

a) Choose A and B so that the particle is in the lowest possible energy state.
b) Choose A and B so that the state has parity $= +1$.

Solution

a) For a particle in a box with sides at $x = \pm a$ the (unnormalized) eigenfunctions are:
parity $= +1$

$$\psi = \cos kx \quad |x| \le a \quad , \quad ka = (n+1/2)\pi \quad , \quad E_n = \frac{\hbar^2 k^2}{2m} \qquad (4.7.42)$$

or for parity $= -1$

$$\psi = \sin kx \quad |x| \le a \quad , \quad ka = n\pi \quad , \quad E_n = \frac{\hbar^2 k^2}{2m} . \qquad (4.7.43)$$

Thus, for the wavefunction

$$f(x) = A\cos(\pi x/2a) + B\sin(\pi x/a) \qquad (4.7.44)$$

the lowest energy is obtained with $B = 0$.
b) The solution with positive parity requires that $f(-x) = f(x)$. This also requires $B = 0$.

4.8   Particles Incident on a Potential

A beam of free particles with intensity N particles per second and energy $E > V_0$ is incident on a potential

$$V(x) = \begin{cases} 0 & x < 0 \\ V_0 & 0 < x < a \\ 0 & x > a \end{cases} .$$

Find the number of particles that are reflected in one second. How many particles per second are transmitted?

Solution

For a beam of particles with energy $E > V_0$ incident from the left we have only a transmitted beam on the right and the wavefunction is given by

$$\psi = \begin{cases} A\left(e^{ikx} + Re^{-ikx}\right) & x < 0 \\ A\left(Ce^{iKx} + De^{-iKx}\right) & 0 < x < a \\ ATe^{ikx} & x > a \end{cases} \qquad (4.8.45)$$

The constant A is determined by the condition that the incident beam $A e^{ikx}$ has a current density of N particles per second. This means that

$$N = \frac{\hbar k}{m} |A|^2 .$$

(4.8.46)

The wavevector k is defined by the energy of the incident beam

$$E = \frac{\hbar^2 k^2}{2m}$$

(4.8.47)

and K is defined by

$$E = \frac{\hbar^2 K^2}{2m} + V_0 .$$

(4.8.48)

Matching the wavefunction and derivatives at $x = 0$ and $x = a$ we get

$$
\begin{aligned}
1 + R &= C + D \\
ik(1 - R) &= iK(C - D) \\
Ce^{iKa} + De^{-iKa} &= Te^{ika} \\
iK\left(Ce^{iKa} - De^{-iKa}\right) &= ikTe^{ika}.
\end{aligned}
$$

(4.8.49)

Solving for R and T we first find

$$
\begin{aligned}
C &= \frac{1}{2K}[K + k + R(K - k)] \\
D &= \frac{1}{2K}[K - k + R(K + k)] .
\end{aligned}
$$

(4.8.50)

Then, we get two expressions for $T e^{ika}$

$$
\begin{aligned}
T e^{ika} &= \frac{1}{2K}[K + k + R(K - k)]e^{iKa} \\
&+ \frac{1}{2K}[K - k + R(K + k)]e^{-iKa} \\
T e^{ika} &= \frac{1}{2k}[K + k + R(K - k)]e^{iKa} \\
&- \frac{1}{2k}[K - k + R(K + k)]e^{-iKa} .
\end{aligned}
$$

(4.8.51)

Thus, by subtracting these two equations, we finally get that

$$
\begin{aligned}
R &= \frac{(k^2 - K^2)\sin Ka}{(k^2 + K^2)\sin Ka + 2ikK\cos Ka} \\
T &= \frac{2ikK\, e^{-i(k-2K)a}}{(k^2 + K^2)\sin Ka + 2ikK\cos Ka} .
\end{aligned}
$$

(4.8.52)

So, the number of particles per second that are reflected are given by

$$N|R|^2 = N \frac{(k^2 - K^2)^2 \sin^2 Ka}{(k^2 + K^2)^2 \sin^2 Ka + 4k^2 K^2 \cos^2 Ka}$$

(4.8.53)

and the number of particles per second that are transmitted are given by

$$N|T|^2 = N \frac{4k^2 K^2}{(k^2 + K^2)^2 \sin^2 Ka + 4k^2 K^2 \cos^2 Ka} \quad . \qquad (4.8.54)$$

Notice that

$$N|R|^2 + N|T|^2 = N \qquad (4.8.55)$$

as conservation of number of particles requires.

4.9   Two Beams Incident on a Potential

A. beam of free particles with intensity N_L particles per second and energy $E_L > V_0$ is incident from the left on a potential

$$V(x) = \begin{cases} 0 & x < 0 \\ V_0 & 0 < x < a \\ 0 & x > a \end{cases} \quad .$$

At the same time another beam of free particles with intensity N_R particles per second and energy $E_R > V_0$ is incident from the right. Calculate the total particle current travelling to the right. [4.1]
Hint: The waves from the left and right are completely independent and their scattering from the potential can be handled independently.

Solution

To solve this problem we simply use the results of problem 4.8. Then the total particle current to the right is given by the transmitted part of the current incident from the left plus the reflected part of the current incident from the right. Corresponding to these two currents we have a transmission amplitude T_L for the current from the left given by

$$T_L = \frac{2ik_L K_L \, e^{-i(k_L - 2K_L)a}}{(k_L^2 + K_L^2) \sin K_L a + 2ik_L K_L \cos K_L a} \qquad (4.9.56)$$

where

$$E_L = \frac{\hbar^2 k_L^2}{2m} \qquad (4.9.57)$$

and also

$$E_L = \frac{\hbar^2 K_L^2}{2m} + V_0 \quad . \qquad (4.9.58)$$

Similarly, we have a reflection amplitude R_R for the current incident from the right given by

$$R_R = \frac{(k_R^2 + K_R^2) \sin K_R a}{(k_R^2 + K_R^2) \sin K_R a + 2ik_R K_R \cos K_R a} \qquad (4.9.59)$$

where

$$E_R = \frac{\hbar^2 k_R^2}{2m} \qquad (4.9.60)$$

and also

$$E_R = \frac{\hbar^2 K_R^2}{2m} + V_0 . \qquad (4.9.61)$$

Then, the total particle current j_R to the right is given by

$$j_R = N_L |T_L|^2 + N_R |R_R|^2 . \qquad (4.9.62)$$

4.10 Ramsauer-Townsend Effect

Consider scattering of a particle with energy $E > V_0 > 0$ from a potential

$$V(x) = \begin{cases} 0 & \text{if } x < 0 \\ V_0 & \text{if } 0 < x < a \\ 0 & \text{if } x > a \end{cases} .$$

Show that if the wavelength of the particle in the region $0 < x < a$ is such that $n\lambda = a$ then no reflection occurs. Give a physical explanation for this result.

The actual Ramsauer-Townsend Effect was observed in the scattering of electrons off atoms of the noble gases since due to their closed shell configurations these atoms have very sharp outer boundaries.

Solution

This is exactly the case of resonance transmission discussed in [4.3]. The Schrödinger equation reads

$$\frac{d^2\psi}{dx^2} = \begin{cases} -k^2\psi & x < 0 \\ -K^2\psi & 0 < x < a \\ -k^2\psi & x > a \end{cases} \qquad (4.10.63)$$

where

$$k^2 = \frac{2mE}{\hbar^2} \qquad K^2 = \frac{2m(E - V_0)}{\hbar^2} . \qquad (4.10.64)$$

Assuming that we have an incoming wave from the left so that for $x > a$ we have only a transmitted wave, the solution may be written

$$\psi = \begin{cases} e^{ikx} + R e^{-ikx} & x < 0 \\ A e^{iKx} + B e^{-iKx} & 0 < x < a \\ T e^{-ikx} & x > a \end{cases} . \qquad (4.10.65)$$

Matching the wavefunction and first derivative at $x = 0$ and $x = a$ we find

$$
\begin{aligned}
1 + R &= A + B \\
ik(1 - R) &= iK(A - B) \\
A e^{iKa} + B e^{-iKa} &= T e^{ika} \\
iK \left(A e^{iKa} - B e^{-iKa} \right) &= ikT e^{ika} .
\end{aligned}
\tag{4.10.66}
$$

Solving these equations for the transmission amplitude T and computing the transmission probability $\mathcal{T} = |T|^2$ we get

$$
\mathcal{T} = \left[1 + \frac{V_0^2}{4E(E - V_0)} \sin^2 Ka \right]^{-1} .
\tag{4.10.67}
$$

Clearly for $\sin Ka = 0$ we have $\mathcal{T} = 1$. Thus, if

$$
Ka = n\pi
\tag{4.10.68}
$$

which is the same as

$$
n\lambda = 2a
\tag{4.10.69}
$$

we get total transmission.

A physical picture of what happens is that the wave inside the barrier matches exactly onto the wave outside the barrier. We have nodes at the edges of the barrier.

4.11   Wronskian and Non-degeneracy in 1 Dimension

a) For the one-dimensional Schrödinger equation with potential $V(x)$ and any two independent solutions $u(x)$ and $v(x)$, corresponding to the same energy, show that the Wronskain

$$
W(x) = u'(x)v(x) - u(x)v'(x)
$$

is a constant.

Hint: Write out the Schrödinger equation for the two solutions and multiply each of the equations by the other solution.

b) Use the result of part a) to show that in one dimension the energy eigenvalues of a Hamiltonian

$$
H = \frac{p^2}{2m} + V(x)
$$

defined on $\mathcal{L}_2(-\infty, \infty)$ are non-degenerate.

Hint: Assume that one of the eigenvalues is degenerate and show that the Wronskian of the eigenfunctions corresponding to this eigenvalue vanishes. Use this fact to obtain a contradiction.

Solution

a) Since the eigenvalue E is degenerate the two solutions corresponding to this eigenvalue satisfy

$$\frac{d^2 u}{dx^2} + (k^2 + U(x)) u = 0$$

$$\frac{d^2 v}{dx^2} + (k^2 + U(x)) v = 0 , \qquad (4.11.70)$$

where, as always,

$$k^2 = \frac{2mE}{\hbar^2} , \quad U(x) = \frac{2V(x)}{\hbar^2} .$$

If we multiply the first of (4.11.70) by v and the second of these by u and subtract, we get

$$v \frac{d^2 u}{dx^2} - u \frac{d^2 v}{dx^2} = 0 . \qquad (4.11.71)$$

This may be rewritten as

$$\frac{dW}{dx} = \frac{d}{dx} \left[v \frac{du}{dx} - u \frac{dv}{dx} \right] = 0 . \qquad (4.11.72)$$

Therefore, W is a constant and we have shown the desired result.

b) Since, the Wronskian is a constant, we can evaluate it at any point such as $x = \pm\infty$ where we have that

$$u(\pm\infty) = v(\pm\infty) = 0 .$$

Thus, we have that $W = 0$. From this it follows that we can write

$$\frac{1}{u} \frac{du}{dx} = \frac{1}{v} \frac{dv}{dx} . \qquad (4.11.73)$$

Integrating this equation we obtain that

$$\ln u = \ln v + \ln A \qquad (4.11.74)$$

where A is an integration constant. Thus,

$$u = A v \qquad (4.11.75)$$

so that u and v are linearly dependent, contrary to our assumption that there are two linearly independent eigenfunctions corresponding to the same energy E.

4.12   Symmetry of Reflection

Consider a repulsive potential $V(x) > 0$ such that it vanishes outside a finite region $a < x < b$. Show that the reflection coefficient is the same for a particle incident from either the left or the right. Assume that the solutions in the interval $a < x < b$ have been normalized such that their Wronskian (see problem 4.11) is 1.

Solution

Let $u(x)$ and $v(x)$ be two linearly independent solutions of the Schrödinger equation in the interval $a < x < b$. For a wave incident from the left the desired scattering solution of the Schrödinger equation is

$$\psi = \begin{cases} e^{ikx} + Re^{-ikx} & x < a \\ Au(x) + Bv(x) & a < x < b \\ Te^{ikx} & x > b \end{cases} \tag{4.12.76}$$

Imposing as usual the condition of continuity of ψ and ψ' we find that

$$\begin{aligned} e^{ika} + Re^{-ika} &= Au(a) + Bv(a) \\ ik\left(e^{ika} - Re^{-ika}\right) &= Au'(a) + Bv'(a) \\ Te^{ikb} &= Au(b) + Bv(b) \\ ikTe^{ikb} &= Au'(b) + Bv'(b) \ . \end{aligned} \tag{4.12.77}$$

Eliminating A and B we find from the last two equations

$$\begin{aligned} A &= Te^{ikb}\left[v'(b) - ik\,v(b)\right] \\ B &= -Te^{ikb}\left[u'(b) - ik\,u(b)\right] \ . \end{aligned} \tag{4.12.78}$$

Here, we have used the fact that the Wronskian $u(x)v'(x) - v(x)u'(x) = 1$. After substituting these into the first pair of equations (4.12.77) we find

$$\begin{aligned} e^{ika} &+ Re^{-ika} \\ &= Te^{ikb}\left[v'(b)u(a) - ik\,v(b)u(a) - u'(b)v(a) - ik\,v(b)u(a)\right] \\ ik&\left(e^{ika} - Re^{-ika}\right) \\ &= Te^{ikb}\left[v'(b)u'(a) - ik\,v(b)u'(a) - u'(b)v'(a) + iku(b)v'(a)\right]. \tag{4.12.79} \end{aligned}$$

Now, eliminating T we get

$$R = -e^{2ika}\frac{E + k^2D + ik(F - C)}{E - k^2D + ik(F + C)} \tag{4.12.80}$$

where

$$\begin{aligned} C &= u(a)v'(b) - v(a)u'(b) \\ D &= v(a)u(b) - u(a)v(b) \\ E &= u'(a)v'(b) - u'(b)v'(a) \\ F &= u(b)v'(a) - u'(a)v(b) \end{aligned} \tag{4.12.81}$$

are real numbers. Therefore,

$$|R|^2 = -\frac{(E + k^2D)^2 + k^2(F - C)^2}{(-Ek^2D)^2 + k^2(F + C)^2} \ . \tag{4.12.82}$$

Similarly, for waves incident from the right we have

$$\bar{\psi} = \begin{cases} e^{ikx} + \bar{R}e^{-ikx} & x > b \\ \bar{A}u(x) + \bar{B}v(x) & a < x < b \\ \bar{T}e^{ikx} & x < a \end{cases} \tag{4.12.83}$$

Again imposing the boundary conditions at $x = a$ and $x = b$ and solving for $|\bar{R}|^2$ we get

$$|\bar{R}|^2 = -\frac{(\bar{E} + k^2\bar{D})^2 + k^2(\bar{F} - \bar{C})^2}{(\bar{E} - k^2\bar{D})^2 + k^2(\bar{F} + \bar{C})^2} \ . \tag{4.12.84}$$

The numbers $\bar{C}$, $\bar{D}$, $\bar{E}$, $\bar{F}$ are obtained from the numbers C, D, E, F respectively by replacing a by b and changing k to $-k$. Under this transformation we find that

$$\begin{aligned} C &\leftrightarrow F \\ D &\to -D \\ E &\to -E \end{aligned} \tag{4.12.85}$$

so that

$$\begin{aligned} \bar{C} &= F \\ \bar{F} &= C \\ \bar{D} &= -D \\ \bar{E} &= -E \ . \end{aligned} \tag{4.12.86}$$

The transformation (4.12.85) clearly leaves $|R|^2$ unchanged and thus proves our result.

4.13 Parity and Electric Dipole Moment

Prove Laporte's rule, that is, show that a system of N particles with charges $q_1, q_2 \ldots, q_N$ in a state of definite parity <u>cannot</u> have an electric dipole moment.

Solution

If the wavefunction for the state of the system is

$$\psi(\vec{x}_1, \vec{x}_2, \ldots, \vec{x}_N) \tag{4.13.87}$$

and if P is the parity operator then, by assumption

$$P\psi = \pm 1 \ \psi \ . \tag{4.13.88}$$

The expression for the electric dipole moment is

$$\vec{d} = \left(\psi, \sum_{i=1}^{N} q_i \vec{x}_i \ \psi \right) \ . \tag{4.13.89}$$

We also have that

$$P\vec{x}P = -\vec{x}P^2 = -\vec{x} \ . \tag{4.13.90}$$

If we write out the expectation value on the right hand side of (4.13.89) and apply the operator $P^2 = 1$ we have

$$
\begin{aligned}
\vec{d} &= \left(\psi, P^2 \sum_{i=1}^{N} q_i \vec{x}_i P^2 \psi \right) \\
&= \left(P\psi, P \sum_{i=1}^{N} q_i \vec{x}_i PP\psi \right) \\
&= \left(\pm\psi, \sum_{i=1}^{N} q_i (P\vec{x}_i P)(\pm)\psi \right) \\
&= -\left(\psi, \sum_{i=1}^{N} q_i \vec{x}_i \psi \right) \\
&= -\vec{d} \ .
\end{aligned}
\tag{4.13.91}
$$

Thus, the electric dipole moment must vanish.

4.14   Bound State Degeneracy and Current

a) Show that a single-particle system, having a non-zero current and evolving according to a Hamiltonian H that commutes with the time-reversal operator T, must be in a degenerate state of this Hamiltonian.
b) Furthermore show that if all the states of the Hamiltonian H are non-degenerate then the system cannot have a non-zero current and may consequently be described by a real wavefunction. (See also problem 4.5.)

Solution

a) The current is given by

$$
\vec{j} = \frac{\hbar}{2im} [\psi^* \nabla \psi - \nabla \psi^* \psi] \ .
\tag{4.14.92}
$$

Now suppose that ψ is an eigenstate of the Hamiltonian H. Thus,

$$
H\psi = E\psi \ .
\tag{4.14.93}
$$

Next we use the fact that the time-reversal operator commutes with H to act on (4.14.93) from the left with T and write (remember that the energy eigenvalue E is real)

$$
THT^{-1}T\psi = ET\psi \ .
\tag{4.14.94}
$$

Using the properties of the time reversal operator we can rewrite this equation as

$$
H(T\psi) = H\psi^* = E\psi^* \ .
\tag{4.14.95}
$$

Thus, ψ^* is again an eigenfunction of H. This leaves only two possibilities
1) The state is non-degenerate so that apart from a phase ψ and ψ^* are the same

$$\psi^* = e^{i\alpha}\,\psi \ . \tag{4.14.96}$$

In this case the current (4.14.92) clearly vanishes. Thus, we are left with the second possibility.
2) The state is degenerate so that

$$\psi^* \neq e^{i\alpha}\,\psi \ . \tag{4.14.97}$$

In this case we have a non-zero current.
b) As already seen, if the state is non-degenerate we have

$$\psi^* = e^{i\alpha}\,\psi \ . \tag{4.14.98}$$

Now, since the phase of ψ is arbitrary we can multiply ψ by $e^{-i\alpha/2}$. In that case we get

$$\psi^*\,e^{i\alpha/2} = e^{i\alpha}\,e^{-i\alpha/2}\,\psi = e^{i\alpha/2}\,\psi \tag{4.14.99}$$

so that, after cancelling the factors of $e^{i\alpha/2}$ on both sides we see that

$$\psi^* = \psi \ . \tag{4.14.100}$$

4.15   Car Reflected from a Cliff

Suppose a car with a mass of 1000 kg travelling very slowly at 1.0 cm/s approaches a very sharp drop. Compute the probability that the car is reflected from this cliff.

Solution

To solve this problem we need the potential energy $V(x)$ of the car that is travelling in the x-direction. We assume that the edge of the cliff is located at $x = 0$, $y = 0$. Then for $x \leq 0$ the potential energy of the car is a constant which we set equal to zero. For $x \geq 0$ the potential energy is mgy. But, the classical trajectory of the car as it goes over the cliff is given by

$$x = vt \quad , \quad y = -\frac{1}{2}gt^2 \tag{4.15.101}$$

or

$$y = -\frac{g}{2v^2}x^2 \ . \tag{4.15.102}$$

Thus, the potential energy for $x \geq 0$ is given by

$$V(x) = -\frac{mg^2}{2v^2}x^2 \quad x \geq 0 \ . \tag{4.15.103}$$

So, the Hamiltonian is

$$H = \begin{cases} \frac{p^2}{2m} & x \leq 0 \\ \frac{p^2}{2m} - \frac{mg^2}{2v^2}x^2 & x \geq 0 \end{cases} \tag{4.15.104}$$

If we now set

$$k^2 = \frac{2mE}{\hbar^2} \quad , \quad \alpha^2 = \frac{m^2 g^2}{\hbar^2 v^2} \tag{4.15.105}$$

then the Schrödinger equation for the effective Hamiltonian becomes

$$\frac{d^2\psi}{dx^2} + k^2\psi = 0 \quad x \ \leq \ 0$$

$$\frac{d^2\psi}{dx^2} + \left(k^2 + \alpha^2 x^2\right)\psi = 0 \quad x \ \geq \ 0 \ . \tag{4.15.106}$$

For $x \leq 0$ we have an incoming wave plus the reflected wave. So, we have

$$\psi(x) = e^{ikx} + R e^{-ikx} \quad x \leq 0 \ . \tag{4.15.107}$$

For $x \geq 0$ we have a purely transmitted wave. So, we try a solution of the form

$$\psi = e^{ikx} f(x) \quad x \geq 0 \ . \tag{4.15.108}$$

Then, $f(x)$ satisfies

$$f'' + 2ik f' + \alpha^2 x^2 f = 0 \ . \tag{4.15.109}$$

The solutions of this equation can be expressed in terms of the confluent hypergeometric function. However, since we are only interested in obtaining the reflection amplitude R and not the entire solution, we can simplify the problem even further at this stage. Thus, we only need enough of the solution for $f(x)$ to be able to match the wavefunction and its drivative at $x = 0$. So, we only need the solution near $x = 0$ and we can try

$$f(x) = a_0 + a_1 x + a_2 x^2 + a_3 x^3 + \cdots \ . \tag{4.15.110}$$

After substituting this expansion into (4.15.109) we get to order x^2

$$2a_2 + 6a_3 x + 2ik(a_1 + 2a_2 x + 3a_3 x^2) + \alpha^2 a_0 x^2 = 0 \ . \tag{4.15.111}$$

Therefore,

$$\begin{aligned} 2a_2 + 2ika_1 &= 0 \\ 6a_3 + 4ika_2 &= 0 \\ 6ika_3 + \alpha^2 a_0 &= 0 \ . \end{aligned} \tag{4.15.112}$$

Thus, we find

$$\begin{aligned} a_1 &= -\frac{i\alpha^2}{4k^3}a_0 \\ a_2 &= -\frac{\alpha^2}{4k^2}a_0 \ . \end{aligned} \tag{4.15.113}$$

So, the wavefunction near $x = 0$ is given by

$$\psi(x) = \begin{cases} e^{ikx} + R\,e^{-ikx} & x \leq 0 \\ a_0\,e^{ikx}\left(1 - \frac{i\alpha^2}{4k^3}x + \cdots\right) & x \geq 0 \end{cases} \qquad (4.15.114)$$

We next match the wavefunction and derivative at $x = 0$ and get

$$\begin{aligned} 1 + R &= a_0 \\ ik(1 - R) &= ika_0\left[1 - \frac{\alpha^2}{4k^4}\right] \end{aligned} \qquad (4.15.115)$$

So, we immediately get

$$a_0 \approx 1 \quad , \quad R = \frac{\alpha^2}{4k^4} \quad . \qquad (4.15.116)$$

The reflection probability is therefore given by

$$\mathcal{R} = |R|^2 = \frac{\alpha^4}{16k^8} \quad . \qquad (4.15.117)$$

Next we replace α and k in terms of

$$\alpha^2 = \frac{m^2 g^2}{\hbar^2 v^2} \quad \text{and} \quad k^2 = \frac{2m(mv^2/2)}{\hbar^2}$$

to get

$$\mathcal{R} = \frac{1}{16}\left(\frac{g\hbar}{mv^3}\right)^4 \quad . \qquad (4.15.118)$$

An order of magnitude calculation now shows that

$$\mathcal{R} < 10^{-128} \quad . \qquad (4.15.119)$$

This shows why classical mechanics works so well when applied to macroscopic objects.

Bibliography

[4.1] A.Z. Capri, *Nonrelativistic Quantum Mechanics* 3rd edition, World Scientific Publishing Co. Pte. Ltd., section 4.3 and 4.4, (2002) .

[4.2] ibid, section 4.5.

[4.3] ibid, section 4.6.1 .

Chapter 5

More One-Dimensional Problems

5.1   Motion of a Wavepacket

A wavefunction

$$\Psi(t, x) = \int_{-\infty}^{\infty} f(k) e^{i[kx - \omega(k)t]} \, dk$$

is normalized such that

$$\int_{-\infty}^{\infty} |\Psi(t, x)|^2 \, dx = 1 \ .$$

Assume $f(k)$ is a smooth function vanishing rapidly at infinity. Show that the velocity of the centre of mass $\bar{x}$ defined by

$$\bar{x} = \int_{-\infty}^{\infty} x |\Psi(t, x)|^2 \, dx$$

is given by

$$\frac{d\bar{x}}{dt} = 2\pi \int_{-\infty}^{\infty} \frac{d\omega}{dk} |f(k)|^2 \, dk \ .$$

Hint: Use the fact that

$$\int_{-\infty}^{\infty} e^{i(k-q)x} \, dx = 2\pi \delta(k - q) \ .$$

Show furthermore that if $\omega(k)$ is adequately approximated by

$$\omega(k) \approx \omega(0) + (k - k_0) \left. \frac{d\omega}{dk} \right|_{k_0}$$

then

$$\frac{d\bar{x}}{dt} \approx \frac{d\omega}{dk}\Big|_{k_0} \quad,$$

the group velocity.

Solution

We first write out $|\Psi|^2$ in detail.

$$
\begin{aligned}
|\Psi|^2 &= \int_{-\infty}^{\infty} f^*(q) e^{-i[qx-\omega(q)t]} \, dq \int_{-\infty}^{\infty} f(k) e^{i[kx-\omega(k)t]} \, dk \\
&= \int_{-\infty}^{\infty}\int_{-\infty}^{\infty} dk\,dq\, f^*(q) f(k) e^{i(k-q)x} e^{-i[\omega(k)-\omega(q)]t} \quad .
\end{aligned}
\tag{5.1.1}
$$

Now,

$$
\begin{aligned}
\bar{x} &= \int_{-\infty}^{\infty} x\,dx \int_{-\infty}^{\infty}\int_{-\infty}^{\infty} dk\,dq\, f^*(q) f(k) e^{i(k-q)x} e^{-i[\omega(k)-\omega(q)]t} \\
&= \int_{-\infty}^{\infty}\int_{-\infty}^{\infty} dk\,dq\, f^*(q) f(k) e^{-i[\omega(k)-\omega(q)]t} \int_{-\infty}^{\infty} x\, e^{i(k-q)x}\,dx \quad .
\end{aligned}
\tag{5.1.2}
$$

But,

$$x\, e^{i(k-q)x} = i\frac{d}{dq} e^{i(k-q)x} \quad .\tag{5.1.3}$$

Therefore,

$$
\begin{aligned}
\bar{x} &= \int_{-\infty}^{\infty}\int_{-\infty}^{\infty} dk\,dq\, f^*(q) f(k) e^{-i[\omega(k)-\omega(q)]t} i\frac{d}{dq} \int_{-\infty}^{\infty} e^{i(k-q)x}\,dx \\
&= \int_{-\infty}^{\infty}\int_{-\infty}^{\infty} dk\,dq\, f^*(q) f(k) e^{-i[\omega(k)-\omega(q)]t} i\frac{d}{dq} 2\pi\delta(k-q) \quad .
\end{aligned}
\tag{5.1.4}
$$

If we differentiate under the integral we now get

$$
\begin{aligned}
\frac{d\bar{x}}{dt} &= 2\pi \int_{-\infty}^{\infty}\int_{-\infty}^{\infty} dk\,dq\, f^*(q) f(k)[\omega(q)-\omega(k)] e^{-i[\omega(k)-\omega(q)]t} \\
&\quad \times \frac{d}{dq}\delta(k-q) \quad .
\end{aligned}
\tag{5.1.5}
$$

Now integrate by parts to get the desired result

$$\frac{d\bar{x}}{dt} = 2\pi \int_{-\infty}^{\infty} dk|f(k)|^2 \frac{d\omega(k)}{dk} \quad .\tag{5.1.6}$$

If

$$\omega(k) \approx \omega_0 + (k-k_0)\frac{d\omega}{dk}\Big|_{k_0}\tag{5.1.7}$$

then,

$$\frac{d\omega}{dk} \approx 0 + \frac{d\omega}{dk}\Big|_{k_0} \tag{5.1.8}$$

so that

$$\frac{d\bar{x}}{dt} = 2\pi \frac{d\omega}{dk}\Big|_{k_0} \int_{-\infty}^{\infty} dk|f(k)|^2 \ . \tag{5.1.9}$$

Now we use the fact that

$$\int |\psi|^2 dx = 2\pi \int |f(k)|^2 dk = 1$$

to get the final result,

$$\frac{d\bar{x}}{dt} = \frac{d\omega}{dk}\Big|_{k_0} \ . \tag{5.1.10}$$

5.2   Lowest Energy States

A particle is in the potential

$$V(x) = V_0 \exp[ax^2/2] \ .$$

Estimate the energy of the 2 lowest eigenstates. What are the parities of these states?

Solution

The potential is

$$V = V_0 \exp\left[\frac{1}{2}ax^2\right] \ . \tag{5.2.11}$$

The lowest lying states will be centred near $x = 0$. In the vicinity of this point we have

$$V \approx V_0 \left[1 + \frac{1}{2}ax^2\right] \ . \tag{5.2.12}$$

This approximate potential is just the potential for a simple harmonic oscillator with its zero shifted and with a spring constant $k = V_0a$. Therefore,

$$\omega = \sqrt{k/m} = \sqrt{V_0a/m} \ . \tag{5.2.13}$$

Then,

$$E_0 \approx V_0 + \frac{1}{2}\hbar\omega \tag{5.2.14}$$

corresponding to the ground state of a simple hamonic oscillator with parity $= +1$, and

$$E_1 \approx V_0 + \frac{3}{2}\hbar\omega \tag{5.2.15}$$

corresponding to the first excited state of a simple hamonic oscillator with parity $= -1$.

5.3 Particle at Rest

Find the wavefunction for a particle at rest at the origin of a coordinate system fixed in space.

Solution

We have to solve the eigenvalue problem

$$x f(x) = 0 f(x) \ . \tag{5.3.16}$$

The only function with this property is

$$f(x) = \delta(x) \ . \tag{5.3.17}$$

5.4 Scattering from Two Delta Functions

Calculate the transmission probability for a particle of mass m incident on a potential

$$V(x) = \Lambda[\delta(x+a) + \delta(x-a)] \ .$$

Compute also the phase shifts.

Solution

The potential is an even function of x so parity considerations may be used for the phase shifts. But first we need the correct boundary conditions. These are:
1) $\psi(x)$ is continuous at both $x = \pm a$.
2) The δ-functions indicate that there is a discontinuity in $d\psi/dx$ at $x = \pm a$. As usual, we find this discontinuity by integrating the Schrödinger equation over a tiny interval that includes the discontinuity. The Schrödinger equation in terms of the reduced variables

$$k^2 = \frac{2mE}{\hbar^2} \quad , \quad K = \frac{2m\Lambda_0}{\hbar^2} \tag{5.4.18}$$

reads

$$\frac{d^2\psi}{dx^2} + k^2\psi = K[\delta(x+a) + \delta(x-a)]\psi \ . \tag{5.4.19}$$

The discontinuity in the derivative at $x = \pm a$ is therefore given by

$$\left.\frac{d\psi}{dx}\right|_{\pm a+0} - \left.\frac{d\psi}{dx}\right|_{\pm a-0} = K\psi(\pm a) \ . \tag{5.4.20}$$

The solution may be written

$$\psi = \begin{cases} e^{ikx} + R\,e^{-ikx} & x \le -a \\ A\,e^{ikx} + B\,e^{-ikx} & -a \le x \le a \\ T\,e^{ikx} & x \ge a \end{cases} \ . \tag{5.4.21}$$

Here we have already imposed the physical boundary conditions that for $x \leq -a$ we have an incoming and a reflected wave and for $x \geq a$ we have a purely transmitted wave. For $-a \leq x \leq a$ we have a combination of waves travelling to the left and to the right.

Applying the boundary conditions at $x = -a$ we get

$$e^{-ika} + R e^{ika} = A e^{-ika} + B e^{ika} \tag{5.4.22}$$

and

$$ik \left(e^{-ika} - R e^{ika} \right) - ik \left(A e^{-ika} - B e^{ika} \right) = -K \left(e^{-ika} + R e^{ika} \right) . \tag{5.4.23}$$

Solving for A and B we find

$$A = \frac{K + 2ik}{2ik} + R \frac{K}{2ik} e^{2ika} \tag{5.4.24}$$

and

$$B = -\frac{K}{2ik} e^{2ika} + R \frac{-K + 2ik}{2ik} . \tag{5.4.25}$$

Similarly, applying the boundary conditions at $x = a$ we get

$$A e^{ika} + B e^{-ika} = T e^{ika} \tag{5.4.26}$$

and

$$ik \left(A e^{ika} + B e^{-ika} \right) + ikT e^{ika} = KT e^{ika} . \tag{5.4.27}$$

Again, solving for A and B we find

$$A = \frac{K + 2ik}{2ik} T \quad , \quad B = -\frac{K}{2ik} T e^{2ika} . \tag{5.4.28}$$

Equating the two expressions for A and B we get two equations for R and T.

$$(K - 2ik)T + K R e^{2ika} = -(K + 2ik) \tag{5.4.29}$$

and

$$KT e^{2ika} + (K + 2ik)R = K e^{-2ika} . \tag{5.4.30}$$

After solving these for T and R we get

$$T = \frac{\left[K^2 + (K + 2ik)^2 \right] e^{-2ika}}{2iK^2 \sin(2ka) - 4k^2 e^{-2ika}} \tag{5.4.31}$$

and

$$R = -\frac{K \left[(K - 2ik) e^{-4ika} + (K + 2ik) \right]}{2iK^2 \sin(2ka) - 4k^2 e^{-2ika}} . \tag{5.4.32}$$

To compute the phase shifts we divide the solutions into positive parity solutions ψ_+ and negative parity solutions ψ_-

$$\psi_+ = \begin{cases} \cos(kx - \delta_+) & x \leq -a \\ A\cos(kx) & -a \leq x \leq a \\ \cos(kx + \delta_+) & x \geq a \end{cases} \tag{5.4.33}$$

and

$$\psi_- = \begin{cases} i\sin(kx - \delta_-) & x \leq -a \\ B\sin(kx) & -a \leq x \leq a \\ i\sin(kx + \delta_-) & x \geq a \end{cases} \tag{5.4.34}$$

Writing out the boundary conditions at either $x = a$ or else $x = -a$ we get

$$A\cos(ka) = \cos(ka + \delta_+) \tag{5.4.35}$$

$$-k\sin(ka + \delta_+) + k\sin(ka) = K\cos(ka + \delta_+) . \tag{5.4.36}$$

Therefore, we get

$$\tan(ka) = \tan(ka + \delta_+) + \frac{K}{k} . \tag{5.4.37}$$

So,

$$\delta_+ = \arctan[\tan(ka) - K/k] - ka . \tag{5.4.38}$$

Proceeding in the same way for the negative parity solution we find

$$B\sin(ka) = i\sin(ka + \delta_-) \tag{5.4.39}$$

$$ik\cos(ka + \delta_-) - kB\cos(ka) = iK\sin(ka + \delta_-) . \tag{5.4.40}$$

Therefore,

$$\cot(ka) = \cot(ka + \delta_-) - \frac{K}{k} . \tag{5.4.41}$$

So,

$$\delta_- = \text{arccot}[\cot(ka) + K/k] - ka . \tag{5.4.42}$$

5.5 Reflection and Transmission Amplitudes: Phase Shifts

Complete all the steps in going from

$$R = \frac{1}{2}\left[e^{2i\delta_+} - e^{2i\delta_-}\right] ,$$

$$T = \frac{1}{2}\left[e^{2i\delta_+} + e^{2i\delta_-}\right] ,$$

$$\delta_+ = \arctan\left[\frac{K}{k}\tan KA\right] - ka ,$$

and

$$\delta_- = \cot^{-1}\left[\frac{K}{k}\cot KA\right] - ka$$

to

$$T = \frac{(k^2 - K^2)e^{-2ika}}{(k^2 + K^2) + ikK(\cot Ka - \tan Ka)}$$

and

$$R = \frac{ikK(\cot Ka + \tan Ka)e^{-2ika}}{(k^2 + K^2) + ikK(\cot Ka - \tan Ka)} .$$

Solution

Our starting point is

$$T = \frac{1}{2}\left(e^{2i\delta_+} + e^{2i\delta_-}\right) , \quad R = \frac{1}{2}\left(e^{2i\delta_+} - e^{2i\delta_-}\right) . \tag{5.5.43}$$

Also,

$$\delta_+ = \arctan[\frac{K}{k}\tan(Ka)] - ka \tag{5.5.44}$$

and

$$\delta_- = \text{arccot}[\frac{K}{k}\cot(Ka)] - ka . \tag{5.5.45}$$

From (5.5.44) we get

$$\tan(\delta_+ + ka) = \frac{K}{k}\tan Ka . \tag{5.5.46}$$

But,

$$\begin{aligned}
\tan(\delta_+ + ka) &= \frac{e^{i(\delta_+ + ka)} - e^{-i(\delta_+ + ka)}}{i[e^{i(\delta_+ + ka)} + e^{-i(\delta_+ + ka)}]}\\
&= \frac{e^{2i(\delta_+ + ka)} - 1}{i[e^{2i(\delta_+ + ka)} + 1]} .
\end{aligned} \tag{5.5.47}$$

Therefore,

$$e^{2i(\delta_+ + ka)} - 1 = i[e^{2i(\delta_+ + ka)} + 1]\frac{K}{k}\tan Ka . \tag{5.5.48}$$

So,

$$e^{2i\delta_+} = \frac{k + iK\tan Ka}{k - iK\tan Ka}e^{-2ika} . \tag{5.5.49}$$

Similarly, starting with (5.5.45) we get

$$\cot(\delta_- + ka) = \frac{i[e^{2i(\delta_- + ka)} + 1]}{e^{2i(\delta_- + ka)} - 1]}$$

$$= \frac{K}{k} \cot Ka .$$ (5.5.50)

Therefore,

$$e^{2i\delta_-} = -\frac{k - iK \cot Ka}{k + iK \cot Ka} e^{-2ika} .$$ (5.5.51)

Combining these results we have

$$T = \frac{1}{2} \left(e^{2i\delta_+} + e^{2i\delta_-} \right)$$

$$= \frac{1}{2} \left[\frac{k + iK \tan Ka}{k - iK \tan Ka} - \frac{k - iK \cot Ka}{k + iK \cot Ka} \right] e^{-2ika}$$

$$= \frac{k^2 - K^2}{k^2 + K^2 + ikK[\cot Ka - \tan Ka]} e^{-2ika}$$ (5.5.52)

and

$$R = \frac{1}{2} \left(e^{2i\delta_+} - e^{2i\delta_-} \right)$$

$$= \frac{1}{2} \left[\frac{k + iK \tan Ka}{k - iK \tan Ka} + \frac{k - iK \cot Ka}{k + iK \cot Ka} \right] e^{-2ika}$$

$$= \frac{ikK[\cot Ka + \tan Ka]}{k^2 + K^2 + ikK[\cot Ka - \tan Ka]} e^{-2ika} .$$ (5.5.53)

5.6 Oscillator Against a Solid Wall

Find the energy levels and normalized wave functions of the stationary states
of a particle moving in the potential

$$V(x) = \begin{cases} \infty & \text{if } x < 0 \\ \frac{1}{2}kx^2 & \text{if } x > 0 \end{cases} .$$

Compare the zero point energy in this potential with that for the simple harmonic oscillator with the same force constant k. Explain any differences you
find.

Solution

For $x > 0$ we again have just the simple harmonic oscillator solutions

$$\psi_n(x) = H_n(y)e^{-1/2 y^2} , \quad y = \sqrt{\frac{m\omega}{\hbar}} x , \quad E_n = (n + 1/2)\hbar\omega .$$ (5.6.54)

However, we now have to impose the additional boundary condition that the wavefunction should vanish at $x = 0$. The easiest way to do this is to take the solutions of the simple harmonic oscillator above and keep only the odd parity solutions since they must vanish at $x = 0$. This requires that n should be an odd integer

$$n = 2p + 1 \quad , \quad p = 0, 1, 2, 3, \dots . \tag{5.6.55}$$

This means that the lowest energy is now the same as that of the first excited state of the simple harmonic oscillator. Since the particle is more confined, the ground state energy is greater. Furthermore all the even parity states are missing.

5.7   Periodic Potential

Find the equation that determines the allowed energy bands for a particle moving in the periodic potential (see figure 5.1)

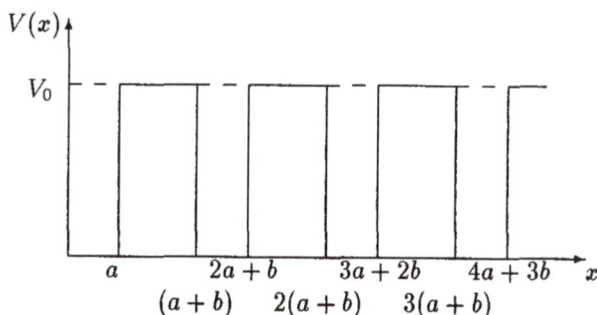

Figure 5.1: Periodic potential with period $(a + b)$.

$$V = \begin{cases} 0 & 0 < x < a \\ V_0 & a < x < a + b \end{cases}$$

where $V(x) = V(x + (a + b)) = V(x + 2(a + b)) = \dots$.

Solution

We first solve the problem in the interval $-b < x < a$ and then use Bloch's Theorem [5.1] to extend the solution to $a < x < a + b$.

For $0 < x < a$ we have with

$$k_1^2 = \frac{2mE}{\hbar^2}$$

that

$$\psi = c_1 e^{ik_1 x} + c_2 e^{-ik_1 x} \quad 0 < x < a . \tag{5.7.56}$$

For $-b < x < 0$ we have with

$$k_2^2 = \frac{2m(E - V_0)}{\hbar^2}$$

that

$$\psi = c_3 e^{ik_2 x} + c_4 e^{-ik_2 x} \qquad -b < x < 0 \ . \tag{5.7.57}$$

Now using Bloch's Theorem we get that

$$\psi = e^{iK(a+b)} \left[c_3 e^{ik_2(x-a-b)} + c_4 e^{-ik_2(x-a-b)} \right] \qquad a < x < a+b \ . \tag{5.7.58}$$

The wavefunction and its derivative must be continuous. Imposing these conditions at $x = 0$ and $x = a$ we get

$$
\begin{aligned}
c_1 + c_2 &= c_3 + c_4 \ . \\
k_1(c_1 - c_2) &= k_2(c_3 - c_4) \ . \\
c_1 e^{ik_1 a} + c_2 e^{-ik_1 a} &= e^{iK(a+b)} \left[c_3 e^{-ik_2 b} + c_4 e^{ik_2 b} \right] \ . \\
k_1 \left[c_1 e^{ik_1 a} - c_2 e^{-ik_1 a} \right] &= k_2 e^{iK(a+b)} \left[c_3 e^{-ik_2 b} - c_4 e^{ik_2 b} \right] \ . \tag{5.7.59}
\end{aligned}
$$

To obtain a nontrivial solution for the parameters c_i we require that the determinant of their coefficients vanish

$$
\begin{vmatrix}
1 & 1 & -1 & -1 \\
k_1 & -k_1 & -k_2 & k_2 \\
e^{ik_1 a} & e^{-ik_1 a} & -e^{iK(a+b)}e^{-ik_2 b} & -e^{iK(a+b)}e^{ik_2 b} \\
k_1 e^{ik_1 a} & -k_1 e^{-ik_1 a} & -k_2 e^{iK(a+b)}e^{-ik_2 b} & k_2 e^{iK(a+b)}e^{ik_2 b}
\end{vmatrix} = 0 \ .
$$

$$\tag{5.7.60}$$

Expanding this out and after some tedious simplifications we get that

$$\cos K(a + b) = \cos k_1 a \cos k_2 b - \frac{k_1^2 + k_2^2}{2k_1 k_2} \sin k_1 a \sin k_2 b \ . \tag{5.7.61}$$

If $E > V_0$ this equation determines the energy bands since it requires that

$$-1 \leq \cos k_1 a \cos k_2 b - \frac{k_1^2 + k_2^2}{2k_1 k_2} \sin k_1 a \sin k_2 b \leq 1 \quad \text{for} \quad E > V_0 \ . \tag{5.7.62}$$

For $E < V_0$ we have to replace k_2 by $i\kappa_2$. In this case the energy bands are determined by

$$-1 \leq \cos k_1 a \cosh \kappa_2 b - \frac{\kappa_2^2 - k_1^2}{2k_1 \kappa_2} \sin k_1 a \sinh \kappa_2 b \leq 1 \quad \text{for} \quad E < V_0 \ . \tag{5.7.63}$$

5.8 Reflection and Transmission Through a Barrier

A particle is incident on a smooth, one-dimensional potential barrier whose maximum height is greater than the energy of the particle. Furthermore, the potential goes to zero rapidly at $x \to \pm\infty$. Verify unitarity. That is, show that the sum of the reflection and transmission probabilities is unity.
Hint: Use the equation of continuity for the probability current.

Solution

Since the potential vanishes rapidly for large values of $|x|$ we can look in the asymptotic regions. If we assume that the particle is incident from the left, then for $x \to -\infty$ we have the incident wave plus a reflected wave, say

$$\psi = e^{ikx} + R e^{-ikx} \ . \tag{5.8.64}$$

For $x \to \infty$ we have only a transmitted wave

$$\psi = T e^{ikx} \ . \tag{5.8.65}$$

The probability current in the two regions is

$$
\begin{aligned}
j &= \frac{\hbar k}{m}[1 - |R|^2] \quad x \to -\infty \\
j &= \frac{\hbar k}{m}|T|^2 \quad x \to +\infty
\end{aligned}
\tag{5.8.66}
$$

The equation of continuity states that

$$\frac{dj}{dx} = 0 \ . \tag{5.8.67}$$

If we integrate this equation from $-L$ to L for very large L we get

$$\int_{-L}^{L} \frac{dj}{dx} dx = 0 = j(L) - j(-L) \ . \tag{5.8.68}$$

Writing this out using the asymptotic forms for j we get

$$\frac{\hbar k}{m}[1 - |R|^2] = \frac{\hbar k}{m}|T|^2 \ . \tag{5.8.69}$$

Therefore,

$$|R|^2 + |T|^2 = 1 \ . \tag{5.8.70}$$

5.9 Hermite Polynomials: Integral Representation

Show that if we define the Hermite polynomials by the integral representation

$$H_n(x) = \frac{2^n}{\sqrt{\pi}} \int_{-\infty}^{\infty} (x + is)^n e^{-s^2} ds \tag{5.9.71}$$

then the following results hold.

a)

$$H_n'(x) = 2n\, H_{n-1}(x)$$

b)

$$2x\, H_n(x) = H_{n+1}(x) + 2n\, H_{n-1}(x)$$

and finally
c)

$$H_n''(x) - 2x\,H_n'(x) + 2n\,H_n(x) = 0 \ .$$

This last equation shows that the functions $H_n(x)$ defined by the integral (5.9.71) satisfy the differential equation for the Hermite polynomials.

Solution

a) If we differentiate the integral representation under the integral sign we immediately get

$$H_n'(x) = \frac{2^n}{\sqrt{\pi}} \int_{-\infty}^{\infty} n(x+is)^{n-1}\,e^{-s^2}\,ds = 2n\,H_{n-1}(x) \ . \tag{5.9.72}$$

b) Also, we immediately see that

$$
\begin{aligned}
2x\,H_n(x) &= \frac{2^{n+1}}{\sqrt{\pi}} \int_{-\infty}^{\infty} x(x+is)^n\,e^{-s^2}\,ds \\
&= H_{n+1}(x) - \frac{2^{n+1}}{\sqrt{\pi}} \int_{-\infty}^{\infty} is(x+is)^n\,e^{-s^2}\,ds \ .
\end{aligned}
\tag{5.9.73}
$$

But,

$$\int_{-\infty}^{\infty} (x+is)^n\,de^{-s^2} = \int_{-\infty}^{\infty} -2s(x+is)^n\,e^{-s^2}\,ds \ . \tag{5.9.74}$$

After an integration by parts this yields

$$\int_{-\infty}^{\infty} is(x+is)^n\,e^{-s^2}\,ds = -\frac{n}{2} \int_{-\infty}^{\infty} (x+is)^{n-1}\,e^{-s^2}\,ds \ . \tag{5.9.75}$$

So, combining this with (5.9.73) we have that

$$2x\,H_n(x) = H_{n+1}(x) + 2n\,H_{n-1}(x) \tag{5.9.76}$$

as desired.

c) Now, using (5.9.72) together with (5.9.76) we have that

$$H_n'(x) = 2n\,H_{n-1}(x) = 2x\,H_n(x) - H_{n+1}(x) \ . \tag{5.9.77}$$

From this it follows, after another application of (5.9.72) that

$$
\begin{aligned}
H_n''(x) &= 2\,H_n(x) + 2x\,H_n'(x) - H_{n+1}'(x) \\
&= 2H_n(x) + 2x\,H_n'(x) - 2(n+1)H_n(x) \\
&= 2x\,H_n'(x) - 2n\,H_n(x) \ .
\end{aligned}
\tag{5.9.78}
$$

5.10   Matrix Element Between Degenerate States

Let ϕ_1 and ϕ_2 be two degenerate eigenfunctions of the Hamiltonian

$$H = \frac{p^2}{2m} + V(x) .$$

That is they correspond to the same eigenvalue E. Show that the matrix element

$$(\phi_1, (xp + px)\phi_2) = \int \phi_1^*(xp + px)\phi_2 \, dx = 0 .$$

Hint: Use the fact that $xp + px$ is self-adjoint.

Solution

Using the hint we have that

$$(\phi_1, (xp + px)\phi_2) = ((xp + px)\phi_1, \phi_2) . \tag{5.10.79}$$

Writing this out we get

$$\frac{\hbar}{i} \int \phi_1^* [x\frac{d\phi_2}{dx} + \frac{d}{dx}(x\phi_2)] \, dx = -\frac{\hbar}{i} \int [x\frac{d\phi_1^*}{dx} + \frac{d}{dx}(x\phi_1^*)]\phi_2 \, dx . \tag{5.10.80}$$

Using this we can rewrite the matrix element as

$$(\phi_1, (xp + px)\phi_2)$$
$$= \frac{1}{2}\frac{\hbar}{i} \int \left[\phi_1^* x\frac{d\phi_2}{dx} - x\frac{d\phi_1^*}{dx}\phi_2 + \phi_1^*\frac{d}{dx}(x\phi_2) - \frac{d}{dx}(x\phi_1^*)\phi_2 \right] \, dx$$
$$= \frac{\hbar}{i} \int x \left[\phi_1^*\frac{d\phi_2}{dx} - \frac{d\phi_1^*}{dx}\phi_2 \right] \, dx . \tag{5.10.81}$$

Furthermore, using the fact that the two eigenfunctions are degenerate we have

$$-\frac{\hbar^2}{2m}\frac{d^2\phi_1^*}{dx^2} + V\phi_1^* = E\phi_1^*$$
$$-\frac{\hbar^2}{2m}\frac{d^2\phi_2}{dx^2} + V\phi_2 = E\phi_2 . \tag{5.10.82}$$

So, multiplying the first of these equations by ϕ_2 and the second by ϕ_1^* and subtracting we get

$$-\frac{\hbar^2}{2m}\left(\phi_2\frac{d^2\phi_1^*}{dx^2} - \phi_1^*\frac{d^2\phi_2}{dx^2} \right) = 0 . \tag{5.10.83}$$

This may be rewritten to read

$$\frac{d}{dx}\left(\phi_2\frac{d\phi_1^*}{dx} - \phi_1^*\frac{d\phi_2}{dx} \right) = 0 . \tag{5.10.84}$$

From this it immediately follows that

$$x^2 \frac{d}{dx}\left(\phi_2 \frac{d\phi_1^*}{dx} - \phi_1^* \frac{d\phi_2}{dx}\right) = 0 \ . \tag{5.10.85}$$

If we now consider the expression

$$\int \frac{d}{dx}\left[x^2\left(\phi_2 \frac{d\phi_1^*}{dx} - \phi_1^* \frac{d\phi_2}{dx}\right)\right] dx \tag{5.10.86}$$

we see that it vanishes since both ϕ_1^* and ϕ_2 vanish rapidly at $\pm\infty$. If we furthermore carry out the indicated differentiation and use (5.10.85) we get the desired result that

$$\int 2x\left(\phi_2 \frac{d\phi_1^*}{dx} - \phi_1^* \frac{d\phi_2}{dx}\right) dx = 0 \ . \tag{5.10.87}$$

5.11 Hellmann-Feynman Theorem

Consider a Hamiltonian $H(\lambda)$ that depends on a parameter λ and let A be some operator such that $[A, H]$ is well defined. Prove that for any stationary state $\psi(\lambda)$ of this Hamiltonian we have that

$$\frac{\partial E(\lambda)}{\partial \lambda}\langle A\rangle = \left\langle A \frac{\partial H(\lambda)}{\partial \lambda}\right\rangle + \left(\psi, [A, H]\frac{\partial \psi}{\partial \lambda}\right) \ .$$

This is known as the generalized Hellmann-Feynman theorem. For $A = 1$ it is the Hellmann-Feynman theorem.

Solution

The equation for the stationary states reads

$$H(\lambda)\psi(\lambda) = E(\lambda)\psi(\lambda) \ . \tag{5.11.88}$$

If we multiply this equation from the left by A, take the expectation value of this equation, and differentiate the resultant equation with respect to λ we get

$$E(\frac{\partial \psi}{\partial \lambda}, A\psi) + (\psi, A\frac{\partial H}{\partial \lambda}\psi) + (\psi, AH\frac{\partial \psi}{\partial \lambda})$$
$$= \frac{\partial E}{\partial \lambda}(\psi, A\psi) + E(\frac{\partial \psi}{\partial \lambda}, A\psi) + E(\psi, A\frac{\partial \psi}{\partial \lambda}) \ . \tag{5.11.89}$$

But,

$$E(\psi, A\frac{\partial \psi}{\partial \lambda}) = (H\psi, A\frac{\partial \psi}{\partial \lambda})$$
$$= (\psi, HA\frac{\partial \psi}{\partial \lambda}) \ . \tag{5.11.90}$$

Hence, we get the required result

$$\frac{\partial E}{\partial \lambda}\langle A\rangle = \langle A\frac{\partial H}{\partial \lambda}\rangle + (\psi, [A, H]\frac{\partial \psi}{\partial \lambda}) \ . \tag{5.11.91}$$

Bibliography

[5.1] A.Z. Capri, *Nonrelativistic Quantum Mechanics* 3rd edition, World Scientific Publishing Co. Pte. Ltd., section 5.8, (2002) .

Chapter 6

Mathematical Foundations

6.1 Cauchy Sequence in a Finite Vector Space

Show that every Cauchy sequence in a finite dimensional vector space converges strongly (converges using the norm).

Solution

We are given a sequence of vectors $\{v_j \ \ j = 1, 2, 3, \ldots\}$ where each of these vectors is an element of an r-dimensional vector space and may therefore be written for a fixed basis in terms of its components as

$$v_j = (v_{j1}, v_{j2}, \ldots, v_{jr}) \ . \tag{6.1.1}$$

Furthermore, we are told that these vectors form a Cauchy sequence. This means that given an $\epsilon > 0$ we can always find a number N such that for $n, m > N$ we have

$$\left(|v_{n1} - v_{m1}|^2 + |v_{n2} - v_{m2}|^2 + \cdots + |v_{nr} - v_{mr}|^2 \right) < \epsilon \ . \tag{6.1.2}$$

As a consequence, we can bound each term $|v_{ns} - v_{ms}|$ where $1 \leq s \leq r$ by ϵ. This means that each component of the sequence converges to some finite limit u_s. Hence, it follows that the limit vector is

$$u = (u_1, u_2, \ldots, u_r) \tag{6.1.3}$$

and the norm of this limit vector is finite.

Notice that if our space were infinite dimensional, and that even if every component u_s were finite, we could not conclude that

$$|u|^2 = \sum_{s=1}^{\infty} |u_s|^2 \tag{6.1.4}$$

is finite since we would now have to sum over an infinite number of components. This is what makes completeness in an infinite dimensional space non-trivial.

6.2   Nonuniqueness of Schrödinger Representation

Show that if

$$[x, p] = i$$

then there exists a unitary operator S such that for a given a real function $f(x)$ we have

$$SpS^{-1} = p + f(x) \ .$$

Solution

The proof is by displaying the operator S. In the Schrödinger representation we have that

$$p = \frac{1}{i}\frac{d}{dx} \quad , \quad x = x. \tag{6.2.5}$$

Thus, we try with a real differentiable function $h(x)$,

$$S = e^{-ih(x)} \quad , \quad S^{-1} = S^\dagger = e^{ih(x)} \ . \tag{6.2.6}$$

Then, for a function g in the domain of p we have

$$\begin{aligned}
pS^{-1}g &= \frac{1}{i}\frac{d}{dx}\left(e^{ih(x)}g(x)\right) \\
&= e^{ih(x)}[p + h'(x)]g(x)
\end{aligned} \tag{6.2.7}$$

where

$$h'(x) = \frac{dh}{dx}(x) \ .$$

Thus,

$$SpS^{-1} = p + h'(x) \ . \tag{6.2.8}$$

So, we simply have to choose

$$h'(x) = f(x) \ . \tag{6.2.9}$$

The unitary operator S serves to give the same additional phase $e^{ih(x)}$ to every state in the hilbert space and thus has no observable effect. This means, we are free to use the Schrödinger representation and don't have to worry about this nonuniqueness.

6.3   Degeneracy and Commutator

Show that if all eigenvalues of the self-adjoint operator A are non-degenerate and that if the self-adjoint operator B commutes with A, i.e.

$$[A, B] = 0$$

then B is a function of A.

Solution

We have

$$A|n\rangle = a_n|n\rangle \ . \tag{6.3.10}$$

The kets $|n\rangle$ form a complete set. Also, using the commutator we have

$$BA|n\rangle = a_n B|n\rangle = AB|n\rangle \ . \tag{6.3.11}$$

Thus,

$$A(B|n\rangle) = a_n(B|n\rangle) \tag{6.3.12}$$

and because the eigenvalues a_n are non-degenerate we have that

$$B|n\rangle = c_n|n\rangle \ . \tag{6.3.13}$$

Now, corresponding to any ket $|n\rangle$ there is a unique eigenvalue a_n. Conversely, corresponding to any eigenvalue a_n there is a unique ket $|n\rangle$. This means that there is a one-one correspondence between n and a_n. So, we can write

$$c_n = c(a_n) \ . \tag{6.3.14}$$

Using (6.3.13) it therefore follows that

$$B|n\rangle = c(a_n)|n\rangle = c(A)|n\rangle \tag{6.3.15}$$

and since, as stated, the $|n\rangle$ form a complete set we have that

$$B = c(A) \ . \tag{6.3.16}$$

6.4 von Neumann's Example

Consider the operator

$$A_n = px^{2n+1} + x^{2n+1}p \ , \quad n = 1, 2, 3, \ldots$$

where

$$p = \frac{\hbar}{i}\frac{d}{dx} \ .$$

Find the eigenvalues and eigenfunctions of A_n. What are the deficiency indices of A_n? The Hilbert space in this case is $\mathcal{L}_2(-\infty, \infty)$. For $n = 1$ this example is due to von Neumann.

Solution

We write the eigenvalue as $\lambda\hbar$. Then, the eigenvalue equation for A_n reads

$$\frac{\hbar}{i}\frac{d}{dx}(x^{2n+1}f) + \frac{\hbar}{i}x^{2n+1}\frac{df}{dx} = \lambda\hbar f \ . \tag{6.4.17}$$

After rearranging this becomes

$$\frac{df}{dx} - \frac{i\lambda}{2}x^{-(2n+1)}f + \frac{2n+1}{2x}f = 0 \ . \tag{6.4.18}$$

Separating we get

$$\frac{df}{f} = \frac{i\lambda}{2}x^{-(2n+1)}dx - \frac{2n+1}{2x}dx \ . \tag{6.4.19}$$

For $x \neq 0$ this can be integrated to yield

$$f = \frac{A}{x^{(n+1/2)}} \exp\left[\frac{-i\lambda}{4n}x^{-2n}\right] \ . \tag{6.4.20}$$

Since this function must be square integrable we consider

$$|f|^2 = \frac{|A|^2}{x^{(2n+1)}} \exp\left[\frac{\Im(\lambda)}{2n}x^{-2n}\right] \ . \tag{6.4.21}$$

This requires that $\Im(\lambda) < 0$ so that the singularity at $x = 0$ does not cause the function to blow up too strongly at the origin. Therefore, all eigenvalues are of the form $\lambda\hbar$ with

$$\lambda\hbar = \alpha - i\beta \ , \quad \beta > 0 \ . \tag{6.4.22}$$

So we see that we get exactly one solution of the equation

$$A_n^\dagger f = -if \tag{6.4.23}$$

and no solutions of the equation

$$A_n^\dagger f = +if \ . \tag{6.4.24}$$

This means that the deficiency indices are $(0,1)$ and there are no self-adjoint extensions of this operator. This example shows that, given two observables $\mathcal{A}$ and $\mathcal{B}$, with corresponding self-adjoint operators A and B, it is not generally true that the operator $AB + BA$ is self-adjoint and corresponds to an obervable.

6.5 Projection Operator

A projection operator is a self-adjoint, non-negative operator P satisfying

$$P^2 = P \ .$$

Let ϕ_n be a normalized eigenfunction of a self-adjoint operator A with only discrete eigenvalues λ_n .
a) Show that the operator $P_n\psi = \phi_n(\phi_n, \psi)$ is a projection operator.
b) Show that A can be written

$$A\psi = \sum_n \int \lambda_n \phi_n(x)\phi_n^*(y)\psi(y)\,dy = \sum_n \lambda_n P_n\psi \ .$$

This is called the spectral resolution of the operator A.
Hint: Assume completeness of the eigenfunctions.

Solution

We are given

$$P_n\psi = \phi_n(\phi_n, \psi). \tag{6.5.25}$$

Hence, using that $(\phi_n, \phi_n) = 1$ we get

$$
\begin{aligned}
P_n^2\psi &= P_n(P_n\psi) \\
&= \phi_n(\phi_n, \phi_n(\phi_n, \psi)) \\
&= \phi_n(\phi_n, \phi_n)(\phi_n, \psi) \\
&= \phi_n(\phi_n, \psi) \\
&= P_n\psi
\end{aligned}
\tag{6.5.26}
$$

b) Assume (using completeness) that

$$\psi = \sum_n \alpha_n \phi_n \tag{6.5.27}$$

so that

$$\alpha_n = (\phi_n, \psi) \ . \tag{6.5.28}$$

Then,

$$\psi = \sum_n \phi_n(\phi_n, \psi) \tag{6.5.29}$$

and

$$
\begin{aligned}
A\psi &= \sum_n A\phi_n(\phi_n, \psi) \\
&= \sum_n \lambda_n \phi_n(\phi_n, \psi) \\
&= \sum_n \lambda_n P_n\psi \ .
\end{aligned}
\tag{6.5.30}
$$

6.6 Spectral Resolution

Find the spectral resolution (see problem 6.5) of the operator

$$A = \begin{pmatrix} a_3 & a_1 + ia_2 \\ a_1 - ia_2 & -a_3 \end{pmatrix} .$$

Solution

The operator is

$$A = \begin{pmatrix} a_3 & a_1 + ia_2 \\ a_1 - ia_2 & -a_3 \end{pmatrix} . \tag{6.6.31}$$

The eigenvalues are given by

$$-(a_3 - x)(a_3 + x) - (a_1^2 + a_2^2) = 0 \tag{6.6.32}$$

or

$$x^2 = a_1^2 + a_2^2 + a_3^2 = a^2 . \tag{6.6.33}$$

So that

$$x = \pm a . \tag{6.6.34}$$

The corresponding normalized eigenvectors are

$$f_+ = \frac{1}{\sqrt{2a(a + a_3)}} \begin{pmatrix} a + a_3 \\ a_1 - ia_2 \end{pmatrix}$$

$$f_- = \frac{1}{\sqrt{2a(a - a_3)}} \begin{pmatrix} a - a_3 \\ -(a_1 - ia_2) \end{pmatrix} . \tag{6.6.35}$$

Thus, the projection operators are

$$P_+ = f_+ f_+^\dagger = \frac{1}{2a} \begin{pmatrix} a + a_3 & a_1 + ia_2 \\ a_1 - ia_2 & a - a_3 \end{pmatrix}$$

$$P_- = f_- f_-^\dagger = \frac{1}{2a} \begin{pmatrix} a - a_3 & -(a_1 + ia_2) \\ -(a_1 - ia_2) & a + a_3 \end{pmatrix} . \tag{6.6.36}$$

So, it is easy to see that

$$aP_+ + (-a)P_- = A . \tag{6.6.37}$$

6.7 Resolvent Operator

For any operator A the corresponding operator

$$R(z) = (A - z1)^{-1} ,$$

where z is a complex number and 1 stands for the unit operator, is called the resolvent operator. Show that for any square matrix A, $R(z)$ is analytic in z with poles at the eigenvalues of A.

Solution

If A is a square $n \times n$ matrix then

$$R(z) = (A - z)^{-1} \tag{6.7.38}$$

may be written as

$$R(z) = \frac{\text{matrix of cofactors of}(A - z)}{\det (A - z)}. \tag{6.7.39}$$

The cofactors are polynomials of degree $\leq n - 1$. The determinant is a polynomial of degree n. Thus, $R(z)$ is the ratio of such polynomials and is analytic except at the zeroes of the denominator which occur when

$$\det (A - z) = 0. \tag{6.7.40}$$

All of these singularities are poles.

6.8 Deficiency Indices

Find the deficiency indices and hence all self-adjoint extensions of the Hamiltonian

$$H = -\frac{\hbar^2}{2m} \frac{d^2}{dx^2}$$

defined on the interval (a, b).
Hint: It may be useful to express the boundary conditions on a function $f \in D_H$ in terms of 2-component quantities

$$F(a) = \begin{pmatrix} f(a) \\ f'(a) \end{pmatrix} \quad \text{and} \quad F(b) = \begin{pmatrix} f(b) \\ f'(b) \end{pmatrix}$$

and assume that $F(b) = U F(a)$ where U is a non-singular 2×2 matrix.

Solution

The Hamiltonian is

$$H = -\frac{\hbar^2}{2m} \frac{d^2}{dx^2}. \tag{6.8.41}$$

To find the deficiency indices when the Hilbert space is $L^2(a, b; dx)$ we have to find solutions of

$$H f_\pm = \pm i f_\pm \tag{6.8.42}$$

belonging to $L^2(a, b; dx)$. All solutions of these differential equations are given by

$$f_\pm = A \exp(k_\pm x) \tag{6.8.43}$$

where

$$-\frac{\hbar^2 k_{\pm}^2}{2m} = \pm i \ . \tag{6.8.44}$$

Therefore, we find

$$k_+ = \frac{\sqrt{2m}}{\hbar} e^{i3\pi/4} \ , \quad -\frac{\sqrt{2m}}{\hbar} e^{i3\pi/4} \tag{6.8.45}$$

$$k_- = \frac{\sqrt{2m}}{\hbar} e^{i\pi/4} \ , \quad -\frac{\sqrt{2m}}{\hbar} e^{i\pi/4} \ . \tag{6.8.46}$$

All four solutions belong to $L^2(a, b; dx)$ and thus, the deficiency indices are $(2, 2)$.

To find the self-adjoint extensions we consider

$$(f, Hg) = -\frac{\hbar^2}{2m} \int_a^b f^* \frac{d^2 g}{dx^2} \, dx$$

$$= -\frac{\hbar^2}{2m} \left(f^* \frac{dg}{dx} \Big|_a^b - \frac{df^*}{dx} g \Big|_a^b \right) + (Hf, g) \ . \tag{6.8.47}$$

So, we need

$$f^*(b)g'(b) - f^*(a)g'(a) - f^{*\prime}(b)g(b) + f^{*\prime}(a)g(a) = 0 \ . \tag{6.8.48}$$

The same boundary conditions must apply to both $f(x)$ and $g(x)$. We therefore define

$$F(x) = \begin{pmatrix} f(x) \\ (b-a)f'(x) \end{pmatrix} \ , \quad G(x) = \begin{pmatrix} g(x) \\ (b-a)g'(x) \end{pmatrix} \tag{6.8.49}$$

(the factors of $b - a$ are simply to have both components with the same dimension) and look for boundary conditions of the form

$$F(b) = AF(a) \ , \quad G(b) = AG(a) \tag{6.8.50}$$

where A is a 2×2 matrix. The condition for self-adjointness when now written out reads

$$F^\dagger(b) \begin{pmatrix} 0 & -1 \\ 1 & 0 \end{pmatrix} G(b) = F^\dagger(a) \begin{pmatrix} 0 & -1 \\ 1 & 0 \end{pmatrix} G(a) \ . \tag{6.8.51}$$

In terms of the matrix A this requires:

$$A^\dagger \begin{pmatrix} 0 & -1 \\ 1 & 0 \end{pmatrix} A = \begin{pmatrix} 0 & -1 \\ 1 & 0 \end{pmatrix} \ . \tag{6.8.52}$$

A general 2×2 matrix involves 8 real parameters and we have 4 conditions. Thus, we will find 4 real parameters. The most general 2×2 matrix may be written

$$A = e^{i\alpha} \begin{pmatrix} a_0 + a_3 & a_1 - ia_2 \\ a_1 + ia_2 & a_0 - a_3 \end{pmatrix} \ . \tag{6.8.53}$$

Since we need only 4 parameters we begin by choosing all 5 parameters in A to be real. Now, writing out the condition on A we find

$$A^\dagger \begin{pmatrix} 0 & -1 \\ 1 & 0 \end{pmatrix} A = \begin{pmatrix} 0 & -a_0^2 + a_1^2 + a_2^2 + a_3^2 \\ a_0^2 - a_1^2 - a_2^2 - a_3^2 & 0 \end{pmatrix}$$
$$= \begin{pmatrix} 0 & -1 \\ 1 & 0 \end{pmatrix}. \tag{6.8.54}$$

Therefore, we need only choose

$$a_0 = \sqrt{1 + a_1^2 + a_2^2 + a_3^2}. \tag{6.8.55}$$

This fixes the matrix A and since it has exactly 4 parameters as required we are finished. Thus, the domain of H consists of all twice differentiable functions $f(x)$ on the interval (a,b) such that

$$\begin{pmatrix} f(b) \\ (b-a)f'(b) \end{pmatrix} = e^{i\alpha} \begin{pmatrix} a_0 + a_3 & a_1 - ia_2 \\ a_1 + ia_2 & a_0 - a_3 \end{pmatrix} \begin{pmatrix} f(a) \\ (b-a)f'(a) \end{pmatrix} \tag{6.8.56}$$

where α, a_1, a_2, a_3 are real and $a_0 = \sqrt{1 + a_1^2 + a_2^2 + a_3^2}$.

6.9 Adjoint Operator

Given an orthonormal basis set $\{u_n : n = 0, 1, 2, \ldots\}$ and an operator a which has the following action on this basis:

$$a u_n = \sqrt{n}\, u_{n-1} \quad n \geq 0.$$

Find the adjoint operator $a^\dagger$ by explicitly giving its action on this basis set. Also find the commutator $[a, a^\dagger]$.

Solution

We have that the set $\{u_n\}$ is complete and orthonormal. Also,

$$a\, u_n = \sqrt{n}\, u_{n-1}. \tag{6.9.57}$$

Therefore, we get

$$(u_m, a\, u_n) = \sqrt{n}(u_m, u_{n-1}) = \sqrt{n}\, \delta_{m,n-1}. \tag{6.9.58}$$

This last term on the right may be rewritten as $\sqrt{m+1}\, \delta_{m+1,n}$ Using the definition of the adjoint operator we therefore have that

$$(a^\dagger u_m, u_n) = \sqrt{m+1}\, \delta_{m+1,n}. \tag{6.9.59}$$

Therefore,

$$a^\dagger u_m = \sqrt{m+1}\, u_{m+1}. \tag{6.9.60}$$

The commutator $[a, a^\dagger]$ is now computed using

$$aa^\dagger u_n = \sqrt{n+1}\, a\, u_{n+1} = (n+1)\, u_n \tag{6.9.61}$$

and

$$a^\dagger a\, u_n = \sqrt{n}\, a^\dagger\, u_{n-1} = n\, u_n \ . \tag{6.9.62}$$

Therefore,

$$[a, a^\dagger]u_n = (aa^\dagger - a^\dagger a)u_n = (n+1-n)\, u_n = u_n \ . \tag{6.9.63}$$

Thus,

$$[a, a^\dagger] = 1 \ . \tag{6.9.64}$$

6.10   Projection Operator

Let A be an operator on a Hilbert space $\mathcal{H}$ and let $\{a_n\}$ be the set of all eigenvalues of A. Show that if A has a complete set of eigenstates $\{|n\rangle\}$ such that

$$A|n\rangle = a_n|n\rangle$$

then

$$P_n = \prod_{m \neq n} \frac{A - a_m}{a_n - a_m} \tag{6.10.65}$$

is a projection operator onto the eigenspace of a_n, that is it projects onto the space spanned by the states belonging to the eigenvalue a_n.
Hint: Prove that

$$P_n |l\rangle = \delta_{nl} |l\rangle \ . \tag{6.10.66}$$

Solution

To use the hint we consider a eigenstate $|l\rangle$, $l \neq n$. In that case

$$P_n |l\rangle = \prod_{m \neq n} \frac{a_l - a_m}{a_n - a_m} |l\rangle = 0 \tag{6.10.67}$$

since in the product we arrive at a term with $m = l$. Now consider the eigenstate $|n\rangle$. Then,

$$P_n |n\rangle = \prod_{m \neq n} \frac{a_n - a_m}{a_n - a_m} |n\rangle = |n\rangle \ . \tag{6.10.68}$$

Thus, applied to an eigenstate we have

$$P_n |l\rangle = \delta_{nl} |l\rangle \ . \tag{6.10.69}$$

It now follows that

$$
\begin{aligned}
P_n^2 |l\rangle &= \delta_{nl} P_n |l\rangle \\
&= \delta_{nl} |l\rangle \\
&= P_n |l\rangle \ .
\end{aligned}
\tag{6.10.70}
$$

Since the eigenstates of A form a complete set we may expand any state in terms of these

$$
|\phi\rangle = \sum_l b_l |l\rangle \ .
\tag{6.10.71}
$$

Then,

$$
\begin{aligned}
P_n |\phi\rangle &= \sum_l b_l P_n |l\rangle \\
&= \sum_l b_l \delta_{nl} |l\rangle \\
&= b_n |n\rangle \ .
\end{aligned}
\tag{6.10.72}
$$

Applying P_n to this equation once more, we get that

$$
\begin{aligned}
P_n^2 |\phi\rangle &= b_n P_n |n\rangle \\
&= b_n |n\rangle \\
&= P_n |\phi\rangle \ .
\end{aligned}
\tag{6.10.73}
$$

Thus, we have shown that

$$
P_n^2 = P_n \ .
\tag{6.10.74}
$$

This shows that P_n is idempotent as required for a projection operator. If we now consider any state $|\phi\rangle$ as in (6.10.71) and apply $\sum_n P_n$ we get

$$
\begin{aligned}
\sum_n P_n |\phi\rangle &= \sum_n \sum_l b_l P_n |l\rangle \\
&= \sum_n \sum_l b_l \delta_{nl} |l\rangle \\
&= \sum_l b_l |l\rangle = |\phi\rangle \ .
\end{aligned}
\tag{6.10.75}
$$

Thus, we have shown that

$$
\sum_n P_n = 1
\tag{6.10.76}
$$

and P_n is indeed the projection operator that was claimed.

6.11   Commutator of L_z and φ

Formally one can derive the relation

$$[L_z, \varphi] = i\hbar$$

and deduce from it that

$$\Delta L_z \, \Delta\varphi \geq \hbar/2 \ .$$

Now, $\Delta\varphi$ is of necessity $\leq 2\pi$, and in an eigenstate of L_z we have $\Delta L_z = 0$. This violates

$$\Delta L_z \, \Delta\varphi \geq \hbar/2 \ .$$

Explain this apparent paradox.

Hint: Examine the domain of L_z on which it is self-adjoint. A similar argument also holds for $[p, x] = i\hbar$ and a particle confined to a finite interval on the line. See also [6.1] and [6.2].

Solution

L_z is defined as $(\hbar/i) \, (d/d\varphi)$ for functions $f(\varphi)$ differentiable and periodic in φ. This defines the domain of L_z and means

$$f(\varphi + 2\pi) = f(\varphi) \qquad\qquad (6.11.77)$$

So if f is such a function then $\varphi L_z f(\varphi)$ is well defined, but $g(\varphi) = \varphi f(\varphi)$ is not periodic and therefore does not belong to the domain of L_z. Thus, we cannot even write $L_z[\varphi f(\varphi)]$. This means that the commutator

$$[L_z, \varphi] = -i\hbar \qquad\qquad (6.11.78)$$

is incorrect .

6.12   Uncertainty Relation: L_z and $\cos\varphi$, $\sin\varphi$

In the previous problem, if instead of φ we choose periodic functions of φ such as $\cos\varphi$ or $\sin\varphi$ we have

$$[L_z, \sin\varphi] = -i\hbar \, \cos\varphi \qquad\qquad (6.12.79)$$

$$[L_z, \cos\varphi] = +i\hbar \, \sin\varphi \ . \qquad\qquad (6.12.80)$$

These equations involve well-defined quantities. Use these commutators to compute the uncertainty relations between L_z and $\cos\varphi$ as well as $\sin\varphi$.

Solution

We know from general considerations [6.3] that

$$\Delta L_z \, \Delta \sin \varphi \geq \frac{\hbar}{2} \langle \cos \varphi \rangle \qquad\qquad (6.12.81)$$

$$\Delta L_z \, \Delta \cos \varphi \geq \frac{\hbar}{2} \langle \sin \varphi \rangle \ . \qquad\qquad (6.12.82)$$

This does not lead to any contradictions. For example, if we are in an eigenstate of L_z so that the wavefunction is

$$\psi(\varphi) = \frac{1}{\sqrt{2\pi}} e^{im\varphi} \qquad\qquad (6.12.83)$$

and $\Delta L_z = 0$ we see that we also have

$$\langle \cos \varphi \rangle = \langle \sin \varphi \rangle = 0 \ . \qquad\qquad (6.12.84)$$

This time, however, no contradiction occurs. The results simply mean that the uncertainty relations read $0 \geq 0$, a statement that is clearly true.

6.13 Domain of Kinetic Energy: Polar Coordinates

a) Show that if $\psi(r, \theta, \varphi)$ is to be in the domain of the kinetic energy operator

$$T = -(\hbar^2/2m)\nabla^2$$

then, near $r = 0$ the function $\psi(r, \theta, \varphi)$ must be bounded by $A\,r^\alpha$, $\alpha > -1/2$ where A is independent of r.
Hint: Examine the condition that

$$T\psi(r, ., .) \in \mathcal{L}_2(0, \infty)$$

near $r = 0$ by writing

$$\psi(r, \theta, \varphi) = r^\alpha \left[a_0 + a_1 r + a_2 r^2 + \cdots \right] \qquad\qquad (6.13.85)$$

where the a_j are independent of r, i.e. they depend only on θ and φ.

Solution

Using the hint we have that near $r = 0$

$$
\begin{aligned}
T\psi &= -\frac{\hbar^2}{2m} \left[\frac{\partial^2 \psi}{\partial r^2} + \frac{2}{r} \frac{\partial \psi}{\partial r} + \frac{L^2 \psi}{r^2} \right] \\
&= r^{\alpha-2} \left[b_0 + b_1 r + b_2 r^2 + \cdots \right]
\end{aligned}
\qquad\qquad (6.13.86)
$$

where the b_j are new functions of θ and φ and are independent of r just like the a_j. Since the integration over r has the measure $r^2\,dr$ it now follows that near $r = 0$ we have

$$(\psi,\ T\psi) = \int_0 dr\, r^{2\alpha}\, [a_0^* b_0 + (a_1^* b_0 + a_0^* b_1)r + \cdots]\ . \tag{6.13.87}$$

For this integral to converge at the lower limit we therefore need that

$$\alpha > -\frac{1}{2}\ . \tag{6.13.88}$$

This is the required condition for $\psi(r, ., .) \in D_T$.

Notice that if in a Hamiltonian

$$H = T + V$$

we have a potential $V(r)$ such that $V(r)$ is less singular than r^{-2} near $r = 0$ then the condition (6.13.88) obtained also suffices for ψ to be in the domain of this Hamiltonian, at least as far as the behaviour near $r = 0$ is concerned.

6.14   Self-Adjoint Extensions of p^4

For the operator p^4 defined on $\mathcal{L}_2(0, \infty)$ find the deficiency indices and conditions to make the operator self-adjoint.

Solution

For these computations it is convenient to choose units such that $\hbar = 1$. Then,

$$p^2 = -\frac{d^2}{dx^2}$$

$$p^4 = \frac{d^4}{dx^4}\ . \tag{6.14.89}$$

To find the deficiency indices we need to find the solutions belonging to $\mathcal{L}_2(0, \infty)$ of

$$\frac{d^4 f_\pm}{dx^4} = e^{\pm i\pi/2}\, f_\pm\ . \tag{6.14.90}$$

We try

$$f_\pm = e^{\alpha \pm x} \tag{6.14.91}$$

and find that

$$\alpha_+^4 = e^{i\pi/2}\ ,\quad \alpha_-^4 = e^{-i\pi/2}\ . \tag{6.14.92}$$

Thus,

$$\alpha_\pm = e^{i\pm\pi/8}\ ,\ e^{i\pm\pi/8+i\pi/2}\ ,\ e^{i\pm\pi/8+i\pi}\ ,\ e^{i\pm\pi/8+i3\pi/2}\ . \tag{6.14.93}$$

In order that these solutions belong to $\mathcal{L}_2(0, \infty)$ we need that $\Re(\alpha_\pm) < 0$. So we find that acceptable values of $\alpha_\pm$ are

$$\alpha_+ = e^{i\pi/8 + i\pi/2}, \ e^{i\pi/8 + i\pi} \tag{6.14.94}$$

and

$$\alpha_- = e^{-i\pi/8 + i\pi}, \ e^{-i\pi/8 + i3\pi/2} \ . \tag{6.14.95}$$

So, the deficiency indices are $(2, 2)$ and we have a $2 \times 2 = 4$ parameter family of self-adjoint extensions.

To find the self-adjoint extensions we consider the matrix element of p^4 and integrate by parts several times. Thus,

$$
\begin{aligned}
\int_0^\infty \phi^* \frac{d^4\psi}{dx^4} dx &= \left. \phi^* \frac{d^3\psi}{dx^3} \right|_0^\infty - \int_0^\infty \frac{d\phi^*}{dx} \frac{d^3\psi}{dx^3} dx \\
&= \left. \phi^* \psi^{(3)} \right|_0^\infty - \left. \phi^{(1)*} \psi^{(2)} \right|_0^\infty + \int_0^\infty \frac{d^2\phi^*}{dx^2} \frac{d^2\psi}{dx^2} dx \\
&= \left. \phi^* \psi^{(3)} \right|_0^\infty - \left. \phi^{(1)*} \psi^{(2)} \right|_0^\infty + \left. \phi^{(2)*} \psi^{(1)} \right|_0^\infty - \int_0^\infty \frac{d^3\phi^*}{dx^3} \frac{d\psi}{dx} dx \\
&= \left. \phi^* \psi^{(3)} \right|_0^\infty - \left. \phi^{(1)*} \psi^{(2)} \right|_0^\infty + \left. \phi^{(2)*} \psi^{(1)} \right|_0^\infty - \left. \phi^{(3)*} \psi \right|_0^\infty \\
&+ \int_0^\infty \frac{d^4\phi^*}{dx^4} \psi \, dx \ .
\end{aligned}
\tag{6.14.96}
$$

Here we have used the notation

$$\psi^{(n)} = \frac{d^n \psi}{dx^n} \ .$$

Since both ψ and ϕ as well as their derivatives have to vanish at $x = \infty$ we are left with the following condition for self-adjointness of p^4.

$$\phi^*(0)\psi^{(3)}(0) - \phi^{(1)*}(0)\psi^{(2)}(0) + \phi^{(2)*}(0)\psi^{(1)}(0) - \phi^{(3)*}(0)\psi(0) = 0 \ . \tag{6.14.97}$$

To satisfy this condition we insist that all $\psi \in \mathcal{L}_2(0, \infty)$ should be of the form

$$\psi^{(3)}(0) = a\psi^{(2)}(0) + b\psi^{(1)}(0) + c\psi(0) \ . \tag{6.14.98}$$

Then, we also have that

$$\phi^{(3)*}(0) = a^*\phi^{(2)*}(0) + b^*\phi^{(1)*}(0) + c^*\phi(0) \ . \tag{6.14.99}$$

Now writing out (6.14.97) we find

$$
\begin{aligned}
&\phi^{(2)*}(0) \left[\psi^{(1)}(0) - a^*\psi(0) \right] - \phi^{(1)*}(0) \left[\psi^{(2)}(0) + b^*\psi(0) \right] \\
&+ \ \phi^*(0) \left[a\psi^{(2)}(0) + b\psi^{(1)}(0) + (c - c^*)\psi(0) \right] = 0 \ .
\end{aligned}
\tag{6.14.100}
$$

If we now cause all three of the square brackets to vanish separately we have
that

$$\begin{aligned}
\psi^{(1)}(0) &= a^*\psi(0) \\
\psi^{(2)}(0) &= -b^*\psi(0) \\
-ab^* + a^*b + c - c^* &= 0 \ .
\end{aligned}$$

(6.14.101)

This last equation is easily satified by choosing

$$c = ab^* \ .$$

(6.14.102)

Writing,

$$a = |a|\,e^{i\beta} \ , \quad b = |b|\,e^{i\gamma}$$

(6.14.103)

we have that

$$c = |ab|\,e^{i(\beta-\gamma)} \ .$$

(6.14.104)

Thus, finally the conditions for self adjointness are

$$\begin{aligned}
\psi^{(1)}(0) &= |a|\,e^{i\beta}\psi(0) \\
\psi^{(2)}(0) &= -|b|\,e^{-i\gamma}\psi(0) \\
\psi^{(3)}(0) &= -|ab|\,e^{-i(\beta-\gamma)}\psi(0) \ .
\end{aligned}$$

(6.14.105)

It is also clear that the four real parameters $|a|$, $|b|$, β, and γ label the self-adjoint
extensions.

Bibliography

[6.1] P. Carruthers and M.M. Nieto - Phase and Angle Variables in Quantum
Mechanics, Rev. Mod. Phys. **40**, 411, (1968) .

[6.2] A.Z. Capri, *Nonrelativistic Quantum Mechanics* 3rd edition, World Scientific Publishing Co. Pte. Ltd., chapter 6, (2002) .

[6.3] ibid, section 7.6.

Chapter 7

Physical Interpretation

7.1 Tetrahedral Die

An unbiased tetrahedral die is thrown and shows the number n. If this die is thrown very many times calculate $\langle n \rangle$ as well as Δn.

Solution

Since each side of the die is equally probable and the sides are numbered from 1 to 4 we get

$$\langle n \rangle = \frac{1}{4}(1 + 2 + 3 + 4) = \frac{5}{2} \, . \qquad (7.1.1)$$

Similarly,

$$
\begin{aligned}
\Delta n &= \frac{1}{4} \left[(1 - 5/2)^2 + (2 - 5/2)^2 + (3 - 5/2)^2 + (4 - 5/2)^2 \right] \\
&= \frac{5}{4} \, . \qquad (7.1.2)
\end{aligned}
$$

Notice that we could also calculate Δn from

$$
\begin{aligned}
\Delta n &= \langle n^2 \rangle - (\langle n \rangle)^2 \\
&= \frac{1}{4} \left[1^2 + 2^2 + 3^2 + 4^2 \right] - \left[\frac{5}{2} \right]^2 \\
&= \frac{5}{4} \, . \qquad (7.1.3)
\end{aligned}
$$

7.2 Probabilities, Expectation Values, Evolution

A particle is in a state given at $t = 0$ by:

$$\psi = \frac{1}{3} u_0(x) + \frac{i\sqrt{2}}{3} u_1(x) - \frac{\sqrt{6}}{3} u_2(x)$$

where u_0 , u_1 , u_2 are simple harmonic oscillator eigenfunctions corresponding to the energies $1/2\ \hbar\omega$, $3/2\ \hbar\omega$ and $5/2\ \hbar\omega$, respectively.

a) What is the most likely value of the energy that will be found in a single observation on this system? What is the probability of finding this value?

b) What is the average of the energy that would be obtained if the experiment in part a) could be repeated many times? What is the probability of getting this value?

c) A measurement of the energy yields a value $3/2\ \hbar\omega$. The measurement is immediately repeated. What is the resultant value of the energy? What is the wavefunction immediately after the second measurement?

d) What is the wavefunction of the undisturbed system after a time t has elapsed?

Solution

Since u_0, u_1, u_2 are normalized and the moduli of their coefficients add up to unity, the whole wavefunction is normalized.

a) The most likely value of the energy corresponds to the eigenfunction with the largest coefficient ie. u_2. This value is $5/2\ \hbar\omega$ with a probability of

$$|-\sqrt{6}/3|^2 = 2/3\ .$$

b) The average energy is

$$\langle E \rangle = \left[\frac{1}{9} \times \frac{1}{2}\hbar\omega + \frac{2}{9} \times \frac{3}{2}\hbar\omega + \frac{6}{9} \times \frac{5}{2}\hbar\omega\right] = \frac{37}{18}\hbar\omega\ . \tag{7.2.4}$$

The probability of observing this value is zero since it is not one of the eigenvalues.

c) The resultant energy value obtained is $3/2\ \hbar\omega$. The wavefunction is

$$u_1(x)\exp(-3i/2\,\omega t)\ .$$

d) The wavefunction is

$$\Psi = \frac{1}{3}u_0(x)\,e^{-i\omega t/2} + \frac{i\sqrt{2}}{3}u_1(x)\,e^{-i3\omega t/2} - \frac{\sqrt{6}}{3}u_2(x)\,e^{-i5\omega t/2}\ . \tag{7.2.5}$$

7.3 $\langle L_x \rangle$ and $\langle L_y \rangle$ in an Eigenstate of L_z

Show that in an eigenstate of L_z the expectation values $\langle L_x \rangle$ and $\langle L_y \rangle$ vanish. Hint: Use the commutation relations for the angular momentum operators.

Solution

The commutation relations for the angular momentum operators are

$$[L_y,\ L_z]\ =\ i\hbar L_x\ . \tag{7.3.6}$$
$$[L_z,\ L_x]\ =\ i\hbar L_y\ . \tag{7.3.7}$$

If we assume that the system is in an eigenstate $|m\rangle$ of L_z such that

$$L_z|m\rangle = m\hbar|m\rangle \qquad\qquad (7.3.8)$$

then, taking the expectation value of (7.3.7) we get

$$
\begin{aligned}
i\hbar\langle m|L_x|m\rangle &= \langle m|[L_y\,,\,L_z]|m\rangle \\
&= m\hbar\langle m|L_y|m\rangle - m\hbar\langle m|L_y|m\rangle \\
&= 0 \; . \qquad\qquad\qquad\qquad\qquad (7.3.9)
\end{aligned}
$$

In exactly the same manner, taking the expectation value of (7.3.7) we find that

$$\langle m|L_y|m\rangle = 0 \; . \qquad\qquad (7.3.10)$$

7.4   Free Particle Propagator

A free particle is located at $x = a$ at $t = 0$; i.e. its wavefunction at $t = 0$ is given by

$$\Psi(0, x) = \delta(x - a) \; .$$

Find the wavefunction for $t > 0$. This solution is called the free particle propagator.

Hint: To evaluate an integral of the form

$$\int_{-\infty}^{\infty} \exp\{i\lambda x^2 + i\beta x\}\, dx$$

pretend that λ is $\lambda + i\epsilon$ with $\epsilon > 0$ so that the integral is convergent. Then complete the square in the exponent of the exponential and change variables. The $i\epsilon$ in your answer will allow you to decide whether to take the positive or negative square root. Finally let $\epsilon \to 0$.

Solution

The eigenfunctions for a free particle are

$$u_k(x) = \frac{1}{\sqrt{2\pi}}\, e^{ikx} \qquad\qquad (7.4.11)$$

with energy

$$E_k = \frac{\hbar^2 k^2}{2m} \; . \qquad\qquad (7.4.12)$$

The initial condition is

$$\Psi(0, x) = \delta(x - a) = \int A(k)u_k(x)\, dk = \frac{1}{\sqrt{2\pi}}\int A(k)\, e^{ikx}\, dk \; . \qquad (7.4.13)$$

Therefore,

$$A(k) = \frac{1}{\sqrt{2\pi}} \int_{-\infty}^{\infty} \delta(x-a) \, e^{-ikx} \, dx = \frac{e^{-ika}}{\sqrt{2\pi}} \ . \tag{7.4.14}$$

Then, we find that

$$
\begin{aligned}
\Psi(t,x) &= \int_{-\infty}^{\infty} u_k(x) \, e^{-iE_k t/\hbar} \, dk \\
&= \frac{1}{2\pi} \int_{-\infty}^{\infty} e^{ik(x-a)} \, e^{-i\hbar k^2 t/2m} \, dk \ .
\end{aligned}
\tag{7.4.15}
$$

This integral is not obviously convergent. To remedy this we replace the co-efficient $-i\hbar t/2m$ of k^2 in the exponential by $-i\hbar t/2m - \epsilon$, $\epsilon > 0$, with the intention of letting ϵ go to zero after the computation is done. This process is called, "regularization". Thus,

$$
\begin{aligned}
\Psi(t,x) &= \frac{1}{2\pi} \int_{-\infty}^{\infty} e^{ik(x-a)} \, e^{-i(\hbar t/2m - i\epsilon)k^2} \, dk \\
&= \frac{1}{2\pi} \sqrt{\frac{\pi}{\epsilon + i\hbar t/2m}} \, \exp\left[i\frac{m(x-a)^2}{2\hbar t}\right] \\
&= \frac{1-i}{2\pi} \, \exp\left[i\frac{m(x-a)^2}{2\hbar t}\right] \ .
\end{aligned}
\tag{7.4.16}
$$

It is in the last step that we used the fact that $\epsilon \to 0+$ to obtain the correct square root .

7.5   Minimum Uncertainty Wavefunction

We have seen that

$$\Delta x \Delta p \geq \hbar/2 \ .$$

Assume

$$\langle x \rangle = \langle p \rangle = 0 \ .$$

Now use the Schwarz inequality

$$\| \, f \, \|^2 \| \, g \, \|^2 \geq |(f, g)|^2$$

and put

$$f = x\psi \ , \quad g = p\psi \ .$$

Show that the equality in the uncertainty will hold only if

$$p = \lambda \, x\psi$$

with λ a constant and

$$(\psi, (xp + px)\psi) = 0 \ .$$

Hence derive an equation for ψ and solve it explicitly.

Solution

We start with $f(x) = x\psi(x)$, $g(x) = p\psi(x)$. For the Schwarz inequality to become an equality requires that

$$g = \lambda f \tag{7.5.17}$$

where λ is a constant. This means that

$$p\psi(x) = \lambda x \psi(x) . \tag{7.5.18}$$

Furthermore,

$$|(\psi, [x, p]\psi)| = |2i \, \Im(x\psi, p\psi)| \leq 2|(x\psi, p\psi)| . \tag{7.5.19}$$

Therefore in order to get equality we further require that

$$\Re(x\psi, p\psi) = 0 . \tag{7.5.20}$$

This means

$$(x\psi, p\psi) + (p\psi, x\psi) = 0 \tag{7.5.21}$$

or

$$(\psi, xp\psi) + (\psi, px\psi) = (\psi, (xp + px)\psi) = 0 . \tag{7.5.22}$$

Now,

$$(\psi, xp\psi) = \lambda(x\psi, x\psi) \quad , \quad (\psi, px\psi) = \lambda^*(x\psi, x\psi) . \tag{7.5.23}$$

Adding we get

$$(\psi, xp\psi) + (\psi, px\psi) = (\lambda + \lambda^*)(x\psi, x\psi) . \tag{7.5.24}$$

This means that λ has to be pure imaginary so that $\lambda = i\alpha$ with α real. Thus, we find

$$p\psi(x) = i\alpha x \psi(x) . \tag{7.5.25}$$

Therefore,

$$\frac{\hbar}{i} \frac{d\psi}{dx} = i\alpha x \psi(x) . \tag{7.5.26}$$

So finally,

$$\psi(x) = A \exp\left(-\frac{\alpha x^2}{2\hbar}\right) \tag{7.5.27}$$

and for this to be a square-integrable function further requires that $\alpha > 0$.

7.6 Spreading of a Wave Packet

A free particle of mass m is at $t = 0$ in a state described by

$$\Psi(0, x) = \left[2\pi L^2\right]^{-1/4} \exp -(x/2L)^2$$

What is the wavefunction for an arbitrary time $t > 0$? Compute the uncertainties Δx and Δp as functions of time. This illustrates the "spreading" of a wave packet.

Solution

We first express $\Psi(0, x)$ as a superposition of free particle states. Since the most general solution for a free particle is

$$\Psi(t, x) = \int_{-\infty}^{\infty} A(k) \, e^{i(kx - \hbar k^2 t/2m)} \, dk \tag{7.6.28}$$

we write

$$\Psi(0, x) = \int_{-\infty}^{\infty} A(k) \, e^{ikx} \, dk \ . \tag{7.6.29}$$

Then, the coefficient of $A(k)$ is determined from the explicit expression for $\Psi(0, x)$.

$$A(k) = \frac{1}{2\pi} \int_{-\infty}^{\infty} \Psi(0, x) \, e^{-ikx} \, dx \ . \tag{7.6.30}$$

Inserting this in the integral and completing the square in the exponent, we can evaluate $A(k)$.

$$A(k) = \frac{\sqrt{2}}{2\pi} (2\pi L^2)^{1/4} \, e^{-k^2 L^2} \ . \tag{7.6.31}$$

Hence, we find (upon again completing the square in the exponent) that

$$\Psi(t, x) \quad = \quad \frac{(8\pi L^2)^{1/4} \exp\left[-\frac{x^2}{4(L^2 + i\hbar t/2m)}\right]}{2\pi} \times$$

$$\times \int_{-\infty}^{\infty} \exp\left\{-(L^2 + i\hbar t/2m)\left[k - \frac{ix}{2(L^2 + i\hbar t/2m)}\right]^2\right\}$$

$$= \quad \frac{(8\pi L^2)^{1/4} \exp\left[-\frac{x^2}{4(L^2 + i\hbar t/2m)}\right]}{2\pi} \sqrt{\frac{\pi}{L^2 + i\hbar t/2m}} \ . \tag{7.6.32}$$

This is a Gaussian of the form

$$\Psi(t, x) = (2\pi w^2)^{-1/4} \exp\left(-\frac{x^2}{4w^2}\right) e^{iA} \tag{7.6.33}$$

where

$$w^2 = L^2 + \frac{\hbar^2 t^2}{4m^2 L^2} \tag{7.6.34}$$

and

$$A = -\left[\frac{1}{2} \arctan\left(\frac{\hbar t}{4mL^2}\right) + \frac{\hbar t x^2}{8mL^2 w^2}\right] . \tag{7.6.35}$$

So, we can immediately conclude that

$$(\Delta x)^2 = \langle x^2 \rangle = 2w^2 . \tag{7.6.36}$$

Furthermore since this is a Gaussian it is a minimum uncertainty wavefunction. Therefore,

$$(\Delta p)^2 = \frac{\hbar^2}{4(\Delta x)^2} = \frac{\hbar^2}{8w^2} . \tag{7.6.37}$$

7.7 Time-dependent Expectation Values

A free particle is, at $t = 0$, in a state described by the wavefunction

$$\Psi(0, x) = \begin{cases} A \sin^2 \frac{\pi x}{a} & |x| < a \\ 0 & |x| > a \end{cases} .$$

Find for $t > 0$ the following expectation values

$$\langle p \rangle_t , \ \langle x \rangle_t , \ \langle p^2 \rangle_t , \ \langle xp + px \rangle_t , \text{ and } \langle x^2 \rangle_t .$$

Hint: Use the equation

$$\frac{d}{dt}\langle A \rangle_t = \frac{i}{\hbar} \left(\Psi(t, x), [H, A]\Psi(t, x) \right) .$$

If you try to evaluate these results using the time-dependent wave-function $\Psi(t, x)$ you will get some impossible integrals. [7.1]

Solution

In this case we have a free particle so

$$H = \frac{p^2}{2m} . \tag{7.7.38}$$

Also we are using the equation

$$\frac{d}{dt}\langle A \rangle = \frac{i}{\hbar}(\Psi, [H, A]\Psi) . \tag{7.7.39}$$

Since

$$[H, p] = 0 \tag{7.7.40}$$

we therefore have that

$$\frac{d}{dt}\langle p \rangle = 0 \tag{7.7.41}$$

so that

$$\langle p \rangle_t = \langle p \rangle_0 = |A|^2 \int_{-a}^{a} \sin^2(\pi x/a) \frac{\hbar}{i}\frac{d}{dx}\left(\sin^2(\pi x/a)\right)\,dx = 0 \ . \tag{7.7.42}$$

Next, we find that

$$[H, x] = -i\hbar\frac{p}{m} \tag{7.7.43}$$

so that

$$\frac{d}{dt}\langle x \rangle = \langle \frac{p}{m} \rangle = 0 \ . \tag{7.7.44}$$

Therefore,

$$\langle x \rangle_t = \langle x \rangle_0 = |A|^2 \int_{-a}^{a} \sin^4(\pi x/a)x\,dx = 0 \ . \tag{7.7.45}$$

Similarly, we see that

$$\langle p^2 \rangle_t = \langle p^2 \rangle_0 - \hbar^2|A|^2 \int_{-a}^{a} \sin^2(\pi x/a)\frac{d^2}{dx^2}\left(\sin^2(\pi x/a)\right)\,dx$$

$$= -\hbar^2|A|^2 \int_{-a}^{a} \sin^2(\pi x/a)\frac{2\pi^2}{a^2}\left(\cos^2(\pi x/a) - \sin^2(\pi x/a)\right)\,dx \ . \tag{7.7.46}$$

Now, we rewrite $\cos^2$ as $1 - \sin^2$ and use the fact that

$$|A|^2 \int_{-a}^{a} \sin^4(\pi x/a)\,dx = 1 \ . \tag{7.7.47}$$

So, we find that

$$\langle p^2 \rangle_t = \frac{4\hbar^2\pi^2}{a^2} - \frac{2\hbar^2\pi^2}{a^2}a|A|^2 \ . \tag{7.7.48}$$

Next, we need to evaluate $|A|^2$. This turns out to be $4/(3a)$. Therefore,

$$\langle p^2 \rangle_t = \frac{2\hbar^2\pi^2}{a^2}(2 - 4/3) = \frac{4\hbar^2\pi^2}{3a^2} \ . \tag{7.7.49}$$

We now find

$$[H, xp + px] = -2i\hbar\frac{p^2}{2m} \ . \tag{7.7.50}$$

Therefore,

$$\frac{d}{dt}\langle xp + px \rangle = -\frac{2}{m}\langle p^2 \rangle = \frac{2(2\pi\hbar)^2}{3ma^2} \ . \tag{7.7.51}$$

So, we get

$$\langle xp + px \rangle_t = \langle xp + px \rangle_0 + \frac{2(2\pi\hbar)^2}{3ma^2} t \ . \tag{7.7.52}$$

We now compute

$$
\begin{aligned}
&\langle xp + px \rangle_0 \\
&= |A|^2 \frac{\hbar}{i} \int_{-a}^{a} \sin^2(\pi x/a) \left(2x \frac{d}{dx} + 1 \right) \sin^2(\pi x/a) \, dx = 0 \ .
\end{aligned} \tag{7.7.53}
$$

Therefore,

$$\langle xp + px \rangle_t = \frac{2(2\pi\hbar)^2}{3ma^2} t \ . \tag{7.7.54}$$

Finally we find

$$[H, x^2] = x[H, x] + [H, x]x = -\frac{i\hbar}{m}(xp + px) \ . \tag{7.7.55}$$

Hence,

$$\frac{d}{dt} \langle x^2 \rangle = \frac{i}{\hbar} \frac{-i\hbar}{m} \langle xp + px \rangle_t = \frac{1}{m} \frac{2(2\pi\hbar)^2}{3ma^2} t \ . \tag{7.7.56}$$

Therefore,

$$\langle x^2 \rangle_t = \langle x^2 \rangle_0 + \frac{(2\pi\hbar)^2}{3m^2a^2} t^2 \ . \tag{7.7.57}$$

Now,

$$\langle x^2 \rangle_0 = |A|^2 \int_{-a}^{a} x^2 \sin^4(\pi x/a) \, dx = \frac{4}{3a} \left(\frac{a}{\pi} \right)^3 \frac{\pi}{4} (\pi - 15/8) \ . \tag{7.7.58}$$

Thus, we finally have

$$\langle x^2 \rangle_t = \frac{4}{3a} \left(\frac{a}{\pi} \right)^3 \frac{\pi}{4} (\pi - 15/8) + \frac{(2\pi\hbar)^2}{3m^2a^2} t^2 \ . \tag{7.7.59}$$

7.8 Ehrenfest Theorem

Consider a particle under the influence of a Hamiltonian

$$H = \frac{p^2}{2m} + V(x)$$

so that

$$i\hbar \frac{\partial \Psi}{\partial t} = H\Psi \ .$$

Show that if $\langle x \rangle$ is the "centre of mass" of the wave packet and $\langle p \rangle$ the average momentum of the particle then

$$\frac{d\langle x \rangle}{dt} = \frac{\langle p \rangle}{m}$$

and

$$\frac{d\langle p \rangle}{dt} = \langle F(x) \rangle = -\langle \frac{dV(x)}{dx} \rangle \ .$$

These are known as the *Ehrenfest equations*. To be equivalent to Newton's equations requires that

$$\langle F(x) \rangle = F(\langle x \rangle) \ .$$

Discuss under what circumstances this condition is approximately valid.

Solution

We have

$$H = \frac{p^2}{2m} + V(x) \ . \tag{7.8.60}$$

Therefore,

$$[H, x] = -i\hbar \frac{p}{m} \tag{7.8.61}$$

and

$$[H, p] = i\hbar \frac{dV}{dx} \ . \tag{7.8.62}$$

From this we find that

$$\frac{d}{dt}\langle x \rangle = \frac{i}{\hbar}\langle [H, x] \rangle = \frac{i}{\hbar}(-i\hbar)\langle \frac{p}{m} \rangle = \frac{\langle p \rangle}{m} \tag{7.8.63}$$

and

$$\frac{d}{dt}\langle p \rangle = \frac{i}{\hbar}\langle [H, p] \rangle = \frac{i}{\hbar}i\hbar\langle \frac{dV}{dx} \rangle = -\langle \frac{dV}{dx} \rangle = \langle F(x) \rangle \ . \tag{7.8.64}$$

For these results to be equivalent to the classical Newton's equations requires that

$$\langle F(x) \rangle = F(\langle x \rangle) \ . \tag{7.8.65}$$

To see what this means, we Taylor expand $F(x)$ about $\langle x \rangle$.

$$F(x) = F(\langle x \rangle) + \sum_{n=1}^{\infty} \frac{(x - \langle x \rangle)^n}{n!} \frac{d^n}{d\langle x \rangle^n} F(\langle x \rangle) \ . \tag{7.8.66}$$

Clearly we require that

$$\langle (x - \langle x \rangle)^n \rangle \approx 0 \quad n \geq 1 \ . \tag{7.8.67}$$

This is exactly true if the potential is a quadratic function of x. It is also approximately true if the dispersion in x, that is Δx is not too large. This means that the wavefunction $\Psi(t, x)$ must be fairly smooth and narrow.

7.9 Compatibility Theorem

Prove the compatibility theorem which states:
The following statements are equivalent for any pair of observables A, B.
i) A and B are compatible.
ii) The operators A and B, representing the observables A and B, possess a common eigenbasis.
iii) A and B commute, that is

$$[A, B] = AB - BA = 0 .$$

Solution

The compatibility theorem states that, given two observables A , B represented respectively by the self-adjoint operators A , B the following statements are equivalent.
i) A and B are compatible observables.
ii) A and B possess a common eigenbasis.
iii) $[A, B] = 0$.
We begin by recalling that two operators A , B are compatible if a measurement of A followed immediately by a measurement of B and another measurement of A yields the same result for the value of A in both cases. To prove the theorem we now show that
i) $\rightarrow$ ii) $\rightarrow$ iii) $\rightarrow$ i).
i) $\rightarrow$ ii)
Suppose the measurement of A yields α. Then, immediately after the measurement the state of the system has to be an eigenstate of A say ϕ_α. To guarantee that after the measurement of B when we again measure A we again get the value α requires that after the measurement of B the state be $\phi_{\alpha,\beta}$ where

$$B\phi_{\alpha,\beta} = \beta\phi_{\alpha,\beta} \tag{7.9.68}$$

and

$$A\phi_{\alpha,\beta} = \alpha\phi_{\alpha,\beta} . \tag{7.9.69}$$

This shows that A and B have a common eigenbasis.
ii) $\rightarrow$ iii)
Since, A and B have a common eigenbasis we have that operating on any common eigenstate

$$AB\phi_{\alpha,\beta} = \beta A\phi_{\alpha,\beta} = \beta\alpha\phi_{\alpha,\beta} = BA\phi_{\alpha,\beta} . \tag{7.9.70}$$

This means that when acting on an eigenbasis $[A, B] = 0$. Since the eigenbasis forms a complete set this means that quite generally

$$[A, B] = 0 . \tag{7.9.71}$$

iii) $\rightarrow$ i)
Since $[A, B] = 0$, the two operators may be diagonalized simultaneously. Then,

a measurement of A followed by a measurement of B will yield a state $\phi_{\alpha,\beta}$ as above. Thus, if a measurement of A is now performed, we are guaranteed to get the result α. This proves the desired result.

7.10   Constant of the Motion

Show that if $[H, A] = 0$ then ΔA does not change in time.

Solution

Since,

$$[H, A] = 0 \tag{7.10.72}$$

it follows that

$$[H, A^2] = A[H, A] + [H, A]A = 0 . \tag{7.10.73}$$

Hence, we can conclude that

$$\frac{d}{dt}\langle A \rangle = 0 \tag{7.10.74}$$

as well as that

$$\frac{d}{dt}\langle A^2 \rangle = 0 . \tag{7.10.75}$$

This immediately implies that

$$\frac{d}{dt}(\Delta A)^2 = \frac{d}{dt}\langle (A - \langle A \rangle)^2 \rangle = 0 . \tag{7.10.76}$$

7.11   Spreading of a Gaussian Wavepacket

Consider a free electron described at time $t = 0$ by the Gaussian wavepacket

$$\Psi(0, x) = \left[2\pi L^2\right]^{-1/4} \exp -(x/2L)^2 .$$

Using the experience from problem 7.6 answer the following questions:
a) What is $\langle x \rangle$ at any time?
b) What is Δx as a function of time?
This is known as the spreading of a wave packet. If this packet corresponds to an electron with 20 eV energy and a width of 1.0 Å at $t = 0$, what is its width after travelling 100 m?

Solution

We are given as initial condition

$$\Psi(0, x) = (2\pi L^2)^{-1/4} \exp\left[-\frac{x^2}{4L^2} + ikx\right] . \qquad (7.11.77)$$

Now following the steps in problem 7.6 we first Fourier transform $\Psi(0, x)$.

$$F(p) = \int_{-\infty}^{\infty} \Psi(0, x) e^{-ipx} \, dx = \frac{\sqrt{2}}{2\pi} (2\pi L^2)^{1/4} e^{-(k-p)^2 L^2} \qquad (7.11.78)$$

and

$$\Psi(t, x) = \int_{-\infty}^{\infty} F(p) e^{i(px - \hbar p^2 t/2m)} \, dp . \qquad (7.11.79)$$

Thus, carrying out the integral, we see that the wavefunction after a time t is again a shifted Gaussian of the form

$$\Psi(t, x) = (2\pi L^2)^{-1/4} \frac{e^{-(k-p)^2 L^2}}{[1 + i\hbar t/2mL^2]} \exp\left[\frac{-x^2 + 4ikL^2 x + 4k^2 L^2}{4(L^2 + i\hbar t/2m)}\right] . \qquad (7.11.80)$$

From this we compute that

$$\langle x \rangle = \frac{\hbar k t}{2m} \qquad (7.11.81)$$

and

$$(\Delta x)^2 = \langle x^2 \rangle - \langle x \rangle^2 = L^2 \left(1 + \frac{\hbar^2 t^2}{4m^2 L^4}\right) . \qquad (7.11.82)$$

For the problem under consideration we have $L = 1.0$ Å$= 10^{-10}$ m . The speed of the electron is given by

$$E = \frac{1}{2} mv^2 . \qquad (7.11.83)$$

Substituting 20 eV for the energy we find that $v = 8.4 \times 10^6$ m/s. Therefore the time to travel 100 m is

$$t = 1.2 \times 10^{-5} \text{ s} . \qquad (7.11.84)$$

Hence, we get that

$$\Delta x \approx \frac{\hbar t}{2mL} = 69 \text{ m} . \qquad (7.11.85)$$

This means that in going down a 100 m beam tube the wavepacket has spread to macroscopic dimensions. The reason for this is that originally it was extremely narrow and therefore had a very large spread in momentum. If instead the wavepacket originally had a width of 1.0 mm the width after travelling down the tube would have increased by a negligible 6.9×10^{-3} mm.

7.12   Incorrect Time Operator

Consider the space of functions of E belonging to $\mathcal{L}_2(0, \infty)$. The relationship

$$[E, t] = i\hbar$$

can be represented on this space by

$$E\, f(E) = E\, f(E)$$

and

$$t\, f(E) = -i\hbar\, \frac{df(E)}{dE}\ .$$

Show that the operator t so defined <u>cannot</u> be an observable, i.e. that it has no self-adjoint extensions. This proves that if the energy has a lower bound, a relationship such as

$$[E, t] = i\hbar\ .$$

cannot hold if time is to be an observable. [7.2]

Solution

We first find the deficiency indices for the putative time operator $-i\hbar\, d/dE$. Thus, we look at

$$-i\hbar \frac{d}{dE} f(E) = \pm iT f(E)\ . \tag{7.12.86}$$

The solutions are

$$f_+(E) = e^{-ET/\hbar}\ , \quad f_-(E) = e^{ET/\hbar}\ . \tag{7.12.87}$$

Only the first of these is square integrable on the interval $(0, \infty)$. Thus, the deficiency indices are $(0,1)$ and this operator has no self adjoint extensions.

Since for any physical system the energy must have a lower bound we can always shift the minimum of energy to $E = 0$. This allows us to consider the space $\mathcal{L}_2(0, \infty)$ as the appropriate space for this problem. Our result means that the relationship

$$[E, t] = i\hbar$$

is <u>false</u>.

7.13   Probability to Find a Particle

A particle is in a state described by the unnormalized wavefunction:

$$f(x) = A e^{-a|x|} \quad a > 0\ .$$

Find the length of an interval around the origin such that the probability of finding the particle in this interval is 40 % .

Solution

We have the unnormalized wavefunction

$$f(x) = Ae^{-a|x|} \quad a > 0 .$$ (7.13.88)

To normalize it we require that

$$\int_{-\infty}^{\infty} |f(x)|^2 \, dx = 2|A|^2 \int_0^{\infty} e^{-2ax} \, dx = 1$$ (7.13.89)

or

$$A = \sqrt{a} .$$ (7.13.90)

Now, the probability that the corresponding particle is found between $-L < x < L$ is given by

$$\int_{-L}^{L} |f(x)|^2 \, dx = 2|A|^2 \int_0^L e^{-2ax} \, dx = 1 - e^{-2aL} .$$ (7.13.91)

We want this probability to be 40%. Thus, we get

$$1 - e^{-2aL} = 0.4 .$$ (7.13.92)

Hence, we find that

$$L = -\frac{ln(0.6)}{2a} .$$ (7.13.93)

7.14 Sphere Bouncing on Sphere

A 1.0 g sphere of radius $r = 1$ cm is dropped from a height of $l = 1.2$ m (separation of the two centres) onto a similar ball rigidly attached to the ground. If the collisions are perfectly elastic, find the number of times the ball could be expected to bounce if the only limitation in the precision of releasing the ball to fall directly on top of the lower ball is due to the uncertainty principle.

Solution

Assume the ball falls in the y-direction and is deflected in the x-direction. The ball actually falls a distance of 1 m. Its terminal momentum just before contact is

$$\begin{aligned} p_y &= m\sqrt{2g(l - 2r)} \approx 10^{-3} \times \sqrt{2 \times 10 \times 1} \text{ kg m/s} \\ &= 4.47 \times 10^{-3} \text{ kg m/s} . \end{aligned}$$ (7.14.94)

Suppose that on contact the vertical makes an angle $\Delta\varphi$ with the line connecting the two centres (see figure 7.1). Then, on bouncing, the angle increases to $2\Delta\varphi$. After n bounces the angle will be $n\Delta\varphi$.

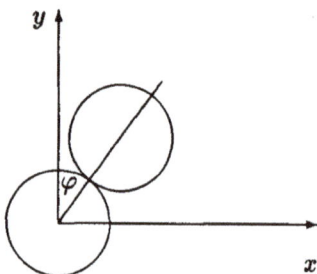

Figure 7.1: Sphere bouncing on a sphere. The angle φ is the same as the angle $\Delta\varphi$ in the text.

Now, if the deflection is $\Delta\varphi$, then the momentum in the x-direction after the collision is

$$\Delta p_x \approx p_y \Delta\varphi \tag{7.14.95}$$

since $\Delta\varphi$ is a very small angle. The angular momentum of the upper ball with respect to the point from which it was dropped is

$$\Delta L = p_x(l - 2r) \approx (l - 2r)p_y \Delta\varphi \ . \tag{7.14.96}$$

On the other hand, for very small angles we have the uncertainty relation

$$\Delta L \Delta\varphi \geq \hbar/2 \ . \tag{7.14.97}$$

So, if this uncertainty relation is the limiting factor we get

$$\Delta L \Delta\varphi \approx \hbar/2 \approx (l - 2r)m\sqrt{2g(l - 2r)}(\Delta\varphi)^2 \ . \tag{7.14.98}$$

Therefore,

$$(\Delta\varphi)^2 \approx \frac{\hbar}{2(l - 2r)m\sqrt{2g(l - 2r)}} \approx 1.1 \times 10^{-32} \ . \tag{7.14.99}$$

So,

$$\Delta\varphi \approx 1 \times 10^{-16} \ . \tag{7.14.100}$$

The last bounce will occur when the angle is approximatly $\varphi \approx r/l = 10^{-2}$ radians. Thus, the number of bounces is given by

$$\varphi = n\Delta\varphi \tag{7.14.101}$$

or

$$n = \frac{\varphi}{\Delta\varphi} \approx \frac{10^{-2}}{10^{-16}} \approx 10^{14} \ . \tag{7.14.102}$$

7.15 Cloud Chamber Tracks

The tracks made by an electron in a cloud chamber consist of small droplets of about 1 μ diameter. Use this fact to show that the track made by a 1.0 keV electron is essentially indistinguishable from the classical trajectory.

Solution

The momentum of the electron in the direction of the track (say x) is given by

$$
\begin{aligned}
p_x &= \sqrt{2mE} \\
&= \sqrt{2 \times 9.1 \times 10^{-31} \times 1.6 \times 10^{-19} \times 1.0 \times 10^3} \\
&= 1.71 \times 10^{-23} \text{ kg m/s} .
\end{aligned}
\tag{7.15.103}
$$

This is also the momentum for the classical trajectory. Quantum mechanically there is also a transverse component (say p_y) of the momentum given by the Heisenberg uncertainty principle.

$$
\Delta p_y \Delta y \approx \hbar .
\tag{7.15.104}
$$

Here, Δy is given by the size of the droplets. Thus,

$$
\Delta p_y \approx \frac{\hbar}{\Delta y} = \frac{1.05 \times 10^{-34}}{10^{-6}} = 1.05 \times 10^{-28} \text{ kg m/s} .
\tag{7.15.105}
$$

This means that the fractional change in the momentum is of the order of

$$
\frac{\Delta p_y}{p_x} \approx 6 \times 10^{-6}
\tag{7.15.106}
$$

and hence, the trajectory is essentially the same as the classical.

7.16 Spin 1 Measurement in Two Directions

A particle has spin $\hbar$. A measurement of the spin of this particle along a given direction yields the largest possible value. Next, a measurement of the spin is made along a new direction making an angle θ with the original direction.
a) What are the possible results of this measurement?
b) What are the probabilities for the results obtained in part a)?

Solution

a) Since the particle has spin $\hbar$ the outcome for the measurement of spin along any direction must be one of $\hbar$, 0, or $-\hbar$.
b) Since we obtain the largest possible value, in the initial measurement, we

obtain $\hbar$. To compute the probabilities for the new measurement we call the direction of the first measurement the z-direction. In that case the corresponding eigenfunction is

$$\psi = \begin{pmatrix} 1 \\ 0 \\ 0 \end{pmatrix} . \tag{7.16.107}$$

We let the new direction lie in the x-z plane. In that case the appropriate angular momentum operator for this measurement is

$$L_n = L_z \cos\theta + L_x \sin\theta = \hbar \begin{pmatrix} \cos\theta & \frac{1}{\sqrt{2}}\sin\theta & 0 \\ \frac{1}{\sqrt{2}}\sin\theta & 0 & \frac{1}{\sqrt{2}}\sin\theta \\ 0 & \frac{1}{\sqrt{2}}\sin\theta & -\cos\theta \end{pmatrix} . \tag{7.16.108}$$

We now compute the eigenfunctions of L_n for the eigenvalues $\hbar$, 0, and $-\hbar$ respectively. In every case we assume the eigenfunction is of the form

$$\psi = \begin{pmatrix} a \\ b \\ c \end{pmatrix} . \tag{7.16.109}$$

The equations for the coefficients a, b, c then read
$\hbar$:

$$a\cos\theta + \frac{1}{\sqrt{2}}\sin\theta\, b = a$$
$$\frac{1}{\sqrt{2}}\sin\theta\, b - \cos\theta\, c = c . \tag{7.16.110}$$

Thus, after normalization, we get

$$\psi_1 = \begin{pmatrix} \frac{1}{2}(1+\cos\theta) \\ \frac{1}{\sqrt{2}}\sin\theta \\ \frac{1}{2}(1-\cos\theta) \end{pmatrix} . \tag{7.16.111}$$

$0\hbar$:

$$a\cos\theta + \frac{1}{\sqrt{2}}\sin\theta\, b = 0$$
$$\frac{1}{\sqrt{2}}\sin\theta\, b - \cos\theta\, c = 0 . \tag{7.16.112}$$

Hence, after normalization, we get

$$\psi_0 = \begin{pmatrix} \frac{1}{\sqrt{2}}\sin\theta \\ \cos\theta \\ \frac{1}{\sqrt{2}}\sin\theta \end{pmatrix} . \tag{7.16.113}$$

$-\hbar$:

$$a \cos\theta + \frac{1}{\sqrt{2}} \sin\theta\, b = -a$$

$$\frac{1}{\sqrt{2}} \sin\theta\, b - \cos\theta\, c = -c \qquad (7.16.114)$$

and again, after normalization, we get

$$\psi_{-1} = \begin{pmatrix} -\frac{1}{2}(1 - \cos\theta) \\ \frac{1}{\sqrt{2}} \sin\theta \\ -\frac{1}{2}(1 + \cos\theta) \end{pmatrix} . \qquad (7.16.115)$$

The desired probabilities are now given by

$$P_s = |(\psi, \psi_s)|^2 \qquad (7.16.116)$$

and are

$$P_1 = \frac{1}{4}(1 + \cos\theta)^2$$

$$P_0 = \frac{1}{2}\sin^2\theta$$

$$P_{-1} = \frac{1}{4}(1 - \cos\theta)^2 . \qquad (7.16.117)$$

7.17 Particle in a Box: Probabilities and Evolution

The wavefunction for a particle in a box $0 \le x \le L$ is given at $t = 0$ by

$$\Psi(x,0) = \begin{cases} \frac{1}{\sqrt{L}} & 0 \le x \le L \\ 0 & \text{otherwise} \end{cases} .$$

a) Find the probability $P_n(0)$ that at $t = 0$ a measurement of the energy yields the value

$$E_n = \frac{\hbar^2 n^2 \pi^2}{2mL^2} .$$

b) Find the corresponding probability $P_n(t)$ for a time t.

Solution

The eigenvalues and eigenfunctions for a particle in a box of length L are

$$E_n = \frac{\hbar^2 n^2 \pi^2}{2mL^2} \qquad (7.17.118)$$

and

$$u_n = \sqrt{\frac{2}{L}} \sin(n\pi x / L) . \qquad (7.17.119)$$

The probability that at $t = 0$ a measurement of the energy yields the value

$$E_n = \frac{\hbar^2 n^2 \pi^2}{2mL^2}$$

is given by

$$
\begin{aligned}
P_n(0) &= |(u_n, \Psi(x,0))|^2 \\
&= \frac{2}{L^2} \left| \int_0^L \sin(n\pi x/L)\, dx \right|^2 \\
&= \frac{2}{n^2 \pi^2} [1 - \cos(n\pi)]^2 \\
&= \begin{cases} \frac{8}{n^2 \pi^2} & n = 1, 3, 5, \dots \\ 0 & n = 2, 4, 6, \dots \end{cases}
\end{aligned}
\tag{7.17.120}
$$

b) To find the probability at time t we can proceed in two different ways.
1) We can find the wavefunctions at time t and use this. Thus, each eigenstate evolves according to

$$u_n(x)\, e^{-iE_n t/\hbar}$$

and the total state evolves according to

$$\Psi(x,t) = \sum_n c_l u_l(x)\, e^{-iE_l t/\hbar} \tag{7.17.121}$$

where, as just calculated,

$$c_l = (\Psi(x,0), u_l(x)) = \frac{2\sqrt{2}}{l\pi} \quad l = 1, 3, 5, \dots . \tag{7.17.122}$$

Then,

$$
\begin{aligned}
P_n(t) &= |(u_l\, e^{-iE_l t/\hbar}, \Psi(x,t))|^2 \\
&= \left| \sum_{l=1,3,5,\dots} c_l(u_n, u_l) \right|^2 \\
&= \left| \sum_{l=1,3,5,\dots} c_l \delta_{nl} \right|^2 \\
&= \begin{cases} \frac{8}{n^2 \pi^2} & n = 1, 3, 5, \dots \\ 0 & n = 2, 4, 6, \dots \end{cases}
\end{aligned}
\tag{7.17.123}
$$

as before.
2) A simper method is to realize that the evolution is unitary. This means that both the eigenstates u_n and the total state Ψ evolve with the unitary operator

$$U(t) = \exp\{-iHt/\hbar\} \tag{7.17.124}$$

where H is the total Hamiltonian for this system. Then,

$$
\begin{aligned}
P_n(t) &= |(U(t)\Psi(x,0), U(t)u_n(x))|^2 \\
&= |(u_n, \Psi(x,0))|^2
\end{aligned}
\tag{7.17.125}
$$

which is the same result as for $t = 0$.

7.18 Free Wave Equation: Translation Invariance

a) Show that the Schrödinger equation for a free particle

$$
i\hbar \frac{\partial \Psi}{\partial t} = -\frac{\hbar^2}{2m} \nabla^2 \Psi
\tag{7.18.126}
$$

is form invariant under a translation

$$
\begin{aligned}
x &\rightarrow x' = x - vt \\
y &\rightarrow y' = y \\
z &\rightarrow z' = z \ .
\end{aligned}
\tag{7.18.127}
$$

That is, show that if the wavefunction in the new coordinate system is $\Psi'(x',y',z',t)$ then it satisfies the Schrödinger equation

$$
i\hbar \frac{\partial \Psi'}{\partial t} = -\frac{\hbar^2}{2m} \nabla'^2 \Psi' \ .
\tag{7.18.128}
$$

Hint: The wavefunction transforms as a scalar with an additional phase transformation.

b) Apply this result to a plane wave and discuss the results so obtained.

Solution

a) Using the hint we write the wavefunction in the new coordinate system as $\Psi'(x',y',z',t)$ and relate it to the wavefunction in the old coordinate system by an additional phase factor. Thus,

$$
\Psi(\vec{x},t) = e^{-if(\vec{x}',t)} \Psi'(\vec{x}',t) \ .
\tag{7.18.129}
$$

This may be rewritten to read

$$
\Psi(x'+vt,y,z,t) = e^{-if(x',y,z,t)} \Psi'(\vec{x}',t) \ .
\tag{7.18.130}
$$

Since the transformation is only on the x coordinate, we need not consider the y and z coordinates and so we choose f independent of y and z. Then,

$$
\begin{aligned}
\frac{\partial \Psi}{\partial t} &= e^{-if} \left[\frac{\partial \Psi'}{\partial x'} \frac{\partial x'}{\partial t} + \frac{\partial \Psi'}{\partial t} - i \left(\frac{\partial f}{\partial x'} \frac{\partial x'}{\partial t} + \frac{\partial f}{\partial t} \right) \Psi' \right] \\
&= e^{-if} \left[-v \frac{\partial \Psi'}{\partial x'} + \frac{\partial \Psi'}{\partial t} - i \left(-v \frac{\partial f}{\partial x'} + \frac{\partial f}{\partial t} \right) \Psi' \right] \ .
\end{aligned}
\tag{7.18.131}
$$

Also,

$$\frac{\partial \Psi}{\partial x} = e^{-if} \left[\frac{\partial \Psi'}{\partial x'} - i \frac{\partial f}{\partial x'} \Psi' \right] \tag{7.18.132}$$

and

$$\frac{\partial^2 \Psi}{\partial x^2} = e^{-if} \left[\frac{\partial^2 \Psi'}{\partial x'^2} - \left(i \frac{\partial^2 f}{\partial x'^2} + (\frac{\partial f}{\partial x'})^2 \right) \Psi' - 2i \frac{\partial f}{\partial x'} \frac{\partial \Psi'}{\partial x'} \right] . \tag{7.18.133}$$

Substituting all this into the Schrödinger equation we get

$$i\hbar \left[-v \frac{\partial \Psi'}{\partial x'} + \frac{\partial \Psi'}{\partial t} - i \left(-v \frac{\partial f}{\partial x'} + \frac{\partial f}{\partial t} \right) \Psi' \right]$$

$$= -\frac{\hbar^2}{2m} \left[\frac{\partial^2 \Psi'}{\partial x'^2} - \left(i \frac{\partial^2 f}{\partial x'^2} + (\frac{\partial f}{\partial x'})^2 \right) \Psi' - 2i \frac{\partial f}{\partial x'} \frac{\partial \Psi'}{\partial x'} \right] . \tag{7.18.134}$$

Therefore, to recover the Schrödinger equation in the new coordinate system we need

$$-i\hbar v = 2i \frac{\hbar^2}{2m} \frac{\partial f}{\partial x'} \tag{7.18.135}$$

$$\hbar \left(-v \frac{\partial f}{\partial x'} + \frac{\partial f}{\partial t} \right) = \frac{\hbar^2}{2m} \left[(\frac{\partial f}{\partial x'})^2 + i \frac{\partial^2 f}{\partial x'^2} \right] . \tag{7.18.136}$$

From (7.18.135) we get that

$$\frac{\partial f}{\partial x'} = -\frac{mv}{\hbar} \tag{7.18.137}$$

so that

$$f(x',t) = -\frac{mv}{\hbar} x' + a(t) \tag{7.18.138}$$

where $a(t)$ is an arbitrary function of t. After substituting this result into (7.18.136) we get

$$\frac{mv^2}{\hbar} + \frac{da}{dt} = \frac{mv^2}{2\hbar} . \tag{7.18.139}$$

Thus,

$$a(t) = -\frac{mv^2}{2\hbar} t \tag{7.18.140}$$

where we have set the arbitrary constant of integration equal to zero since it would only yield an irrelevant constant phase factor. So, finally

$$f(x',t) = -\frac{mv}{\hbar} (x' + vt/2) . \tag{7.18.141}$$

The wavefunction in the primed coordinate system is therefore given by

$$\Psi(x',y',z',t) = e^{-i\frac{mv}{\hbar}(x'+vt/2)} \Psi(x' + vt, y, z, t) \tag{7.18.142}$$

and satisfies the Schrödinger equation

$$i\hbar\frac{\partial\Psi'(\vec{x}',t)}{\partial t} = -\frac{\hbar^2}{2m}\nabla'^2\Psi'(\vec{x}',t) \ . \tag{7.18.143}$$

b) For a plane wave we have

$$\Psi(\vec{x},t) = e^{-i[\omega t - \vec{k}\cdot\vec{x}]} \ . \tag{7.18.144}$$

According to (7.18.143) the wavefunction in the primed coordinate system is

$$\Psi'(\vec{x}',t) = e^{-i\frac{mv}{\hbar}(x'+vt/2)} e^{-i[\omega t - \vec{k}\cdot(\vec{x}'+\vec{v}t)]}$$

$$= \ \exp -i\left[\left(\omega + \frac{mv^2}{2\hbar}\right)t - \left(k_x + \frac{mv}{\hbar}\right)x' - k_y y - k_z z\right] \ . \tag{7.18.145}$$

This means that in the moving frame (primed system) the frequency and wavenumbers are given by

$$\begin{aligned}
\omega' &= \omega + \frac{mv^2}{2\hbar} \\
k_x' &= k_x + \frac{mv}{\hbar} \\
k_y' &= k_y \\
k_z' &= k_z \ .
\end{aligned} \tag{7.18.146}$$

If we express these results in terms of the energy $E = \hbar\omega$ and momentum $\vec{p} = \hbar\vec{k}$ we get

$$\begin{aligned}
E' &= E + \frac{1}{2}mv^2 \\
p_x' &= p_x + mv \\
p_y' &= p_y \\
p_z' &= p_z \ .
\end{aligned} \tag{7.18.147}$$

These are exactly the same results that would be obtained classically.

7.19 Free Wave Equation: Accelerated Frame

Show that if in the Schrödinger equation for a free particle

$$i\hbar\frac{\partial\Psi}{\partial t} = -\frac{\hbar^2}{2m}\frac{\partial^2\Psi}{\partial x^2} \tag{7.19.148}$$

we go to a uniformly accelerated frame

$$x \to x' = x + \frac{1}{2}at^2 \tag{7.19.149}$$

then the wavefunction $\Psi'(x',t)$ in the new coordinate system satisfies a Schrödinger equation of the form

$$i\hbar\frac{\partial\Psi'(x',t)}{\partial t} = -\frac{\hbar^2}{2m}\frac{\partial^2\Psi'(x',t)}{\partial x'^2} + V(x')\Psi'(x',t) \tag{7.19.150}$$

where

$$V(x') = -max' \tag{7.19.151}$$

is the potential for a constant force that produces a uniform acceleration a.
Hint: In transforming the wavefunction to the accelerated frame allow for a
phase factor so that wavefunction in the stationary frame $\Psi(x,t)$ and the wavefunction in the accelerated frame are related by

$$\Psi(x,t) = e^{-if(x',t)}\,\Psi(x',t) \ . \tag{7.19.152}$$

See also problem 7.18.

Solution

We begin with equation (7.19.152) and rewrite it using (7.19.149) as

$$\Psi(x' - 1/2\,at^2, t) = e^{-if(x',t)}\,\Psi'(x',t) \ . \tag{7.19.153}$$

Then,

$$
\begin{aligned}
\frac{\partial \Psi}{\partial t} &= e^{-if}\left[\frac{\partial \Psi'}{\partial t} + \frac{\partial \Psi'}{\partial x'}\frac{\partial x'}{\partial t} - i\left(\frac{\partial f}{\partial x'}\frac{\partial x'}{\partial t} + \frac{\partial f}{\partial t}\right)\Psi'\right] \\
&= e^{-if}\left[\frac{\partial \Psi'}{\partial t} + at\frac{\partial \Psi'}{\partial x'} - i\left(at\frac{\partial f}{\partial x'} + \frac{\partial f}{\partial t}\right)\Psi'\right] \ .
\end{aligned}
\tag{7.19.154}
$$

Also,

$$\frac{\partial \Psi}{\partial x} = e^{-if}\left[\frac{\partial \Psi'}{\partial x'} - i\frac{\partial f}{\partial x'}\Psi'\right] \tag{7.19.155}$$

and

$$\frac{\partial^2 \Psi}{\partial x^2} = e^{-if}\left[\frac{\partial^2 \Psi'}{\partial x'^2} - 2i\frac{\partial f}{\partial x'}\frac{\partial \Psi'}{\partial x'} - \left(i\frac{\partial^2 f}{\partial x'^2} + (\frac{\partial f}{\partial x'})^2\right)\Psi'\right] \ . \tag{7.19.156}$$

Substituting this into the Schrödinger equation we get

$$
\begin{aligned}
&i\hbar\left[\frac{\partial \Psi'}{\partial t} + at\frac{\partial \Psi'}{\partial x'} - i\left(at\frac{\partial f}{\partial x'} + \frac{\partial f}{\partial t}\right)\Psi'\right] \\
&= -\frac{\hbar^2}{2m}\left[\frac{\partial^2 \Psi'}{\partial x'^2} - 2i\frac{\partial f}{\partial x'}\frac{\partial \Psi'}{\partial x'} - \left(i\frac{\partial^2 f}{\partial x'^2} + (\frac{\partial f}{\partial x'})^2\right)\Psi'\right] \ .
\end{aligned}
\tag{7.19.157}
$$

Therefore, to recover a Schrödinger equation, with the possibility of a potential,
in the new coordinate system we need

$$i\hbar at = 2i\frac{\hbar^2}{2m}\frac{\partial f}{\partial x'} \tag{7.19.158}$$

as well as

$$\frac{\hbar^2}{2m}\left[i\frac{\partial^2 f}{\partial x'^2} + \left(\frac{\partial f}{\partial x'}\right)^2\right] - \hbar\left[at\frac{\partial f}{\partial x'} + \frac{\partial f}{\partial t}\right] = V(x') \ . \tag{7.19.159}$$

From (7.19.158) we get

$$f(x',t) = \frac{mat}{\hbar}x' + b(t) \tag{7.19.160}$$

where $b(t)$ is an arbitrary function of t. Substituting this result into (7.19.159) we find

$$-\frac{ma^2t^2}{2\hbar} - \frac{max'}{\hbar} - \frac{db}{dt} = V(x')/\hbar \ . \tag{7.19.161}$$

Therefore we must take

$$V(x') = -max' \tag{7.19.162}$$

and

$$\frac{db}{dt} = -\frac{ma^2t^2}{2\hbar} \ . \tag{7.19.163}$$

So,

$$b(t) = -\frac{ma^2t^3}{6\hbar} \tag{7.19.164}$$

where we have dropped an irrelevant constant of integration. Hence,

$$f(x',t) = \frac{mat}{\hbar}\left(x' - \frac{1}{6}at^2\right) \ . \tag{7.19.165}$$

Furthermore the wavefunction $\Psi'(x',t)$ satisfies the Schrödinger equation

$$i\hbar\frac{\partial\Psi'(x',t)}{\partial t} = -\frac{\hbar^2}{2m}\frac{\partial^2\Psi'(x',t)}{\partial x'^2} + V(x')\Psi'(x',t) \tag{7.19.166}$$

where, as stated,

$$V(x') = -max' \tag{7.19.167}$$

is the potential due to a constant force that produces a uniform acceleration a.

7.20   The Wigner Function

Show that if for a given wavefunction $\Psi(\vec{x}, t)$ one defines the "Wigner function" [7.3]

$$W(\vec{x}, \vec{p}, t) = \frac{1}{h^3}\int d^3y\, \Psi(\vec{x} + \vec{y}/2, t)\Psi^*(\vec{x} - \vec{y}/2, t)\, e^{-i\vec{p}\cdot\vec{y}/\hbar}$$

then it satisfies the following properties.

a)

$$W(\vec{x}, \vec{p}, t) = \frac{1}{h^6}\int d^3q\, \tilde{\Psi}(\vec{p} + \vec{q}/2, t)\tilde{\Psi}^*(\vec{p} - \vec{q}/2, t)\, e^{i\vec{q}\cdot\vec{x}/\hbar} \ .$$

b)

$$\int d^3p\, W(\vec{x},\vec{p},t) = |\Psi(\vec{x},t)|^2 \ .$$

c)

$$\int d^3x\, W(\vec{x},\vec{p},t) = \left(\frac{2\pi}{\hbar}\right)^3 |\tilde{\Psi}(\vec{p},t)|^2 \ .$$

Hint: These properties depend only on the properties of Fourier integrals.

Solution

We begin by writing out the Fourier transforms.

$$
\begin{aligned}
\Psi(\vec{x}+\vec{y}/2,t) &= \frac{1}{\hbar^3}\int d^3q\, e^{i\vec{q}\cdot(\vec{x}+\vec{y}/2)/\hbar}\,\tilde{\Psi}(\vec{q},t) \\
\Psi^*(\vec{x}-\vec{y}/2,t) &= \frac{1}{\hbar^3}\int d^3k\, e^{-i\vec{k}\cdot(\vec{x}-\vec{y}/2)/\hbar}\,\tilde{\Psi}^*(\vec{k},t) \ .
\end{aligned}
\qquad (7.20.168)
$$

Therefore,

$$
\begin{aligned}
&W(\vec{x},\vec{p},t) \\
&= \frac{1}{\hbar^9}\int d^3y\, d^3q\, d^3k\, e^{-i\vec{p}\cdot\vec{y}/\hbar}\, e^{i\vec{q}\cdot(\vec{x}+\vec{y}/2)/\hbar}\, e^{-i\vec{k}\cdot(\vec{x}-\vec{y}/2)/\hbar}\,\tilde{\Psi}(\vec{q},t)\tilde{\Psi}^*(\vec{k},t) \\
&= \frac{(2\pi\hbar)^3}{\hbar^9}\int d^3q\, d^3k\, \delta(\vec{p}-\vec{q}/2-\vec{k}/2)\, e^{i(\vec{q}-\vec{k})\cdot\vec{x}/\hbar}\,\tilde{\Psi}(\vec{q},t)\tilde{\Psi}^*(\vec{k},t) \\
&= \frac{8}{\hbar^6}\int d^3k\, e^{2i(\vec{p}-\vec{k})\cdot\vec{x}/\hbar}\,\tilde{\Psi}(2\vec{p}-\vec{k},t)\tilde{\Psi}^*(\vec{k},t) \ .
\end{aligned}
\qquad (7.20.169)
$$

We now let

$$\vec{p}-\vec{k} = \vec{q}/2$$

so that

$$\vec{k} = \vec{p}-\vec{q}/2 \quad \text{and} \quad \frac{1}{8}d^3q = d^3k \ .$$

Then,

$$W(\vec{x},\vec{p},t) = \frac{1}{\hbar^6}\int d^3q\,\tilde{\Psi}(\vec{p}+\vec{q}/2,t)\tilde{\Psi}^*(\vec{p}-\vec{q}/2,t)\, e^{i\vec{q}\cdot\vec{x}/\hbar} \ . \qquad (7.20.170)$$

b) In this case we simply integrate W over $\vec{p}$ and get

$$
\begin{aligned}
\int d^3p\, W(\vec{x},\vec{p},t) &= \frac{1}{\hbar^3}\int d^3y\, d^3p\,\Psi(\vec{x}+\vec{y}/2,t)\Psi^*(\vec{x}-\vec{y}/2,t)\, e^{-i\vec{p}\cdot\vec{y}/\hbar} \\
&= \frac{(2\pi\hbar)^3}{\hbar^3}\int d^3y\,\Psi(\vec{x}+\vec{y}/2,t)\Psi^*(\vec{x}-\vec{y}/2,t)\delta(\vec{y}) \\
&= \Psi(\vec{x},t)\Psi^*(\vec{x},t) = |\Psi(\vec{x},t)|^2 \ .
\end{aligned}
\qquad (7.20.171)
$$

This result shows that

$$\int d^3p\, W(\vec{x},\, \vec{p},\, t)$$

gives the probability density for finding a particle at the point $\vec{x}$ at the time t.

c) Here, just as in part b) above we simply integrate over $\vec{x}$. Thus,

$$\begin{aligned}
\int d^3x\, W(\vec{x},\, \vec{p},\, t) &= \frac{1}{h^6} \int d^3q\, d^3x\, \tilde{\Psi}(\vec{p}+\vec{q}/2,\, t)\tilde{\Psi}^*(\vec{p}-\vec{q}/2,\, t)\, e^{i\vec{q}\cdot\vec{x}/\hbar} \\
&= \frac{1}{h^3} \int d^3q\, \tilde{\Psi}(\vec{p}+\vec{q}/2,\, t)\tilde{\Psi}^*(\vec{p}-\vec{q}/2,\, t)\, \delta(\vec{q}) \\
&= \frac{1}{h^3} \tilde{\Psi}(\vec{p},\, t)\tilde{\Psi}^*(\vec{p},\, t) = \frac{1}{h^3}|\tilde{\Psi}(\vec{p},\, t)|^2 \ . \qquad (7.20.172)
\end{aligned}$$

This result shows that

$$\int d^3x\, W(\vec{x},\, \vec{p},\, t)$$

now gives the probability density for finding a particle with momentum $\vec{p}$ at the time t.

7.21 Properties of the Wigner Function

Show that although the Wigner function (see the previous problem) seems to behave like a classical probability density for $\vec{x}$ and $\vec{p}$ it cannot be a true probability distribution since it is not positive definite.

Hint: Try the wavefunction

$$\Psi(\vec{x}) = A\, x\, e^{-r^2/2a^2}$$

where A is a normalization constant and $r^2 = \vec{x}\cdot\vec{x}$.

Solution

Using the suggested wavefunction we have

$$\begin{aligned}
&W(\vec{x},\vec{p},t) \\
&= \frac{|A|^2}{h^3} \int d^3p\, e^{i\vec{p}\cdot\vec{\rho}/\hbar}(x+\xi/2)\, e^{-(x+\xi/2)^2/2a^2} \\
&\times\ e^{-(y+\eta/2)^2/2a^2}\, e^{-(z+\zeta/2)^2/2a^2} \\
&\times\ (x-\xi/2)\, e^{-(x-\xi/2)^2/2a^2}\, e^{-(y-\eta/2)^2/2a^2}\, e^{-(z-\zeta/2)^2/2a^2}\ . \qquad (7.21.173)
\end{aligned}$$

Here,

$$\vec{\rho} = (\xi,\, \eta,\, \zeta) \ .$$

After some simplification we find

$$
W(\vec{x}, \vec{p}, t)
$$

$$
= \frac{|A|^2}{h^3} \int d\xi d\eta d\zeta \, e^{i\vec{p}\cdot\vec{\rho}/\hbar} \left(x^2 - \xi^2/4\right) e^{-2(x^2+\xi^2/4)/2a^2}
$$

$$
e^{-2(y^2+\eta^2/4)/2a^2} \, e^{-2(z^2+\zeta^2/4)/2a^2}
$$

$$
= \frac{|A|^2}{h^3} \int d\xi \, e^{-ip_x\xi/\hbar} \left(x^2 - \xi^2/4\right) e^{-x^2/a^2} e^{-\xi^2/4a^2}
$$

$$
\times \int d\eta \, e^{-ip_y\eta/\hbar} \, e^{-y^2/a^2} e^{-\eta^2/4a^2} \int d\zeta \, e^{-ip_z\zeta/\hbar} \, e^{-z^2/a^2} e^{-\zeta^2/4a^2}
$$

$$
= \frac{|A|^2}{h^3} e^{-r^2/a^2} e^{-p^2 a^2/\hbar^2} \pi^{3/2} 4a^5 \left(\frac{2x^2}{a^2} + \frac{2a^2 p_x^2}{\hbar^2} - 1 \right) . \qquad (7.21.174)
$$

So, for

$$
\frac{2x^2}{a^2} + \frac{2a^2 p_x^2}{\hbar^2} < 1
$$

we get negative values. This means that, in this case, $W(\vec{x}, \vec{p}, t)$ cannot be a true probability density. Notice the region for which we get negative values is the interior of an ellipse in the $x - p_x$ plane with semi-major and semi-minor axes $a/\sqrt{2}$ and $\hbar/(\sqrt{2}a)$. Inside this region we would have

$$
x p_x < a/\sqrt{2} \times \hbar/(\sqrt{2}a) = \hbar/2 .
$$

7.22 Uncertainty Relation and Wigner Function

Show that if $\Psi(\vec{x}, t)$ is a properly normalized wavefunction then

$$
|W(\vec{x}, \vec{p}, t)|^2
$$

$$
\leq \frac{1}{h^6} \int d^3y \, |\Psi(\vec{x} + \vec{y}/2, t)|^2 \int d^3z \, |\Psi(\vec{x} - \vec{z}/2, t)|^2
$$

$$
= \left(\frac{2}{h} \right)^6 .
$$

$$
(7.22.175)
$$

This means that

$$
|W| \leq \left(\frac{2}{h} \right)^3
$$

and states that a particle can not be localized in a cell of phase space smaller than $(h/2)^3$.

Hint: Use the Schwarz inequality.

Solution

The Schwarz inequality states that for any two square integrable functions f and g we have

$$|(f, g)|^2 \le (f, f)(g, g) \tag{7.22.176}$$

where

$$(f, g) = \int f^*(\vec{x})g(\vec{x})\, d^3x \ .$$

We now choose

$$
\begin{aligned}
f(\vec{y}) &= \frac{1}{h^{3/2}}\Psi(\vec{x} - \vec{y}/2, t)\, e^{i\vec{p}\cdot\vec{y}/h} \\
g(\vec{y}) &= \frac{1}{h^{3/2}}\Psi(\vec{x} + \vec{y}/2, t) \ .
\end{aligned} \tag{7.22.177}
$$

Then,

$$
\begin{aligned}
(f, g) &= \frac{1}{h^3}\int d^3y\, \Psi(\vec{x} + \vec{y}/2, t)\Psi^*(\vec{x} - \vec{y}/2, t)\, e^{-i\vec{p}\cdot\vec{y}/h} \\
&= W(\vec{x}, \vec{p}, t) \ .
\end{aligned} \tag{7.22.178}
$$

So,

$$
\begin{aligned}
(f, g)(g, f) &= |W(\vec{x}, \vec{p}, t)|^2 \\
&\le \frac{1}{h^6}\int d^3y|\Psi(\vec{x} + \vec{y}/2, t)|^2 \int d^3z|\Psi^*(\vec{x} - \vec{z}/2, t)|^2 \\
&= \frac{2^6}{h^6}\int d^3(y/2)|\Psi(\vec{x} + \vec{y}/2, t)|^2 \int d^3(z/2)|\Psi^*(\vec{x} - \vec{z}/2, t)|^2 \\
&= \left(\frac{2}{h}\right)^6 \ .
\end{aligned} \tag{7.22.179}
$$

Bibliography

[7.1] A.Z. Capri, *Nonrelativistic Quantum Mechanics* 3rd edition, World Scientific Publishing Co. Pte. Ltd., section 7.9, (2002) .

[7.2] ibid section 7.10 .

[7.3]E.P. Wigner, Phys. Rev., **40**, 749, (1932).

Chapter 8

Distributions and Fourier Transforms

8.1 Properties of the Delta Function

Use the representation

$$\delta(x) = \lim_{\epsilon \to 0+} h_\epsilon(x)$$

where

$$h_\epsilon(x) = \begin{cases} \frac{1}{2\epsilon} & \text{if } |x| < \epsilon \\ 0 & \text{if } |x| > \epsilon \end{cases}$$

to verify the following equations

a)

$$\int_{-\infty}^{\infty} f(x)\delta(x-a)\,dx = f(a) \tag{8.1.1}$$

b)

$$\delta(ax)\,dx = \frac{1}{a}\delta(x) \tag{8.1.2}$$

c)

$$f(x)\delta(x-a) = f(a)\delta(x-a) \tag{8.1.3}$$

d)

$$\int_{-\infty}^{\infty} \delta(x-y)\delta(y-z)\,dy = \delta(x-z) \tag{8.1.4}$$

e)

$$\delta(x^2 - a^2) = \frac{1}{2|a|}[\delta(x-a) + \delta(x+a)]\,. \tag{8.1.5}$$

Hint: Integrate both sides of the equation with a well-behaved function $f(x)$.

Solution

a) We begin with

$$h_\epsilon(x) = \begin{cases} \frac{1}{2\epsilon} & |x| < \epsilon \\ 0 & |x| > \epsilon \end{cases} \tag{8.1.6}$$

Therefore,

$$\lim_{\epsilon \to 0+} \int_{-\infty}^{\infty} f(x) h_\epsilon(x-a)\, dx = \lim_{\epsilon \to 0+} \int_{-\epsilon}^{\epsilon} \frac{f(x+a)}{2\epsilon}\, dx \ . \tag{8.1.7}$$

Now, for $f(x)$ sufficiently smooth near $x = a$ we can use a mean value theorem to write (with $-1 < \theta < 1$) the above expression as

$$\lim_{\epsilon \to 0+} \int_{-\epsilon}^{\epsilon} \frac{f(x+a)}{2\epsilon}\, dx = \lim_{\epsilon \to 0+} f(a + \theta\epsilon) = f(a) \ . \tag{8.1.8}$$

b) In this case we consider

$$\begin{aligned} \lim_{\epsilon \to 0+} \int_{-\infty}^{\infty} f(x) h_\epsilon(ax)\, dx &= \lim_{\epsilon \to 0+} \int_{-\epsilon/a}^{\epsilon/a} \frac{f(z/a)}{2\epsilon} \frac{dz}{a} \\ &= \lim_{\epsilon \to 0+} 2\frac{\epsilon}{a} \frac{f(\theta\epsilon/a)}{2\epsilon} \\ &= \frac{1}{a} f(0) \ . \end{aligned} \tag{8.1.9}$$

Here we have changed variables from x to $z = ax$ in going to the second line and again invoked a mean value theorem to get to the third line where θ is a number bewtween 0 and 1. So, as required,

$$\delta(ax) = \frac{1}{a}\delta(x) \ . \tag{8.1.10}$$

c) We evaluated $\delta(x - a)$ in part a) so now we simply compute

$$\lim_{\epsilon \to 0+} \int_{-\epsilon}^{\epsilon} \frac{f(a)}{2\epsilon}\, dx = f(a) \ . \tag{8.1.11}$$

This equals the result from part a) and therefore we have verified that

$$f(x)\delta(x-a) = f(a)\delta(x-a) \ . \tag{8.1.12}$$

d) In this case we first write out the two functions whose limit we need to take.

$$\begin{aligned} h_\epsilon(x-y) &= \begin{cases} \frac{1}{2\epsilon} & x - \epsilon < y < x + \epsilon \\ 0 & \text{otherwise} \end{cases} \\ h_\epsilon(y-z) &= \begin{cases} \frac{1}{2\epsilon} & z - \epsilon < x < z + \epsilon \\ 0 & \text{otherwise} \end{cases} \end{aligned} \tag{8.1.13}$$

Therefore,

$$\lim_{\epsilon_1 \to 0+} \lim_{\epsilon_2 \to 0+} \int_{-\infty}^{\infty} f(x) h_{\epsilon_1}(x-y) h_{\epsilon_2}(y-z) \, dx \, dy$$

$$= \lim_{\epsilon_1 \to 0+} \lim_{\epsilon_2 \to 0+} \frac{1}{2\epsilon_1} \frac{1}{2\epsilon_2} \int_{z-\epsilon_2}^{z+\epsilon_2} dx \, f(x) \int_{x-\epsilon_1}^{x+\epsilon_1} dy$$

$$= \lim_{\epsilon_1 \to 0+} \lim_{\epsilon_2 \to 0+} \frac{1}{2\epsilon_1} \frac{1}{2\epsilon_2} 2\epsilon_1 \int_{z-\epsilon_2}^{z+\epsilon_2} dx \, f(x)$$

$$= f(z) \ . \tag{8.1.14}$$

Therefore, we have again verified that

$$\int_{-\infty}^{\infty} \delta(x-y)\delta(y-z) \, dy = \delta(x-z) \ . \tag{8.1.15}$$

e) In this case we have to write out the limiting function

$$h_{\epsilon}(x^2 - a^2) = \begin{cases} \frac{1}{2\epsilon} & \sqrt{a^2 - \epsilon} < x < \sqrt{a^2 + \epsilon} \\ \frac{1}{2\epsilon} & -\sqrt{a^2 + \epsilon} < x < -\sqrt{a^2 - \epsilon} \\ 0 & \text{otherwise} \end{cases} \ . \tag{8.1.16}$$

Therefore,

$$\lim_{\epsilon \to 0+} \int_{-\infty}^{\infty} f(x) h_{\epsilon}(x^2 - a^2) f(x) \, dx$$

$$= \lim_{\epsilon \to 0+} \frac{1}{2\epsilon} \left[\int_{-\sqrt{a^2+\epsilon}}^{-\sqrt{a^2-\epsilon}} f(x) \, dx + \int_{\sqrt{a^2-\epsilon}}^{\sqrt{a^2+\epsilon}} f(x) \, dx \right]$$

$$= \lim_{\epsilon \to 0+} \frac{1}{2\epsilon} \left[\int_{-(a+\epsilon/2|a|)}^{-(a-\epsilon/2|a|)} f(x) \, dx + \int_{(a-\epsilon/2|a|)}^{(a+\epsilon/2|a|)} f(x) \, dx \right]$$

$$= \lim_{\epsilon \to 0+} \frac{1}{2\epsilon} \left[\int_{-\epsilon/2|a|}^{\epsilon/2|a|} f(x+a) \, dx + \int_{-\epsilon/2|a|}^{\epsilon/2|a|} f(x-a) \, dx \right]$$

$$= \frac{1}{2|a|} [f(a) + f(-a)] \ . \tag{8.1.17}$$

Thus, we have also verified that

$$\delta(x^2 - a^2) = \frac{1}{2|a|} [\delta(x-a) + \delta(x+a)] \ . \tag{8.1.18}$$

8.2   Representation of Delta Function

Repeat the previous problem (problem 8.1) using the representation

$$\delta(x) = \lim_{\epsilon \to 0+} \frac{1}{2\pi} \int_{-\infty}^{\infty} e^{ikx - \epsilon k^2} \, dk \ .$$

Solution

a) In this case we again consider

$$I = \lim_{\epsilon \to 0} \frac{1}{2\pi} \int_{-\infty}^{\infty} dx \int_{-\infty}^{\infty} f(x) e^{ikx - \epsilon k^2} dk . \tag{8.2.19}$$

First, we evaluate the integral over k (by completing the square in the argument of the exponential and changing variables).

$$\int_{-\infty}^{\infty} e^{ikx - \epsilon k^2} dk = \sqrt{\frac{\pi}{\epsilon}} \exp\left(-\frac{x^2}{4\epsilon}\right) . \tag{8.2.20}$$

Then we get that

$$I = \lim_{\epsilon \to 0} \frac{1}{2\pi} \int_{-\infty}^{\infty} \sqrt{\frac{\pi}{\epsilon}} \exp\left(-\frac{x^2}{4\epsilon}\right) f(x) \, dx . \tag{8.2.21}$$

Now, as $\epsilon \to 0$ the exponential in the integrand is non-zero only near $x = 0$ i.e. within a distance $\theta\epsilon$ where θ is some finite number. So we can use a mean value theorem to write

$$
\begin{aligned}
I &= \lim_{\epsilon \to 0} \frac{1}{2\pi} \sqrt{\frac{\pi}{\epsilon}} f(\theta\epsilon) \int_{-\infty}^{\infty} \exp\left(-\frac{x^2}{4\epsilon}\right) dx \\
&= \lim_{\epsilon \to 0} \frac{1}{2\pi} \sqrt{\frac{\pi}{\epsilon}} f(\theta\epsilon) \sqrt{4\pi\epsilon} = f(0) .
\end{aligned} \tag{8.2.22}
$$

b) For this part we proceed exactly as in part a) but we change the integration variable from x to $y = ax$. The desired result then follows immediately from part a).

c) This result is also obvious from the calculation in part a).

d) In this case we first consider the expression

$$I = \lim_{\epsilon_1 \to 0+} \lim_{\epsilon_2 \to 0+} \frac{1}{(2\pi)^2} \int dy \, dk \, dq \, e^{ik(x-y) - \epsilon_1 k^2} e^{iq(x-y) - \epsilon_2 q^2} \tag{8.2.23}$$

and integrate over k and q to get

$$I = \lim_{\epsilon_1 \to 0+} \lim_{\epsilon_2 \to 0+} \frac{1}{(2\pi)^2} \int dy \sqrt{\frac{\pi^2}{\epsilon_1 \epsilon_2}} \exp\left[-\frac{(x-y)^2}{4\epsilon_1} - \frac{(y-z)^2}{4\epsilon_2}\right] . \tag{8.2.24}$$

The argument of the exponential simplifies as follows

$$\frac{\epsilon_2(x-y)^2 + \epsilon_1(y-z)^2}{4\epsilon_1\epsilon_2} = \frac{(x-z)^2}{4(\epsilon_1 + \epsilon_2)} + \frac{\epsilon_1 + \epsilon_2}{4\epsilon_1\epsilon_2}\left[y - \frac{\epsilon_2 x + \epsilon_1 z}{\epsilon_1 + \epsilon_2}\right]^2 . \tag{8.2.25}$$

If we now substitute this result back into (8.2.24) and integrate over y we get

$$I = \frac{1}{2\pi} \sqrt{\frac{\pi}{\epsilon_1 + \epsilon_2}} \exp\left[-\frac{(x-z)^2}{4(\epsilon_1 + \epsilon_2)}\right] . \tag{8.2.26}$$

Now taking the limit $\epsilon_2 \to 0+$ we immediately obtain an expression that we recognize from part a) to be the desired result.

e) In this case we consider the expression

$$
\begin{aligned}
I &= \lim_{\epsilon \to 0+} \frac{1}{2\pi} \int f(x)\, dx \int e^{ik(x^2 - a^2) - \epsilon k^2}\, dk \\
&= \lim_{\epsilon \to 0+} \frac{1}{2\pi} \int f(x)\, dx \sqrt{\frac{\pi}{\epsilon}} \exp\left[-\frac{(x^2 - a^2)^2}{4\epsilon}\right] .
\end{aligned}
\tag{8.2.27}
$$

In the limit as $\epsilon \to 0+$ the integrant vanishes everywhere except in the vicinity of $x^2 = a^2$ which is the same as in the vicinity of $x = \pm a$. So we can use a mean value theorem and write

$$
\begin{aligned}
I &= \lim_{\epsilon \to 0+} \frac{1}{2\pi} \left\{ f(a + \theta\epsilon) \int \sqrt{\frac{\pi}{\epsilon}} f(x)\, dx \exp\left[-\frac{(x - a)^2 4a^2}{4\epsilon}\right] \right. \\
&\quad + \left. f(-a + \theta\epsilon) \int \sqrt{\frac{\pi}{\epsilon}} f(x)\, dx \exp\left[-\frac{(x + a)^2 4a^2}{4\epsilon}\right] \right\} \\
&= \lim_{\epsilon \to 0+} \frac{1}{2\pi} [f(a + \theta\epsilon) + f(-a + \theta\epsilon)] \sqrt{\frac{\pi}{\epsilon}} \sqrt{\frac{\pi\epsilon}{a^2}} \\
&= \frac{1}{2|a|} [f(a) + f(-a)] .
\end{aligned}
\tag{8.2.28}
$$

8.3   Normalization of Scattering Solution

Show that the appropriate normalization for the positive parity solution for scattering from a square well

$$
\psi_+(x) = C \begin{cases}
\cos(kx - \delta_+) & \text{if } x < -a \\
A \cos Kx & \text{if } |x| < a \\
\cos(kx + \delta_+) & \text{if } x > a
\end{cases}
$$

to yield δ-function normalization is $1/\sqrt{\pi}$. You will have to use the continuity of the wavefunction at $x = \pm a$ as well as the equation

$$
k \tan(ka + \delta_+) = K \tan Ka .
$$

Solution

We begin with the given solution.

$$
\psi_+(x) = C \begin{cases}
\cos(kx - \delta_+(k)) & x \le a \\
A(k) \cos(Kx) & -a \le x \le a \\
\cos(kx + \delta_+(k)) & x \ge a
\end{cases}
\tag{8.3.29}
$$

where throughout we have $k > 0$ and, from continuity of the wavefunction, that

$$
A(k) = \frac{\cos(ka + \delta_+(k))}{\cos(Ka)}
\tag{8.3.30}
$$

as well as

$$k \tan(ka + \delta_+(k)) = K \tan(Ka)$$
$$q \tan(qa + \delta_+(q)) = Q \tan(Qa) \tag{8.3.31}$$

with

$$K^2 = k^2 + \frac{2mV_0}{\hbar^2} \quad , \quad Q^2 = q^2 + \frac{2mV_0}{\hbar^2} \tag{8.3.32}$$

The normalization is given by the integral

$$\frac{1}{C^2} I = \frac{1}{C^2} \int \psi_+(k, x)\psi_+(q, x)\, dx \tag{8.3.33}$$

or

$$
\begin{aligned}
I &= \int_{-\infty}^{-a} dx \, \cos[kx - \delta_+(k)] \cos[qx - \delta_+(q)] \\
&+ A(k)A(q) \int_{-a}^{a} dx \, \cos(Kx) \cos(Qx) \\
&+ \int_{a}^{\infty} dx \, \cos[kx + \delta_+(k)] \cos[qx + \delta_+(q)] \\
&= 2 \int_{0}^{\infty} dx \, \cos[kx + \delta_+(k)] \cos[qx + \delta_+(q)] \\
&+ 2A(k)A(q) \int_{-a}^{a} dx \, \cos(Kx) \cos(Qx) \\
&- 2 \int_{0}^{a} dx \, \cos[kx + \delta_+(k)] \cos[qx + \delta_+(q)] \\
&= \pi\delta(k - q) - \frac{2}{k^2 - q^2}[k \sin(\delta_+(k)) - q \sin(\delta_+(q))] \\
&+ \frac{\cos[ka + \delta_+(k)] \cos[qa + \delta_+(q)]}{\cos(Ka) \cos(Qa)} \\
&\times \frac{(K + Q) \sin((K - Q)a) + (K - Q) \sin((K + Q)A)}{k^2 - q^2} \\
&- \frac{1}{k^2 - q^2} \{(k + q) \sin[(k - q)a + \delta_+(k) - \delta_+(q)] \\
&+ (k - q) \sin[(k + q)a + \delta_+(k) + \delta_+(q)]\} \ . \tag{8.3.34}
\end{aligned}
$$

We now use equation (8.3.31) to simplify and get

$$I = \pi\delta(k - q) \ . \tag{8.3.35}$$

Therefore, we find that

$$C = \frac{1}{\sqrt{\pi}} \tag{8.3.36}$$

8.4   Tempered Distribution

Show that T is a tempered distribution if T is defined by

$$T(f) = \sum_{k=0}^{m} \int_{-\infty}^{\infty} F_k(x)\, \frac{d^k f(x)}{dx^k}\, dx$$

where F_k are continuous functions bounded by

$$|F_k(x)| \le C_k(1 + |x|^j)$$

for some C_k and j depending on k. As a matter of fact every tempered distribution can be written in this form. Symbolically one then writes

$$T = \sum_{k=0}^{m} (-1)^k\, \frac{d^k F_k(x)}{dx^k}\ .$$

This formula cannot be taken literally however since the $F_k(x)$ need not be differentiable. It arises from a formal integration by parts of the first equation above.

Solution

Let $f \in \mathcal{S}$. Then,

$$\frac{d^k f}{dx^k} \in \mathcal{S} \quad \text{for all } k\ . \tag{8.4.37}$$

Furthermore, $|x|^j f(x) \in \mathcal{S}$ for all j. Therefore, if we define

$$T(f) = \sum_{k=0}^{m} \int_{-\infty}^{\infty} F_k(x) \frac{d^k f}{dx^k}\, dx \tag{8.4.38}$$

we find that

$$|T(f)| \le \sum_{k=0}^{m} \int_{-\infty}^{\infty} \left| C_k(1 + |x|^{j(k)}) \right| \left| \frac{d^k f}{dx^k} \right| dx\ . \tag{8.4.39}$$

Each term on the right can be bounded by one of the semi-norms $\| f \|_{r,n}$ and hence, $T(f)$ is a continuous functional on $\mathcal{S}$.

8.5   Fourier Transform of $\mathcal{D}$

The test function space $\mathcal{D}$ consists of the space of $C^{(\infty)}$ functions of bounded support. The support of a function f, (supp f) is the complement of the largest open set on which the function vanishes. Show that if $f \in \mathcal{FD}$ then $\tilde{f}$ is an entire function.

Solution

Consider

$$\tilde{f}(p) = \int_{-\infty}^{\infty} e^{-ipx} f(x)\, dx \ . \tag{8.5.40}$$

Since the support of f, supp$f$, is bounded, the integral on the right converges uniformly for all complex values of p. We may therefore differentiate under the the integral sign with respect to the complex variable p to get

$$\frac{d\tilde{f}(p)}{dp} = -i \int_{-\infty}^{\infty} e^{-ipx}\, x f(x)\, dx < \infty \ . \tag{8.5.41}$$

Thus, $\tilde{f}$ is an entire function.

8.6 Tempered Distribution of Fast Decrease

Prove the Theorem: The Fourier transform of a *tempered distribution of fast decrease* is a $C^{(\infty)}$ function bounded by a polynomial. A tempered distribution of fast decrease F is of the form

$$F = fT$$

where $f \in \mathcal{S}$ and T is also a tempered distribution.
Hint: To prove that the Fourier transform of F is bounded by a polynomial use the result of problem 8.2.

Solution

We have that a tempered distribution of fast decrease is of the form fT where $f \in \mathcal{S}$ and T is a tempered distribution. Now, $f(x)\, e^{ikx}$ is itself a test function belonging to $\mathcal{S}$ and therefore

$$T\left(f(x)\, e^{ikx}\right) = \int_{-\infty}^{\infty} T(x)\, f(x)\, e^{ikx}\, dx \tag{8.6.42}$$

is uniformly convergent for all real values of k. Using the representation of T obtained in problem 8.2, namely

$$T(f(x)e^{ikx}) = \sum_{n=0}^{m} \int_{-\infty}^{\infty} T_n(x)\frac{d^n\left(f(x)e^{ikx}\right)}{dx^n}\, dx \tag{8.6.43}$$

we can perform the differentiation and obtain that the highest possible power of k that occurs is k^m. Furthermore, since the integral is uniformly convergent we may differentiate under the integral sign any number of times to obtain that the Fourier transform is a C^{∞} function.

8.7   A Useful Identity

Let $f(z)$ be an entire function vanishing rapidly at large $|\Re(z)|$. Show that

$$\lim_{\epsilon \to 0+} \frac{1}{2} \int_{-\infty}^{\infty} \left[\frac{1}{x - a + i\epsilon} + \frac{1}{x - a - i\epsilon} \right] f(x) \, dx = P \int_{-\infty}^{\infty} \frac{f(x)}{x - a} \, dx$$

where the principle value integral is defined by

$$P \int_{-\infty}^{\infty} \frac{f(x)}{x - a} \, dx = \lim_{\epsilon \to 0+} \left[\int_{-\infty}^{a-\epsilon} \frac{f(x)}{x - a} \, dx + \int_{a+\epsilon}^{\infty} \frac{f(x)}{x - a} \, dx \right] .$$

Furthermore show that

$$\lim_{\epsilon \to 0+} \frac{\epsilon/\pi}{x^2 + \epsilon^2} = \delta(x) .$$

Hence conclude that, considered as distributions,

$$\lim_{\epsilon \to 0+} \frac{1}{x - a \pm i\epsilon} = P \frac{1}{x - a} \mp i\pi\delta(x - a)$$

that is,

$$\lim_{\epsilon \to 0+} \int_{-\infty}^{\infty} \frac{f(x) \, dx}{x - a \pm i\epsilon} = P \int_{-\infty}^{\infty} \frac{f(x) \, dx}{x - a} \mp i\pi f(a) .$$

Solution

To study

$$I = \lim_{\epsilon \to 0+} \frac{1}{2} \int_{-\infty}^{\infty} \left(\frac{1}{x - a - i\epsilon} + \frac{1}{x - a + i\epsilon} \right) f(x) \, dx \tag{8.7.44}$$

we consider the contour integral

$$\frac{1}{2} \int_{-R}^{R} \left(\frac{1}{x - a - i\epsilon} + \frac{1}{x - a + i\epsilon} \right) f(x) \, dx$$
$$+ \frac{1}{2} \int_{C_R} \left(\frac{1}{z - a - i\epsilon} + \frac{1}{z - a + i\epsilon} \right) f(z) \, dz$$
$$= \frac{1}{2} 2\pi i \, f(a + i\epsilon) \tag{8.7.45}$$

around the contour shown in figure 8.1. We could equally well have closed the contour in the lower half plane. Next we consider the integral

$$0 = \int_{-R}^{a-\epsilon} \frac{f(x)}{x - a} \, dx + \int_{a+\epsilon}^{R} \frac{f(x)}{x - a} \, dx + \int_{C_1} \frac{f(z)}{z - a} \, dz + \int_{C_R} \frac{f(z)}{z - a} \, dz$$
$$= P \int_{-R}^{R} \frac{f(x)}{x - a} \, dx + \int_{C_R} \frac{f(z)}{z - a} \, dz + \int_{C_1} \frac{f(z)}{z - a} \, dz , \tag{8.7.46}$$

where the contour is shown in figure 8.2. To evaluate the integral around the

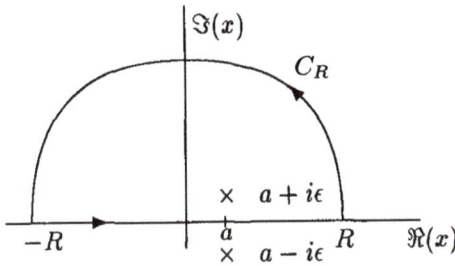

Figure 8.1: The contour used to evaluate the integral I.

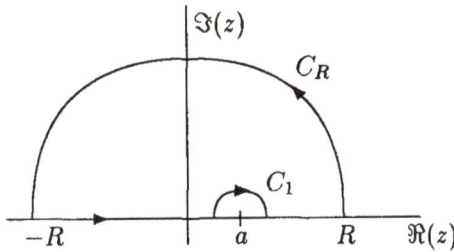

Figure 8.2: The contours C_1 and C_R

semicircle C_1 of radius ϵ we let

$$z = a + \epsilon\, e^{i\theta} \ . \tag{8.7.47}$$

We then obtain

$$\int_{C_1} \frac{f(z)}{z-a}\, dz = \int_\pi^0 \frac{f(a + \epsilon\, e^{i\theta})}{\epsilon\, e^{i\theta}} i\, \epsilon\, e^{i\theta}\, d\theta = -i\pi f(a) \ . \tag{8.7.48}$$

Here we have already taken the limit of $\epsilon \to 0$ under the integral sign since the integral converges uniformly. This means we can conclude that

$$P \int_{-R}^R \frac{f(x)}{x-a}\, dx = -\int_{C_R} \frac{f(z)}{z-a}\, dz + i\pi f(a) \ . \tag{8.7.49}$$

Next taking the limit $\epsilon \to 0+$ of equation (8.7.46) and comparing with (8.7.49) we conclude that

$$P \int_{-R}^R \frac{f(x)}{x-a}\, dx = \lim_{\epsilon \to 0+} \frac{1}{2} \int_{-R}^R \left(\frac{1}{x - a - i\epsilon} + \frac{1}{x - a + i\epsilon} \right) f(x)\, dx \ . \tag{8.7.50}$$

After taking the $\lim_{R\to\infty}$ we have the desired result.

For the next part we simply do a contour integration by closing the path of integration with a large semicircle (at infinity) in either the upper or lower half plane. In either case we get

$$\lim_{\epsilon \to 0+} \int_{-\infty}^{\infty} \frac{\epsilon/\pi}{x^2 + \epsilon^2} f(x)\, dx = \lim_{\epsilon \to 0+} \frac{\epsilon}{\pi} 2\pi i \frac{f(i\epsilon)}{2i\epsilon} = f(0)\ . \tag{8.7.51}$$

Now a simple partial fraction decomposition yields that

$$\frac{\epsilon/\pi}{x^2 + \epsilon^2} = \frac{1}{2\pi}\left[\frac{1}{x - a - i\epsilon} - \frac{1}{x - a + i\epsilon}\right]\ . \tag{8.7.52}$$

Combining this with our previous computation we again have the desired result.

8.8   A Representation of $\delta(x)$

Using the result of problem 8.4 and defining

$$\frac{1}{2\pi}\int_{-\infty}^{\infty} e^{ikx}\, dk = \lim_{\epsilon \to 0+} \frac{1}{2\pi}\left[\int_0^{\infty} e^{ik(x+i\epsilon)}\, dk + \int_{-\infty}^0 e^{ik(x-i\epsilon)}\, dk\right]\ .$$

Prove that

$$\int_{-\infty}^{\infty} e^{ikx}\, dk = \delta(x)\ .$$

Solution

The desired result follows immediately by writing out the integrals. Thus,

$$\lim_{\epsilon \to 0+} \frac{1}{2\pi}\left(\int_0^{\infty} e^{ik(x+i\epsilon)}\, dk + \int_{-\infty}^0 e^{ik(x-i\epsilon)}\, dk\right)$$

$$= \lim_{\epsilon \to 0+} \frac{1}{2\pi i}\left(\frac{-1}{x + i\epsilon} + \frac{1}{x - i\epsilon}\right) = \lim_{\epsilon \to 0+} \frac{\epsilon/\pi}{x^2 + \epsilon^2}\ . \tag{8.8.53}$$

8.9   Fourier Transform of $\delta^{(n)}(x)$

Calculate the Fourier transform of $\delta^{(n)}(x)$.

Solution

In this case we can proceed quite formally

$$\int \delta^{(n)}(x)\, e^{ikx}\, dx = (-1)^n \int \delta(x)\frac{d^n}{dx^n} e^{ikx}\, dx$$

$$= (-ik)^n \int \delta(x)\, e^{ikx}\, dx = (-ik)^n\ . \tag{8.9.54}$$

8.10 Value of $x^m \delta^{(n)}(x)$

Show that

$$x^m \delta^n(x) = \begin{cases} 0 & \text{if } n < m \\ (-1)^m \, m! \, \delta(x) & \text{if } n = m \\ (-1)^m \frac{n!}{(n-m)!} \delta^{(n-m)}(x) & \text{if } n > m . \end{cases}$$

Solution

Let $f(x)$ be a test function. Then,

$$\int x^m \delta^{(n)}(x) f(x)\, dx = (-1)^n \frac{d^n}{dx^n} (x^m f(x))|_{x=0} . \tag{8.10.55}$$

Now we consider the three cases separately and remember that $f(0)$ is finite. Thus, we find:

i) for $n < m$ the righthand side vanishes. So,

$$x^m \delta^{(n)}(x) = 0 \quad n < m . \tag{8.10.56}$$

ii) for $n = m$ the righthand side yields $(-1)^n n! f(0)$. So,

$$x^m \delta^{(n)}(x) = (-1)^n n! \delta(x) \quad n = m . \tag{8.10.57}$$

iii) for $n > m$ the righthand side yields $(-1)^n \frac{n!}{(n-m)!} f^{(n-m)}(0)$. So,

$$\begin{aligned} x^m \delta^{(n)}(x) &= (-1)^n \frac{n!}{(n-m)!} (-1)^{(n-m)} \delta^{(n-m)}(x) \\ &= (-1)^m \frac{n!}{(n-m)!} \delta^{(n-m)}(x) \quad n > m . \end{aligned} \tag{8.10.58}$$

8.11 Distribution Occurring in Fermi's Golden Rule

Show that the function

$$\phi(\omega, t) = \frac{1}{\omega^2 t} |e^{i\omega t} - 1|^2 \tag{8.11.59}$$

that occurs in the derivation of Fermi's Golden Rule satisfies the following properties.

a)

$$\int_{-\infty}^{\infty} \phi(\omega, t)\, d\omega = 2\pi . \tag{8.11.60}$$

b)

$$\lim_{t \to \infty} \phi(\omega, t) = 2\pi \delta(\omega) . \tag{8.11.61}$$

Solution

a) We begin by rewriting ϕ

$$\phi(\omega, t) = \frac{1}{\omega^2 t} \left(e^{i\omega t} - 1 \right) \left(e^{-i\omega t} - 1 \right)$$

$$= \frac{1}{\omega^2 t} \left(2 - e^{i\omega t} - e^{-i\omega t} \right)$$

$$= \frac{2}{\omega^2 t} (1 - \cos \omega t) \,. \qquad (8.11.62)$$

So,

$$\int_{-\infty}^{\infty} \phi(\omega, t) \, d\omega = \frac{2}{t} \int_{-\infty}^{\infty} \frac{1 - \cos \omega t}{\omega^2} \, d\omega$$

$$= 2 \int_{-\infty}^{\infty} \frac{1 - \cos x}{x^2} \, dx \,. \qquad (8.11.63)$$

After an integration by parts this integral reduces to

$$2 \int_{-\infty}^{\infty} \frac{\sin x}{x} \, dx = 2 \Im \int_{-\infty}^{\infty} \frac{e^{ix}}{x} \, dx \qquad (8.11.64)$$

and may be evaluated by contour integration. To do this we realize that this integral is to be understood as a prinipal value integral so that what we want is

$$\lim_{\epsilon \to 0+} \left[\int_{-\infty}^{-\epsilon} \frac{e^{ix}}{x} \, dx + \int_{\epsilon}^{\infty} \frac{e^{ix}}{x} \, dx \right] \,.$$

Thus, to evaluate this integral we close the contour with a large semicircle C in the upper half plane and a small semicircle C_0 of radius ϵ around the origin as shown in figure 8.3. Since there are no poles inside the contour the total contour

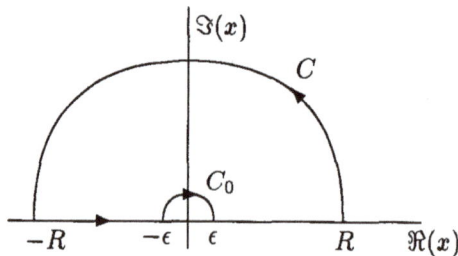

Figure 8.3: Contour used to evaluate the integral $\int_{-\infty}^{\infty} \frac{e^{ix}}{x} dx$.

integral vanishes and we get

$$0 = \lim_{\epsilon \to 0+} \left[\int_{-\infty}^{-\epsilon} \frac{e^{ix}}{x} \, dx + \int_{\epsilon}^{\infty} \frac{e^{ix}}{x} \, dx \right.$$

$$\left. + \int_{C} \frac{e^{ix}}{x} \, dx + \int_{C_0} \frac{e^{ix}}{x} \, dx \right] \,. \qquad (8.11.65)$$

The contribution from the large semicircle vanishes and after taking the limit for the first two terms we see that

$$
\begin{aligned}
\int_{-\infty}^{\infty} \frac{e^{ix}}{x}\, dx &= -\lim_{\epsilon \to 0+} \int_{C_0} \frac{e^{ix}}{x}\, dx \\
&= -\lim_{\epsilon \to 0+} \int_{\pi}^{0} \frac{\exp(i\epsilon\, e^{i\theta})}{\epsilon\, e^{i\theta}} i\epsilon\, e^{i\theta}\, d\theta \\
&= -i \int_{\pi}^{0} d\theta = i\pi \ .
\end{aligned}
\tag{8.11.66}
$$

Therefore,

$$
\int_{-\infty}^{\infty} \phi(\omega, t)\, d\omega = 2\pi \ .
\tag{8.11.67}
$$

b) To prove that

$$
\lim_{t \to \infty} \phi(\omega, t) = 2\pi \delta(\omega)
\tag{8.11.68}
$$

we integrate the left hand side with a test function $f(\omega)$ before taking the limit. Thus, we consider

$$
\begin{aligned}
\lim_{t \to \infty} \int_{-\infty}^{\infty} \phi(\omega, t) f(\omega)\, d\omega &= \lim_{t \to \infty} \int_{-\infty}^{\infty} \frac{2}{\omega^2 t}(1 - \cos \omega t) f(\omega)\, d\omega \\
&= \lim_{t \to \infty} \int_{-\infty}^{\infty} \frac{2}{x^2}(1 - \cos x) f(x/t)\, dx \\
&= f(0) \int_{-\infty}^{\infty} \frac{2}{x^2}(1 - \cos x)\, dx \\
&= 2\pi f(0) \ .
\end{aligned}
\tag{8.11.69}
$$

Bibliography

[8.1] A.Z. Capri, *Nonrelativistic Quantum Mechanics* 3rd edition, World Scientific Publishing Co. Pte. Ltd., chapter 8, (2002) .

Chapter 9

Algebraic Methods

9.1 An Operator Identity

Show that for any two operators A, B such that

$$[A, B] = c \quad \text{a c-number}$$

it is true that

$$[A, e^B] = c\, e^B \ .$$

Hint: Expand e^B and use the fact that

$$[A, B^n] = nc B^{n-1} \ .$$

Solution

We are told that

$$[A, B] = c \quad \text{a c-number} \tag{9.1.1}$$

and thus it follows that

$$[A, B^n] = nc B^{n-1} \ . \tag{9.1.2}$$

But,

$$
\begin{aligned}
[A, e^B] &= \sum_{n=0}^{\infty} \frac{1}{n!}[A, B^n] \\
&= \sum_{n=0}^{\infty} \frac{1}{n!} nc B^{n-1} \\
&= c \sum_{n=1}^{\infty} \frac{1}{(n-1)!} B^{n-1} \\
&= c\, e^B \ .
\end{aligned}
\tag{9.1.3}
$$

9.2 Expectation Values: Simple Harmonic Oscillator

Use algebraic techniques to evaluate the following expectation values as a function of time for a simple harmonic oscillator state which at $t = 0$ is given by

$$\psi(0, x) = Au_0(x) + Bu_1(x) + Cu_3(x)$$

where $u_0(x)$ is the ground state, $u_1(x)$ is the first excited state and $u_3(x)$ is the third excited state of a S.H.O.:

$$\langle H \rangle , \ \langle p^2/2m \rangle , \ \langle 1/2 \ kx^2 \rangle , \ \langle p \rangle , \ \langle x \rangle , \ (\Delta p)^2 , \ (\Delta x)^2 \ .$$

Solution

The wavefunction at $t = 0$ is given by

$$\Psi = Au_0 + Bu_1 + Cu_3 \qquad (9.2.4)$$

where $|A|^2 + |B|^2 + |C|^2 = 1$ and the Hamiltonian is

$$H = \frac{p^2}{2m} + \frac{1}{2}kx^2 , \qquad (9.2.5)$$

so that

$$Hu_n = (n + 1/2)\hbar\omega u_n \quad \omega = \sqrt{k/m} . \qquad (9.2.6)$$

We want $\langle H \rangle$, $\langle p^2/2m \rangle$, $\langle 1/2 \ kx^2 \rangle$, $\langle p \rangle$, and $\langle x \rangle$ as functions of time. Therefore,

$$\langle H \rangle = \hbar\omega \left[\frac{1}{2}|A|^2 + \frac{3}{2}|B|^2 + \frac{7}{2}|C|^2 \right] . \qquad (9.2.7)$$

Now, using $p = i\sqrt{(m\hbar\omega)/2}(a^\dagger - a)$ we have

$$\frac{p^2}{2m} = -\frac{1}{4}\hbar\omega(a - a^\dagger)^2 = -\frac{1}{4}\hbar\omega(a^2 + a^{\dagger 2} - aa^\dagger - a^\dagger a) . \qquad (9.2.8)$$

Also the state at time t is

$$\Psi(t) = Au_0 e^{-i\omega t/2} + Bu_1 e^{-3i\omega t/2} + Cu_3 e^{-7i\omega t/2} . \qquad (9.2.9)$$

Next, using that

$$au_n = \sqrt{n}u_{n-1} , \quad a^\dagger u_n = \sqrt{n + 1}u_{n+1} \qquad (9.2.10)$$

we get

$$\begin{aligned} a^{\dagger 2}\Psi(t) &= \sqrt{2}Au_2 e^{-i\omega t/2} + \sqrt{6}u_3 e^{-3i\omega t/2} + \sqrt{20}Cu_5 e^{-7i\omega t/2} \\ a^2 \ \Psi(t) &= \sqrt{6}Cu_1 e^{-7i\omega t/2} \\ a^\dagger a\Psi(t) &= Bu_1 e^{-3i\omega t/2} + 3Cu_3 e^{-7i\omega t/2} . \end{aligned} \qquad (9.2.11)$$

Furthermore,

$$aa^\dagger = a^\dagger a + 1 \ . \tag{9.2.12}$$

So, we find

$$\langle \frac{p^2}{2m} \rangle$$

$$= -\frac{\hbar\omega}{4} \left[B^* C\sqrt{6}\, e^{-4i\omega t/2} + C^* B\sqrt{6}\, e^{4i\omega t/2} - 2(B^* B + 3C^* C) - 1 \right]$$

$$= \frac{\hbar\omega}{4} \left[2(|B|^2 + 3|C|^2) + 1 - 2\sqrt{6}|B^* C| \cos(2\omega t + \phi) \right] \tag{9.2.13}$$

where

$$B^* C = |B^* C|\, e^{i\phi} \ . \tag{9.2.14}$$

Since

$$\langle H \rangle = \langle \frac{p^2}{2m} \rangle + \langle \frac{1}{2} kx^2 \rangle \tag{9.2.15}$$

it follows that

$$\langle \frac{1}{2} kx^2 \rangle = \langle H \rangle - \langle \frac{p^2}{2m} \rangle \ . \tag{9.2.16}$$

Next, we again use that

$$p = i\sqrt{\frac{m\hbar\omega}{2}}(a^\dagger - a) \quad \text{as well as} \quad x = \sqrt{\frac{\hbar}{2m\omega}}(a^\dagger + a) \tag{9.2.17}$$

and find

$$\langle p \rangle = -\sqrt{2m\hbar\omega}\,|A^* B| \sin(\omega t + \alpha) \tag{9.2.18}$$

$$\langle x \rangle = \sqrt{\frac{2\hbar}{m\omega}}\,|A^* B| \cos(\omega t + \alpha) \tag{9.2.19}$$

where

$$A^* B = |A^* B|\, e^{i\alpha} \ . \tag{9.2.20}$$

9.3 Angular Momentum Matrices

a) Compute the three 3×3 matrices

$$(L_j)_{m,m'} = (Y_{1m}, L_j Y_{1m'})$$

where $j = x, y, z$.

b) Show that they satisfy the cyclic commutation relation

$$[L_x, L_y] = i\hbar L_z \quad \text{etc.}$$

c) Furthermore show that each matrix L_j satisfies the characteristic equation

$$\left(L_j^2 - \hbar^2 1\right) L_j = 0 .$$

d) Evaluate in closed form the expression for the matrix $\exp(iL_z\theta)$.
Hint: Write out the series and resum it after using the characteristic equation to simplify. Compare this result to a rotation matrix corresponding to a rotation through an angle θ about the z-axis for the quantities corresponding to

$$\left(\frac{x+iy}{\sqrt{2}} , z , \frac{x-iy}{\sqrt{2}}\right) .$$

Solution

a) We begin with

$$L_z Y_{1,m} = m\hbar Y_{1,m} . \tag{9.3.21}$$

Therefore,

$$(L_z)_{m,m'} = (Y_{1,m}, L_z Y_{1,m'}) = m\hbar \delta_{m,m'} . \tag{9.3.22}$$

As a matrix this has the following form

$$L_z = \hbar \begin{pmatrix} 1 & 0 & 0 \\ 0 & 0 & 0 \\ 0 & 0 & -1 \end{pmatrix} . \tag{9.3.23}$$

Also

$$(L_\pm)_{m,m'} = (Y_{1,m}, L_\pm Y_{1,m'}) = \sqrt{2 - m'(m' \pm 1)}\hbar \delta_{m,m'\pm1} . \tag{9.3.24}$$

In matrix form this becomes

$$L_+ = \sqrt{2}\hbar \begin{pmatrix} 0 & 1 & 0 \\ 0 & 0 & 1 \\ 0 & 0 & 0 \end{pmatrix} , \quad L_- = \sqrt{2}\hbar \begin{pmatrix} 0 & 0 & 0 \\ 1 & 0 & 0 \\ 0 & 1 & 0 \end{pmatrix} . \tag{9.3.25}$$

So we finally get

$$L_x = \frac{1}{2}(L_+ + L_-) = \frac{\hbar}{\sqrt{2}} \begin{pmatrix} 0 & 1 & 0 \\ 1 & 0 & 1 \\ 0 & 1 & 0 \end{pmatrix} \tag{9.3.26}$$

$$L_y = \frac{-i}{2}(L_+ - L_-) = \frac{\hbar}{\sqrt{2}} \begin{pmatrix} 0 & -i & 0 \\ i & 0 & -i \\ 0 & i & 0 \end{pmatrix} . \tag{9.3.27}$$

b) By straightforward matrix multiplication we now find that

$$[L_x, L_y] = i\hbar L_z , \quad [L_y, L_z] = i\hbar L_x , \quad [L_z, L_x] = i\hbar L_y . \tag{9.3.28}$$

c) In a similar manner we find that these matrices satisfy the characteristic equation

$$(L_x^2 - \hbar^2)L_x = (L_y^2 - \hbar^2)L_y = (L_z^2 - \hbar^2)L_z = 0 \ . \tag{9.3.29}$$

d) Writing simply L for L_x or L_y or L_z we now have that

$$L^3 = \hbar^2 L \ , \quad L^4 = \hbar^2 L^2 \ , \quad L^5 = \hbar^4 L \ \text{ etc.} \tag{9.3.30}$$

Thus,

$$\begin{aligned}
\exp\left(\frac{iL\theta}{\hbar}\right) &= 1 + \frac{iL\theta}{\hbar} + \frac{1}{2!}\left(\frac{iL\theta}{\hbar}\right)^2 + \frac{1}{3!}\left(\frac{iL\theta}{\hbar}\right)^3 + \cdots \\
&= 1 + \frac{L}{\hbar}\left[i\theta + \frac{(i\theta)^3}{3!} + \frac{(i\theta)^5}{5!} + \cdots\right] \\
&\quad + \left(\frac{L}{\hbar}\right)^2\left[\frac{(i\theta)^2}{2!} + \frac{(i\theta)^4}{4!} + \frac{(i\theta)^6}{6!} + \cdots\right] \\
&= 1 - \frac{L^2}{\hbar^2} + \frac{iL}{\hbar}\sin\theta + \frac{L^2}{\hbar^2}\cos\theta \ . \tag{9.3.31}
\end{aligned}$$

Now,

$$\frac{L_z}{\hbar} = \begin{pmatrix} 1 & 0 & 0 \\ 0 & 0 & 0 \\ 0 & 0 & -1 \end{pmatrix}, \quad \frac{L_z^2}{\hbar^2} = \begin{pmatrix} 1 & 0 & 0 \\ 0 & 0 & 0 \\ 0 & 0 & 1 \end{pmatrix} \tag{9.3.32}$$

and

$$1 - \frac{L_z^2}{\hbar^2} = \begin{pmatrix} 0 & 0 & 0 \\ 0 & 1 & 0 \\ 0 & 0 & 0 \end{pmatrix} \ . \tag{9.3.33}$$

Therefore writing (9.3.31) out for $L = L_z$ and $\theta = \gamma$ we find

$$\exp\left(\frac{iL_z\gamma}{\hbar}\right) = \begin{pmatrix} e^{i\gamma} & 0 & 0 \\ 0 & 1 & 0 \\ 0 & 0 & e^{-i\gamma} \end{pmatrix} \ . \tag{9.3.34}$$

To see what this means, consider a rotation about the z-axis

$$\begin{aligned}
x' &= x\cos\theta + y\sin\theta \\
y' &= -x\sin\theta + y\cos\theta \\
z' &= z \ . \tag{9.3.35}
\end{aligned}$$

Instead of x, y, z we now introduce

$$\begin{aligned}
x_1 &= \frac{x+iy}{\sqrt{2}} \\
x_0 &= z \\
x_{-1} &= \frac{x-iy}{\sqrt{2}} \tag{9.3.36}
\end{aligned}$$

since then

$$x_m = \frac{\sqrt{4\pi}}{3} r\, Y_{1,m} \,. \tag{9.3.37}$$

For these variables the transformation properties are

$$
\begin{aligned}
x_1' &= x_1 e^{i\gamma} \\
x_0' &= x_0 \\
x_{-1}' &= x_{-1} e^{-i\gamma} \,.
\end{aligned}
\tag{9.3.38}
$$

This allows us to write

$$x_m' = \sum_{m'} \exp\left(\frac{iL_z\theta}{\hbar}\right)_{m,m'} x_{m'} \,. \tag{9.3.39}$$

9.4 Displaced Oscillator

Diagonalize the following Hamiltonian, that is, find all of its energy eigenvalues.

$$H = E\, a^\dagger a + V(a + a^\dagger)$$

where E and V are constants and

$$[a, a^\dagger] = \beta^2 \,,$$

a positive constant.
Hint: Try to transform to operators

$$b = ua + v \,, \quad b^\dagger = u^* a^\dagger + v^*$$

where u and v are complex numbers, and recall the simple harmonic oscillator.

Solution

We are given

$$H = E\, a^\dagger a + V(a^\dagger + a) \tag{9.4.40}$$

where

$$[a, a^\dagger] = \beta^2 \tag{9.4.41}$$

So we try as suggested

$$b = ua + v \,, \quad b^\dagger = u^* a^\dagger + v^* \,. \tag{9.4.42}$$

Then,

$$[b, b^\dagger] = u^2 [a, a^\dagger] = \beta^2 |u|^2 \,. \tag{9.4.43}$$

This suggests that we should choose $u = 1/\beta$. Then we get

$$[b, b^\dagger] = 1 \tag{9.4.44}$$

and we find

$$a = \beta(b - v) \quad , \quad a^\dagger = \beta(b^\dagger - v^*) \ . \tag{9.4.45}$$

So,

$$H = \beta^2 E(b^\dagger b - vb^\dagger - v^* b + |v|^2) + \beta V (b^\dagger + b - v - v^*) \ . \tag{9.4.46}$$

Therefore, if we now choose v real and such that $\beta^2 Ev = \beta V$ then the terms linear in b and $b^\dagger$ go away and we have just a simple harmonic oscillator Hamiltonian.

$$H = \beta^2 E b^\dagger b - \frac{V^2}{E} \ . \tag{9.4.47}$$

The energy eigenvalues are therefore given by

$$E_n = (n + 1/2)\beta^2 E - \frac{V^2}{E} \ . \tag{9.4.48}$$

9.5   Dipole Matrix Elements

Evaluate matrix elements of the form $(Y_{l,m}, z Y_{l',m'})$. These occur in the evaluation of dipole radiation rates.

Hint: Use the following recurrence relation:

$$(2l + 1)x P_l^m(x) = (l + m)P_{l-1}^m(x) + (l - m + 1)P_{l+1}^m(x) \ .$$

Solution

In spherical coordinates we have $z = r \cos \theta$. So we have to evaluate the integral

$$I = \int \sin \theta \, d\theta \, d\varphi \, Y_{l,m}^*(\theta, \varphi) \cos \theta \, Y_{l',m'}(\theta, \varphi) \ . \tag{9.5.49}$$

Using the explicit form for the spherical harmonics

$$Y_{l,m}(\theta, \varphi) = \left[\frac{2l + 1}{4\pi} \frac{(l - m)!}{(l + m)!} \right]^{1/2} P_l^m(\cos \theta) \, e^{im\varphi} \tag{9.5.50}$$

we see that the φ-integration yields $\delta_{m,m'}$. After we make the substitution $u = \cos \theta$ the integral becomes:

$$\begin{aligned} I &= \frac{\sqrt{(2l + 1)(2l' + 1)}}{4\pi} \frac{(l - m)!(l' - m)!}{(l + m)!(l' + m)!} \delta_{m,m'} \\ &\quad \times \int_{-1}^1 du \, P_l^m(u) \, u \, P_{l'}^m(u) \ . \end{aligned} \tag{9.5.51}$$

But,

$$u \, P_{l'}^m(u) = \frac{1}{2l' + 1} \left[(l' - m + 1)P_{l'+1}^m + (l' + m)P_{l'-1}^m \right] \ . \tag{9.5.52}$$

Therefore we find that

$$
\begin{aligned}
I &= \frac{\sqrt{(2l+1)(2l'+1)}}{4\pi}\frac{(l-m)!(l'-m)!}{(l+m)!(l'+m)!}\delta_{m,m'} \\
&\times \frac{2}{(2l+1)(2l'+1)}\left[(l-m)\delta_{l,l'+1}+(l+1+m)\delta_{l,l'-1}\right] \\
&= \frac{1}{2\pi\sqrt{(2l+1)(2l'+1)}}\frac{(l-m)!(l'-m)!}{(l+m)!(l'+m)!}\delta_{m,m'} \\
&\times \left[(l-m)\delta_{l,l'+1}+(l+1+m)\delta_{l,l'-1}\right] .
\end{aligned}
\tag{9.5.53}
$$

9.6   Scalar Operator

Let A be an operator such that

$$[A, L_x] = [A, L_y] = 0 .$$

Calculate $[A, L^2]$.
Hint: First compute $[A, L_z]$.

Solution

We have

$$[L_x, L_y] = -i\hbar\, L_z .$$

Therefore,

$$
\begin{aligned}
i\hbar[A, L_z] &= [A, [L_x, L_y]] = [A, L_x L_y] - [A, L_y L_x] \\
&= [A, L_x]L_y + L_x[A, L_y] - [A, L_y]L_x - L_y A, L_x \\
&= 0 .
\end{aligned}
\tag{9.6.54}
$$

Thus, $[A, L_z] = 0$ and hence it follows that

$$[A, L^2] = 0 . \tag{9.6.55}$$

9.7   Probability to Obtain l, m

Let a wavefunction be given by

$$
\psi = A\left[\frac{x^2 - y^2 + 2ixy}{x^2 + y^2 + z^2} + \frac{3z}{\sqrt{x^2 + y^2 + z^2}} + 5\right]\exp(-\alpha\sqrt{x^2 + y^2 + z^2})
$$

Find the probability of obtaining any (l, m) value.

Solution

The wavefunction is

$$\psi = A\left[\frac{x^2 - y^2 + 2ixy}{x^2 + y^2 + z^2} + \frac{3z}{\sqrt{x^2 + y^2 + z^2}} + 5\right] e^{-\alpha\sqrt{x^2+y^2+z^2}} . \quad (9.7.56)$$

In spherical coordinates this can be rewritten

$$\psi = A\left[\sin^2\theta\, e^{2i\varphi} + 3\cos\theta + 5\right] e^{-\alpha r} . \quad (9.7.57)$$

In terms of spherical harmonics (see for example table (9.1) of [9.1]) we find

$$\psi = A\left[\sqrt{\frac{32\pi}{15}}Y_{2,2} + 3\sqrt{\frac{4\pi}{3}}Y_{1,0} + 5\sqrt{4\pi}\,Y_{0,0}\right] e^{-\alpha r} . \quad (9.7.58)$$

Thus, a measurement of L^2, L_z will only yield the values (2,2), (1,0), (0,0). The corresponding probabilities are

$$
\begin{aligned}
P_{2,2} &= \frac{32\pi}{15}\frac{1}{32\pi/15+9(4\pi/3)+25(4\pi)} = \frac{32}{1712}\\
P_{1,0} &= \frac{12\pi(15)}{1712\pi} = \frac{180}{1712}\\
P_{0,0} &= \frac{100\pi(15)}{1712\pi} = \frac{1500}{1712} .
\end{aligned}
\quad (9.7.59)
$$

9.8 Probability to Obtain l, m Along Different Axis

a) A measurement of the z-component of angular momentum is made on a particle in a state of total angular momentum 1 and x-component 1. What are the probabilities for obtaining the values $1, 0, -1$?
b) For a particle in a state of total angular momentum 1 and z-component 1 a measurement is made of the component of angular momentum along an axis lying in the $x - z$ plane and making an angle θ with the z-axis. What is the probability of getting the values $1, 0, -1$?

Solution

a) We have

$$L_x\psi = \hbar\psi , \quad L^2\psi = 2\hbar^2\psi . \quad (9.8.60)$$

So, the total angular momentum is $\hbar\cdot 1$. We can therefore represent the angular momentum operators by the following matrices.

$$L_x = \frac{\hbar}{\sqrt{2}}\begin{pmatrix} 0 & 1 & 0 \\ 1 & 0 & 1 \\ 0 & 1 & 0 \end{pmatrix}$$

$$L_y = \frac{\hbar}{\sqrt{2}} \begin{pmatrix} 0 & -i & 0 \\ i & 0 & -i \\ 0 & i & 0 \end{pmatrix}$$

$$L_z = \hbar \begin{pmatrix} 1 & 0 & 0 \\ 0 & 0 & 0 \\ 0 & 0 & -1 \end{pmatrix} . \tag{9.8.61}$$

The eigenstate of L_x corresponding to $1\hbar$ is

$$\psi = \frac{1}{2} \begin{pmatrix} 1 \\ \sqrt{2} \\ 1 \end{pmatrix} . \tag{9.8.62}$$

The eigenstates of L_z are

$$\psi_1 = \begin{pmatrix} 1 \\ 0 \\ 0 \end{pmatrix} , \quad \psi_0 = \begin{pmatrix} 0 \\ 1 \\ 0 \end{pmatrix} , \quad \psi_{-1} = \begin{pmatrix} 0 \\ 0 \\ 1 \end{pmatrix} . \tag{9.8.63}$$

Thus, the required probabilities are: $P_m = |(\psi, \psi_m)|^2$. This yields,

$$P_{+1} = \frac{1}{4} , \quad P_0 = \frac{1}{2} , \quad P_{-1} = \frac{1}{4} . \tag{9.8.64}$$

b) In this case the state corresponding to $L_z = 1\hbar$ is

$$\psi = \begin{pmatrix} 1 \\ 0 \\ 0 \end{pmatrix} . \tag{9.8.65}$$

We want the eigenstates of the operator

$$L_\theta = L_z \cos\theta + L_x \sin\theta = \hbar \begin{pmatrix} \cos\theta & \frac{1}{\sqrt{2}}\sin\theta & 0 \\ \frac{1}{\sqrt{2}}\sin\theta & 0 & \cos\theta \\ 0 & \frac{1}{\sqrt{2}}\sin\theta & -\cos\theta \end{pmatrix} . \tag{9.8.66}$$

The eigenvalues are, of course, still $\pm\hbar$, 0. The eigenstates are found in the standard way to be (up to an arbitrary phase factor)

$$\psi_{\pm 1} = \frac{1}{2} \begin{pmatrix} \pm(1 \pm \cos\theta) \\ \sqrt{2}\sin\theta \\ \pm(1 \mp \cos\theta) \end{pmatrix} , \quad \psi_0 = \frac{1}{2} \begin{pmatrix} \sqrt{2}\sin\theta \\ -2\cos\theta \\ -\sqrt{2}\sin\theta \end{pmatrix} . \tag{9.8.67}$$

The probabilities are as always given by $P_m = |(\psi, \psi_m)|^2$. So we get

$$P_{+1} = \frac{1}{4}(1 + \cos\theta)^2 , \quad P_0 = \frac{1}{2}\sin^2\theta , \quad P_{-1} = \frac{1}{4}(1 - \cos\theta)^2 . \tag{9.8.68}$$

9.9 Commutators of $\vec{x}$ and $\vec{p}$ with $\vec{L}$

Compute the commutators $[\vec{x}, \vec{L}]$ and $[\vec{p}, \vec{L}]$ and compare the results with $[\vec{L}, \vec{L}]$. What does this suggest about the commutator $[\vec{A}, \vec{L}]$ where $\vec{A}$ is an arbitrary vector operator? Note that each of the commutator brackets above constitutes nine commutators.

Solution

By definition

$$\vec{L} = \vec{r} \times \vec{p} . \tag{9.9.69}$$

This means

$$L_x = yp_z - zp_y \ , \quad L_y = zp_x - xp_z \ , \quad L_z = xp_y - yp_x \ . \tag{9.9.70}$$

Then we find

$$
\begin{aligned}
{[x, L_x]} &= 0 \\
{[x, L_y]} &= z[x, p_x] = i\hbar z \\
{[x, L_z]} &= -y[x, p_x] = -i\hbar y \\
{[y, L_x]} &= -z[y, p_y] = -i\hbar z \\
{[y, L_y]} &= 0 \\
{[y, L_z]} &= -x[y, p_y] = i\hbar x \\
{[z, L_x]} &= y[z, p_z] = i\hbar y \\
{[z, L_y]} &= -x[z, p_z] = -i\hbar x \\
{[z, L_z]} &= 0 .
\end{aligned}
\tag{9.9.71}
$$

This means that $\vec{r}$ and $\vec{L}$ satisfy the same sort of cyclic commutation relations as $\vec{L}$ and $\vec{L}$. Similarly we find that $\vec{p}$ and $\vec{L}$ satisfy

$$[p_x, L_y] = i\hbar p_z \quad \text{etc. in cyclic fashion} . \tag{9.9.72}$$

It is therefore quite reasonable to guess that for a vector operator $\vec{A}$ we will have

$$[A_x, L_y] = i\hbar A_z \quad \text{etc. in cyclic fashion} . \tag{9.9.73}$$

9.10 Some Eigenfunctions of Angular Momentum

Let

$$\psi_x = x\, f(r) \ , \quad \psi_y = y\, f(r) \ , \quad \psi_{(z)} = z\, f(r) \ .$$

Show, by explicit calculation, that they are respectively eigenfunctions of L_x, L_y, and L_z as well as of L^2 and find the corresponding eigenvalues.

Solution

By straightforward computation we have

$$
\begin{aligned}
L_x \psi_x &= (y p_z - z p_y) x f(r) \\
&= \frac{\hbar}{i} \left[xy \frac{z}{r} - xz \frac{y}{r} \right] \frac{df}{dr} \\
&= 0 .
\end{aligned}
\tag{9.10.74}
$$

Similarly,

$$
\begin{aligned}
L_y \psi_y &= (z p_x - x p_z) y f(r) \\
&= \frac{\hbar}{i} \left[yz \frac{x}{r} - yx \frac{z}{r} \right] \frac{df}{dr} \\
&= 0
\end{aligned}
\tag{9.10.75}
$$

and

$$
\begin{aligned}
L_z \psi_z &= (x p_y - y p_x) z f(r) \\
&= \frac{\hbar}{i} \left[zx \frac{y}{r} - zy \frac{x}{r} \right] \frac{df}{dr} \\
&= 0 .
\end{aligned}
\tag{9.10.76}
$$

Thus, all three are eigenfunctions of the corresponding component of angular momentum with eigenvalue 0.

Continuing we find

$$
\begin{aligned}
L_y \psi_x &= (z p_x - x p_z) x f(r) \\
&= \frac{\hbar}{i} \left[xz \frac{x}{r} - x^2 \frac{z}{r} \right] \frac{df}{dr} + \frac{\hbar}{i} z f(r) \\
&= \frac{\hbar}{i} \psi_z
\end{aligned}
\tag{9.10.77}
$$

and

$$
\begin{aligned}
L_z \psi_x &= (x p_y - y p_x) x f(r) \\
&= \frac{\hbar}{i} \left[x^2 \frac{y}{r} - xy \frac{x}{r} \right] \frac{df}{dr} - \frac{\hbar}{i} y f(r) \\
&= -\frac{\hbar}{i} \psi_y .
\end{aligned}
\tag{9.10.78}
$$

In exactly the same way we find

$$
\begin{aligned}
L_x \psi_y &= -\frac{\hbar}{i} \psi_z \\
L_z \psi_y &= \frac{\hbar}{i} \psi_x
\end{aligned}
$$

$$
\tag{9.10.79}
$$

and

$$L_x \psi_z = \frac{\hbar}{i} \psi_y$$

$$L_y \psi_z = -\frac{\hbar}{i} \psi_x \ .$$ (9.10.80)

Therefore,

$$L^2 \psi_x = \frac{\hbar}{i} L_y \psi_z - \frac{\hbar}{i} L_z \psi_y$$

$$= \frac{\hbar}{i} [-2 \frac{\hbar}{i}] \psi_x$$

$$= 2 \hbar^2 \psi_x \ ,$$ (9.10.81)

$$L^2 \psi_y = -\frac{\hbar}{i} L_x \psi_z + \frac{\hbar}{i} L_z \psi_x$$

$$= 2 \hbar^2 \psi_y$$

and (9.10.82)

$$L^2 \psi_z = \frac{\hbar}{i} L_x \psi_y - \frac{\hbar}{i} L_y \psi_x$$

$$= 2 \hbar^2 \psi_z \ .$$ (9.10.83)

9.11   Expectation Value of L_x

A system is in a state of angular momentum given by

$$\Psi = a Y_{1,1} + b Y_{1,0} + c Y_{1,-1}$$

where

$$|a|^2 + |b|^2 + |c|^2 = 1 \ .$$

a) Compute the expectation value of L_x.
Hint: The equation

$$L_\pm Y_{l,m} = \sqrt{l(l+1) - m(m \pm 1)} \hbar Y_{l,m\pm1}$$

may be useful.
b) Compute the expectation value of L^2.
c) What are the possible values of the coefficients a, b, c in order that

$$L_x \Psi = \hbar \Psi \ \ ?$$

It may again be useful to recall that

$$L_x = \frac{1}{2}(L_+ + L_-) \ .$$

Solution

The wavefunction is

$$\psi = aY_{1,1} + bY_{1,0} + cY_{1,-1} \tag{9.11.84}$$

where

$$|a|^2 + |b|^2 + |c|^2 = 1 \tag{9.11.85}$$

a) we use

$$\langle L_x \rangle = \frac{1}{2}\langle L_+ \rangle + \frac{1}{2}\langle L_- \rangle \tag{9.11.86}$$

and

$$L_+\psi = \sqrt{2}\hbar \, bY_{1,1} + \sqrt{2}\hbar \, cY_{1,0} \tag{9.11.87}$$

$$L_-\psi = \sqrt{2}\hbar \, aY_{1,0} + \sqrt{2}\hbar \, bY_{1,-1} \ . \tag{9.11.88}$$

So,

$$\langle L_+ \rangle = (\psi, L_+\psi) = \sqrt{2}\hbar(a^*b + b^*c) \tag{9.11.89}$$

$$\langle L_- \rangle = (\psi, L_+\psi) = \sqrt{2}\hbar(b^*a + c^*b) \ . \tag{9.11.90}$$

Therefore,

$$\langle L_x \rangle = \frac{\sqrt{2}\hbar}{2}(2\Re(a^*b) + 2\Re(b^*c)) = \sqrt{2}\hbar(\Re(a^*b) + \Re(b^*c)) \ . \tag{9.11.91}$$

b) Since ψ is an eigenstate of L^2 we have that

$$\langle L^2 \rangle = 1(1+1)\hbar^2 = 2\hbar^2 \ . \tag{9.11.92}$$

c) We want an eigenstate of L_x with eigenvalue $\hbar$. This means

$$L_x \, \psi = \hbar \, \psi \ . \tag{9.11.93}$$

But, writing this out we get

$$
\begin{aligned}
L_x \, \psi &= \frac{1}{2}(L_+ + L_-)\psi = \frac{1}{\sqrt{2}}\hbar \, (bY_{1,1} + (c+a)Y_{1,0} + bY_{1,-1}) \\
&= \hbar(aY_{1,1} + bY_{1,0} + cY_{1,-1}) \ .
\end{aligned} \tag{9.11.94}
$$

This requires that

$$b = \sqrt{2}\,a \ , \quad c + a = \sqrt{2}\,b \ , \quad b = \sqrt{2}\,c \ . \tag{9.11.95}$$

Therefore,

$$\psi = \frac{b}{|b|}\left(\frac{1}{2}Y_{1,1} + \frac{1}{\sqrt{2}}Y_{1,0} + \frac{1}{2}Y_{1,-1}\right) \ . \tag{9.11.96}$$

9.12   Rotation Invariance of the Hamiltonian

Starting from the equation,

$$U_R \, H \, U_R^{-1} = H$$

and using

$$U = \exp\{\frac{i\epsilon}{\hbar}\hat{n} \cdot \vec{L}\}$$

with $|\epsilon| << 1$ obtain to first order in ϵ that

$$[H, \hat{n} \cdot \vec{L}] = 0 \,.$$

Hence conclude that

$$[H, \vec{L}] = 0 \quad.$$

Solution

We are given

$$U_R(\theta) = \exp\left[\frac{i}{\hbar}\theta\hat{n} \cdot \vec{L}\right] \tag{9.12.97}$$

$$U_R(\theta)^{-1} = \exp\left[-\frac{i}{\hbar}\theta\hat{n} \cdot \vec{L}\right] \quad. \tag{9.12.98}$$

For $\theta \to \epsilon$ we get

$$U_R(\theta) \approx 1 + \frac{i}{\hbar}\epsilon\hat{n} \cdot \vec{L} \tag{9.12.99}$$

$$U_R^{-1}(\theta) \approx 1 - \frac{i}{\hbar}\epsilon\hat{n} \cdot \vec{L} \quad. \tag{9.12.100}$$

Therefore, the equation

$$U_R H U_R^{-1} = H \tag{9.12.101}$$

becomes to this approximation

$$\left(1 + \frac{i}{\hbar}\epsilon\hat{n} \cdot \vec{L}\right) H \left(1 - \frac{i}{\hbar}\epsilon\hat{n} \cdot \vec{L}\right) = H \quad, \tag{9.12.102}$$

or to order ϵ^2

$$[\hat{n} \cdot \vec{L}, H] = 0 \quad. \tag{9.12.103}$$

Since $\hat{n}$ is an arbitrary direction, this means that

$$[\vec{L}, H] = 0 \quad. \tag{9.12.104}$$

9.13   Uncertainty Relation for SHO

Show that for any eigenstate of the simple harmonic oscillator

$$(\Delta x)^2 = \langle x^2 \rangle$$

$$(\Delta p)^2 = \langle p^2 \rangle$$

and also show for the state with quantum number n that

$$(\Delta x)(\Delta p) = (n + 1/2)\hbar \ .$$

Solution

We have

$$x = \sqrt{\frac{\hbar}{2m\omega}}(a^\dagger + a) \tag{9.13.105}$$

$$p = i\sqrt{\frac{m\hbar\omega}{2}}(a^\dagger - a) \ . \tag{9.13.106}$$

Thus,

$$
\begin{aligned}
\langle x \rangle &= \langle n|x|n \rangle = \sqrt{\frac{\hbar}{2m\omega}}\langle n|(a^\dagger + a)|n \rangle \\
&= \sqrt{\frac{\hbar}{2m\omega}}\left[\sqrt{n+1}\langle n|n+1 \rangle + \sqrt{n}\langle n|n-1 \rangle\right] = 0 \ . \tag{9.13.107}
\end{aligned}
$$

Here we have used the fact that $\langle n|m \rangle = \delta_{n,m}$. For the very same reason we find that $\langle p \rangle = \langle n|p|n \rangle = 0$. Therefore,

$$(\Delta x)^2 = \langle x^2 \rangle - \langle x \rangle^2 = \langle x^2 \rangle \tag{9.13.108}$$

and

$$(\Delta p)^2 = \langle p^2 \rangle - \langle p \rangle^2 = \langle p^2 \rangle \ . \tag{9.13.109}$$

Now, computing explicitly we have

$$
\begin{aligned}
(\Delta x)^2 &= \langle x^2 \rangle = \frac{\hbar}{2m\omega}\langle n|a^{\dagger\,2} + a^2 + a^\dagger a + aa^\dagger|n \rangle \\
&= \frac{\hbar}{2m\omega}\langle n|a^\dagger a + aa^\dagger|n \rangle \\
&= \frac{\hbar}{2m\omega}(2n+1) \ . \tag{9.13.110}
\end{aligned}
$$

Similarly,

$$
\begin{aligned}
(\Delta p)^2 &= \langle p^2 \rangle = -\frac{m\hbar\omega}{2}\langle n|a^{\dagger\,2} + a^2 - a^\dagger a - aa^\dagger|n \rangle \\
&= \frac{m\hbar\omega}{2}(2n+1) \ . \tag{9.13.111}
\end{aligned}
$$

So, we get that

$$(\Delta x)^2 (\Delta p)^2 = \frac{\hbar^2}{4}(2n+1)^2 \ . \tag{9.13.112}$$

Therefore,

$$\Delta x \, \Delta p = \hbar(n+1/2) \ . \tag{9.13.113}$$

9.14   Baker-Campbell-Hausdorff Formula

Show that if you have two operators A, B such that

$$[A,[A,B]] = [B,[A,B]] = 0$$

then

$$e^{A+B} = e^A \, e^B \, e^{-1/2[A,B]} \ .$$

Hint: Consider the operator

$$f(x) = e^{xA} \, e^{xB}$$

and show that

$$\frac{df}{dx} = (A + B + [A,B]x) \, f(x) \ .$$

Integrate this equation and obtain the desired result. This result is a special case of the *Baker-Campbell-Hausdorff formula*.

Solution

The following derivation is due to Glauber. We have

$$[A,[A,B]] = [B,[A,B]] = 0 \ . \tag{9.14.114}$$

Now, consider the operator

$$f(x) = e^{xA} \, e^{xB} \ . \tag{9.14.115}$$

Then,

$$\begin{aligned}
\frac{df}{dx} &= A \, e^{xA} \, e^{xB} + e^{xA} \, B \, e^{xB} \\
&= (A + e^{xA} \, B \, e^{-xA}) f(x) \ .
\end{aligned} \tag{9.14.116}$$

We also have that

$$[B, A^n] = nA^{n-1}[B,A] \tag{9.14.117}$$

so that

$$[B, e^{-xA}] = -x e^{-xA}[B,A] \tag{9.14.118}$$

and

$$e^{xA} B e^{-xA} = B - [B, A]x .$$ (9.14.119)

Therefore,

$$\frac{df}{dx} = (A + B + [A, B]x)f(x)$$ (9.14.120)

with $f(0) = 1$. Furthermore, since

$$[A + B, [A, B]] = 0$$ (9.14.121)

we can integrate this like an equation for a c-number function to get

$$f(x) = e^{x(A+B)} e^{1/2 x^2 [A,B]} .$$ (9.14.122)

If we now set $x = 1$ and multiply from the right by $e^{-1/2[A,B]}$ we get the desired result

$$e^{A+B} = e^A e^B e^{-1/2[A,B]} .$$ (9.14.123)

9.15 A Useful Commutator

Show that if

$$[A, B] = \lambda A$$

then

$$A e^B = e^\lambda e^B A .$$

This formula can be used, for example, if the operator B is proportional to the Hamiltonian $\hbar\omega(a^\dagger a + 1/2)$ and the operator A is proportional to either a or $a^\dagger$.

Solution
Given

$$[A, B] = \lambda A$$ (9.15.124)

we find, using a proof by induction, that

$$AB^n = (\lambda + B)^n A .$$ (9.15.125)

Thus,

$$\begin{aligned}
A e^B &= \sum_n \frac{1}{n!}(\lambda + B)^n A \\
&= e^{\lambda+B} A \\
&= e^\lambda e^B A .
\end{aligned}$$ (9.15.126)

9.16   Uncertainty in L_z

A particle is in a state described by the wavefunction

$$\psi = \frac{1}{\sqrt{6}}[Y_{l,l} + 2iY_{l,l-1} - Y_{l,l-2}] \ .$$

a) What are the most probable values that would be obtained in a single measurement of L^2 and L_z? What are the possible values of L_z and what are their corresponding probabilities?
b) Compute $(\Delta L_z)^2$ the uncertainty in L_z for this state.
c) If the state is now simply $Y_{l,m}$ find the uncertainty $(\Delta L_z)^2$ in L_z.

Solution

The particle is in the state given by

$$\psi = \frac{1}{\sqrt{6}} (Y_{l,l} + 2iY_{l,l-1} - Y_{l,l-2}) \ . \tag{9.16.127}$$

a) A measurement is then sure to yield that the value of

$$L^2 = l(l+1)\hbar^2 \tag{9.16.128}$$

since the state is an eigenstate of L^2 with this eigenvalue. The possible values of L_z are:
$l\hbar \ , \quad (l-1)\hbar \ , \quad (l-2)\hbar$
with the corresponding probabilities:
$1/6 \ , \ 4/6 \ , \ 1/6$.
Thus, the most likely value to be obtained is $(l-1)\hbar$.
b) $(\Delta L_z)^2$ is given by

$$
\begin{aligned}
(\Delta L_z)^2 &= \langle L_z^2 \rangle - (\langle L_z \rangle)^2 \\
&= \frac{\hbar^2}{6}[l^2 + 4(l-1)^2 + (l-2)^2 - \{\frac{\hbar}{6}[l + 4(l-1) + (l-2)]\}^2 \\
&= \frac{1}{3}\hbar^2 \ . \tag{9.16.129}
\end{aligned}
$$

c) In this case the state is an eigenstate of L_z and thus we are certain to get the eigenvalue namely $m\hbar$. So, $\Delta L_z = 0$.

9.17   Expectation Values of Angular Momentum

Given the state $|l \ , m\rangle$
a) compute the expectation values $\langle L_x^2 \rangle$, $\langle L_y^2 \rangle$,and $\langle L_z^2 \rangle$ as well as the uncertainties ΔL_x, ΔL_y,and ΔL_z.
b) Discuss how the uncertainties computed in part a) conform with the commutator

$$[L_x, L_y] = i\hbar L_z \ .$$

Solution

a) We begin with the step-up and step-down operators

$$L_\pm = L_x \pm iL_y \ . \tag{9.17.130}$$

So,

$$L_x = \frac{1}{2}(L_+ + L_-) \ , \quad L_y = \frac{1}{2i}(L_+ - L_-) \ . \tag{9.17.131}$$

Therefore,

$$
\begin{aligned}
L_x^2 &= \frac{1}{4}\left(L_+^2 + L_-^2 + L_+L_- + L_-L_+\right) \\
L_y^2 &= -\frac{1}{4}\left(L_+^2 + L_-^2 - L_+L_- - L_-L_+\right) \ .
\end{aligned}
\tag{9.17.132}
$$

Now,

$$\langle l, m|L_\pm|l, m\rangle = \langle l, m|L_\pm^2|l, m\rangle = 0 \ . \tag{9.17.133}$$

Therefore,

$$\langle l, m|L_x^2|l, m\rangle = \langle l, m|L_y^2|l, m\rangle = \frac{1}{4}\langle l, m|L_+L_- + L_-L_+|l, m\rangle \ . \tag{9.17.134}$$

But,

$$L_\pm|l, m\rangle = \sqrt{l(l+1) - m(m \pm 1)}\,\hbar|l, m\rangle \ . \tag{9.17.135}$$

Therefore,

$$
\begin{aligned}
L_\mp L_\pm|l, m\rangle &= \sqrt{l(l+1) - m(m \pm 1)}\,\hbar L_\mp|l, m \pm 1\rangle \\
&= \sqrt{l(l+1) - m(m \pm 1)}\sqrt{l(l+1) - (m \pm 1)m}\,\hbar^2|l, m\rangle \\
&= [l(l+1) - m(m \pm 1)]\hbar^2|l, m\rangle \ .
\end{aligned}
\tag{9.17.136}
$$

So,

$$
\begin{aligned}
\langle L_x^2\rangle &= \langle L_y^2\rangle = \frac{\hbar^2}{4}[2l(l+1) - m(m+1) - m(m-1)] \\
&= \frac{\hbar^2}{2}[l(l+1) - m^2] \ .
\end{aligned}
\tag{9.17.137}
$$

Also, since the state $|l, m\rangle$ is an eigenstate of L_z we immediately have that

$$
\begin{aligned}
\langle L_z\rangle &= m\hbar \\
\langle L_z^2\rangle &= m^2\hbar^2 \ .
\end{aligned}
\tag{9.17.138}
$$

Also, as already seen,

$$\langle L_x\rangle = \langle L_y\rangle = 0 \ . \tag{9.17.139}$$

Therefore,

$$\Delta L_x = \Delta L_y = \frac{\hbar}{\sqrt{2}}\sqrt{l(l+1) - m^2} \qquad (9.17.140)$$

and

$$\Delta L_z = 0 . \qquad (9.17.141)$$

b) From the commutator

$$[L_x, L_y] = i\hbar L_z$$

we obtain on general grounds the uncertainty relation

$$\Delta L_x \Delta L_y \geq \frac{1}{2}|\langle[L_x, L_y]\rangle|$$

$$= \frac{\hbar}{2}\langle L_z\rangle = \frac{\hbar^2}{2}m . \qquad (9.17.142)$$

So, consistency requires that

$$\frac{\hbar^2}{2}[l(l+1) - m^2] \geq \frac{\hbar^2}{2}m . \qquad (9.17.143)$$

But, quite generally

$$l \geq m . \qquad (9.17.144)$$

Therefore the inequality is satisfied and the results are consistent. Notice that for a state with $l = 0$ the right hand side of (9.17.142) vanishes so that the product of the uncertainties

$$\Delta L_x \Delta L_y = 0 .$$

This is also consistent since the individual uncertainties ΔL_x and ΔL_y vanish separately.

9.18   Validity of Ehrenfest's Theorem

Ehrenfest's theorem states that

$$\frac{d}{dt}\langle p\rangle = -\langle\frac{dV}{dx}\rangle = \langle F(x)\rangle .$$

Suppose that

$$F(x) = Ax^2 , \quad A = \text{constant} .$$

Given that the system of interest is a simple harmonic oscillator and the wave function at $t = 0$ is

$$\Psi(0, x) = \frac{1}{2}[\sqrt{3}u_n(x) + iu_{n+1}(x)]$$

compute the difference between $F(\langle x\rangle)$ and $\langle F(x)\rangle$ as a function of time.

Solution

As a first step we need to compute the wavefunction for all times. This is given by

$$\psi(t,x) = \frac{1}{2}\left[\sqrt{3}u_n(x) + iu_{n+1}(x)\,e^{-i\omega t}\right]e^{-i(n+1/2)\omega t} . \qquad (9.18.145)$$

Now,

$$F(\langle x\rangle) = A\langle x\rangle^2 . \qquad (9.18.146)$$

Thus, we need

$$\langle x\rangle = \sqrt{\frac{\hbar}{2m\omega}}\langle(a^\dagger + a)\rangle . \qquad (9.18.147)$$

But,

$$(\psi(t,x),(a^\dagger + a)\psi(t,x)) = \frac{1}{4}\Big(\{\sqrt{3}u_n(x) + iu_{n+1}(x)\,e^{-i\omega t}\},$$

$$\{\sqrt{3}[\sqrt{n}u_{n-1} + \sqrt{n+1}u_{n+1}] + i[\sqrt{n+1}u_n + \sqrt{n+2}u_{n+2}]e^{-i\omega t}\}\Big)$$

$$= \sqrt{\frac{3(n+1)}{4}}\,\sin\omega t . \qquad (9.18.148)$$

Thus,

$$F(\langle x\rangle) = A\frac{3(n+1)}{4}\frac{\hbar}{2m\omega}\sin^2\omega t . \qquad (9.18.149)$$

Similarly,

$$\langle F(x)\rangle = A\langle x^2\rangle = \frac{\hbar}{2m\omega}(\psi(t,x),(a^\dagger + a)^2\psi(t,x))$$

$$= \frac{A\hbar}{8m\omega}\Big(\{\sqrt{3}u_n + iu_{n+1}e^{-i\omega t}\},$$

$$\{\sqrt{3}[\sqrt{n(n-1)}u_{n-2} + (2n+1)u_n + \sqrt{(n+1)(n+2)}u_{n+2}]$$

$$+i[\sqrt{n(n+1)}u_{n-1} + (2n+3)u_{n+1} + \sqrt{(n+2)(n+3)}u_{n+3}]e^{-i\omega t}\}\Big)$$

$$= (2n+3/2)\frac{A\hbar}{2m\omega} . \qquad (9.18.150)$$

9.19 Wigner Problem: Annihilation and Creation

This problem is based on a paper by Wigner [9.2]. For more discussion of that paper see also the following problem.

Consider the Hamiltonian

$$H = \hbar\omega(ba + 1/2) . \qquad (9.19.151)$$

Show that the Heisenberg equations

$$\dot{a} = i[H, a] = -i\hbar\omega a \qquad (9.19.152)$$
$$\dot{b} = i[H, b] = i\hbar\omega b \qquad (9.19.153)$$

imply that

$$[a, b] = 1 \qquad (9.19.154)$$

so that a is an annihilation operator and b is a creation operator. Assume that all the energy eigenvalues are positive and non-degenerate.
Hint: Work with matrix elements in the energy representation.

Solution

If we take matrix elements of (9.19.152) and (9.19.153) in the energy representation where H is diagonal and the energies are labelled E_0, E_1, E_2, ... we get

$$-i\hbar\omega a_{nm} = i(E_n - E_m)a_{nm}$$
$$i\hbar\omega b_{nm} = i(E_n - E_m)b_{nm} \ .$$

These equations imply

$$E_n - E_m = -\hbar\omega \qquad (9.19.155)$$
$$E_n - E_m = \hbar\omega \ . \qquad (9.19.156)$$

Thus, we have that

$$E_n = E_m \pm \hbar\omega \ . \qquad (9.19.157)$$

Since the energies are all positive we have

$$E_1 = E_0 + \hbar\omega$$
$$E_2 = E_1 + \hbar\omega = E_0 + 2\hbar\omega$$
$$\cdots$$
$$E_n = E_0 + n\hbar\omega \ . \qquad (9.19.158)$$

Furthermore, we see that the only non-zero matrix elements are: $a_{n,n+1}$ and $b_{n,n-1}$. We now impose that the diagonal Hamiltonian should have the eigenvalues that we obtained. Thus, we get

$$H = \hbar\omega(b_{n,n-1}a_{n-1,n} + 1/2)$$
$$= \hbar\omega\left(n + \frac{E_0}{\hbar\omega}\right) \ . \qquad (9.19.159)$$

Therefore,

$$b_{n,n-1}a_{n-1,n} = (n - 1/2) + \frac{E_0}{\hbar\omega} \ . \qquad (9.19.160)$$

We can now compute the matrix elements of the commutator.

$$\langle n|[a,b]|n\rangle = \sum_m (a_{n,m}b_{m,n} - b_{n,m}a_{m,n})$$

$$= a_{n,n+1}b_{n+1,n} - b_{n,n-1}a_{n-1,n}$$

$$= (n+1/2) + \frac{E_0}{\hbar\omega} - (n-1/2) - \frac{E_0}{\hbar\omega}$$

$$= 1 . \qquad\qquad (9.19.161)$$

Thus,

$$[a,b] = 1 \qquad\qquad (9.19.162)$$

and a and b satisfy the commutation relations for an annihilation and creation operator, respectively.

9.20 Wigner Problem: SHO

The following problem was first considered by Wigner [9.2]. Consider the Hamiltonian

$$H = \frac{1}{2}(p^2 + x^2) \qquad\qquad (9.20.163)$$

with units such that $\hbar = 1$. Show that the Heisenberg equations

$$p = \dot{x} = i[H,x] \qquad\qquad (9.20.164)$$
$$-x = \dot{p} = i[H,p] \qquad\qquad (9.20.165)$$

imply that

$$([p,x] + i)^2 = -(2E_0 - 1) \qquad\qquad (9.20.166)$$

where E_0 is the ground state energy. You may assume that all the energy eigenvalues are positive and non-degenerate as well as that all the matrix elements of x in the energy representation are real.
Hint: Work with matrix elements in the energy representation.

Solution

Since we are in the energy representation, H is diagonal and has eigenvalues $E_0, E_1, E_2, \ldots$ where E_0 is the lowest eigenvalue and the eigenvalues are ordered by increasing size. Taking matrix elements of (9.20.164) and (9.20.165) we get

$$p_{nm} = i(E_n - E_m)x_{nm}$$
$$-x_{nm} = i(E_n - E_m)p_{nm} . \qquad\qquad (9.20.167)$$

Combining these equations we find

$$x_{nm} = (E_n - E_m)^2 x_{nm} . \qquad\qquad (9.20.168)$$

Thus, to have a solution we need

$$(E_n - E_m)^2 = 1 \tag{9.20.169}$$

or

$$(E_n - E_m) = \pm 1 \ . \tag{9.20.170}$$

So,

$$E_n = E_0 + n \ . \tag{9.20.171}$$

It also follows that the only non-zero matrix elements of x and p are of the form $x_{n,n+1}$, $x_{n+1,n}$ and $p_{n,n+1}$ and $p_{n+1,n}$. By hermiticity we also have that

$$
\begin{aligned}
x_{n,n+1} &= x_{n+1,n} \\
p_{n,n+1} &= p^*_{n+1,n} \ .
\end{aligned}
\tag{9.20.172}
$$

Also, from the Heisenberg equations we have

$$p_{n,n+1} = i x_{n,n+1} \ , \quad p_{n+1,n} = i x_{n+1,n} = -i x_{n,n+1} \ . \tag{9.20.173}$$

At this stage we have satisfied the Heisenberg equations. We still have to impose the condition that the Hamiltonian (9.20.163) has only the diagonal matrix elements $E_n = E_0 + n$. So, for $n \neq 0$ we get

$$
\begin{aligned}
E_0 + n &= \frac{1}{2} \sum_m (x_{nm} x_{mn} + p_{nm} p_{mn}) \\
&= \frac{1}{2} (x_{n,n+1} x_{n+1,n} + x_{n,n-1} x_{n-1,n} + p_{n,n+1} p_{n+1,n} + p_{n,n-1} p_{n-1,n}) \\
&= \frac{1}{2} (x^2_{n,n+1} + x^2_{n,n-1} + x^2_{n,n+1} + x^2_{n,n-1}) \\
&= x^2_{n,n+1} + x^2_{n,n-1} \ .
\end{aligned}
\tag{9.20.174}
$$

For $n = 0$ we get

$$E_0 = x^2_{01} \ . \tag{9.20.175}$$

For the next few values of n we find

$$
\begin{aligned}
E_0 + 1 &= x^2_{1,2} + x^2_{1,0} &&\implies x^2_{1,2} = 1 \\
E_0 + 2 &= x^2_{2,3} + x^2_{1,2} &&\implies x^2_{2,3} = E_0 + 1 \\
E_0 + 3 &= x^2_{3,4} + x^2_{2,3} &&\implies x^2_{3,4} = 2 \\
E_0 + 4 &= x^2_{4,5} + x^2_{3,4} &&\implies x^2_{4,5} = E_0 + 2 \\
& \quad \cdots && \quad \cdots \ .
\end{aligned}
\tag{9.20.176}
$$

These results mean that

$$
\begin{aligned}
n &= \text{odd} & x^2_{n,n+1} &= \frac{1}{2}(n+1) \\
n &= \text{even} & x^2_{n,n+1} &= E_0 + \frac{n}{2} \ .
\end{aligned}
\tag{9.20.177}
$$

As a consequence we find that the commutator $[p, x]$ is also diagonal due to the relationships (9.20.173). For $n \neq 0$ the matrix elements of this commutator are

$$
\begin{aligned}
&\langle n|[p, x]|n\rangle \\
&= p_{n,n+1}x_{n+1,n} - x_{n,n+1}p_{n+1,n} + p_{n,n-1}x_{n-1,n} - x_{n,n-1}p_{n-1,n} \\
&= 2i(x_{n,n+1} - x_{n-1,n}) \ .
\end{aligned}
\tag{9.20.178}
$$

For the first few values of n these become

$$
-2iE_0, \ -2i(1 - E_0), \ -2iE_0, \ -2i(1 - E_0), \ \text{etc.}
$$

This can be summarized in the single equation

$$
([p, x] + i)^2 = -(2E_0 - 1)^2
\tag{9.20.179}
$$

as required.

For further discussions of how the $[p, x]$ commutation relations are determined by the Heisenberg equations and the Hamiltonian the reader should read the paper by Wigner [9.2].

9.21 Identity for Pauli Matrices

Show that if $\vec{A}$ and $\vec{B}$ are two vector operators that commute with the Pauli matrices $\vec{\sigma}$ then the following equation holds

$$
(\vec{\sigma} \cdot \vec{A})(\vec{\sigma} \cdot \vec{B}) = \vec{A} \cdot \vec{B} + i\vec{\sigma} \cdot (\vec{A} \times \vec{B}) \ .
$$

Solution

We want to show that if $\vec{A}$ and $\vec{B}$ are two vector operators that commute with the Pauli matrices $\vec{\sigma}$ then the following equation holds

$$
(\vec{\sigma} \cdot \vec{A})(\vec{\sigma} \cdot \vec{B}) = \vec{A} \cdot \vec{B} + i\vec{\sigma} \cdot (\vec{A} \times \vec{B}) \ .
$$

To see this we write out the left hand side

$$
(\vec{\sigma} \cdot \vec{A})(\vec{\sigma} \cdot \vec{B}) = \begin{pmatrix} A_z & A_x - iA_y \\ A_x + iA_y & A_z \end{pmatrix} \begin{pmatrix} B_z & B_x - iB_y \\ B_x + iB_y & B_z \end{pmatrix} \ .
\tag{9.21.180}
$$

After explicitly multiplying out these two matrices we get

$$
\begin{pmatrix} \vec{A} \cdot \vec{B} + i(A_xB_y - A_yB_x) & A_zB_x - A_xB_z + i(A_yB_z - A_zB_y) \\ A_xB_z - A_zB_x + i(A_yB_z - A_zB_y) & \vec{A} \cdot \vec{B} - i(A_xB_y - A_yB_x) \end{pmatrix}
$$

which establishes the required identity

$$
(\vec{\sigma} \cdot \vec{A})(\vec{\sigma} \cdot \vec{B}) = \vec{A} \cdot \vec{B} + i\vec{\sigma} \cdot (\vec{A} \times \vec{B}) \ .
\tag{9.21.181}
$$

9.22   Operator Identity - Spin Rotation

a) Verify to order λ^3 that for any two operators G, A

$$e^{i\lambda G} A e^{-i\lambda G} = A + i\lambda[G,]A + \frac{(i\lambda)^2}{2!}[G,[G,A]] + \cdots \quad .$$

b) Use this result to show that if S_x, S_y, S_z are spin operators then

$$e^{iS_z\varphi/\hbar} S_x e^{-iS_z\varphi/\hbar} = S_x \cos\varphi - S_y \sin\varphi$$

$$e^{iS_z\varphi/\hbar} S_y e^{-iS_z\varphi/\hbar} = S_y \cos\varphi + S_x \sin\varphi \quad .$$

Solution

a) The proof follows simply by expanding the exponentials. Thus,

$$e^{i\lambda G} A e^{-i\lambda G} = [1 + i\lambda G + \frac{(i\lambda G)^2}{2!} + \frac{(i\lambda G)^3}{3!} + \cdots]$$

$$\times \quad A[1 - i\lambda G + \frac{(-i\lambda G)^2}{2!} + \frac{(-i\lambda G)^3}{3!} + \cdots]$$

$$= \quad A + i\lambda[G,A] + \frac{(i\lambda)^2}{2!}[G^2 A - 2GAG + AG^2]$$

$$+ \quad \frac{(i\lambda)^3}{3!}[G^3 A + 3GAG^2 - 3G^2 AG - AG^3] + \cdots \quad . \tag{9.22.182}$$

But,

$$[G,[G,A]] = [G, GA - AG]$$

$$= \quad G^2 A - 2GAG + AG^2 \quad . \tag{9.22.183}$$

Similarly,

$$[G,[G,[G,A]]] = [G, G^2 A - 2GAG + AG^2]$$

$$= \quad G^3 A + 3GAG^2 - 3G^2 AG - AG^3 \quad . \tag{9.22.184}$$

Substituting these results back into 9.22.182 we have the desired verification.
b) Since S_x, S_y, S_z are spin operators we have that

$$[S_z, S_x] = i\hbar S_y$$

$$[S_z, S_y] = -i\hbar S_x \quad . \tag{9.22.185}$$

Using this we have

$$e^{iS_z\varphi/\hbar} S_x e^{-iS_z\varphi/\hbar} = S_x + \frac{i\varphi}{\hbar}i\hbar S_y + \frac{1}{2!}\left(\frac{i\varphi}{\hbar}\right)^2 \hbar^2 S_x$$

$$+ \quad \frac{1}{3!}\left(\frac{i\varphi}{\hbar}\right)^3 i\hbar^3 S_y + \cdots$$

$$= \quad S_x\left[1 - \frac{\varphi^2}{2!} + \frac{\varphi^4}{4!} - + \cdots\right] - S_y\left[\varphi - \frac{\varphi^3}{3!} + \frac{\varphi^5}{5!} - + \cdots\right]$$

$$= \quad S_x \cos\varphi - S_y \sin\varphi \quad . \tag{9.22.186}$$

Similarly, we get

$$e^{iS_z\varphi/\hbar} S_y e^{-iS_z\varphi/\hbar} = S_y + \frac{i\varphi}{\hbar}(-i\hbar)S_x + \frac{1}{2!}\left(\frac{i\varphi}{\hbar}\right)^2 \hbar^2 S_y$$

$$+ \frac{1}{3!}\left(\frac{i\varphi}{\hbar}\right)^3 (-i\hbar)^3 S_x + \cdots$$

$$= S_y\left[1 - \frac{\varphi^2}{2!} + \frac{\varphi^4}{4!} - + \cdots\right] + S_x\left[\varphi - \frac{\varphi^3}{3!} + \frac{\varphi^5}{5!} - + \cdots\right]$$

$$= S_y \cos\varphi + S_x \sin\varphi \ . \tag{9.22.187}$$

9.23 An Operator Identity

Let A and B be two arbitrary operators and define

$$Z(\alpha) = e^{\alpha A} B e^{-\alpha A} \ .$$

Show that

$$\frac{dZ}{d\alpha} = [A, Z] \ .$$

Use this result to derive the operator identity

$$e^{\alpha A} B e^{-\alpha A} = B + \alpha[A, B] + \frac{\alpha^2}{2!}[A, [A, B]] + \frac{\alpha^3}{3!}[A[A, [A, B]]] + \cdots \ .$$

Hint: The right hand side is a Taylor expansion.

Solution

To prove the first part we simply differentiate $Z(\alpha)$, keeping in mind that the operator order must be maintained. Then,

$$\frac{dZ}{d\alpha} = \left(\frac{de^{\alpha A}}{d\alpha}\right) B e^{-\alpha A} + e^{\alpha A} B \left(\frac{de^{-\alpha A}}{d\alpha}\right)$$

$$= A e^{\alpha A} B e^{-\alpha A} - e^{\alpha A} B e^{-\alpha A} A$$

$$= [A, Z] \ . \tag{9.23.188}$$

We now use induction. Thus, we assume that

$$\frac{d^n Z}{d\alpha^n} = \underbrace{[A, [A, \ldots [A, Z]]] \ldots]}_{n} \ . \tag{9.23.189}$$

Differentiating this equation it follows that

$$\frac{d^{n+1} Z}{d\alpha^{n+1}} = \underbrace{[A, [A, \ldots [A, \frac{dZ}{d\alpha}]]] \ldots]}_{n}$$

$$= \underbrace{[A, [A, \ldots [A, Z]]] \ldots]}_{n+1} \ . \tag{9.23.190}$$

Thus, we have that

$$\frac{d^n Z}{d\alpha^n}\bigg|_{\alpha=0} = [A,[A,\ldots \underbrace{[A,B]]]\ldots]}_{n}$$
(9.23.191)

and therefore the Taylor expansion of Z becomes

$$e^{\alpha A} B e^{-\alpha A} = B + \alpha[A,B] + \frac{\alpha^2}{2!}[A,[A,B]] + \frac{\alpha^3}{3!}[A[A,[A,B]]] + \cdots (9.23.192)$$

as desired.

9.24 Commutator with Inverse Operator

Show that

$$[A,B^{-1}] = -B^{-1}[A,B]B^{-1} .$$

Hint: Use the fact that

$$[A,BB^{-1}] = 0 .$$

Solution

Using the hint we have that

$$[A,BB^{-1}] = 0 = B[A,B^{-1}] + [A,B]B^{-1} .$$
(9.24.193)

Therefore,

$$B[A,B^{-1}] = -[A,B]B^{-1} .$$
(9.24.194)

So, acting with B^{-1} on the left of this equation we obtain the desired result

$$[A,B^{-1}] = -B^{-1}[A,B]B^{-1} .$$
(9.24.195)

9.25 Schwinger Method for Angular Momenta

Let a_1, $a_1^\dagger$ and a_2, $a_2^\dagger$ be simple harmonic oscillator annihilation and creation operators. Define

$$
\begin{aligned}
J_1 &= \frac{\hbar}{2}\left(a_2^\dagger a_1 + a_1^\dagger a_2\right) \\
J_2 &= \frac{i\hbar}{2}\left(a_2^\dagger a_1 - a_1^\dagger a_2\right) \\
J_3 &= \frac{\hbar}{2}\left(a_1^\dagger a_1 - a_2^\dagger a_2\right) \\
j &= \frac{\hbar}{2}\left(a_1^\dagger a_1 + a_2^\dagger a_2\right) .
\end{aligned}
$$
(9.25.196)

Show that J_1, J_2, J_3 obey the angular momentum commutation relations

$$[J_1, J_2] = i\hbar J_3 \quad \text{and cyclic permutations}$$

and that

$$[j, \vec{J}] = 0$$

as well as that

$$\vec{J}^2 = j(j+1)\hbar^2 \ .$$

Solution

The transformations given are due to Schwinger [9.3]. We first verify the angular momentum commutation relations using that

$$[a_i, a_j^\dagger] = \delta_{ij} \ . \tag{9.25.197}$$

Then,

$$
\begin{aligned}
[J_1, J_2] &= \frac{i\hbar^2}{4}[a_2^\dagger a_1 + a_1^\dagger a_2, a_2^\dagger a_1 - a_1^\dagger a_2] \\
&= \frac{i\hbar^2}{4}\left([a_1^\dagger a_2, a_2^\dagger a_1] - [a_2^\dagger a_1, a_1^\dagger a_2]\right) \\
&= \frac{i\hbar^2}{2}\left([a_1^\dagger, a_2^\dagger a_1]a_2 + a_1^\dagger[a_2, a_2^\dagger a_1]\right) \\
&= \frac{i\hbar^2}{2}\left(-a_2^\dagger a_2 + a_1^\dagger a_1\right) \\
&= i\hbar J_3 \ . \tag{9.25.198}
\end{aligned}
$$

Similarly we find

$$
\begin{aligned}
[J_2, J_3] &= \frac{i\hbar^2}{4}[a_2^\dagger a_1 - a_1^\dagger a_2, a_1^\dagger a_1 - a_2^\dagger a_2] \\
&= \frac{i\hbar^2}{4}\left(a_2^\dagger a_1 + a_1 a_2^\dagger + a_2 a_1^\dagger + a_1^\dagger a_2\right) \\
&= i\hbar J_1 \ . \tag{9.25.199}
\end{aligned}
$$

And finally

$$
\begin{aligned}
[J_3, J_1] &= \frac{\hbar^2}{4}[a_1^\dagger a_1 - a_2^\dagger a_2, a_2^\dagger a_1 + a_1^\dagger a_2] \\
&= \frac{\hbar^2}{4}\left(-a_2^\dagger a_1 + a_1^\dagger a_2 - a_2^\dagger a_1 + a_2 a_1^\dagger\right) \\
&= i\hbar J_2 \ . \tag{9.25.200}
\end{aligned}
$$

Next we compute $\vec{J}^2$.

$$
\begin{aligned}
\vec{J}^2 &= \frac{\hbar^2}{4}\Big[(a_2^\dagger)^2 a_1^2 + (a_1^\dagger)^2 a_2^2 + a_2^\dagger a_1 a_2 a_1^\dagger + a_1^\dagger a_2 a_2^\dagger a_1 \\
&\quad - (a_2^\dagger)^2 a_1^2 - (a_1^\dagger)^2 a_2^2 + a_2^\dagger a_1 a_2 a_1^\dagger + a_1^\dagger a_2 a_2^\dagger a_1 \\
&\quad + (a_1^\dagger a_1 + a_2^\dagger a_2)^2 - 4a_1^\dagger a_1 a_2^\dagger a_2\Big] \\
&= \frac{\hbar^2}{4}\Big[(a_1^\dagger a_1 + a_2^\dagger a_2)^2 - 4a_1^\dagger a_1 a_2^\dagger a_2 \\
&\quad + 2a_2^\dagger a_1(a_1^\dagger a_2 + 1) + a_1^\dagger a_1(a_2^\dagger a_2 + 1)\Big] \\
&= \frac{\hbar^2}{4}\Big[(a_1^\dagger a_1 + a_2^\dagger a_2)^2 + 2(a_1^\dagger a_1 + a_2^\dagger a_2)\Big] \\
&= j(j+1) \ .
\end{aligned}
\tag{9.25.201}
$$

This last equation also shows that

$$
[\vec{J}^2, j] = 0 \ .
\tag{9.25.202}
$$

9.26 Minimum Uncertainty in J_x

Given the eigenstates $|j, m\rangle$ of J^2, J_z find the value of m that minimizes the uncertainty in J_x.

Solution

We first write

$$
J_x = \frac{1}{2}(J_+ + J_-)
\tag{9.26.203}
$$

where

$$
J_\pm|j, m\rangle = \sqrt{j(j+1) - m(m \pm 1)}|j, m \pm 1\rangle \ .
\tag{9.26.204}
$$

It thus follows that

$$
\langle j, m|J_x|j, m\rangle = 0 \ .
\tag{9.26.205}
$$

Also,

$$
J_x^2 = \frac{1}{4}(J_+^2 + J_-^2 + J_+ J_- + J_- J_+) \ .
\tag{9.26.206}
$$

But,

$$
\begin{aligned}
J_+ J_- &= J_x^2 + J_y^2 - i[J_x, J_y] = J^2 - J_z^2 + \hbar J_z \\
J_- J_+ &= J_x^2 + J_y^2 + i[J_x, J_y] = J^2 - J_z^2 - \hbar J_z \ .
\end{aligned}
\tag{9.26.207}
$$

Combining all these results we find that

$$
(\Delta J_x)^2 = \langle J_x^2 \rangle - \langle J_x \rangle^2 = \frac{1}{2}[j(j+1) - m^2] \ .
\tag{9.26.208}
$$

This result is clearly minimized by $|m| = j$.

9.27 Unsöld's Theorem and its Application

a) Prove that

$$\sum_{m=-l}^{l} Y_{lm}^{*}(\theta, \varphi) Y_{lm}(\theta, \varphi)$$

is a constant. This is Unsöld's Theorem. [9.4] Also evaluate the constant.
Hint: Use the addition theorem for the spherical harmonics.
b) Use this result to show that the probability of finding a particle at the point $\vec{r}$, if it is acted on by a potential $V(r)$ and is in a state of definite angular momentum l, is only a function of $r = |\vec{r}|$.
Hint: Assume that there is no special alignment so that all the z-components (m-values of the angular momentum) are equally probable.

Solution

a) The addition theorem for the sperical harmonics states that

$$\sum_{m=-l}^{l} Y_{lm}^{*}(\theta, \varphi) Y_{lm}(\theta', \varphi') = \frac{2l+1}{4\pi} P_{l}(\cos \gamma) \qquad (9.27.209)$$

where

$$\cos \gamma = \sin \theta \sin \theta' \cos(\varphi - \varphi') + \cos \theta \cos \theta' . \qquad (9.27.210)$$

But, setting $\theta = \theta'$ and $\varphi = \varphi'$ in (9.27.209) we get

$$\cos \gamma = \sin^2 \theta + \cos^2 \theta = 1 . \qquad (9.27.211)$$

This means that

$$P_{l}(\cos \gamma) = P_{l}(1) = 1 . \qquad (9.27.212)$$

Therefore,

$$\sum_{m=-l}^{l} Y_{lm}^{*}(\theta, \varphi) Y_{lm}(\theta, \varphi) = \frac{2l+1}{4\pi} . \qquad (9.27.213)$$

b) Since we have a state of definite l the wavefunction must be of the form

$$\psi_{n,l}(\vec{r}) = \sum_{m=-l}^{l} c_m R_{nl}(r) Y_{lm}(\theta, \varphi) . \qquad (9.27.214)$$

But, all values of m are equally probable. This means that all the states $R_{nl}(r) Y_{lm}(\theta, \varphi)$ are equally probable. So, we can take

$$c_m = \frac{1}{\sqrt{2l+1}} \qquad (9.27.215)$$

so that all states

$$\psi_{n,l,m}(\vec{r}) = \frac{1}{\sqrt{2l+1}} R_{nl}(r) Y_{lm}(\theta, \varphi) \tag{9.27.216}$$

are equally probable. Therefore, the probability density for finding the particle at $\vec{r}$ is the sum over all m values of these states

$$
\begin{aligned}
|\psi_{n,l}(\vec{r})|^2 &= \frac{1}{2l+1} \sum_{m=-l}^{l} \frac{1}{2l+1} |R_{nl}(r)|^2 Y_{lm}^*(\theta, \varphi) Y_{lm}(\theta, \varphi) \\
&= \frac{|R_{nl}(r)|^2}{2l+1} \frac{2l+1}{4\pi} \\
&= \frac{|R_{nl}(r)|^2}{4\pi} \tag{9.27.217}
\end{aligned}
$$

and is a function in terms of r only.

9.28   Rotation Matrix for $j=1$

In problem 9.3 we showed that using the matrix representations for $j = 1$

$$J_x = \frac{\hbar}{\sqrt{2}} \begin{pmatrix} 0 & 1 & 0 \\ 1 & 0 & 1 \\ 0 & 1 & 0 \end{pmatrix} \tag{9.28.218}$$

$$J_y = \frac{\hbar}{\sqrt{2}} \begin{pmatrix} 0 & -i & 0 \\ i & 0 & -i \\ 0 & i & 0 \end{pmatrix} \tag{9.28.219}$$

$$J_z = \hbar \begin{pmatrix} 1 & 0 & 0 \\ 0 & 0 & 0 \\ 0 & 0 & -1 \end{pmatrix} \tag{9.28.220}$$

it follows that for a unit vector $\hat{n}$

$$R = \exp\left[i(\vec{J}/\hbar) \cdot \hat{n}\theta\}\right] = 1 + \frac{\vec{J} \cdot \hat{n}}{\hbar} \sin\theta + \left(\frac{\vec{J} \cdot \hat{n}}{\hbar}\right)^2 (\cos\theta + 1) . \tag{9.28.221}$$

Use this result to show that the trace of this expression determines the rotation angle θ and that the antisymmetric part of this expression determines the rotation axis $\hat{n}$.

Solution

Again using the matrix representations (9.28.218) - (9.28.220) we find that

$$\frac{\vec{J} \cdot \hat{n}}{\hbar} = \hbar \begin{pmatrix} n_z & \frac{n_x - i n_y}{\sqrt{2}} & 0 \\ \frac{n_x + i n_y}{\sqrt{2}} & 0 & \frac{n_x - i n_y}{\sqrt{2}} \\ 0 & \frac{n_x + i n_y}{\sqrt{2}} & -n_z \end{pmatrix} . \tag{9.28.222}$$

Also, recalling that

$$(\hat{n})^2 = n_x^2 + n_y^2 + n_z^2 = 1 \qquad (9.28.223)$$

we find that we can write

$$\left(\frac{\vec{J} \cdot \hat{n}}{\hbar}\right)^2$$

$$= \hbar^2 \begin{pmatrix} \frac{1}{2}(1 + n_z^2) & \frac{n_+}{\sqrt{2}}(n_x - in_y) & \frac{1}{2}(n_x - in_y)^2 \\ \frac{n_+}{\sqrt{2}}(n_x + in_y) & 1 - n_z^2 & -\frac{n_+}{\sqrt{2}}(n_x - in_y) \\ \frac{1}{2}(n_x + in_y)^2 & -\frac{n_+}{\sqrt{2}}(n_x + in_y) & \frac{1}{2}(1 + n_z^2) \end{pmatrix} . \qquad (9.28.224)$$

Taking the trace of the right hand side of (9.28.221) we now obtain the result that

$$\text{Trace} = 3 + 2(\cos\theta + 1) = 4 + 2\cos\theta . \qquad (9.28.225)$$

Thus, the trace determines the rotation angle θ. Also, taking the three different antisymmetric parts of the right hand side of (9.28.221) we get

$$\begin{aligned} R_{21} - R_{12} &= \sqrt{2}\, i[n_y \sin\theta + n_z n_y(\cos\theta + 1)] \\ R_{31} - R_{13} &= 2\, i n_x n_y(\cos\theta + 1) \\ R_{23} - R_{32} &= \sqrt{2}\, i[n_y \sin\theta - n_z n_y(\cos\theta + 1)] . \end{aligned} \qquad (9.28.226)$$

The three components n_x, n_y, n_z are now easily solved from these expressions. Thus, the antisymmetric part determines the rotation axis $\hat{n}$.

9.29 Algebra and Constants of the Motion

Consider the Hamiltonian for a symmetric two-dimensional harmonic oscillator

$$H = \frac{1}{2M}(p_x^2 + p_y^2) + \frac{1}{2}M\omega^2(x^2 + y^2)$$

as well as the two operators

$$Q = \frac{1}{2M}(p_x^2 - p_y^2) + \frac{1}{2}M\omega^2(x^2 - y^2)$$

and

$$L_z = x p_y - y p_x .$$

a) Show that

$$[L_z, H] = 0$$

b) Show that if

$$[A, H] = 0 \quad \text{and} \quad [B, H] = 0$$

then

$$[[A, B], H] = 0$$

and hence deduce that

$$[R, H] = 0 \quad \text{where} \quad R = [L_z, Q] \ .$$

c) Finally show that the algebra of the operators H, L_z, Q, R closes. In other words show that no other new operators can be obtained by continuing in this way. This shows that, by proceeding in this way, one can construct at most a finite number of independent constants of the motion of the Hamiltonian H.

Solution

a) Since the given Hamiltonian H is invariant under rotations about the z-axis it follows that

$$[L_z, H] = 0 \ . \tag{9.29.227}$$

This result also follows, of course, from a brute force computation which yields

$$
\begin{aligned}
[L_z, H] &= [xp_y - yp_x, \frac{1}{2M}(p_x^2 + p_y^2) + \frac{1}{2}M\omega^2(x^2 + y^2) \\
&= \frac{i\hbar}{2M}[2p_xp_y - 2p_xp_y] + \frac{i\hbar}{2}M\omega^2[-2xy + 2xy] \\
&= 0 \ .
\end{aligned}
\tag{9.29.228}
$$

To evaluate $[Q, H]$ we first rewrite the Hamiltonian as

$$
\begin{aligned}
H &= \frac{1}{2}\left[\frac{1}{2M}(p_x + p_y)^2 + \frac{1}{2}M\omega^2(x + y)^2 \right. \\
&+ \left. \frac{1}{2M}(p_x - p_y)^2 + \frac{1}{2}M\omega^2(x - y)^2 \right]
\end{aligned}
\tag{9.29.229}
$$

and Q as

$$Q = \frac{1}{2M}(p_x - p_y)(p_x + p_y) + \frac{1}{2}M\omega^2(x - y)(x + y) \ . \tag{9.29.230}$$

Now, we easily find that

$$
\begin{aligned}
[p_x - p_y, x + y] &= 0 \\
[p_x + p_y, x - y] &= 0 \ .
\end{aligned}
\tag{9.29.231}
$$

So, we get

$$
\begin{aligned}
[(p_x - p_y)(p_x + p_y), (x + y)^2] &= (p_x - p_y)[(p_x + p_y), (x + y)^2] \\
&= -4i\hbar(p_x - p_y)(x + y) \ .
\end{aligned}
\tag{9.29.232}
$$

Similarly,

$$
\begin{aligned}
[(p_x - p_y)(p_x + p_y), (x - y)^2] &= [(p_x - p_y), (x - y)^2](p_x + p_y) \\
&= -4i\hbar(x - y)(p_x + p_y) \ .
\end{aligned}
\tag{9.29.233}
$$

In an exactly similar manner we find

$$
\begin{aligned}
[(x - y)(x + y), (p_x + p_y)^2] &= (x - y)[(x + y), (p_x + p_y)^2] \\
&= 4i\hbar(x - y)(p_x + p_y)
\end{aligned}
\tag{9.29.234}
$$

and

$$
\begin{aligned}
[(x - y)(x + y), (p_x - p_y)^2] &= [(x - y), (p_x - p_y)^2](x + y) \\
&= 4i\hbar(p_x - p_y)(x + y) \ .
\end{aligned}
\tag{9.29.235}
$$

Therefore, combining these results we find that

$$
[Q, H] = 0 \ .
\tag{9.29.236}
$$

b) Now, consider operators A, B, H such that

$$
[A, H] = 0 \quad \text{and} \quad [B, H] = 0 \ .
\tag{9.29.237}
$$

Then,

$$
\begin{aligned}
[[A, B], H] &= [AB, H] - [BA, H] \\
&= A[B, H] + [A, H]B - B[A, H] - [B, H]A \\
&= 0
\end{aligned}
\tag{9.29.238}
$$

since each of the four commutators vanishes. Therefore, we also have that

$$
\begin{aligned}
[R, H] &= [[L_z, Q], H] \\
&= 0 \ .
\end{aligned}
\tag{9.29.239}
$$

c) To show that the algebra closes we explicitly calculate the operator R.

$$
\begin{aligned}
R &= [L_z, Q] \\
&= [xp_y - yp_x, \frac{1}{2M}(p_x^2 - p_y^2) + \frac{1}{2}M\omega^2(x^2 - y^2)] \\
&= 4i\hbar \left(\frac{p_x p_y}{2M} + \frac{1}{2}M\omega^2 xy \right) \ .
\end{aligned}
\tag{9.29.240}
$$

Proceeding in the same spirit we compute

$$
\begin{aligned}
[L_z, R/(4i\hbar)] &= [xp_y - yp_x, \frac{p_x p_y}{2M} + \frac{1}{2}M\omega^2 xy] \\
&= -i\hbar \left(\frac{1}{2M}(p_x^2 - p_y^2) + \frac{1}{2}M\omega^2(x^2 - y^2) \right) \\
&= -i\hbar Q \ .
\end{aligned}
\tag{9.29.241}
$$

Also,

$$\begin{aligned}
[Q, R/(4i\hbar)] &= [\frac{1}{2M}(p_x^2 - p_y^2) + \frac{1}{2}M\omega^2(x^2 - y^2), \frac{p_x p_y}{2M} + \frac{1}{2}M\omega^2 xy] \\
&= i\hbar\omega^2 (xp_y - yp_x) \\
&= i\hbar L_z .
\end{aligned} \tag{9.29.242}$$

This shows that the algebra closes and no further constants of the motion can be produced in this manner.

9.30   Coherent State and Normal Ordering

In many cases it is convenient to "normal order" operators constructed from annihilation and creation operators. An operator $F(a, a^\dagger)$ is said to be normal ordered if all the annihilation operators a appear to the right of all the creation operators $a^\dagger$. In that case the operator is written with colons on each side : $F(a, a^\dagger)$:.

Show that for any normal ordered operator : $F(a, a^\dagger)$: and any two coherent states $|z\rangle$ $|w\rangle$ we have that

$$\langle z| : F(a, a^\dagger) : |w\rangle = F(z^*, w)\langle z|w\rangle .$$

Solution

For the coherent states we have

$$a|w\rangle = w|w\rangle \quad , \quad \langle z|a^\dagger = \langle z|z^* . \tag{9.30.243}$$

Since : $F(a, a^\dagger)$: is, as indicated by the colon, normal order we have that if F is written out as a series that

$$: F(a, a^\dagger) := \sum_{nm} c_{nm}(a^\dagger)^n a^m . \tag{9.30.244}$$

But, using (9.30.243)

$$\langle z|c_{nm}(a^\dagger)^n a^m|w\rangle = c_{nm}(z^*)^n w^m \langle z|w\rangle . \tag{9.30.245}$$

Hence,

$$\langle z| : F(a, a^\dagger) : |w\rangle = F(z^*, w)\langle z|w\rangle . \tag{9.30.246}$$

9.31   Normal Ordering of x^n

Suppose we have units such that

$$x = \frac{1}{\sqrt{2}}(a + a^\dagger) .$$

Show that the normal ordered operator (see the previous problem) $: x^n :$ is given by

$$: x^n : = \frac{1}{2^n} H_n(x)$$

where $H_n(x)$ are the Hermite polynomials.
Hint: It may be convenient to write

$$: x^n : = f_n(x)$$

and show that $f_n(x)$ satisfies a certain recursion relation.
Hint: The Hermite polynomials satisfy

$$H_{n+1}(x) = 2x H_n(x) - 2n H_{n-1}(x) \ .$$

Solution

We start with the algebra of the annihilation and creation operators

$$[a, a^\dagger] = 1 \ . \tag{9.31.247}$$

It then follows immediately that

$$
\begin{aligned}
a(a^\dagger)^n &= (a^\dagger)^n a + n(a^\dagger)^{n-1} \\
a^n a^\dagger &= a^\dagger a^n + n a^{n-1} \ .
\end{aligned} \tag{9.31.248}
$$

Now, following the hint we assume that

$$: x^n : = f_n(x) \ . \tag{9.31.249}$$

It is also clear that

$$: x^n : = 2^{-n/2} \sum_{r=0}^{n} \binom{n}{r} (a^\dagger)^{n-r} a^r \ . \tag{9.31.250}$$

Therefore, using (9.31.248)

$$
\begin{aligned}
: x^n : x &= 2^{-(n+1)/2} \sum_{r=0}^{n} \binom{n}{r} [(a^\dagger)^{n-r} a^{r+1} + (a^\dagger)^{n-r} a^r a^\dagger] \\
&= 2^{-(n+1)/2} \sum_{r=0}^{n} \binom{n}{r} [(a^\dagger)^{n-r} a^{r+1} + (a^\dagger)^{n+1-r} a^r \\
&\quad + r(a^\dagger)^{n-r} a^{r-1}] \ . \tag{9.31.251}
\end{aligned}
$$

But,

$$
\begin{aligned}
&\sum_{r=0}^{n} \binom{n}{r} [(a^\dagger)^{n-r} a^{r+1} + (a^\dagger)^{n+1-r} a^r] \\
&= (a^\dagger)^{n+1} + a^{n+1}
\end{aligned}
$$

$$+ \sum_{r=1}^{n} \left[\frac{r}{n+1} \frac{(n+1)!}{(n+1-r)!r!} + \frac{n!}{(n-r)!r!} \right] (a^\dagger)^{n+1-r} a^r$$

$$= (a^\dagger)^{n+1} + a^{n+1} \sum_{r=1}^{n} \left[\frac{n!(r+n+1-r)}{(n+1-r)!r!} (a^\dagger)^{n+1-r} a^r \right]$$

$$= \sum_{r=0}^{n+1} \binom{n+1}{r} (a^\dagger)^{n+1-r} a^r \ . \tag{9.31.252}$$

Also,

$$\sum_{r=0}^{n} \binom{n}{r} r (a^\dagger)^{n-r} a^{r-1}$$

$$= \sum_{r=1}^{n} n \binom{n-1}{r-1} (a^\dagger)^{n-1-(r-1)} a^{r-1}$$

$$= n \sum_{r=0}^{n-1} n \binom{n-1}{r} (a^\dagger)^{n-1-r} a^r \ . \tag{9.31.253}$$

Therefore,

$$: x^n : x =: x^{n+1} : + n : x^{n-1} : \ . \tag{9.31.254}$$

So, we have obtained that

$$x f_n(x) = f_{n+1}(x) + \frac{n}{2} f_{n-1}(x) \ . \tag{9.31.255}$$

However, according to the hint, the Hermite polynomials satisfy

$$x H_n(x) = \frac{1}{2} H_{n+1}(x) + n H_{n-1}(x) \tag{9.31.256}$$

and we see immediately that

$$g_n(x) = 2^n f_n(x)$$

satisfy the same recursion relation as $H_n(x)$. Therefore,

$$: x^n : \ = 2^{-n} H_n(x) \ . \tag{9.31.257}$$

Bibliography

[9.1] A.Z. Capri, *Nonrelativistic Quantum Mechanics* 3rd edition, World Scientific Publishing Co. Pte. Ltd., section 9.5, (2002) .

[9.2] E.P. Wigner, Phys. Rev. **77**, 711, (1950) .

[9.3] J. Schwinger, "On Angular Momentum". US Atomic Energy Commission NYO - 3071, (1952)

[9.4] A. Unsöld, Ann. Phys., **82**, 355, (1927).

[9.5]P. Carruthers and M.M. Nieto - Phase and Angle Variables in Quantum Mechanics, Rev. Mod. Phys. **40**, 411, (1968).

Chapter 10

Central Force Problems

10.1 Isotropic SHO in Two Dimensions

Solve the isotropic simple harmonic oscillator problem in two dimensions in both Cartesian and cylindrical coordinates.

Hint: L_z commutes with the Hamiltonian.

Solution

We are given

$$H = \frac{\vec{p}^2}{2m} + \frac{1}{2}k\vec{r}^2 \quad , \quad \vec{r}^2 = x^2 + y^2 \quad , \quad \vec{p}^2 = p_x^2 + p_y^2 . \tag{10.1.1}$$

a) Cartesian coordinates

$$H = H_x + H_y \qquad [H_x, H_y] = 0 . \tag{10.1.2}$$

Therefore, we simply have two one-dimensional simple harmonic oscillators and the energy levels are

$$E_{n_1,n_2} = (n_1 + 1/2)\hbar\omega + (n_2 + 1/2)\hbar\omega = (n_1 + n_2 + 1)\hbar\omega \tag{10.1.3}$$

where, as always, $\omega^2 = k/M$. Also the eigenstates are (here we are setting $\alpha^2 = M\omega/\hbar$)

$$\phi_{n_1,n_2}(\vec{r}) = \frac{\alpha}{\sqrt{\pi}} \frac{1}{\sqrt{2^{n_1}n_1!}} \frac{1}{\sqrt{2^{n_2}n_2!}} H_{n_1}(\alpha x) H_{n_2}(\alpha y) e^{-\alpha^2 r^2/2} . \tag{10.1.4}$$

b) Cylindrical coordinates: The Schrödinger equation reads

$$-\frac{\hbar^2}{2M}\left(\frac{\partial^2}{\partial r^2} + \frac{1}{r}\frac{\partial}{\partial r} + \frac{1}{r^2}\frac{\partial}{\partial\varphi^2} + \frac{1}{2}kr^2\right)\psi = E\psi . \tag{10.1.5}$$

Since $[L_z, H] = 0$ we can now set

$$\psi = R(r) e^{im\varphi} \tag{10.1.6}$$

and $\omega^2 = k/M$. Then,

$$\left(\frac{d^2}{dr^2} + \frac{1}{r}\frac{d}{dr} - \frac{m^2}{r^2} - \frac{M^2\omega^2}{\hbar^2}r^2 + \frac{2ME}{\hbar^2}\right)R = 0 . \tag{10.1.7}$$

Now let

$$y^2 = \frac{M\omega}{\hbar}r^2 \; , \quad \lambda = \frac{2E}{\hbar\omega} \;\; \text{and} \;\; R(r) = G(y) \; . \tag{10.1.8}$$

Then,

$$\left(\frac{d^2}{dy^2} + \frac{1}{y}\frac{d}{dy} - \frac{m^2}{y^2} - y^2 + \lambda\right)G = 0 . \tag{10.1.9}$$

We first examine the asymptotic behaviour of the solutions of this equation.

$$G \sim e^{\pm y^2/2} \;\; \text{for} \;\; y \to \infty . \tag{10.1.10}$$

$$G \sim y^{\pm m} \;\; \text{for} \;\; y \to 0 . \tag{10.1.11}$$

Therefore we set

$$G = y^m \, e^{-y^2/2} \, F(y) \tag{10.1.12}$$

and find the differential equation satisfied by F. After some simple algebra we find

$$\frac{d^2 F}{dy^2} + \left(\frac{2m+1}{y} - 2y\right)\frac{dF}{dy} + (\lambda - 2m - 2)F = 0 . \tag{10.1.13}$$

Next, we look for a solution as a power series in y.

$$F = \sum_n a_n y^n \; . \tag{10.1.14}$$

Substituting in the equation for F and equating the coefficients of equal powers of y we find

$$a_{n+2} = \frac{2(n+m+1) - \lambda}{(n+2)(2m+2+n)} \; . \tag{10.1.15}$$

If this series does not terminate it leads to behaviour like $\exp(y^2)$ for large y and is therefore unacceptable. Therefore the series must terminate and we require that

$$\lambda = \frac{2E}{\hbar\omega} = 2(n+m+1) . \tag{10.1.16}$$

Thus, the allowed energies are

$$E = (n+m+1)\hbar\omega . \tag{10.1.17}$$

10.2   Attractive Exponential Potential

Consider the attractive potential

$$V(r) = -V_0\, e^{-\alpha r}$$

for $l = 0$. This is one of the few solvable problems.

Hint: Change variables to $u = e^{-\alpha r}$. The resultant equation is Bessel's equation. Discuss carefully the boundary conditions to be obeyed by

$$\phi(u) = R(r)$$

and find the equation that determines the energy eigenvalues.

Solution

The Schrödinger equation, for the s-wave ($l = 0$), reads

$$-\frac{\hbar^2}{2m}\frac{1}{r}\frac{d^2}{dr^2}(r\psi) - V_0\, e^{-\alpha r}\,\psi = E\psi\ . \tag{10.2.18}$$

We now set

$$r\psi = R\ ,\quad \frac{2mE}{\hbar^2} = -\kappa^2\ ,\quad \frac{2mV_0}{\hbar^2} = \beta^2\ . \tag{10.2.19}$$

Then, the equation reads

$$\frac{d^2 R}{dr^2} + \beta^2\, e^{-\alpha r}\, R = \kappa^2 R\ . \tag{10.2.20}$$

Let

$$u = \frac{2\beta}{\alpha}\, e^{-\alpha r/2}\ \ \text{and}\ \ F(u) = R(r)\ . \tag{10.2.21}$$

Then, we find that the equation for F becomes

$$\alpha^2 u^2 \frac{d^2 F}{du^2} + \alpha^2 u \frac{dF}{du} + \alpha^2 u^2 F = \kappa^2 F\ . \tag{10.2.22}$$

This can be rewritten

$$\frac{d^2 F}{du^2} + \frac{1}{u}\frac{dF}{du} + \left(1 - \frac{\kappa^2/\alpha^2}{u^2}\right) F = 0\ . \tag{10.2.23}$$

This is Bessels equation of order κ/α. The boundary conditions derive from the boundary conditions on R. Thus,

$$R(0) = 0\ \ \text{and}\ \ R(r) \to 0\ \text{for}\ r \to \infty\ . \tag{10.2.24}$$

This means that

$$F(2\beta/\alpha) = 0\ \ \text{and}\ \ F(0) = 0. \tag{10.2.25}$$

Therefore the solutions are

$$F(u) = A\, J_{\kappa/\alpha}(u) \tag{10.2.26}$$

with the quantization of energy given by

$$J_{\kappa/\alpha}(2\beta/\alpha) = 0\ . \tag{10.2.27}$$

10.3 Reduction of the Two-body Problem

Show that the Hamiltonian

$$H = \frac{\vec{p}_1^{\,2}}{2m_1} + \frac{\vec{p}_2^{\,2}}{2m_2} + V(\vec{r}_1 - \vec{r}_2)$$

reduces to the Hamiltonian

$$H = \frac{\vec{P}^{\,2}}{2M} + \frac{\vec{p}^{\,2}}{2m} + V(\mathbf{r})$$

under the transformations

$$\vec{R} = \frac{m_1 \vec{r}_1 + m_2 \vec{r}_2}{m_1 + m_2}\ , \tag{10.3.28}$$

$$\vec{r} = \vec{r}_1 - \vec{r}\ , \tag{10.3.29}$$

$$M = m_1 + m_2\ , \tag{10.3.30}$$

and

$$m = \frac{m_1 m_2}{m_1 + m_2}\ . \tag{10.3.31}$$

Solution

The Hamiltonian is

$$H = \frac{\vec{p}_1^{\,2}}{2m_1} + \frac{\vec{p}_2^{\,2}}{2m_2} + V(\vec{r}_1 - \vec{r}_2)\ . \tag{10.3.32}$$

We define

$$\vec{R} = \frac{m_1 \vec{r}_1 + m_2 \vec{r}_2}{M}\ , \quad M = m_1 + m_2 \tag{10.3.33}$$

$$\vec{r} = \vec{r}_1 - \vec{r}_2\ , \quad m = \frac{m_1 m_2}{m_1 + m_2}\ . \tag{10.3.34}$$

Now consider any differentiable function

$$F(\vec{r}_1, \vec{r}_2) = F(x_1, y_1, z_1; x_2, y_2, z_2)\ . \tag{10.3.35}$$

Then,

$$\frac{\partial F}{\partial x_1} = \frac{\partial F}{\partial X}\frac{\partial X}{\partial x_1} + \frac{\partial F}{\partial x}\frac{\partial x}{\partial x_1} = \frac{m_1}{M}\frac{\partial F}{\partial X} + \frac{\partial F}{\partial x} \tag{10.3.36}$$

and

$$\frac{\partial^2 F}{\partial x_1^2} = \left(\frac{m_1}{M}\right)^2 \frac{\partial^2 F}{\partial X^2} + \frac{2m_1}{M}\frac{\partial^2 F}{\partial x \partial X} + \frac{\partial^2 F}{\partial x^2}\ . \tag{10.3.37}$$

Thus we find,

$$\nabla_1^2 F = \left(\frac{m_1}{M}\right)^2 \nabla_X^2 F + \frac{2m_1}{M} \nabla_X \cdot \nabla_x F + \nabla_x^2 F \ . \qquad (10.3.38)$$

Similarly we find

$$\frac{\partial F}{\partial x_2} = \frac{\partial F}{\partial X} \frac{\partial X}{\partial x_2} + \frac{\partial F}{\partial x} \frac{\partial x}{\partial x_2} = \frac{m_2}{M} \frac{\partial F}{\partial X} - \frac{\partial F}{\partial x} \qquad (10.3.39)$$

and

$$\frac{\partial^2 F}{\partial x_2^2} = \left(\frac{m_2}{M}\right)^2 \frac{\partial^2 F}{\partial X^2} - \frac{2m_2}{M} \frac{\partial^2 F}{\partial x \partial X} + \frac{\partial^2 F}{\partial x^2} \qquad (10.3.40)$$

So,

$$\nabla_2^2 F = \left(\frac{m_2}{M}\right)^2 \nabla_X^2 F - \frac{2m_2}{M} \nabla_X \cdot \nabla_x F + \nabla_x^2 F \ . \qquad (10.3.41)$$

Therefore,

$$\begin{aligned}
\left(\frac{1}{2m_1}\nabla_1^2 + \frac{1}{2m_2}\nabla_1^2\right) F &= \frac{1}{2M}\nabla_X^2 F + \frac{1}{2}\left(\frac{1}{m_1} + \frac{1}{m_2}\right)\nabla_x^2 F \\
&= \frac{1}{2M}\nabla_X^2 F + \frac{1}{2m}\nabla_x^2 F \ . \qquad (10.3.42)
\end{aligned}$$

So we see that

$$H = \frac{\vec{P}^2}{2M} + \frac{\vec{p}^2}{2m} + V(\vec{r}) \ . \qquad (10.3.43)$$

10.4   Particle in a Spherical Potential Well

A particle is in a spherical potential well

$$V(r) = \begin{cases} -V_0 & \text{for} \quad r < a \\ 0 & \text{for} \quad r > a \end{cases} \ .$$

Find the transcendental equation which yields the energy eigenvalue for the state with angular momentum l.

What is the minimum degeneracy of this state?

If a proton and neutron are bound in an $l = 0$ state with an energy of 2.2 MeV, determine V_0 given that $a \approx 2 \times 10^{-13}$ cm.

Solution

The radial wave equation becomes

$$\begin{aligned}
-\frac{\hbar^2}{2m}\frac{d^2u}{dr^2} + \frac{l(l+1)\hbar^2}{2mr^2}u - V_0 u &= -Eu \quad \text{for} \quad r < a \\
-\frac{\hbar^2}{2m}\frac{d^2u}{dr^2} + \frac{l(l+1)\hbar^2}{2mr^2}u &= -Eu \qquad\qquad \text{for} \quad r > a \ . \qquad (10.4.44)
\end{aligned}$$

Here we have written $-E$ for the energy since we are looking for bound states. Defining as usual

$$k^2 = \frac{2mE}{\hbar^2} \quad , \quad K^2 = \frac{2m(V_0 - E)}{\hbar^2} \qquad (10.4.45)$$

the equations become

$$\frac{d^2u}{dr^2} + \frac{l(l+1)}{r^2}u + K^2u = 0 \quad \text{for } r < a$$

$$\frac{d^2u}{dr^2} + \frac{l(l+1)}{r^2}u - k^2u = 0 \quad \text{for } r > a . \qquad (10.4.46)$$

The solutions are

$$u = A\sqrt{r}Z_{l+1/2}(Kr) \quad \text{for } r < a$$

$$u = B\sqrt{r}Z_{l+1/2}(ikr) \quad \text{for } r > a \qquad (10.4.47)$$

where $Z_{l+1/2} = J_{l+1/2}$ or $N_{l+1/2}$. For $r < a$ we must choose $J_{l+1/2}$ whereas for $r > a$ we need $H_{l+1/2}^{(1)}(ikr) = J_{l+1/2}(ikr) + iN_{l+1/2}(ikr) = K_{l+1/2}(kr)$ since this damps exponentially. Also at $r = a$ both u and its first derivative are continuous. Therefore, it suffices to match the logarithmic derivatives

$$\frac{\frac{d}{dr}\left[\sqrt{r}J_{l+1/2}(Kr)\right]}{\sqrt{r}J_{l+1/2}(Kr)}\Bigg|_{r=a} = \frac{\frac{d}{dr}\left[\sqrt{r}K_{l+1/2}(kr)\right]}{\sqrt{r}K_{l+1/2}(kr)}\Bigg|_{r=a} \qquad (10.4.48)$$

This is the transcendental equation for the eigenvalue E.

Since this equation does not involve the magnetic quantum number m, the minimum degeneracy of each level $E_{n,l}$ is $2l + 1$ corresponding to the possible m values $m = 0, \pm 1, \pm 2, \ldots, \pm l$. For $l = 0$ we do not need Bessel functions since the differential equations are simply

$$\frac{d^2u}{dr^2} + K^2u = 0 \quad \text{for } r < a$$

$$\frac{d^2u}{dr^2} - k^2u = 0 \quad \text{for } r > a . \qquad (10.4.49)$$

Thus, we get

$$u = A\sin Kr \quad r > a$$

$$u = Be^{-kr} \quad r > a . \qquad (10.4.50)$$

Matching the logarithmic derivatives at $r = a$ we get

$$\tan Ka = -\frac{K}{k} . \qquad (10.4.51)$$

Let

$$\frac{K}{k} = z . \qquad (10.4.52)$$

Then we have

$$\tan kaz = -z \ .$$

Substituting the values appropriate for a deuteron and remembering that we have to use the reduced mass which is about $1/2$ the proton mass we find that $ka = 0.46$. Now a glance at a table of $\pm z$ versus $\tan kaz$ shows that we want z somewhat greater than $\pi/2$. This corresponds to $z = \pi/(2ka) = 3.41$. Choosing a series of slightly larger z values we get the following table. Therefore, to an

z	$\tan kaz$
3.6	−11.71
3.8	−5.58
3.9	−4.40
3.95	−3.97

accuracy of about 1% we have $z = 3.95$. Then, we get

$$V_0 = E + z^2 E = 16.6E = 36.5 \ \text{MeV} \ . \tag{10.4.53}$$

10.5   Particle on Surface of a Cylinder

A particle is free to move on the surface of a circular cylinder of radius R. The Laplacian in cylindrical coordinates is given by:

$$\nabla^2 = \frac{\partial^2}{\partial r^2} + \frac{1}{r}\frac{\partial}{\partial r} + \frac{1}{r^2}\frac{\partial^2}{\partial \varphi^2} + \frac{\partial^2}{\partial z^2} \ .$$

Find the energy eigenvalues and eigenfunctions for this motion.

Solution

Since the particle has to move on the surface of the cylinder we have that $r = R$, the radius of the cylinder. Thus, the Schrödinger equation reads

$$\frac{1}{R^2}\frac{\partial^2 \psi}{\partial \varphi^2} + \frac{\partial^2 \psi}{\partial z^2} = -k^2 \psi \ . \tag{10.5.54}$$

Here we have set

$$k^2 = \frac{2ME}{\hbar^2} \ . \tag{10.5.55}$$

Since L_z commutes with the Hamiltonian for a free particle we can write,

$$\psi = e^{im\varphi} F(z) \tag{10.5.56}$$

where F satisfies

$$\frac{d^2 F}{dz^2} + \left(k^2 - \frac{m^2}{R^2}\right) F = 0 \ . \tag{10.5.57}$$

For a free particle the total energy

$$E \geq \frac{L_z^2}{2MR^2}$$

(10.5.58)

Therefore,

$$k^2 \geq \frac{m^2}{R^2}$$

(10.5.59)

and the eigenfunctions are given by

$$\psi = e^{im\varphi} \left[A \, \exp(iz\sqrt{k^2 - m^2/R^2}) + B \, \exp(-iz\sqrt{k^2 - m^2/R^2}) \right] . \quad (10.5.60)$$

10.6 Expectation Values: Electron in a H-atom

An electron in the Coulomb field of a proton is in a state described by the ket:

$$\frac{1}{6} \left[4|1,0,0\rangle + 3|2,1,1\rangle - |2,1,0\rangle + \sqrt{10}|2,1,-1\rangle \right]$$

where the labelling is $|n, l, m\rangle$. Find
a) The expectation value of the energy.
b) The expectation value of L^2.
c) The expectation value of L_z.

Solution

We are given

$$|\psi\rangle = \frac{1}{6} \left[4|1,0,0\rangle + 3|2,1,1\rangle - |2,1,0\rangle + \sqrt{10}|2,1,-1\rangle \right] . \quad (10.6.61)$$

a) The energy expectation value of the electron, neglecting the motion of the proton-electron center of mass, is given by

$$
\begin{aligned}
\langle H \rangle &= \langle \psi | H | \psi \rangle \\
&= \frac{1}{36} \left[16\frac{1}{1^2} + 9\frac{1}{2^2} + 1\frac{1}{2^2} + 10\frac{1}{2^2} \right] \frac{1}{2}mc^2\alpha^2 \\
&= \frac{7}{24}mc^2\alpha^2 .
\end{aligned}
$$

(10.6.62)

b)

$$\langle L^2 \rangle = \frac{1}{36} [16 \cdot 0 + 9 \cdot 2 + 1 \cdot 2 + 10 \cdot 2] \, \hbar^2 = \frac{10}{9}\hbar^2 . \quad (10.6.63)$$

c)

$$\langle L_z \rangle = \frac{1}{36} [16 \cdot 0 + 9 \cdot 1 + 1 \cdot 0 + 10 \cdot (-1)] \, \hbar = -\frac{1}{36}\hbar . \quad (10.6.64)$$

10.7   Parity in Spherical Coordinates

Show that under a parity transformation

$$\theta \to \theta' = \pi - \theta$$

$$\varphi \to \varphi' = \varphi + \pi \ .$$

Solution

Under a parity transformation we have that

$$
\begin{array}{lll}
x = r \sin\theta \cos\varphi & \to & x' = -x = r\sin\theta'\cos\varphi' \\
y = r \sin\theta \sin\varphi & \to & y' = -y = r\sin\theta'\sin\varphi' \\
z = r \cos\theta & \to & z' = -z = r\cos\theta' \ .
\end{array}
\tag{10.7.65}
$$

Therefore,

$$\cos\theta' = -\cos\theta = \cos(\pi - \theta) \ . \tag{10.7.66}$$

Since $0 \le \theta \le \pi$ and $0 \le \theta' \le \pi$ we therefore get that

$$\theta' = \pi - \theta \ . \tag{10.7.67}$$

Using this result we have that

$$
\begin{aligned}
\cos\varphi' &= -\cos\varphi \\
\sin\varphi' &= -\sin\varphi \ .
\end{aligned}
\tag{10.7.68}
$$

Therefore,

$$\varphi' = \varphi + \pi \ . \tag{10.7.69}$$

10.8   Magnetic Moment due to Orbital Motion

Consider an electron in the state $|n, l, m\rangle$ of a hydrogen atom.
a) Compute the current in the direction φ.
b) Compute the current flowing through an element of area normal to $\hat{e}_\varphi$. Use this result to compute the magnetic moment of the electron due to its orbital motion.
Hint: The magnetic moment due to a current loop is the magnitude of the current times the area enclosed by the current loop and points in a direction given by the right hand rule in tracing the loop around the area.

Solution

a) The electric current in the direction $\hat{e}_\varphi$ due to an electron of charge $-e$ and mass M is given by

$$j_\varphi = \frac{-e\hbar}{2iM}\frac{1}{r\sin\theta}[\psi^*\frac{\partial\psi}{\partial\varphi} - \frac{\partial\psi^*}{\partial\varphi}\psi] \ . \tag{10.8.70}$$

For the state $|n,l,m\rangle$ the wavefunction is of the form

$$\psi_{n,l,m}(r,\theta,\varphi) = R_{n,l}(r)P_l^m(\cos\theta)\,e^{im\varphi} \ . \tag{10.8.71}$$

Therefore carrying out the required differentiation in (10.8.70) we see that

$$j_\varphi = \frac{-e\hbar}{M}m\frac{1}{r\sin\theta}|\psi_{n,l,m}(\vec{r})|^2 \ . \tag{10.8.72}$$

b) Consider an element of area $dA = dz\,d\rho$ normal to $\hat{e}_\varphi$. Here we have written

$$x = \rho\cos\varphi \quad , \quad y = \rho\sin\varphi \ . \tag{10.8.73}$$

Then we have

$$\rho = r\sin\theta \quad , \quad z = r\cos\theta \tag{10.8.74}$$

so that

$$dz\,d\rho = r\,dr\,d\theta \ . \tag{10.8.75}$$

The element of current through this area is

$$dI = j_\varphi dA = \frac{-e\hbar}{M}m\frac{1}{r\sin\theta}|\psi_{n,l,m}(\vec{r})|^2\,r\,dr\,d\theta \ . \tag{10.8.76}$$

The area enclosed by the current loop is

$$\pi\rho^2 = \pi r^2\sin^2\theta$$

and the normal to this area points in the $\hat{e}_z$ direction. Thus, the magnetic moment due to the electron's orbital motion is

$$\begin{aligned}
\vec{\mu} &= \frac{-e\hbar}{M}m\hat{e}_z\int|\psi_{n,l,m}(\vec{r})|^2\,\pi r^2\,dr\,\sin\theta\,d\theta \\
&= -\frac{e\hbar}{2M}m\hat{e}_z \ .
\end{aligned} \tag{10.8.77}$$

Here we have used the fact that the wavefunction is normalized to unity.

10.9   Spherical Square Well

For a particle in an s-state ($l = 0$) acted on by a spherically symmetric square well potential

$$V = \begin{cases} -V_0 & 0 \le r < a \\ 0 & r > a \end{cases} \ .$$

a) Find the equation for the bound state energies.
b) What is the minimum strength for V_0 in order that at least one bound state exists?

Solution

a) The Hamiltonian is

$$H = \frac{\bar{p}^2}{2m} + V(r) \; . \tag{10.9.78}$$

The eigenvalue equation for the $l = 0$ bound states now becomes

$$\frac{d^2 u}{dr^2} + k^2 u \; = \; 0 \quad r < a$$

$$\frac{d^2 u}{dr^2} - \kappa^2 u \; = \; 0 \quad r > a \tag{10.9.79}$$

where

$$k^2 = \frac{2m(E + V_0)}{\hbar^2} \quad , \quad \kappa^2 = \frac{2m|E|}{\hbar^2} \tag{10.9.80}$$

and

$$u = r\psi \tag{10.9.81}$$

so that $u(0) = 0$. The solutions are

$$u \; = \; A \sin kr \quad r < a$$

$$u \; = \; B e^{-\kappa r} \quad r > a \; . \tag{10.9.82}$$

Imposing the condition of continuity of u and its derivative at $r = a$ by equating the logarithmic derivatives we get the equation for the bound state eigenvalues

$$k \cot ka = -\kappa \quad \text{or} \quad \tan ka = -\frac{k}{\kappa} \; . \tag{10.9.83}$$

This completes part a).

b) For a solution to exist requires that the equation

$$\tan ka = -\frac{k}{\kappa} = -\sqrt{\frac{V_0 - |E|}{|E|}} \tag{10.9.84}$$

should have a solution. This requires that

$$ka \geq \frac{\pi}{2} \quad \text{or} \quad \left(\frac{2mV_0}{\hbar^2} - \frac{2m|E|}{\hbar^2} \right) a^2 \geq \frac{\pi^2}{4} \; . \tag{10.9.85}$$

So, for $|E| > 0$ we need

$$V_0 \geq \frac{\pi^2}{4} \frac{\hbar^2}{2ma^2} \; . \tag{10.9.86}$$

This result is also clearly the minimum condition since using the equal sign, by taking the limit from above, yields the eigenvalue $E = 0$.

10.10 Binding Energy and Potential

A spinless particle (mass m) moves in a short range central potential $V(r)$. The wavefunction describing the state of the particle is

$$\psi = A\frac{e^{-\alpha r} - e^{-\beta r}}{r}$$

where A is some constant and $\alpha < \beta$ are both positive constants.
a) What is the angular momentum of this particle?
b) What is the binding energy of this state?
c) What is the potential that produced this state?

Solution

a) Since the wavefunction has no angular dependence it follows that

$$L^2\psi = 0 \tag{10.10.87}$$

so that $l = 0$.
b) To find the binding energy we look at the asymptotic behaviour of this wavefunction for large r. This must be of the form

$$\frac{1}{r}\exp\left[-\sqrt{-2mE}/\hbar\right] \ .$$

Since $\beta > \alpha$ we find that

$$\psi \to A\frac{e^{-\alpha r}}{r} \quad \text{for} \quad r \to \infty \ . \tag{10.10.88}$$

Therefore,

$$E = -\frac{\hbar^2\alpha^2}{2m} \ . \tag{10.10.89}$$

c) To find the potential that produced this state we simply stick the wavefunction and the given energy into the Schrödinger equation and solve for the potential. Thus, we find

$$-\frac{\hbar^2}{2m}\frac{1}{r}\frac{d^2(r\phi)}{dr^2} = -\frac{\hbar^2}{2m}A\frac{\alpha^2 e^{-\alpha r} - \beta^2 e^{-\beta r}}{r}$$

$$= (E - V)\phi = A\frac{e^{-\alpha r} - e^{-\beta r}}{r}$$

$$= \left[-\frac{\hbar^2\alpha^2}{2m} - V\right]A\frac{e^{-\alpha r} - e^{-\beta r}}{r} \ . \tag{10.10.90}$$

Therefore,

$$V(r) = -\frac{\hbar^2}{2m}(\beta^2 - 1)\frac{e^{-\beta r}}{e^{-\alpha r} - e^{-\beta r}} \ . \tag{10.10.91}$$

10.11   Generating Function: Laguerre Polynomials

a) Use the integral representation for the Laguerre polynomials [10.1]

$$L_n(x) = \frac{n!}{2\pi i} \oint e^{-xt} \frac{(t+1)^n}{t^{n+1}} \, dt$$

to derive the generating function

$$f(x,s) = \sum_{n=0}^{\infty} \frac{L_n(x)}{n!} s^n = \frac{1}{1-s} \exp\left[\frac{-sx}{1-s}\right] \qquad (10.11.92)$$

for these polynomials. The contour in the integral is a circle about the origin.
b) By differentiating (10.11.92) l times and using the definition

$$L_{n-l}^l(x) = (-1)^l \frac{d^l}{dx^l} L_n(x) \qquad (10.11.93)$$

obtain the generating formula

$$\sum_{n=0}^{\infty} \frac{L_n^l(x)}{(n+l)!} s^n = \frac{1}{(1-s)^{l+1}} \exp\left[\frac{-sx}{1-s}\right] . \qquad (10.11.94)$$

Solution

a) By changing the integration variable from t to $-xt$ in (10.11.92) we get

$$f(x,s) = \frac{1}{2\pi i} \oint e^t \sum_{n=0}^{\infty} \left(s\frac{t-x}{t} \right)^n \frac{dt}{t} . \qquad (10.11.95)$$

We now take $|s| < 1$ to insure that for sufficiently large t the series converges.
Thus,

$$
\begin{aligned}
f(x,s) &= \frac{1}{2\pi i} \oint e^t \frac{1}{1 - s(t-x)/t} \frac{dt}{t} \\
&= \frac{1}{2\pi i} \oint e^t \frac{dt}{t - s(t-x)} \\
&= \frac{1}{2\pi i} \frac{1}{1-s} \oint e^t \frac{dt}{t + sx/(1-s)} .
\end{aligned}
\qquad (10.11.96)
$$

The integrand has a simple pole at

$$t = -\frac{sx}{1-s}$$

so we immediately get the desired result

$$f(x,s) = \sum_{n=0}^{\infty} \frac{L_n(x)}{n!} s^n = \frac{1}{1-s} \exp\left[\frac{-sx}{1-s}\right] . \qquad (10.11.97)$$

b) If we differentiate formula (10.11.97) l times we get

$$\sum_{n=l}^{\infty} \frac{1}{n!} \frac{d^l L_n(x)}{dx^l} s^n = \left(\frac{-s}{1-s}\right)^l \frac{1}{1-s} \exp[\frac{-sx}{1-s}] \ . \tag{10.11.98}$$

Now using the definition of $L_{n-l}^l(x)$, namely,

$$L_{n-l}^l(x) = (-1)^l \frac{d^l L_n(x)}{dx^l} \tag{10.11.99}$$

we see that we have

$$\sum_{n=l}^{\infty} \frac{L_{n-l}^l(x)}{n!} s^n = \left(\frac{1}{1-s}\right)^{l+1} \exp[\frac{-sx}{1-s}] \ . \tag{10.11.100}$$

10.12 Normalization of Hydrogen Wavefunction

a) Use the generating function for the associated Laguerre polynomials

$$\sum_{n=0}^{\infty} \frac{L_n^l(x)}{(n+l)!} s^n = \left(\frac{1}{1-s}\right)^{l+1} \exp[\frac{-sx}{1-s}]$$

(see problem 10.11) to evaluate the integral

$$\int_0^{\infty} e^{-x} x^p L_{n-l-1}^{2l+1}(x) L_{n'-l-1}^{2l+1}(x) \, dx$$

for $p = 2l + 1$ as well as $p = 2l + 2$.
The value $p = 2l + 2$ yields the normalization for a hydrogenic wavefunction.

Solution

a) We first rewrite the generating formula in the form

$$\sum_{n=l+1}^{\infty} \frac{L_{n-l-1}^{2l+1}(x)}{(n+l)!} s^{n-l-1} = \left(\frac{1}{1-s}\right)^{2l+2} \exp[\frac{-sx}{1-s}] \tag{10.12.101}$$

and then use this given generating formula twice. This gives,

$$\sum_{n,n'=l+1}^{\infty} \frac{s^{n-l-1} t^{n'-l-1}}{(n+l)!\,(n'+l)!} \int_0^{\infty} e^{-x} x^p L_{n-l-1}^{2l+1}(x) L_{n'-l-1}^{2l+1}(x) \, dx$$

$$= \left(\frac{1}{1-s}\right)^{2l+2} \left(\frac{1}{1-t}\right)^{2l+2} \int_0^{\infty} x^p \exp\left[-x \frac{1-st}{(1-s)(1-t)}\right] dx$$

$$= (1-s)^{p-2l-1}(1-t)^{p-2l-1}(1-st)^{-(p+1)} \int_0^{\infty} z^p e^{-z} \, dz$$

$$= (1-s)^{p-2l-1}(1-t)^{p-2l-1}(1-st)^{-(p+1)} p! \ . \tag{10.12.102}$$

If $p = 2l + 1$ then the right hand side reduces to

$$\frac{(2l+1)!}{(1-st)^{2l+2}} = \sum_{n=0}^{\infty} \frac{(n+2l+1)!}{n!}(st)^n$$

$$= \sum_{n=l+1}^{\infty} \frac{(n+l)!}{(n-l-1)!}(st)^{n-l-1} \qquad (10.12.103)$$

where we have used the binomial expansion

$$\frac{1}{(1-x)^{p+1}} = \sum_{n=0}^{\infty} \frac{(n+p)!}{p!\,n!}x^n \quad . \qquad (10.12.104)$$

For this to equal the left hand side for all values of s and t we clearly need $n = n'$ so that the left hand side reduces to

$$\sum_{n=l+1}^{\infty} \frac{(st)^{n-l-1}}{[(n+l)!]^2} \int_0^{\infty} e^{-x}\, x^{2l+1}\, L_{n-l-1}^{2l+1}(x) L_{n-l-1}^{2l+1}(x)\, dx \quad . \qquad (10.12.105)$$

Hence, we find

$$\int_0^{\infty} e^{-x}\, x^{2l+1}\, L_{n-l-1}^{2l+1}(x) L_{n'-l-1}^{2l+1}(x)\, dx = \frac{[(n+l)!]^3}{(n-l-1)!} \delta_{n,n'} \quad . \qquad (10.12.106)$$

If we now put $p = 2l + 2$ and $n = n'$ we find that the right hand side yields

$$RHS = (2l+2)!(1-s)(1-t)(1-st)^{-2l+3}$$

$$= (1-s-t+st) \sum_{n=0}^{\infty} \frac{(n+2l+2)!}{n!}(st)^n$$

$$= (1-s-t+st) \sum_{n=l+1}^{\infty} \frac{(n+l+1)!}{(n-l-1)!}(st)^{n-l-1} \quad . \qquad (10.12.107)$$

On the other hand, the left hand side reads

$$LHS = \sum_{n,n'=l+1}^{\infty} \left\{ \frac{(st)^{n-l-1)}}{[(n+l)!]^2} \delta_{n,n'} + \frac{s(st)^{n-l-1)}}{(n+l+1)[(n+l)!]^2} \delta_{n,n'+1} \right.$$

$$+ \left. \frac{t(st)^{n'-l-1)}}{(n'+l+1)[(n'+l)!]^2} \delta_{n,n'-1} \right\}$$

$$\times \int_0^{\infty} e^{-x}\, x^{2l+2}\, L_{n-l-1}^{2l+1}(x) L_{n'-l-1}^{2l+1}(x)\, dx \quad . \qquad (10.12.108)$$

Now, the right hand side may be further rewritten as

$$RHS = \sum_{n,n'=l+1}^{\infty} \left\{ \frac{(n+l+1)!}{(n-l-1)!}\left[(st)^{n-l-1} + (st)^{n-l}\right]\delta_{n,n'} \right.$$

$$- \frac{(n+l+1)!}{(n-l-1)!}s\left(st^{n-l-1}\delta_{n,n'+1}\right)$$

$$- \left. \frac{(n'+l+1)!}{(n'-l-1)!}t\left(st^{n'-l-1}\delta_{n,n'-1}\right) \right\} \quad . \qquad (10.12.109)$$

Choosing the diagonal terms $n = n'$ we get

$$\sum_{n=l+1}^{\infty} \frac{(st)^{n-l-1}}{[(n+l)!]^2} \int_0^{\infty} e^{-x} x^{2l+2} L_{n-l-1}^{2l+1}(x) L_{n-l-1}^{2l+1}(x)\, dx$$

$$= \sum_{n=l+1}^{\infty} \left[\frac{(n+l+1)!}{(n-l-1)!} + \frac{(n+l)!}{(n-l-2)!} \right] (st)^{n-l-1}$$

$$= \sum_{n=l+1}^{\infty} \frac{(n+l)!}{(n-l-1)!} 2n (st)^{n-l-1} \ . \tag{10.12.110}$$

This immediately yields that

$$\int_0^{\infty} e^{-x} x^{2l+2} L_{n-l-1}^{2l+1}(x) L_{n-l-1}^{2l+1}(x)\, dx = 2n \frac{[(n+l)!]^3}{(n-l-1)!} \ . \tag{10.12.111}$$

Hence, we obtain the desired normalization for the hydrogenic wavefunctions.

$$A_{n,l} = \sqrt{\left(\frac{2Z}{na_0}\right)^3 \frac{(n-l-1)!}{2n[(n+l)!]^3}} \ . \tag{10.12.112}$$

10.13 Kramers' Relation

a) Use the radial equation for the hydrogenic atom

$$R_{nl}'' + \frac{2}{r} R_{nl}' - \left[\frac{l(l+1)}{r^2} - \frac{1}{na} \frac{1}{r} + \frac{1}{n^2 a^2} \right] R_{nl} = 0$$

to derive the recursion relation

$$\frac{p+1}{n^2} \langle r^p \rangle - (2p+1) a \langle r^{p-1} \rangle + \frac{p}{4} \left[(2l+1)^2 - p^2 \right] a^2 \langle r^{p-2} \rangle = 0 \quad p \geq 1 \ .$$

This result is also known as the *Pasternach relation* [10.2]. Here,

$$\langle r^p \rangle = \int_0^{\infty} |R_{nl}|^2 r^{p+2}\, dr \ .$$

b) Use this result and the results of problem 10.12 to calculate

$$\langle r \rangle, \ \langle r^{-1} \rangle, \text{ and } \langle r^2 \rangle \ .$$

Solution

a) If we write the radial equation in the form

$$R_{nl}'' = -\frac{2}{r} R_{nl}' + \left[\frac{l(l+1)}{r^2} - \frac{2}{a} \frac{1}{r} + \frac{1}{n^2 a^2} \right] R_{nl} \tag{10.13.113}$$

and multiply it by $r^{p+2}R_{nl}$ and integrate over all r we get (after dropping the subscripts on R_{nl})

$$
\begin{aligned}
\int_0^\infty R'' r^{p+2} R \, dr &= -2 \int_0^\infty R' r^{p+1} R \, dr + l(l+1)\langle r^{p-2}\rangle \\
&\quad - \frac{2}{a}\langle r^{p-1}\rangle + \frac{1}{n^2 a^2}\langle r^p\rangle .
\end{aligned}
\tag{10.13.114}
$$

If we integrate the left hand side of this equation by parts we obtain

$$
\int_0^\infty R'' r^{p+2} R \, dr = -(p+2)\int_0^\infty R' r^{p+1} R \, dr - \int_0^\infty R' r^{p+2} R' \, dr .
\tag{10.13.115}
$$

After another integration by parts we further find

$$
\begin{aligned}
\int_0^\infty R'' r^{p+2} R \, dr &= (p+2)\int_0^\infty R\left[(p+1)r^p R + r^{p+1} R'\right] dr \\
&\quad + \int_0^\infty R\left[(p+2)r^{p+1} R' + r^{p+2} R''\right] dr .
\end{aligned}
\tag{10.13.116}
$$

Hence, it follows that

$$
\int_0^\infty R' r^{p+1} R \, dr = -\frac{p+1}{2}\langle r^{p-2}\rangle .
\tag{10.13.117}
$$

We now multiply the radial equation (10.13.113) by $r^{p+3} R'$ and again integrate over all r to get

$$
\begin{aligned}
\int_0^\infty R'' r^{p+3} R' \, dr &= \frac{1}{2}\int_0^\infty r^{p+3} d(R')^2 = -\frac{p+3}{2}\int_0^\infty R' r^{p+2} R' \, dr \\
&= -2 \int_0^\infty R' r^{p+2} R' \, dr + l(l+1)\int_0^\infty R' r^{p+1} R \, dr \\
&\quad - \frac{2}{a}\int_0^\infty R' r^{p+2} R \, dr + \frac{1}{n^2 a^2}\int_0^\infty R' r^{p+3} R \, dr \\
&= -2 \int_0^\infty R' r^{p+2} R' \, dr - l(l+1)\frac{p+1}{2}\langle r^{p-2}\rangle \\
&\quad + \frac{2p+2}{a}\frac{1}{2}\langle r^{p-1}\rangle - \frac{1}{n^2 a^2}\frac{p+3}{2}\langle r^p\rangle .
\end{aligned}
\tag{10.13.118}
$$

Therefore,

$$
\begin{aligned}
&\int_0^\infty R' r^{p+2} R' \, dr \\
&= l(l+1)\frac{p+1}{p-1}\langle r^{p-2}\rangle - \frac{2p+2}{a}\frac{1}{p-1}\langle r^{p-1}\rangle + \frac{1}{n^2 a^2}\frac{p+3}{p-1}\langle r^p\rangle .
\end{aligned}
\tag{10.13.119}
$$

Combining the results of (10.13.114), (10.13.115), and (10.13.119) we get

$$
\frac{(p+1)}{2}(p+2)\langle r^{p-2}\rangle - l(l+1)\frac{(p+1)}{p-1}\langle r^{p-2}\rangle
$$

$$+\frac{2}{a}\frac{p+2}{p-1}\langle r^{p-1}\rangle - \frac{1}{n^2 a^2}\frac{p+3}{p-1}\langle r^p\rangle$$
$$= (p+1)\langle r^{p-2}\rangle + l(l+1)\langle r^{p-2}\rangle$$
$$-\frac{2}{a}\langle r^{p-1}\rangle + \frac{1}{n^2 a^2}\langle r^p\rangle \ . \tag{10.13.120}$$

Simplifying this result yields the desired expression

$$\frac{p+1}{n^2}\langle r^p\rangle - (2p+1)a\langle r^{p-1}\rangle + \frac{p}{4}\left[(2l+1)^2 - p^2\right]a^2\langle r^{p-2}\rangle \ . \tag{10.13.121}$$

Throughout the calculation we have dropped the terms resulting from the integration by parts. This required that $p \geq 1$. Hence, that condition.

For an alternate derivation that relies on the generalized Hellmann-Feynman theorem see the article by Balasubramanian [10.3].

b) The radial wavefunctions are

$$R_{n,l}(r) = \sqrt{\frac{(n-l-1)!}{2n[(n+l)!]^3}\frac{8}{(na)^3}}\left(\frac{2r}{na}\right)^l e^{-r/na}L_{n-l-1}^{2l+1}(2r/na). \tag{10.13.122}$$

Also, from problem 10.11 we have that

$$\int_0^\infty e^{-2r/na}\left(\frac{2r}{na}\right)^{2l+2}\left(L_{n-l-1}^{2l+1}(2r/na)\right)^2 d(2r/na)$$
$$= \frac{2n[(n+l)!]}{(n-l-1)!} \tag{10.13.123}$$

as well as

$$\int_0^\infty e^{-2r/na}\left(\frac{2r}{na}\right)^{2l+1}\left(L_{n-l-1}^{2l+1}(2r/na)\right)^2 d(2r/na)$$
$$= \frac{[(n+l)!]}{(n-l-1)!} \ . \tag{10.13.124}$$

Equation (10.13.123) is just the normalization of the wavefunction and states that

$$\langle r^0\rangle = 1 \tag{10.13.125}$$

while equation (10.13.124) states that

$$\langle r^{-1}\rangle = \frac{1}{n^2 a} \ . \tag{10.13.126}$$

Now, using these results and (10.13.121) with $p = 1$ we get

$$\frac{2}{n^2}\langle r\rangle = 3a - \frac{1}{4}\left[(2l+1)^2 - 1\right]a^2\frac{1}{n^2 a} \ . \tag{10.13.127}$$

After simplification this reads

$$\langle r\rangle = \frac{1}{2}a\left[3n^2 - l(l+1)\right] \ . \tag{10.13.128}$$

Similarly, putting $p = 2$ in (10.13.121) we get

$$\frac{3}{n^2}\langle r^2 \rangle = \frac{5}{2}a^2 \left[3n^2 - l(l+1) \right] - \frac{1}{2} \left[(2l+1)^2 - 4 \right] a^2 \ . \qquad (10.13.129)$$

Simplification now yields

$$\langle r^2 \rangle = \frac{a^2}{2}n^2 \left[5n^2 + 1 - 3l(l+1) \right] \ . \qquad (10.13.130)$$

10.14   Quantum Mechanical Virial Theorem

a) Prove the quantum mechanical Virial Theorem for stationary states, namely

$$2\langle T \rangle = \langle \vec{r} \cdot \nabla V \rangle \ . \qquad (10.14.131)$$

Hint: Use the equation

$$i\hbar \frac{d}{dt}\langle A \rangle = \langle [A, H] \rangle \qquad (10.14.132)$$

where A is an operator, corresponding to the radial momentum, and H is the Hamiltonian.

b) Use the result of part a) to show that for a potential of the form

$$V(r) = V_0\, r^n$$

this yields

$$2\langle T \rangle = n\langle V \rangle \ .$$

Solution

a) The proof of the first part is similar to the proof used in classical mechanics. For a stationary state, the expectation value of $\vec{r} \cdot \vec{p}$, the radial component of momentum, is independent of t. Therefore, using (10.14.132) we have that

$$\frac{d}{dt}\langle \vec{r} \cdot \vec{p} \rangle = 0 = \frac{1}{i\hbar}\langle [\vec{r} \cdot \vec{p}, H] \rangle \ . \qquad (10.14.133)$$

But,

$$
\begin{aligned}
[\vec{r} \cdot \vec{p}, H] &= \left[xp_x + yp_y + zp_z, \frac{p_x^2 + p_y^2 + p_z^2}{2m} + V(\vec{r}) \right] \\
&= \frac{i\hbar}{m}\left(p_x^2 + p_y^2 + p_z^2 \right) - i\hbar \left(x\frac{\partial V}{\partial x} + y\frac{\partial V}{\partial y} + z\frac{\partial V}{\partial z} \right) \\
&= i\hbar 2T - i\hbar \vec{r} \cdot \nabla V \ . \qquad (10.14.134)
\end{aligned}
$$

Therefore,

$$2\langle T \rangle = \langle \vec{r} \cdot \nabla V \rangle \ . \qquad (10.14.135)$$

b) If

$$V(r) = V_0 \, r^n \tag{10.14.136}$$

then,

$$\vec{r} \cdot \nabla V(\vec{r}) = n V(\vec{r}) \ . \tag{10.14.137}$$

Therefore the desired result immediately follows and

$$2\langle T \rangle = n\langle V \rangle \ . \tag{10.14.138}$$

10.15 Ehrenfest Theorem for Angular Momentum

a) Show that for a particle evolving according to the Hamiltonian

$$H = \frac{\vec{p}^2}{2m} + V(\vec{r})$$

the rate of change of the expectation value of the angular momentum operator

$$\vec{L} = \vec{r} \times \vec{p}$$

is given by

$$\frac{d}{dt}\langle \vec{L} \rangle = -\langle \vec{r} \times \nabla V \rangle \ .$$

b) Under what conditions is this equation exactly the classical equation of motion for $\vec{L}$?

Solution

a) The result follows immediately if we use the equation for the time evolution of the expectation value of an operator A.

$$\frac{d}{dt}\langle A \rangle = \frac{1}{i\hbar}\langle [A, H] \rangle \ . \tag{10.15.139}$$

Thus,

$$\frac{d}{dt}\langle \vec{L} \rangle = \frac{1}{i\hbar}\langle [\vec{r} \times \vec{p}, H] \rangle \ . \tag{10.15.140}$$

But,

$$\begin{aligned}[\vec{r} \times \vec{p}, H] &= \vec{r} \times [\vec{p}, H] + [\vec{r}, H] \times \vec{p} \\ &= -i\hbar \vec{r} \times \nabla V \ . \end{aligned} \tag{10.15.141}$$

Therefore,

$$\frac{d}{dt}\langle \vec{L} \rangle = -\langle \vec{r} \times \nabla V \rangle \ . \tag{10.15.142}$$

b) For this result to be the same as the classical equation requires that, if

$$\vec{F}(\vec{r}) = -\nabla V$$

then,

$$\langle \vec{r} \times \vec{F} \rangle = \langle \vec{r} \rangle \times \vec{F}(\langle \vec{r} \rangle) \ . \tag{10.15.143}$$

This is possible only if $\vec{F}(\vec{r})$ is a constant vector, or parallel to $\vec{r}$.

10.16   Angular Momentum of a Two-Particle System

For a two-particle system with coordinates $\vec{x}_1$, $\vec{x}_2$ introduce the centre of mass and relative coordinates

$$\vec{X} = \frac{m_1 \vec{x}_1 + m_2 \vec{x}_2}{M} \quad , \quad \vec{x} = \vec{x}_1 - \vec{x}_2$$

where

$$M = m_1 + m_2 \quad \text{and} \quad m = \frac{m_1 m_2}{M}$$

are the total and reduced mass respectively.

Show that the total angular momentum of the two-particle system is

$$\vec{L} = \vec{X} \times \vec{P} + \vec{x} \times \vec{p}$$

where

$$\vec{P} = \frac{\hbar}{i} \nabla_X \quad , \quad \vec{p} = \frac{\hbar}{i} \nabla_x$$

are the centre of mass momentum and relative momentum respectively. Give an interpretation of this result.

Solution

Comment: This calculation is the same as the corresponding classical one if we maintain the order of the operators.

The total angular momentum is given by

$$\vec{L} = \vec{x}_1 \times \vec{p}_1 + \vec{x}_2 \times \vec{p}_2 \ . \tag{10.16.144}$$

But,

$$\begin{aligned}
\vec{x}_1 &= \vec{X} + \frac{m_2}{M} \vec{x} \\
\vec{x}_2 &= \vec{X} - \frac{m_1}{M} \vec{x}
\end{aligned} \tag{10.16.145}$$

and

$$\vec{P} = \vec{p}_1 + \vec{p}_2$$
$$\vec{p} = \frac{m_2}{M}\vec{p}_1 - \frac{m_1}{M}\vec{p}_2 \ . \tag{10.16.146}$$

Inverting these two equations we find

$$\vec{p}_1 = \frac{m_1}{M}\vec{P} + \vec{p}$$
$$\vec{p}_2 = \frac{m_2}{M}\vec{P} - \vec{p} \ . \tag{10.16.147}$$

Combining these results we now get

$$
\begin{aligned}
\vec{L} &= \vec{x}_1 \times \vec{p}_1 + \vec{x}_2 \times \vec{p}_2 \\
&= (\vec{X} + \frac{m_2}{M}\vec{x}) \times (\frac{m_1}{M}\vec{P} + \vec{p}) + (\vec{X} - \frac{m_1}{M}\vec{x}) \times (\frac{m_2}{M}\vec{P} - \vec{p}) \\
&= \frac{m_1}{M}\vec{X} \times \vec{P} + \vec{X} \times \vec{p} + \frac{m_1 m_2}{M}\vec{x} \times \vec{P} + \frac{m_2}{M}\vec{x} \times \vec{p} \\
&\quad + \frac{m_2}{M}\vec{X} \times \vec{P} - \vec{X} \times \vec{p} - \frac{m_1 m_2}{M}\vec{x} \times \vec{P} + \frac{m_1}{M}\vec{x} \times \vec{p} \\
&= \vec{X} \times \vec{P} + \vec{x} \times \vec{p} \ . \tag{10.16.148}
\end{aligned}
$$

This result states that the total angular momentum of a two-particle system consists of the angular momentum of the centre of mass of the system plus the angular momentum of the relative motion.

10.17 Hulthén Potential: Ground State

Find the ground state energy and wavefunction for the Hulthén potential

$$V(r) = -\frac{V_0\, e^{-\alpha r}}{1 - e^{-\alpha r}} \ . \tag{10.17.149}$$

Hint: Try a solution of the form

$$\psi = \frac{A}{r} e^{-\gamma r}\left(1 - e^{-\alpha r}\right) \ . \tag{10.17.150}$$

Give a reason why this is indeed the ground state.

Solution

The ground state is an s-state ($l = 0$) and so setting the radial wavefunction $R(r)$ equal to $1/r\, u(r)$ we get the Schrödinger equation

$$-\frac{\hbar^2}{2m}\frac{d^2 u}{dr^2} - \frac{V_0\, e^{-\alpha r}}{1 - e^{-\alpha r}} u = Eu \ . \tag{10.17.151}$$

If we set

$$\beta^2 = \frac{2mV_0}{\hbar^2} \quad , \quad \kappa^2 = -\frac{2mE}{\hbar^2} \tag{10.17.152}$$

the equation reduces to

$$\frac{d^2 u}{dr^2} - \left[\kappa^2 - \frac{\beta^2 e^{-\alpha r}}{1 - e^{-\alpha r}} \right] u = 0 \ . \tag{10.17.153}$$

We now insert the suggested solution

$$u = A e^{-\gamma r} \left(1 - e^{-\alpha r} \right) \tag{10.17.154}$$

into (10.17.153). This equation reduces to

$$\left[\gamma^2 - \kappa^2 \right] e^{-\gamma r} + \left[-(\gamma + \alpha)^2 + \kappa^2 + \beta^2 \right] e^{-(\gamma + \alpha) r} = 0 \ . \tag{10.17.155}$$

Thus,

$$\gamma^2 = \kappa^2 \quad \text{or} \quad \gamma = \kappa \tag{10.17.156}$$

and

$$(\gamma + \alpha)^2 = \kappa^2 + \beta^2 \ . \tag{10.17.157}$$

Solving this equation for κ we get

$$\kappa = \frac{\beta^2}{2\alpha} - \frac{\alpha}{2} \ . \tag{10.17.158}$$

Therefore,

$$\gamma = \kappa = \frac{mV_0}{\hbar^2 \alpha} - \frac{\alpha}{2} \tag{10.17.159}$$

and

$$E = -\frac{\hbar^2 \kappa^2}{2m} = -\frac{\hbar^2}{2m} \left(\frac{mV_0}{\hbar^2 \alpha} - \frac{\alpha}{2} \right)^2 \ . \tag{10.17.160}$$

This is an exact solution of the $l = 0$ Schrödinger equation and does not have any nodes. Therefore, it is indeed the ground state solution.

10.18 Hydrogenic Atom in Two Dimensions

Solve the hydrogenic atom for its eigenvalues and eigenfunctions in two dimensions. [10.4]

Solution

The Hamiltonian for the hydrogenic atom is

$$H = \frac{\vec{p}^2}{2m} - \frac{Ze^2}{r} \ . \tag{10.18.161}$$

In cylindrical coordinates, and in two dimensions, this leads to the Schrödinger equation

$$-\frac{\hbar^2}{2m}\left[\frac{\partial^2\psi}{\partial r^2} + \frac{1}{r}\frac{\partial\psi}{\partial r} + \frac{1}{r^2}\frac{\partial^2\psi}{\partial\varphi^2}\right] - \frac{Ze^2}{r}\psi = E\psi \ . \tag{10.18.162}$$

The equation separates into

$$\psi(r,\varphi) = R(r)\,\frac{1}{\sqrt{2\pi}}\,e^{im\varphi} \quad m = 0, \pm1, \pm2, \dots \tag{10.18.163}$$

where the radial function satisfies

$$-\frac{\hbar^2}{2m}\left[\frac{d^2R}{dr^2} + \frac{1}{r}\frac{dR}{dr} - \frac{m^2}{r^2}\right] - \frac{Ze^2}{r}R = ER \ . \tag{10.18.164}$$

It is convenient to introduce new variables

$$
\begin{aligned}
x &= \frac{\hbar^2}{2me^2}r \\
\epsilon &= \frac{\hbar^2}{2me^2}\frac{E}{e^2} \\
R(r) &= \frac{1}{\sqrt{x}}y(x) \ .
\end{aligned}
\tag{10.18.165}
$$

The radial equation now becomes

$$\frac{d^2y}{dx^2} - \frac{(m^2 - 1/4)}{x^2}y + \frac{Z}{x}y + \epsilon y = 0 \ . \tag{10.18.166}$$

The asymptotic behaviour near $x = 0$ yields the solutions

$$y \to x^{\pm(m\pm1/2)} \quad \text{as} \quad x \to 0 \ . \tag{10.18.167}$$

The only acceptable (square integrable) solution is therefore the one with the asymptotic behaviour $x^{|m|+1/2}$.

For $x \to \infty$ we get that y satisfies

$$\frac{d^2y}{dx^2} + \epsilon y \approx 0 \quad \text{as} \quad x \to \infty \ . \tag{10.18.168}$$

In this case the acceptable solution is of the form

$$y \to e^{-\sqrt{-\epsilon}\,x} \quad \text{as} \quad x \to \infty \ . \tag{10.18.169}$$

So, we look for a solution of the form

$$y = x^{|m|+1/2}\,e^{-\sqrt{-\epsilon}\,x}\,f(x) \ . \tag{10.18.170}$$

Substituting this into (10.18.166) we get the equation for f

$$\frac{d^2f}{dx^2} + \left[\frac{2|m|+1}{x} - 2\sqrt{-\epsilon}\right]\frac{df}{dx} + \frac{1}{x}\left[Z - (2|m|+1)\sqrt{-\epsilon}\right]f = 0. \tag{10.18.171}$$

As usual we now look for a series solution

$$y = \sum_k a_k x^k \tag{10.18.172}$$

and substitute this into (10.18.171). The resulting equation is

$$\sum_k a_k \left\{ [(k+1)(k+2|m|+1)x^{k-2} \right.$$
$$\left. - [(2(|m|+k)+1)\sqrt{-\epsilon} - Z]x^{k-1} \right\} = 0 \ . \tag{10.18.173}$$

Therefore we obtain the recursion relation

$$a_{k+1} = \frac{[2(|m|+k)+1]\sqrt{-\epsilon} - Z}{(k+1)(k+2|m|+1)} a_k \ . \tag{10.18.174}$$

Unless this series terminates it leads to exponential growth yielding a non square integrable wavefunction. Therefore we need that the series terminates. This is acomplished by setting the numerator equal to zero. The result is

$$\sqrt{-\epsilon} = \frac{Z}{2(|m|+k)+1} \tag{10.18.175}$$

or

$$\epsilon = -\frac{Z^2}{N^2} \tag{10.18.176}$$

where

$$N = 2(|m|+k)+1 = 1, 3, 5, \ldots \ . \tag{10.18.177}$$

Thus the energy eigenvalues are

$$E = \frac{2me^4}{\hbar^2}\epsilon = -\frac{Z^2 e^2}{a_0} \frac{1}{(n+1/2)^2} \tag{10.18.178}$$

where

$$a_0 = \frac{\hbar^2}{me^2} \quad \text{is the Bohr radius} \tag{10.18.179}$$

and

$$n = |m| + k = 1, 2, 3, \ldots \ . \tag{10.18.180}$$

10.19   Runge-Lenz Vector: Constant of the Motion

The classical Kepler (attractive $1/r$ potential) problem has closed orbits. Classically, this means that the vector pointing from the origin along the semi-major axis of the orbit is a constant of the motion. This is called the Runge-Lenz vector. In quantum mechanics the Runge-Lenz vector is given by

$$\vec{R} = \frac{1}{2m}(\vec{p} \times \vec{L} - \vec{L} \times \vec{p}) - e^2 \frac{\vec{r}}{r} \ . \tag{10.19.181}$$

Given the hydrogen Hamiltonian

$$H = \frac{\vec{p}^2}{2m} - \frac{e^2}{r} \tag{10.19.182}$$

1) Verify that $\vec{R}$ is formally hermitian.
2) Show that $\vec{R}$ is a constant of the motion. That is, show that

$$[\vec{R}, H] = 0 \ .$$

Solution

1) It is convenient to work with individual components of the Runge-Lenz vector. This is made even simpler by using the Levi-Civita antisymmetric tensor symbol

$$\epsilon_{ijk} = \begin{cases} 1 & \text{if } ijk \text{ is an even permutation of 123} \\ -1 & \text{if } ijk \text{ is an odd permutation of 123} \\ 0 & \text{if any two of } ijk \text{ are equal} \end{cases} \ . \tag{10.19.183}$$

Now,

$$\vec{R}^\dagger = \frac{1}{2m} \left[(\vec{p} \times \vec{L})^\dagger - (\vec{L} \times \vec{p})^\dagger \right] - e^2 \frac{\vec{r}}{r} \ . \tag{10.19.184}$$

Also,

$$(\vec{p} \times \vec{L})_i = \epsilon_{ijk} p_j L_k \tag{10.19.185}$$

and

$$(\vec{p} \times \vec{L})_i^\dagger = \epsilon_{ijk} L_k p_j = -(\vec{L} \times \vec{p})_i \ . \tag{10.19.186}$$

Similarly,

$$(\vec{L} \times \vec{p})_i = \epsilon_{ijk} L_j p_k \tag{10.19.187}$$

and

$$(\vec{L} \times \vec{p})_i^\dagger = \epsilon_{ijk} p_k L_j = -(\vec{p} \times \vec{L})_i \ . \tag{10.19.188}$$

Therefore,

$$\vec{R}^\dagger = \frac{1}{2m} \left[-(\vec{L} \times \vec{p}) + (\vec{p} \times \vec{L}) \right] - e^2 \frac{\vec{r}}{r} = \vec{R} \ . \tag{10.19.189}$$

2) To show that $\vec{R}$ is a constant of the motion we consider the commutator

$$[R_i, H] = \left[\frac{1}{2m}\epsilon_{ijk}(p_j L_k - L_j p_k) - e^2 \frac{x_i}{r}, \frac{\vec{p}^2}{2m} - \frac{e^2}{r}\right]$$

$$= -\frac{e^2}{2m}\left\{\epsilon_{ijk}\left([p_j, \frac{1}{r}]L_k - L_j[p_k, \frac{1}{r}]\right) - [\frac{x_i}{r}, p^2]\right\} . \qquad (10.19.190)$$

Here we have used the fact that

$$\begin{aligned}[p_j, p^2] &= 0 \\ [L_k, p^2] &= 0 \\ [L_k, r] &= 0 .\end{aligned} \qquad (10.19.191)$$

Continuing, we find

$$\begin{aligned}[R_i, H] &= i\hbar\frac{e^2}{2m}\left\{\epsilon_{ijk}\frac{2}{r^3}x_i x_j p_j - \frac{2}{r}p_i - 2i\hbar\frac{x_i}{r^3}\right. \\ &+ \left.\frac{2}{r}p_i - \epsilon_{ijk}\frac{2}{r^3}x_j p_j x_i + 2i\hbar\frac{x_i}{r^3}\right\} \\ &= 0 .\end{aligned} \qquad (10.19.192)$$

10.20 Runge-Lenz Vector: Hydrogen Spectrum

In the previous problem (problem 10.19) we introduced the Runge-Lenz vector and found that it is a constant of the motion. The "accidental" degeneracy of the hydrogen atom is due to the fact that this vector along with angular momentum is conserved. Use the Runge-Lenz vector

$$\vec{R} = \frac{1}{2m}(\vec{p} \times \vec{L} - \vec{L} \times \vec{p}) - e^2 \frac{\vec{r}}{r} \qquad (10.20.193)$$

together with the hydrogen Hamiltonian

$$H = \frac{\vec{p}^2}{2m} - \frac{e^2}{r} \qquad (10.20.194)$$

to prove the following steps.

1)

$$\vec{R} \cdot \vec{L} = \vec{L} \cdot \vec{R} = 0 . \qquad (10.20.195)$$

2)

$$[R_i, R_j] = i\hbar\frac{-2H}{m}\epsilon_{ijk}L_k . \qquad (10.20.196)$$

3)

$$[R_i, L_j] = i\hbar\epsilon_{ijk}R_k . \qquad (10.20.197)$$

4)

$$R^2 = e^4 + \frac{2H(L^2 + \hbar^2)}{m} \; . \tag{10.20.198}$$

By restricting yourself to the space of bound states (energy $E < 0$) define

$$\vec{K} = \sqrt{-\frac{m}{2H}} \, \vec{R} \tag{10.20.199}$$

and
5) show that the components of $\vec{K}$ satisfy the commutation rules

$$[K_i, L_j] = i\hbar\epsilon_{ijk}L_k \tag{10.20.200}$$

as well as that

$$[K_i, L_j] = i\hbar\epsilon_{ijk}K_k \; . \tag{10.20.201}$$

Next define the two operators

$$\vec{M} = \frac{1}{2}(\vec{L} + \vec{K}) \;\; , \;\;\; \vec{N} = \frac{1}{2}(\vec{L} - \vec{K}) \tag{10.20.202}$$

and verify that they both satisfy angular momentum commutation relations as well as that they commute with each other. Finally,
6) use all these results to deduce the hydrogen atom energy levels

$$E_n = -\frac{me^4}{2\hbar^2 n^2} \tag{10.20.203}$$

as well as the degeneracy of these levels.

Solution

1) In preparation for subsequent computations we list two properties of the Levi-Civita symbol ϵ_{ijk} as well as some commutators. Throughout we assume the Einstein convention that repeated indices are to be summed over.

$$\epsilon_{ijk}\epsilon_{ilm} = \delta_{jl}\delta_{km} - \delta_{jm}\delta_{lk} \tag{10.20.204}$$

$$\epsilon_{ijk}\epsilon_{ijm} = 2\delta_{km} \; . \tag{10.20.205}$$

$$[p_i, L_j] = i\hbar\epsilon_{ijk}p_k \; . \tag{10.20.206}$$

Also,

$$[p_i, x_i/r] = -i\hbar\frac{r^2 - x_i^2}{r^3} \; . \tag{10.20.207}$$

We are now ready to prove the first result. To do this we first compute

$$\vec{r} \cdot \vec{L} = x_i\epsilon_{ijk}x_jp_k = 0 \tag{10.20.208}$$

as well as

$$\begin{aligned} \vec{L} \cdot \vec{r} &= \epsilon_{ijk} x_j p_k x_i \\ &= \epsilon_{ijk} (x_j x_i p_k - i\hbar x_j \delta_{ik}) \\ &= 0 \ . \end{aligned}$$

(10.20.209)

Similarly,

$$\vec{L} \cdot \vec{p} = \epsilon_{ijk} x_j p_k p_i = 0$$

(10.20.210)

and, with a little more algebra

$$\begin{aligned} \vec{p} \cdot \vec{L} &= \epsilon_{ijk} p_i x_j p_k \\ &= \epsilon_{ijk} (x_j p_i p_k - i\hbar p_k \delta_{ij}) \\ &= 0 \ . \end{aligned}$$

(10.20.211)

Therefore,

$$\begin{aligned} \vec{R} \cdot \vec{L} &= \frac{1}{2m} (\vec{p} \times \vec{L} - \vec{L} \times \vec{p}) \cdot \vec{L} \\ &= \frac{1}{2m} \epsilon_{ijk} (p_j L_k - L_j p_k) L_i \\ &= -\frac{1}{2m} \epsilon_{ijk} L_j p_k L_i \\ &= -\frac{1}{2m} \epsilon_{ijk} L_j (L_i p_k - i\hbar \epsilon_{kil} p_l) \\ &= \frac{i\hbar}{2m} 2\delta_{jl} L_j p_l = 0 \end{aligned}$$

(10.20.212)

where we have used that

$$\epsilon_{ijk} p_j L_k L_i = p_j L_j = 0 \ .$$

(10.20.213)

Similarly,

$$\begin{aligned} \vec{L} \cdot \vec{R} &= \frac{1}{2m} \epsilon_{ijk} L_i (p_j L_k - L_j p_k) \\ &= -\frac{1}{2m} \epsilon_{ijk} L_i p_j L_k = 0 \ . \end{aligned}$$

(10.20.214)

2) To compute the commutators between the components of $\vec{R}$ we use our results from above. Then,

$$\begin{aligned} [R_i, R_j] &= \frac{1}{4m^2} [\epsilon_{ikl} (p_k L_l - L_k p_l), \ \epsilon_{jmn} (p_m L_n - L_m p_n)] \\ &\quad - \frac{e^2}{2m} \Big\{ \epsilon_{ilk} [p_l L_k - L_l p_k, \ \frac{x_j}{r}] \\ &\quad - \epsilon_{jlm} [\frac{x_j}{r}, \ p_l L_m - L_l p_m] \Big\} \ . \end{aligned}$$

(10.20.215)

To evaluate this we need several commutators. These are:

$$[p_k L_l , p_m L_n] = i\hbar(\epsilon_{knq}p_m p_q L_l - \epsilon_{lmq}p_k p_q L_n + \epsilon_{lnq}p_k p_m L_q)$$
$$[L_k p_l , L_m p_n] = i\hbar(\epsilon_{lmq}L_k p_q p_n + \epsilon_{kmq}L_q p_n p_l - \epsilon_{knq}L_m p_q p_l)$$
$$[L_k p_l , p_m L_n] = i\hbar(\epsilon_{lnq}L_k p_m p_q - \epsilon_{kmq}p_q L_n p_l + \epsilon_{knq}p_m L_q p_l)$$
$$[p_k L_l , L_m p_n] = i\hbar(\epsilon_{lmq}p_k L_q p_n + \epsilon_{kmq}p_n p_q L_l - \epsilon_{lnq}p_k L_m p_q)$$

$$(10.20.216)$$

as well as

$$[p_l L_k , \frac{x_j}{r}] = i\hbar \left\{ \epsilon_{kjq}p_l \frac{x_q}{r} + \left(\frac{x_j x_l}{r^3} - \frac{\delta_{lj}}{r} \right) L_k \right\}$$
$$[L_l p_k , \frac{x_j}{r}] = i\hbar \left\{ \epsilon_{ljq}\frac{x_q}{r}p_k L_l \left(\frac{x_j x_k}{r^3} - \frac{\delta_{kj}}{r} \right) \right\}$$
$$[\frac{x_i}{r} , p_l L_m] = i\hbar \left\{ -\epsilon_{imq}p_l \frac{x_q}{r} - \left(\frac{x_i x_l}{r^3} - \frac{\delta_{li}}{r} \right) L_m \right\}$$
$$[\frac{x_i}{r} , L_l p_m] = i\hbar \left\{ -\epsilon_{ilq}p_l \frac{x_q}{r} + L_l \left(\frac{x_i x_m}{r^3} - \frac{\delta_{im}}{r} \right) \right\} .$$

$$(10.20.217)$$

Putting all this together we get that

$$[R_i, R_j] = -i\hbar \left\{ \epsilon_{ijk}\frac{p^2}{m^2}L_k - \frac{2e^2}{mr}\epsilon_{ijk}L_k \right\}$$
$$= i\hbar \frac{-2H}{m}\epsilon_{ijk}L_k .$$

$$(10.20.218)$$

3) The next result is

$$[R_i, L_j] = \frac{1}{2m}\epsilon_{ikl}[p_k L_l - L_k p_l , L_j] - \frac{e^2}{r}[x_i , L_j]$$
$$= \frac{i\hbar}{2m}\epsilon_{ikl}\{\epsilon_{ljm}p_k L_m + \epsilon_{kjm}p_m L_l$$
$$- \epsilon_{kjm}L_m p_l - \epsilon_{ljm}L_k p_m\}$$
$$- i\hbar\epsilon_{ijm}e^2 \frac{x_m}{r}$$
$$= \frac{i\hbar}{2m}\{(\delta_{ij}\delta_{km} - \delta_{im}\delta_{jk})(p_k L_m - L_k p_m)$$
$$- (\delta_{ij}\delta_{lm} - \delta_{im}\delta_{jl})(p_m L_l - L_m p_l)\}$$
$$- i\hbar\epsilon_{ijm}e^2 \frac{x_m}{r}$$
$$= i\hbar\{\frac{1}{2m}[-p_j L_i + p_i L_j - L_i p_j + L_j p_i] - \epsilon_{ijm}e^2 \frac{x_m}{r}\}$$
$$= i\hbar\epsilon_{ijm} R_m .$$

$$(10.20.219)$$

4) Now we compute

$$R^2 = \left\{ \frac{1}{2m}\epsilon_{ikl}(p_k L_l - L_k p_l) - e^2 \frac{x_i}{r} \right\}$$

$$\times \quad \left\{\frac{1}{2m}\epsilon_{ikl}(p_k L_l - L_k p_l) - e^2\frac{x_i}{r}\right\}$$

$$= \quad \frac{1}{4m^2}(\delta_{km}\delta_{ln} - \delta_{kn}\delta_{ml})(p_k L_l - L_k p_l)(p_m L_n - L_m p_n)$$

$$- \quad \frac{e^2}{2m}\epsilon_{ikl}[(p_k L_l - L_k p_l)\frac{x_i}{r} + \frac{x_i}{r}(p_k L_l - L_k p_l)]$$

$$+ \quad e^4$$

$$= \quad \frac{1}{4m^2}(p_m L_n - p_n L_m - L_m p_n + L_n p_m)(p_m L_n - L_m p_n)$$

$$- \quad \frac{e^2}{2m}\epsilon_{ikl}[(p_k L_l - L_k p_l)\frac{x_i}{r} + \frac{x_i}{r}(p_k L_l - L_k p_l)]$$

$$+ \quad e^4 \ . \tag{10.20.220}$$

But,

$$
\begin{aligned}
p_m L_n p_m L_n &= p_m(p_m L_n - i\hbar\epsilon_{mnq}p_q)L_n \\
&= p^2 L^2 \\
p_m L_n L_m p_n &= (L_n p_m + i\hbar\epsilon_{mnq}p_q)L_m p_n \\
&= i\hbar\epsilon_{mnq}p_q(p_n L_m - i\hbar\epsilon_{nmr}p_r) \\
&= -2\hbar^2 p^2 \\
p_n L_m p_m L_n &= 0 \\
p_n L_m L_m p_n &= (L_m p_n + i\hbar\epsilon_{nmq}p_q)L_m p_n \\
&= L_m(L_m p_n i\hbar\epsilon_{nmq}p_q)p_n + i\hbar\epsilon_{nmq}(L_m p_q + i\hbar\epsilon_{qmr}p_r)p_r \\
&= L^2 p^2 + 2\hbar^2 p^2 = p^2 L^2 + 2\hbar^2 p^2 \\
L_m p_n p_m L_n &= L_m p_n(L_n p_m + i\hbar\epsilon_{mnq}p_q) = 0 \\
L_m p_n L_m p_n &= L_m(L_m p_n + i\hbar\epsilon_{nmq}p_q)p_n \\
&= L^2 p^2 = p^2 L^2 \\
L_n p_m p_m L_n &= L^2 p^2 = p^2 L^2 \\
L_n p_m L_m p_n &= 0 \ . \tag{10.20.221}
\end{aligned}
$$

Also,

$$
\begin{aligned}
\epsilon_{ikl}p_k\frac{1}{r}L_l x_i &= \epsilon_{ikl}(\frac{1}{r}p_k + i\hbar\frac{x_k}{r^3})(x_i L_l - i\hbar\epsilon_{ilm}x_m) \\
&= \epsilon_{ikl}(\frac{1}{r}x_i p_k L_l - i\hbar\epsilon_{ilm}p_k x_m + \hbar^2\epsilon_{ilm}\frac{x_k x_m}{r^3}) \\
&= \frac{1}{r}L^2 + 2i\hbar\vec{p}\cdot\vec{r} - 2\frac{\hbar^2}{r} \\
&= \frac{1}{r}L^2 + 2i\hbar\vec{r}\cdot\vec{p} + 4\frac{\hbar^2}{r} \\
\epsilon_{ikl}L_k p_l\frac{x_i}{r} &= \epsilon_{ikl}L_k\frac{1}{r}(x_i p_l - i\hbar\delta_{il}) \\
&= \frac{1}{r}\epsilon_{ikl}L_k x_i p_l
\end{aligned}
$$

$$= -\frac{1}{r}L^2$$

$$\epsilon_{ikl}\frac{x_i}{r}p_k L_l = \frac{1}{r}L^2$$

$$\epsilon_{ikl}\frac{x_i}{r}L_k p_l = \frac{1}{r}\epsilon_{ikl}x_i\left(p_l L_k - i\hbar\epsilon_{lkm}p_m\right)$$

$$= -\frac{1}{r}L^2 + 2i\hbar\vec{r}\cdot\vec{p} \ . \tag{10.20.222}$$

Combining these results we have

$$R^2 = e^4 + \left(\frac{p^2}{m^2} - 2\frac{e^2}{mr}\right)(L^2 + \hbar^2)$$

$$= e^4 + \frac{2H}{m}(L^2 + \hbar^2) \ . \tag{10.20.223}$$

We now restrict ourselves to the subspace of bound states of the Hamiltonian and define

$$\vec{K} \equiv \sqrt{-\frac{m}{2H}}\,\vec{R} \ . \tag{10.20.224}$$

5) It now follows immediately from the commutation relations satisfied by the components of $\vec{R}$ that

$$[K_i\,,\ K_j] = i\hbar\epsilon_{ijk}L_k \tag{10.20.225}$$

so that $\vec{K}$ satisfies angular momentum commutation relations. Furthermore from the commutators between R_i and L_j it also follows immediately that

$$[K_i, L_j] = i\hbar\epsilon_{ijk}K_k \ . \tag{10.20.226}$$

Next, we define the two operators

$$\vec{M} = \frac{1}{2}(\vec{L} + \vec{K}) \ , \quad \vec{N} = \frac{1}{2}(\vec{L} - \vec{K}) \ . \tag{10.20.227}$$

It then follows that

$$[M_i\,,\ M_j] = \frac{1}{4}\left([L_i\,,\ L_j] + [L_i\,,\ K_j] + [K_i\,,\ L_j] + [K_i\,,\ K_j]\right)$$

$$= \frac{i\hbar}{4}\epsilon_{ijk}\left(L_k + K_k + K_k + L_k\right)$$

$$= i\hbar\epsilon_{ijk}M_k \ . \tag{10.20.228}$$

Similarly,

$$[N_i\,,\ N_j] = \frac{1}{4}\left([L_i\,,\ L_j] - [L_i\,,\ K_j] - [K_i\,,\ L_j] + [K_i\,,\ K_j]\right)$$

$$= \frac{i\hbar}{4}\epsilon_{ijk}\left(L_k - K_k - K_k + L_k\right)$$

$$= i\hbar\epsilon_{ijk}N_k \ . \tag{10.20.229}$$

Also,

$$
\begin{aligned}
[N_i\,,\ M_j] &= \frac{1}{4}\left([L_i\,,\ L_j]+[L_i\,,\ K_j]-[K_i\,,\ L_j]-[K_i\,,\ K_j]\right) \\
&= \frac{i\hbar}{4}\epsilon_{ijk}\left(L_k+K_k-K_k-L_k\right) \\
&= 0\ .
\end{aligned}
\qquad (10.20.230)
$$

This means we can simultaneously diagonalize $\vec{M}^{\,2}$ and $\vec{N}^{\,2}$. Their respective eigenvalues are

$$\hbar^2\mu(\mu+1) \quad\text{and}\quad \hbar^2\nu(\nu+1)\quad \mu\,,\nu=0,1/2,1,3/2,\dots\ .$$

6) To get the hydrogen atom spectrum we now use (10.20.223)

$$R^2 = e^4 + \frac{2H}{m}(L^2+\hbar^2)\ . \qquad (10.20.231)$$

In terms of K^2 this reads

$$-\frac{2H}{m}K^2 = e^4 + \frac{2H}{m}(L^2+\hbar^2)\ . \qquad (10.20.232)$$

Solving for H we find

$$H = -\frac{me^4}{2(K^2+L^2+\hbar^2)}\ . \qquad (10.20.233)$$

But,

$$K^2 + L^2 = 2(M^2+N^2)\ . \qquad (10.20.234)$$

On the other hand, we also have that

$$\vec{R}\cdot\vec{L} = \vec{L}\cdot\vec{R} = \vec{K}\cdot\vec{L} = \vec{L}\cdot\vec{K} = 0\ . \qquad (10.20.235)$$

Therefore,

$$M^2 = N^2\ . \qquad (10.20.236)$$

So, we can write the spectrum of the hydrogen Hamiltonian as

$$
\begin{aligned}
H &= -\frac{me^4}{2\hbar^2[4\mu(\mu+1)+1]} \\
&= -\frac{me^4}{2\hbar^2(2\mu+1)^2}\ .
\end{aligned}
\qquad (10.20.237)
$$

But,

$$2\mu + 1 = 1,2,3,4,\dots\ . \qquad (10.20.238)$$

So, using more conventional notation we call $2\mu+1 = n$ and find that the energy levels for the hydrogen atom are given by the Bohr formula

$$E_n = -\frac{1}{2}\frac{me^4}{\hbar^2 n^2}\quad n=1,2,3,4,\dots\ . \qquad (10.20.239)$$

The degeneracy is also easy to count since for a given value of $n = 2\mu+1 = 2\nu+1$ we have

$$(2\mu + 1)(2\nu + 1) = n^2$$

different states with the same eigenvalue. Therefore, including the spin, the degeneracy is $2n^2$.

It was by a technique of this sort that Pauli [10.5] first succeeded to use matrix mechanics to find the spectrum of the hydrogen atom before Schrödinger came up with his result.

Bibliography

[10.1] A.Z. Capri, *Nonrelativistic Quantum Mechanics* 3rd edition, World Scientific Publishing Co. Pte. Ltd., section 10.6.1, (2002) .

[10.2] S. Pasternach, Proc. Natl. Acad. Sci. USA, **23**, 91 (1937). ibid **23**, 250, (1937).

[10.3] S. Balasubramanian, Am. J. Phys. **68**, 959, (2000)

[10.4] B. Zaslow and M.E. Zander, Am. J. Phys. **35**, 1118, (1967).

[10.5] W. Pauli, Zeitschrift f. Phys. **36**, 336, (1926).

Chapter 11

Transformation Theory

11.1 Fourier Transform of Hermite Functions

Verify the equation

$$V_{p,n} = \int_{-\infty}^{\infty} dx\, e^{-ipx} u_n(x) = i^n\, u_n(k)$$

where

$$u_n(x) = \frac{1}{(\pi)^{1/4}} \frac{1}{\sqrt{2^n n!}} H_n(x)\, e^{-x^2/2}$$

and $H_n(x)$ is a Hermite polynomial. That is, compute the Fourier transform of the Hermite functions.

Solution

Let $\{u_n(x)\}$ be the set of normalized Hermite functions with $k/(\hbar\omega) = 1$. So that

$$u_n(x) = \frac{(-1)^n}{\sqrt{n!}} 2^{-n/2} \left(\frac{d}{dx} - x\right)^n \frac{1}{\pi^{1/4}} e^{-x^2/2} \tag{11.1.1}$$

and

$$V_{p,n} = \int_{-\infty}^{\infty} e^{-ipx}\, u_n(x)\, dx\ . \tag{11.1.2}$$

Since the Hermite functions damp very rapidly at $\pm\infty$ we can integrate by parts and use the following general results.

$$\int_{-\infty}^{\infty} e^{-ipx} \frac{df}{dx}\, dx = -\int_{-\infty}^{\infty} f(x) \frac{de^{-ipx}}{dx}\, dx = ip \int_{-\infty}^{\infty} e^{-ipx} f(x)\, dx\ . \tag{11.1.3}$$

196

Also,

$$\int_{-\infty}^{\infty} e^{-ipx} \, x f(x) \, dx = i \frac{d}{dp} \int_{-\infty}^{\infty} e^{-ipx} \, f(x) \, dx \ . \tag{11.1.4}$$

Thus,

$$
\begin{aligned}
V_{p,n} &= \frac{(-1)^n}{\sqrt{n!}} 2^{-n/2} \left(\frac{d}{dp} - p \right)^n \frac{1}{\pi^{1/4}} \int_{-\infty}^{\infty} e^{-ipx} \, e^{-x^2/2} \, dx \\
&= (-i)^n \frac{(-1)^n}{\sqrt{n!}} 2^{-n/2} \left(\frac{d}{dp} - p \right)^n \frac{1}{\pi^{1/4}} e^{-p^2/2} \\
&= (-i)^n u_n(p) \ .
\end{aligned}
\tag{11.1.5}
$$

11.2   Schrödinger Equation in Momentum Space

Transform the Schrödinger equation

$$-\frac{\hbar^2}{2m} \nabla^2 \psi(\vec{r}) + V(\vec{r})\psi(\vec{r}) = E\psi(\vec{r}) \tag{11.2.6}$$

to momentum space.

Solution

It is convenient to first rewrite the Schrödinger equation in the form

$$(\nabla^2 + k^2)\psi(\vec{r}) = U(\vec{r})\psi(\vec{r}) \tag{11.2.7}$$

where

$$k^2 = \frac{2mE}{\hbar^2} \quad , \quad U(\vec{r}) = \frac{2mV(\vec{r})}{\hbar^2} \ . \tag{11.2.8}$$

We then define the momentum space wavefunction $\phi(\vec{p})$ by

$$\phi(\vec{p}) = \int e^{i\vec{p}\cdot\vec{r}} \, \psi(\vec{r}) \, d^3r \ . \tag{11.2.9}$$

The inverse of this relation is

$$\psi(\vec{r}) = \frac{1}{(2\pi)^3} \int d^3q \, \phi(\vec{q}) \, e^{-i\vec{q}\cdot\vec{r}} \ . \tag{11.2.10}$$

To proceed we take the Fouier transform of (11.2.7). Thus,

$$\int d^3r \, (\nabla^2 + k^2)\psi(\vec{r}) \, e^{i\vec{p}\cdot\vec{r}} = \int d^3r \, U(\vec{r})\psi(\vec{r}) \, e^{i\vec{p}\cdot\vec{r}} \ . \tag{11.2.11}$$

The left hand side may be integrated by parts twice to become

$$\int d^3r \, (-p^2 + k^2)\psi(\vec{r}) \, e^{i\vec{p}\cdot\vec{r}} = (-p^2 + k^2)\phi(\vec{p}) \ . \tag{11.2.12}$$

To simplify the right hand side we replace $\psi(\vec{r})$ by its expression in terms of $\phi(\vec{p})$. Then,

$$
\begin{aligned}
\int d^3r\, U(\vec{r})\psi(\vec{r})\, e^{i\vec{p}\cdot\vec{r}} &= \frac{1}{(2\pi)^3}\int d^3r\, U(\vec{r})e^{i\vec{p}\cdot\vec{r}}\int d^3q\, \phi(\vec{q})\, e^{-i\vec{q}\cdot\vec{r}} \\
&= \frac{1}{(2\pi)^3}\int d^3q\, \phi(\vec{q})\int U(\vec{r})\, e^{i(\vec{p}-\vec{q})\cdot\vec{r}} \\
&= \frac{1}{(2\pi)^3}\int d^3q\, \tilde{U}(\vec{p}-\vec{q})\phi(\vec{q}) \qquad (11.2.13)
\end{aligned}
$$

where

$$
\tilde{U}(\vec{p}-\vec{q}) = \int d^3r\, U(\vec{r})\, e^{i(\vec{p}-\vec{q})\cdot\vec{r}} \qquad (11.2.14)
$$

is the Fourier transform of $U(\vec{r})$. Thus, the Schrödinger equation in momentum space reads

$$
(-p^2 + k^2)\phi(\vec{p}) = \frac{1}{(2\pi)^3}\int d^3q\, \tilde{U}(\vec{p}-\vec{q})\phi(\vec{q}) \; . \qquad (11.2.15)
$$

11.3   Heisenberg Equation for a Free Particle

Find and solve the Heisenberg equation of motion for a particle with a Hamiltonian

$$
H = \frac{\vec{p}^2}{2m} \; .
$$

If at $t = 0$ the particle is in the state $|0,0,0\rangle$ of a harmonic oscillator basis, find $\langle r^2 \rangle$ as a function of t .

Solution

We are given the Hamiltonian

$$
H = \frac{\vec{p}^2}{2m} \; . \qquad (11.3.16)
$$

The corresponding Heisenberg equations of motion are:

$$
\begin{aligned}
i\hbar\frac{d\vec{p}}{dt} &= [\vec{p}, H] = 0 \\
i\hbar\frac{d\vec{r}}{dt} &= [\vec{r}, H] = \frac{i\hbar}{m}\vec{p} \; . \qquad (11.3.17)
\end{aligned}
$$

Therefore $\vec{p} = \vec{p}_0$ where $\vec{p}_0$ is the momentum operator in the Schrödinger picture

$$
\vec{p}_0 = -i\hbar\nabla \qquad (11.3.18)
$$

and

$$\vec{r} = \frac{\vec{p_0}}{m}t + \vec{x} \tag{11.3.19}$$

where $\vec{x}$ is the position operator in the Schrödinger picture. The wavefunction in the Heisenberg picture is the same as the wavefunction in the Schrödinger picture at time $t = 0$. Thus, we have

$$\psi(\vec{x}) = (k/\pi)^{3/4} e^{-kr^2/2} \tag{11.3.20}$$

where

$$k = \frac{m\omega}{\hbar} \ . \tag{11.3.21}$$

Also,

$$\langle \vec{r}^2 \rangle = \langle \vec{p_0}^2 \rangle \frac{t^2}{m^2} + \langle \vec{p_0} \cdot \vec{x} + \vec{x} \cdot \vec{p_0} \rangle \frac{t}{m} + \langle \vec{x}^2 \rangle \ . \tag{11.3.22}$$

So, we get

$$
\begin{aligned}
\langle \vec{r}^2 \rangle \ = \ & (k/\pi)^{3/2} \frac{t^2}{m^2} 4\pi \int_0^\infty -\hbar^2 \left(k^2 r^2 - k \right) e^{-kr^2} r^2 \, dr \\
& + (k/\pi)^{3/2} \frac{t}{m} 4\pi \int_0^\infty \frac{\hbar}{i} \left(-2kr^2 + 3 \right) e^{-kr^2} r^2 \, dr \\
& + (k/\pi)^{3/2} 4\pi \int_0^\infty r^2 e^{-kr^2} r^2 \, dr \ .
\end{aligned}
\tag{11.3.23}
$$

Evaluating these integrals we get

$$\langle \vec{r}^2 \rangle = 4\pi \frac{k}{\pi} \left[-\frac{\hbar^2 t^2}{m^2}(3/8 - 1/4) + \frac{i\hbar t}{2m} \left(\frac{6k}{2(2k)^2} - \frac{3}{4k} \right) + \frac{3}{8k^2} \right] \ . \tag{11.3.24}$$

Simplifying, we find

$$\langle \vec{r}^2 \rangle = \frac{3}{2k} - \frac{\hbar^2 k t^2}{2m^2} \ . \tag{11.3.25}$$

Substituting the value of k we finally obtain

$$\langle \vec{r}^2 \rangle = \frac{3\hbar}{2m\omega} - \frac{\hbar\omega t^2}{2m} \ . \tag{11.3.26}$$

11.4   Dirac Picture for Displaced SHO

Transform the displaced one-dimensional SHO

$$H = \frac{p^2}{2m} + \frac{1}{2}kx^2 + \beta x = H_0 + \beta x$$

to the Dirac picture. Solve for both $\psi_D(t)$, the wavefunction of the system and $x_D(t)$, the position operator of the system. Assume $\psi_D(0)$ describes the ground state of H_0 in the Schrödinger picture.

Hint: Rewrite everything in terms of the Schrödinger picture annihilation and creation operators and use the results of problem 9.15 . The expression for $\psi_D(t)$ cannot be completely evaluated.

Solution

We have the Hamiltonian in the Schrödinger picture

$$H = \frac{p^2}{2m} + \frac{1}{2}kx^2 + \beta x = H_0 + \beta x . \tag{11.4.27}$$

Then we find the unitary operator $U_0(t)$ that transforms to the Dirac picture

$$U_0(t) = \exp\left[-i\frac{H_0 t}{\hbar}\right] . \tag{11.4.28}$$

If we write H_0 in the form

$$H_0 = \hbar\omega(a^\dagger a + 1/2) \tag{11.4.29}$$

and

$$\beta x = \beta\sqrt{\frac{\hbar}{2m\omega}}(a^\dagger + a) \tag{11.4.30}$$

then

$$U_0(t) = \exp\left[-i\omega t(a^\dagger a + 1/2)\right] . \tag{11.4.31}$$

Next we compute

$$x_D(t) = U_0^\dagger(t) \, x \, U_0(t) \tag{11.4.32}$$

So we need to compute

$$x_D(t) = \sqrt{\frac{\hbar}{2m\omega}} \exp\left[i\omega t(a^\dagger a + 1/2)\right] (a^\dagger + a) \exp\left[-i\omega t(a^\dagger a + 1/2)\right] . \tag{11.4.33}$$

Using the results from problem 9.15 that if

$$[A, B] = \lambda A \tag{11.4.34}$$

then

$$A \, e^B = e^\lambda \, e^B \, A \tag{11.4.35}$$

we find that

$$a \, e^{-i\omega t a^\dagger a} = e^{-i\omega t} \, e^{-i\omega t a^\dagger a} a \tag{11.4.36}$$

and

$$a^\dagger \, e^{-i\omega t a^\dagger a} = e^{i\omega t} \, e^{-i\omega t a^\dagger a} a^\dagger \tag{11.4.37}$$

so that

$$\begin{aligned}
x_D(t) &= \sqrt{\frac{\hbar}{2m\omega}}[e^{-i\omega t} a + e^{i\omega t} a^\dagger] \\
&= x_S \cos\omega t + \frac{p_S}{m\omega}\sin\omega t .
\end{aligned} \tag{11.4.38}$$

An alternate way to solve this part is to use the equations of motion for $x_D(t)$ and $p_D(t)$.

$$ih\frac{dx_D}{dt} = [x_D, H_0] = \frac{i\hbar}{m}p_D \quad \text{or} \quad \frac{dx_D}{dt} = \frac{p_D}{m}$$

$$ih\frac{dp_D}{dt} = [p_D, H_0] = -i\hbar k x_D \quad \text{or} \quad \frac{dp_D}{dt} = -kx_D . \qquad (11.4.39)$$

Therefore,

$$\frac{d^2 x_D}{dt^2} = \frac{1}{m}\frac{dp_D}{dt} = -\frac{k}{m}x_D = -\omega^2 x_D \qquad (11.4.40)$$

and

$$x_D = x_S \cos \omega t + \frac{p_S}{m\omega}\sin \omega t \qquad (11.4.41)$$

just as before.

b) To find the state in the Dirac picture we first solve for the state in the Schrödinger picture. The Hamiltonian is

$$H = \hbar\omega(a^\dagger a + 1/2) + \beta\sqrt{\frac{\hbar}{2m\omega}}(a^\dagger + a) = \hbar\omega[a^\dagger a + c(a^\dagger + a) + 1/2](11.4.42)$$

where

$$c = \beta\sqrt{\frac{\hbar}{2m\omega}} \qquad (11.4.43)$$

Therefore the Hamiltonian may be rewritten as

$$H = \hbar\omega(b^\dagger b + 1/2 - c^2) \qquad (11.4.44)$$

where

$$b = a + c , \quad b^\dagger = a^\dagger + c . \qquad (11.4.45)$$

So, the ground state has an energy

$$E_0 = \hbar\omega(1/2 - c^2) \qquad (11.4.46)$$

and satisfies

$$b|0\rangle = 0 . \qquad (11.4.47)$$

The Schrödinger ground state is then

$$|\Psi_S(t)\rangle = e^{-i\omega t(1/2 - c^2)}|0\rangle . \qquad (11.4.48)$$

So, the equation for the state in the Dirac picture is

$$i\hbar\frac{\partial}{\partial t}\Psi_D(t) = \beta x_D(t)\Psi_D(t) \qquad (11.4.49)$$

where $\Psi_D(0) = \Psi_S(0)$. To proceed we convert this to an integral equation

$$\Psi_D(t) = \Psi_S(t) - \frac{i\beta}{\hbar} \int_0^t x_D(t')\Psi_D(t')\,dt' . \tag{11.4.50}$$

This integral equation is of a type known as a *Volterra* equation and can always be solved by simply iterating. Thus, we find

$$
\begin{aligned}
\Psi_D(t) &= \Psi_S(t) - \frac{i\beta}{\hbar} \int_0^t dt_1 x_D(t_1)\Psi_S(t_1) \\
&+ \left(\frac{-i\beta}{\hbar}\right)^2 \int_0^t dt_1 x_D(t_1) \int_0^{t_1} dt_2 x_D(t_2)\Psi_S(t_2) \\
&= \Psi_S(t) + \sum_{n=1}^{\infty} \left(\frac{-i\beta}{\hbar}\right)^n \int_0^t dt_1 x_D(t_1) \\
&\times \int_0^{t_1} dt_2 x_D(t_2) \cdots \int_0^{t_{n-1}} dt_n x_D(t_n)\psi_S(t_n) . \tag{11.4.51}
\end{aligned}
$$

This is as far as one can go since this last expression cannot be simplified.

11.5   Heisenberg Picture for Displaced SHO

Repeat the previous problem for the Heisenberg picture.

Solution

As in the previous problem, 11.4 above, we have

$$H = \frac{p^2}{2m} + \frac{1}{2}kx^2 + \beta x . \tag{11.5.52}$$

The Heisenberg equations of motion are:

$$i\hbar \frac{dx}{dt} = [x, H] = i\hbar \frac{p}{m} \tag{11.5.53}$$

or

$$\frac{dx}{dt} = \frac{p}{m} \tag{11.5.54}$$

and

$$i\hbar \frac{dp}{dt} = [p, H] = -i\hbar m\omega^2 x - i\hbar\beta \tag{11.5.55}$$

or

$$\frac{dp}{dt} = -m\omega^2 x - \beta . \tag{11.5.56}$$

The solutions are:

$$x = x_0 \cos \omega t + \frac{p_0}{m} \sin \omega t - \frac{\beta}{m\omega^2}$$
$$p = p_0 \cos \omega t - m\omega x_0 \sin \omega t \ . \qquad (11.5.57)$$

Also the Heisenberg state at any time t is the same as the Schrödinger state at time $t = 0$.

$$|\Psi_H(t)\rangle = |\Psi_S(0)\rangle = |0\rangle \ . \qquad (11.5.58)$$

11.6 Heisenberg Picture: SHO and Constant Force

Use the Heisenberg picture to find the expectation values for p and x for a particle in a harmonic oscillator potential and also acted on by a constant force F. Assume that the state of the system is the ground state of the simple harmonic oscillator without the constant force F.

Solution

The Hamiltonian is

$$H = \frac{p^2}{2m} + \frac{1}{2}kx^2 - Fx \qquad (11.6.59)$$

The equations of motion are

$$\frac{dx}{dt} = \frac{p}{m}$$
$$\frac{dp}{dt} = -kx + F \ . \qquad (11.6.60)$$

The state in the Heisenberg picture is $|0\rangle$, the ground state of the simple harmonic oscillator at $t = 0$ in the Schrödinger picture. This state is annihilated by the Schrödinger picture annihilation operator which coincides with the Heisenberg picture annihilation operator at time $t = 0$. Also, in terms of the annihilation and creation operators, the position and momentum are given by

$$x = \sqrt{\frac{\hbar}{2m\omega}}(a^\dagger + a)$$
$$p = i\sqrt{\frac{m\omega\hbar}{2}}(a^\dagger - a) \qquad (11.6.61)$$

or

$$a = \sqrt{\frac{m\omega\hbar}{2}}\left[x + \frac{ip}{m\omega}\right]$$
$$a^\dagger = \sqrt{\frac{m\omega\hbar}{2}}\left[x - \frac{ip}{m\omega}\right] \ . \qquad (11.6.62)$$

The equation of motion for the annihilation operator, in the Heisenberg picture is, by direct substitution

$$\frac{da}{dt} = -i\omega\, a + iF\sqrt{\frac{1}{2m\hbar\omega}} \ .$$
(11.6.63)

Therefore,

$$a = a(0)\, e^{-i\omega t} + iF\sqrt{\frac{1}{2m\hbar\omega^3}}\left(1 - e^{-i\omega t}\right)$$
(11.6.64)

and

$$a^\dagger = a^\dagger(0)\, e^{i\omega t} - iF\sqrt{\frac{1}{2m\hbar\omega^3}}\left(1 - e^{i\omega t}\right) \ .$$
(11.6.65)

Hence using that

$$\langle 0|a(0)|0\rangle = \langle 0|a^\dagger(0)|0\rangle = 0$$
(11.6.66)

we find

$$\langle 0|x|0\rangle = 0$$
(11.6.67)

and

$$\langle 0|p|0\rangle = \frac{F}{\omega}\left(1 - \cos\omega t\right) \ .$$
(11.6.68)

11.7   Heisenberg Picture: Constant Force

A free particle is acted on by a constant force F. Use the Heisenberg picture to find the expectation values of p and x if the state of the particle is described by a Gaussian wavefunction.

Solution

Here, the Hamiltonian is given by

$$H = \frac{p^2}{2m} - Fx \ .$$
(11.7.69)

The state is

$$\psi_0(x) = \frac{1}{\sqrt{\pi}}\, e^{-x^2/2} \ .$$
(11.7.70)

The Heisenberg equations of motion yield

$$\frac{dx}{dt} = \frac{p}{m}$$
$$\frac{dp}{dt} = F \ .$$
(11.7.71)

Therefore, we find

$$p = p_0 + Ft \tag{11.7.72}$$

and

$$\frac{dx}{dt} = \frac{p_0}{m} + \frac{F}{m}t \tag{11.7.73}$$

so that

$$x = x_0 + \frac{p_0}{m}t + \frac{F}{2m}t^2 . \tag{11.7.74}$$

Since the Gaussian wavefunction is an even parity state we immediately find that

$$(\psi_0, p_0\psi_0) = (\psi_0, x_0\psi_0) = 0 . \tag{11.7.75}$$

So, we get

$$(\psi_0, p\psi_0) = Ft \tag{11.7.76}$$

and

$$(\psi_0, x\psi_0) = \frac{F}{2m}t^2 . \tag{11.7.77}$$

11.8 Schrödinger Picture: Constant Force

Repeat problem 11.6 in the Schrödinger picture.
Hint: Do not try to actually find the time-dependent wavefunction, but try to find an alternate method. It may help to review some of the results of [11.1]. The point that this problem illustrates is that in this case (a constant force) it is much simpler to do the calculation in the Heisenberg picture.

Solution

In this case we could try to first solve the time-dependent Schrödinger equation

$$i\hbar\frac{\partial\Psi}{\partial t} = -\frac{\hbar^2}{2m}\frac{\partial^2\Psi}{\partial x^2} - Fx\Psi . \tag{11.8.78}$$

We would do this by first separating out the time

$$\Psi(t,x) = \psi(x)\,e^{-iEt/\hbar} \tag{11.8.79}$$

so that

$$-\frac{\hbar^2}{2m}\frac{d^2\psi}{dx^2} - Fx\psi = E\psi . \tag{11.8.80}$$

This equation leads to Bessel functions of order $\pm 1/3$. We would then have to expand the initial state, namely

$$\psi_0(x) = \frac{1}{\sqrt{\pi}}\,e^{-x^2/2} \tag{11.8.81}$$

in terms of these eigenfunctions to get the time-dependent solution. This is rather complicated. The expectation values of p and x could then be evaluated using this result.

A better way to do this is to use the results discussed in [11.1]

$$\frac{d\langle p\rangle}{dt} = -\frac{i}{\hbar}(\Psi, [p, H]\Psi) = (\Psi, F\Psi) = F \tag{11.8.82}$$

This simply says that

$$\langle p\rangle = Ft . \tag{11.8.83}$$

Therefore,

$$\frac{d\langle x\rangle}{dt} = -\frac{i}{\hbar}(\Psi, [x, H]\Psi) = (\Psi, \frac{p}{m}t\Psi) = \frac{F}{m}t \tag{11.8.84}$$

and hence,

$$\langle x\rangle = \langle x\rangle|_{t=0} + \frac{F}{2m}t^2 . \tag{11.8.85}$$

But at $t = 0$ we have that $\langle x\rangle = 0$. Thus, we recover the results obtained in the Heisenberg picture.

11.9   Dirac Picture: Constant Magnetic Field

An electron in the potential $V(\vec{x})$ is also acted on by a weak constant magnetic field so that the resultant Hamiltonian is

$$H = \frac{1}{2m}\left(\vec{p} - eA(\vec{x})\right)^2 + V(\vec{x})$$

with

$$\vec{A} = -\frac{1}{2}\vec{x} \times \vec{B} .$$

Introduce the interaction picture with

$$H_0 = \frac{\vec{p}^2}{2m} + V(\vec{x})$$

and find the equations of motion for $\vec{p}$ and $\vec{x}$ in this picture. Write also the Schrödinger equation for the wavefunction in the interaction picture.
Hint: Use the fact that the magnetic field is weak as well as, the result that you will prove later (problem 17.8), that if $\vec{B}$ is a constant vector then

$$\vec{p} \cdot \vec{A} + \vec{A} \cdot \vec{p} = \vec{B} \cdot \vec{L} .$$

Solution

In the Interaction picture the equation of motion for any dynamical variable is given by

$$i\hbar \frac{dA_D}{dt} = [A_D, H_0] \; . \tag{11.9.86}$$

Therefore using

$$H_0 = \frac{\vec{p}^2}{2m} + V(\vec{x}) \tag{11.9.87}$$

we get

$$\frac{d\vec{x}_D}{dt} = \frac{\vec{p}_D}{m} \;\; , \quad \frac{d\vec{p}_D}{dt} = -\nabla V(\vec{x}_D) \; . \tag{11.9.88}$$

Here we have used the fact that for any function of momentum

$$F_D(\vec{p}) = U_0^\dagger F(\vec{p}) U_0 = F(U_0^\dagger \vec{p} U_0) = F(\vec{p}_D) \tag{11.9.89}$$

where

$$U_0 = e^{-iH_0t/\hbar} \; . \tag{11.9.90}$$

The Schrödinger equation is

$$i\hbar \frac{\partial}{\partial t} |\Psi_D\rangle = H_D'|\Psi_D\rangle \tag{11.9.91}$$

where

$$H_D' = U_0^\dagger H' U_0 \; . \tag{11.9.92}$$

Now for a weak magnetic field we can drop the term $(e^2/2m)\, \vec{A} \cdot \vec{A}$. Thus,

$$H_D' = e^{iH_0t/\hbar} \left[-\frac{e}{2m} (\vec{p} \cdot \vec{A} + \vec{A} \cdot \vec{p}) \right] e^{-iH_0t/\hbar} \; . \tag{11.9.93}$$

But, using the hint that

$$\vec{p} \cdot \vec{A} + \vec{A} \cdot \vec{p} = \vec{B} \cdot \vec{L} \tag{11.9.94}$$

we see (since $\vec{B} \cdot \vec{L}$ commutes with H_0) that

$$H_D' = -\frac{e}{2m} \vec{B} \cdot \vec{L} \; . \tag{11.9.95}$$

So, the Schrödinger equation now reads

$$i\hbar \frac{\partial}{\partial t} |\Psi_D\rangle = -\frac{e}{2m} \vec{B} \cdot \vec{L} |\Psi_D\rangle \; . \tag{11.9.96}$$

11.10   Coherent State

Show that the normalized coherent state

$$|z\rangle = e^{-|z|^2/2}\, e^{z a^\dagger}|0\rangle$$

may also be written

$$|z\rangle = e^{[z a^\dagger - z^* a]}|0\rangle \ .$$

Hint: Use the results of problem 9.14 (the Baker-Campbell-Hausdorff formula).

Solution

We have that

$$|z\rangle = e^{-z^2/2}\, e^{z a^\dagger}|0\rangle \tag{11.10.97}$$

as well as the result of problem 9.14 that if there are two operators A and B such that

$$[A,[A,B]] = [B,[A,B]] = 0 \tag{11.10.98}$$

then

$$e^{A+B} = e^A\, e^B\, e^{-1/2[A,B]} \ . \tag{11.10.99}$$

The operators $z a^\dagger$ and $z^* a$ satisfy this condition. Therefore we have that

$$e^{z a^\dagger - z^* a} = e^{z a^\dagger}\, e^{-z^* a}\, e^{-1/2|z|^2} \ . \tag{11.10.100}$$

And since

$$a|0\rangle = 0 \tag{11.10.101}$$

so that

$$e^{-z^* a}|0\rangle = |0\rangle \ . \tag{11.10.102}$$

It follows that

$$|z\rangle = e^{z a^\dagger - z^* a}|0\rangle \ . \tag{11.10.103}$$

11.11   Coherent State: Overlap of Two States

Verify the equation

$$|\langle w|z\rangle|^2 = \exp - \big[|w|^2 + |z|^2 - w^* z - w z^*\big] = e^{-|w-z|^2}$$

for any two coherent states $|z\rangle$, $|w\rangle$.

Solution

We want to show that

$$|\langle w|z\rangle|^2 = e^{-[|w|^2+|z|^2-w^*z-wz^*]} = e^{-|w-z|^2} \qquad (11.11.104)$$

where

$$|z\rangle = e^{-|z|^2/2} \sum_{n=0}^{\infty} \frac{z^n}{\sqrt{n!}} |n\rangle . \qquad (11.11.105)$$

So, we have

$$
\begin{aligned}
\langle w|z\rangle &= e^{-(|z|^2+|w|^2)/2} \sum_{n,m=0}^{\infty} \frac{z^n}{\sqrt{n!}} \frac{w^{*m}}{\sqrt{m}} \langle m|n\rangle \\
&= e^{-(|z|^2+|w|^2)/2} \sum_{n=0}^{\infty} \frac{(w^*z)^n}{n!} \\
&= e^{-(|z|^2+|w|^2-2w^*z)/2} .
\end{aligned}
\qquad (11.11.106)
$$

Similarly,

$$\langle z|w\rangle = e^{-(|w|^2+|z|^2-2wz^*)/2} . \qquad (11.11.107)$$

Thus,

$$|\langle w|z\rangle|^2 = e^{-[|w|^2+|z|^2-w^*z-wz^*]} = e^{-|w-z|^2} . \qquad (11.11.108)$$

11.12   Coherent State: Wavefunction

Verify the equation

$$|z\rangle = (1-|z|^2)^{1/4} \exp\left(\frac{z}{2}a^{\dagger 2}\right) |0\rangle$$

where

$$|z\rangle = (1-|z|^2)^{1/4} \sum_{n=0}^{\infty} z^n \frac{\sqrt{(2n-1)!!}}{\sqrt{(2n)!!}} |2n\rangle .$$

Solution

We have that

$$|z\rangle = (1-|z|^2)^{1/4} \sum_{n=0}^{\infty} z^n \frac{\sqrt{(2n-1)!!}}{\sqrt{(2n)!!}} |2n\rangle . \qquad (11.12.109)$$

But, it is easy to see that

$$(2n)!! = (2n)(2n-2)(2n-4)\ldots(2) = 2^n n! \qquad (11.12.110)$$

and

$$(2n)! = (2n)!! \, (2n-1)!! \, .$$

(11.12.111)

Also,

$$|2n\rangle = \frac{1}{\sqrt{(2n)!}}(a^\dagger)^{2n}|0\rangle \, .$$

(11.12.112)

Thus,

$$
\begin{aligned}
|z\rangle &= \left(1-|z|^2\right)^{1/4}\sum_{n=0}^{\infty} z^n \frac{1}{\sqrt{2n!!}}\frac{1}{\sqrt{2n!!}}(a^\dagger)^{2n}|0\rangle \\
&= \left(1-|z|^2\right)^{1/4}\sum_{n=0}^{\infty} z^n \frac{1}{2^n n!}(a^\dagger)^{2n}|0\rangle \\
&= \left(1-|z|^2\right)^{1/4}\exp\left[z/2\,(a^\dagger)^2\right]|0\rangle
\end{aligned}
$$

(11.12.113)

as required.

11.13   Squeezing Operator

Use the result of problem 9.15 to derive the equation

$$S^\dagger(r)xS(r) = \sqrt{\frac{\hbar}{2m\omega}}S^\dagger(r)(a+a^\dagger)S(r) = e^r x$$

as well as the result

$$S^\dagger(r)pS(r) = -i\sqrt{\frac{m\omega\hbar}{2}}S^\dagger(r)(a-a^\dagger)S(r) = e^{-r}p$$

where

$$S(r) = \exp\left[\frac{r}{2}(a^{\dagger\,2} - a^2)\right] \, .$$

Solution

From problem 9.15 we know that if

$$[A, B] = \lambda A$$

(11.13.114)

then

$$A\,e^B = e^\lambda\, e^B\, A \, .$$

(11.13.115)

We apply this result to $A = (a \pm a^\dagger)$ and

$$e^B = S(r) = \exp\left[\frac{r}{2}(a^{\dagger\,2} - a^2)\right] \, .$$

(11.13.116)

As a first step we find

$$[a, \frac{r}{2}(a^{\dagger 2} - a^2)] = \frac{r}{2}2a^{\dagger} = ra^{\dagger} \tag{11.13.117}$$

and

$$[a^{\dagger}, \frac{r}{2}(a^{\dagger 2} - a^2)] = \frac{r}{2}2a = ra . \tag{11.13.118}$$

Therefore,

$$
\begin{aligned}
(a \pm a^{\dagger})S(r) &= (a \pm a^{\dagger})\exp\left[\frac{r}{2}(a^{\dagger 2} - a^2)\right] \\
&= \exp\left[\frac{r}{2}(a^{\dagger 2} - a^2)\right] e^{\pm r} (a \pm a^{\dagger}) \\
&= S(r) e^{\pm r} (a \pm a^{\dagger}) .
\end{aligned}
\tag{11.13.119}
$$

This is the desired result since

$$SS^{\dagger} = S^{\dagger}S = 1 . \tag{11.13.120}$$

11.14   No Eigenstates for Creation Operator

Show that the creation operator has no normalizable eigenstates.
Hint: Assume that eigenstates exist and show that they are not normalizable.

Solution

We assume that the creation operator $a^{\dagger}$ has a normalizable eigenstate $|z\rangle$ so that

$$a^{\dagger}|z\rangle = z|z\rangle \quad , \quad \langle z|z\rangle < \infty . \tag{11.14.121}$$

Next, we expand $|z\rangle$ in terms of the harmonic oscillator eigenkets $|n\rangle$ so that

$$|z\rangle = \sum_n c_n |n\rangle . \tag{11.14.122}$$

Then, using (11.14.121) we get

$$\sum_n c_n \sqrt{n+1}|n+1\rangle = z \sum_n c_n |n\rangle . \tag{11.14.123}$$

Therefore,

$$c_n = \frac{\sqrt{n}}{z} c_{n-1} = \frac{\sqrt{n!}}{z^n} c_0 . \tag{11.14.124}$$

Hence,

$$|z\rangle = c_0 \sum_n \frac{\sqrt{n!}}{z^n} |n\rangle \tag{11.14.125}$$

and

$$\langle z|z\rangle = |c_0|^2 \sum_n \frac{n!}{z^{2n}} \ . \tag{11.14.126}$$

This sum clearly diverges for all values of z. Thus, there are no normalizable eigenstates of the creation operator $a^\dagger$.

11.15 Spin Coherent State: Euler Angles

The Euler angle representation of a spin coherent state is defined by

$$|\theta, \varphi, \psi\rangle = e^{-i\varphi J_z/\hbar}\, e^{-i\theta J_y/\hbar}\, e^{-i\psi J_z/\hbar}\, |j, m\rangle \tag{11.15.127}$$

where $|j, m\rangle$ is an eigenstate of J^2 and J_z corresponding to the eigenvalues $j(j+1)\hbar^2$ and $m\hbar$ respectively. Since the action of $e^{-i\psi J_z/\hbar}$ on $|j, m\rangle$ is trivial it is usual to take $\psi = 0$. So we shall do the same and consider the spin coherent states

$$|\theta, \varphi\rangle = e^{-i\varphi J_z/\hbar}\, e^{-i\theta J_y/\hbar}\, |j, m\rangle \ . \tag{11.15.128}$$

For the state $|j, m\rangle$ with $j = 1$ evaluate the spin coherent states corresponding to
a) $m = 1$.
b) $m = 0$.
c) In both cases evaluate the overlap $\langle \theta', \varphi'|\theta, \varphi\rangle$.
Hint: Use the result obtained in problem 9.3 that for any component J_k $k = x,\, y,\, z$ of the angular momentum operator corresponding to $j = 1$ you have

$$e^{-i\theta J_k/\hbar} = i - \frac{J_k^2}{\hbar^2} + \frac{iJ_k}{\hbar}\sin\theta + \frac{J_k^2}{\hbar^2}\cos\theta \ . \tag{11.15.129}$$

Solution

For $j = 1$ we have that

$$\frac{J_z}{\hbar} = \begin{pmatrix} 1 & 0 & 0 \\ 0 & 0 & 0 \\ 0 & 0 & -1 \end{pmatrix} \tag{11.15.130}$$

and

$$\frac{J_y}{\hbar} = \frac{1}{\sqrt{2}}\begin{pmatrix} 0 & -i & 0 \\ i & 0 & -i \\ 0 & i & 0 \end{pmatrix} \ . \tag{11.15.131}$$

Therefore,

$$\frac{J_z^2}{\hbar^2} = \begin{pmatrix} 1 & 0 & 0 \\ 0 & 0 & 0 \\ 0 & 0 & 1 \end{pmatrix} \tag{11.15.132}$$

and

$$\frac{J_y^2}{\hbar^2} = \frac{1}{2} \begin{pmatrix} 1 & 0 & -1 \\ 0 & 2 & 0 \\ -1 & 0 & 1 \end{pmatrix} . \tag{11.15.133}$$

Now, using the hint we have that

$$e^{-i\varphi J_z/\hbar} = \begin{pmatrix} e^{-i\varphi} & 0 & 0 \\ 0 & 0 & 0 \\ 0 & 0 & e^{i\psi} \end{pmatrix} \tag{11.15.134}$$

and

$$e^{-i\theta J_y/\hbar} = \begin{pmatrix} \frac{1}{2}(1+\cos\theta) & -\frac{1}{\sqrt{2}}\sin\theta & \frac{1}{2}(1-\cos\theta) \\ \frac{1}{\sqrt{2}}\sin\theta & \cos\theta & -\frac{1}{\sqrt{2}}\sin\theta \\ \frac{1}{2}(1-\cos\theta) & -\frac{1}{\sqrt{2}}\sin\theta & \frac{1}{2}(1+\cos\theta) \end{pmatrix} . \tag{11.15.135}$$

a) If $m = 1$ then

$$|1,1\rangle = \begin{pmatrix} 1 \\ 0 \\ 0 \end{pmatrix} \tag{11.15.136}$$

and

$$|\theta, \varphi\rangle = \begin{pmatrix} -\frac{1}{2}e^{-i\varphi}(1+\cos\theta) \\ \frac{1}{\sqrt{2}}\sin\theta \\ \frac{1}{2}e^{i\varphi}(1-\cos\theta) \end{pmatrix} . \tag{11.15.137}$$

b) If $m = 0$ then

$$|1,0\rangle = \begin{pmatrix} 0 \\ 1 \\ 0 \end{pmatrix} \tag{11.15.138}$$

and

$$|\theta, \varphi\rangle = \begin{pmatrix} -\frac{1}{\sqrt{2}}e^{-i\varphi}\sin\theta \\ \cos\theta \\ \frac{1}{\sqrt{2}}e^{i\varphi}\sin\theta \end{pmatrix} . \tag{11.15.139}$$

c) For $m = 1$ the overlap is

$$\begin{aligned}
&\langle \theta', \varphi'|\theta, \varphi\rangle \\
&= \frac{1}{4}e^{-i(\varphi-\varphi')}(1+\cos\theta)(1+\cos\theta') + \frac{1}{4}\sin\theta\sin\theta' \\
&+ \frac{1}{4}e^{i(\varphi-\varphi')}(1-\cos\theta)(1-\cos\theta') \\
&= \frac{1}{2}(1+\cos\theta\cos\theta')\cos(\varphi-\varphi') - i(\cos\theta+\cos\theta')\sin(\varphi-\varphi') \\
&+ \frac{1}{2}\sin\theta\sin\theta' . \tag{11.15.140}
\end{aligned}$$

For $m = 0$ the overlap is

$$\langle \theta', \varphi' | \theta, \varphi \rangle$$
$$= \frac{1}{2} e^{-i(\varphi - \varphi')} \sin \theta \sin \theta' + \cos \theta \cos \theta'$$
$$+ \frac{1}{2} e^{i(\varphi - \varphi')} \sin \theta \sin \theta'$$
$$= \cos \theta \cos \theta' + \sin \theta \sin \theta' \cos(\varphi - \varphi') . \tag{11.15.141}$$

11.16 Minimum Uncertainty Spin Coherent States

Show that for any choice of the "fiducial vector" $|j, m\rangle$ not *all* the spin coherent states

$$|\theta, \varphi\rangle = e^{-i\varphi J_z / \hbar} e^{-i\theta J_y / \hbar} |j, m\rangle \tag{11.16.142}$$

can be simultaneously minimum uncertainty states for any pair of non-commuting operators. This is to be contrasted with the coherent states based on the creation and annihilation operators of the simple harmonic oscillator.

Hint: For the required proof first find an equation to be satisfied by any minimum uncertainty state. Then show that this equation does not have enough solutions to encompass *all* the spin coherent states.

Solution

Supppose the operators of interest are A, and B. To derive the Heisenberg uncertainty relation we first define

$$A' = A - \langle A \rangle \quad , \quad B' = B - \langle B \rangle , \tag{11.16.143}$$

where the expectation values are taken with the state ψ. Then,

$$(\Delta A)^2 = (A'\psi, A'\psi) \quad , \quad (\Delta B)^2 = (B'\psi, B'\psi) . \tag{11.16.144}$$

We now consider

$$
\begin{aligned}
|\langle [A, B] \rangle|^2 &= |\langle [A', B'] \rangle|^2 \\
&= (\psi, A'B'\psi) - (\psi, B'A'\psi) \\
&= (A'\psi, B'\psi) - (B'\psi, A'\psi) \\
&= (A'\psi, B'\psi) - (A'\psi, B'\psi)^* \\
&= 2i \Im(A'\psi, B'\psi) .
\end{aligned}
\tag{11.16.145}
$$

Therefore,

$$|\langle [A, B] \rangle| = |2\Im(A'\psi, B'\psi)| \leq 2|(A'\psi, B'\psi)| . \tag{11.16.146}$$

By the Schwarz inequality we get for the last term

$$2|(A'\psi, B'\psi)| \leq (A'\psi, A'\psi)^{1/2}(B'\psi, B'\psi)^{1/2} = 2\Delta A \Delta B \ . \qquad (11.16.147)$$

Thus, we obtain the Heisenberg uncertainty relation

$$\Delta A \Delta B \geq \frac{1}{2}|(\psi, [A, B]\psi)| \ . \qquad (11.16.148)$$

For a minimum uncertainty state this must be an equality and therefore both of the inequalities (in (11.16.146) and (11.16.147)) must also be equalities. The first of these requires that

$$|\Im(A'\psi, B'\psi)| = |(A'\psi, B'\psi)| \qquad (11.16.149)$$

and the second requires that

$$A'\psi = \lambda B'\psi \qquad (11.16.150)$$

where λ is some constant. Substituting this into (11.16.149) we get that

$$|\Im\lambda||(B'\psi, B'\psi)| = |\lambda||(B'\psi, B'\psi)| \qquad (11.16.151)$$

so that λ is pure imaginary

$$\lambda = i\alpha \ , \quad \alpha = \text{real} \ . \qquad (11.16.152)$$

We use these results to write out in detail (11.16.150) (the equation for the minimum uncertainty state ψ).

$$(A - \langle A \rangle)\psi = i\alpha(B - \langle B \rangle)\psi \ . \qquad (11.16.153)$$

This equation can be cast in the form of an eigenvalue equation

$$(A - i\alpha B)\psi = (\langle A \rangle - i\alpha\langle B \rangle)\psi = \beta\psi \ . \qquad (11.16.154)$$

Here,

$$\beta = \langle A \rangle - i\alpha\langle B \rangle \qquad (11.16.155)$$

is the eigenvalue.

Now, for spin coherent states, any operators A and B are simply $(2j + 1) \times (2j + 1)$ dimensional matrices. Therefore, to obtain solutions to the eigenvalue equation (11.16.154) we have to solve the characteristic equation

$$\det(A - i\alpha B - \beta) = 0 \ . \qquad (11.16.156)$$

However, this equation yields at most $2j+1$ distinct complex eigenvalues depending on the real parameter α. The corresponding eigenvectors may be labelled

$$|k(\alpha)\rangle, \beta(\alpha)) \quad k = 1, 2, \cdots, (2j + 1) \ .$$

These states can never represent *all* the states $|\theta, \varphi\rangle$ which are labelled by *two* continuous real variables. So, not all the spin coherent states $|\theta, \varphi\rangle$ can be minimum uncertainty states.

Incidentally, those spin coherent states that are minimum uncertainty states are called "intelligent spin states". [11.2]

11.17   Spin Coherent States: Complex Variables

Consider the spin coherent state

$$|z\rangle = N\, e^{(zJ_-)/(\hbar\sqrt{2})}\, |j,j\rangle , \qquad (11.17.157)$$

where z is a complex number, N is a normalization constant, and

$$J_- = J_x - iJ_y .$$

This is a complex variable representation of a spin coherent state.
a) Find the normalization constant N.
Hint: Use the fact that

$$J_-|j,j\rangle = \sqrt{(j-m+1)(j+m)}\hbar\, |j,j\rangle \qquad (11.17.158)$$

as well as

$$J_-|j,-j\rangle = 0 . \qquad (11.17.159)$$

b) Show that the overlap of two such states is given by

$$\langle z|z'\rangle = \frac{(1+z^*z')^{2j}}{(1+|z|^2)^j(1+|z'|^2)^j} . \qquad (11.17.160)$$

c) Verify the completeness relation

$$\frac{2j+1}{\pi}\int |z\rangle\langle z|\frac{d^2z}{(1+z^*z')^2} = 1 \qquad (11.17.161)$$

where

$$d^2z = dx\, dy , \qquad z = x + iy . \qquad (11.17.162)$$

Hint: To evaluate the integral it may be useful to use polar coordinates. Also the following integral may prove useful.

$$\int_0^\infty \frac{x^n\, dx}{(1+x)^m} = \frac{n!(m-n-2)!}{(m-1)!} \qquad n > -1 , \quad m > n+1 . \qquad (11.17.163)$$

Solution

a) To start we evaluate the action of J_-^n on $|j,j\rangle$ by making repeated use of (11.17.158).

$$
\begin{aligned}
J_-|j,j\rangle &= \sqrt{1(2j)}|j,j-1\rangle \\
J_-^2|j,j\rangle &= \sqrt{1(2j)}\sqrt{2(2j-1)}|j,j-2\rangle \\
J_-^3|j,j\rangle &= \sqrt{1(2j)}\sqrt{2(2j-1)}\sqrt{3(2j-2)}|j,j-3\rangle \\
&\cdots \\
J_-^{2j}|j,j\rangle &= \sqrt{1(2j)}\sqrt{2(2j-1)}\cdots\sqrt{2(2j-1)}\sqrt{1(2j)}|j,-j\rangle \quad (11.17.164)
\end{aligned}
$$

and finally

$$J_-^{2j+1}|j,j\rangle = 0 \ . \tag{11.17.165}$$

Expanding the exponential in the definition of $|z\rangle$ we get

$$\frac{1}{N}|z\rangle = \left(1 + (z/\sqrt{2})J_- + \frac{(z/\sqrt{2})^2}{2!}J_-^2 \right.$$

$$+ \cdots + \left.\frac{(z/\sqrt{2})^{2j}}{(2j)!}J_-^{2j} \right)|j,j\rangle$$

$$= |j,j\rangle + z\sqrt{1(j)}|j,j-1\rangle + \frac{z^2}{2!}\sqrt{j(2j-1)}|j,j-2\rangle$$

$$+ \cdots + z^{2j}|j,-j\rangle \ . \tag{11.17.166}$$

It now follows that

$$\frac{1}{|N|^2}\langle z|z\rangle = 1 + |z|^2 + (|z|^2)^2\frac{j(2j-1)}{2!} + (|z|^2)^3\frac{j(2j-1)(2j-2)}{3!}$$

$$+ \cdots + (|z|^2)^{2j}$$

$$= \sum_{n=0}^{2j}(|z|^2)^n \begin{pmatrix} 2j \\ n \end{pmatrix}$$

$$= \left(1 + |z|^2\right)^{2j} \ . \tag{11.17.167}$$

Therefore,

$$N = \left(1 + |z|^2\right)^{-j} \tag{11.17.168}$$

b) We begin with the definition of $|z\rangle$. Then,

$$\langle z|z'\rangle = \left(1 + |z|^2\right)^{-j}\left(1 + |z'|^2\right)^{-j} \times$$

$$\langle j,j|e^{(z^*J_+)/(\hbar\sqrt{2})}e^{(z'J_-)/(\hbar\sqrt{2})}|j,j\rangle \ . \tag{11.17.169}$$

But, as we saw above,

$$e^{(zJ_-)/(\hbar\sqrt{2})}|j,j\rangle = |j,j\rangle + z\sqrt{1(j)}|j,j-1\rangle + \frac{z^2}{2!}\sqrt{j(2j-1)}|j,j-2\rangle$$

$$+ \cdots + z^{2j}|j,-j\rangle \ . \tag{11.17.170}$$

Therefore,

$$\langle j,j|e^{(z^*J_+)/(\hbar\sqrt{2})}e^{(z'J_-)/(\hbar\sqrt{2})}|j,j\rangle$$

$$= 1 + (z^*z') + (z^*z')^2\frac{j(2j-1)}{2!} + (z^*z')^3\frac{j(2j-1)(2j-2)}{3!}$$

$$+ \cdots + (z^*z')^{2j}$$

$$= \sum_{n=0}^{2j}(z^*z')^n \begin{pmatrix} 2j \\ n \end{pmatrix}$$

$$= \left(1 + z^*z'\right)^{2j} \ . \tag{11.17.171}$$

So,

$$\langle z|z'\rangle = \frac{\left(1+|z|^2\right)^{-j}\left(1+|z'|^2\right)^{-j}}{\left(1+z^*z'\right)^{2j}}\ . \tag{11.17.172}$$

c) To verify the completeness relation we simply write it out.

$$\frac{2j+1}{\pi}\int\frac{d^2z}{(1+|z|^2)^2}|z\rangle\langle z|$$

$$=\ \frac{2j+1}{\pi}\int\frac{d^2z}{(1+|z|^2)^{2j+2}}\left(|j,j\rangle+z\sqrt{j}|j,j-1\rangle+\cdots+z^{2j}|j,-j\rangle\right)$$

$$\times\ \left(\langle j,j|+z^*\sqrt{j}\langle j,j-1|+\cdots+(z^*)^{2j}\langle j,-j|\right)\ . \tag{11.17.173}$$

The typical term involves an integral of the form

$$\int\frac{d^2z\,z^n(z^*)^m}{(1+|z|^2)^{2j+2}}\ .$$

If, in this integral, we let

$$x=r\cos\theta\ ,\quad y=r\sin\theta\ ,\quad dx\,dy=r\,dr\,d\theta \tag{11.17.174}$$

so that

$$z=r\,e^{i\theta}\ ,\quad z^*=r\,e^{-i\theta} \tag{11.17.175}$$

then,

$$\int\frac{d^2z\,z^n(z^*)^m}{(1+|z|^2)^{2j+2}}\ =\ \int_0^\infty\frac{r^{n+m+1}\,dr}{(1+r^2)^{2j+2}}\int_0^{2\pi}e^{i(n-m)\theta}\,d\theta$$

$$=\ 2\pi\,\delta_{n,m}\int_0^\infty\frac{r^{n+m+1}\,dr}{(1+r^2)^{2j+2}}\ . \tag{11.17.176}$$

So, only the diagonal terms survive. But these are now all of the form

$$\frac{2j+1}{\pi}2\pi\left(\begin{array}{c}2j\\n\end{array}\right)\int_0^\infty\frac{r^{2n+1}\,dr}{(1+r^2)^{2j+2}}|j,j-n\rangle\langle j,j-n|$$

$$=\ (2j+1)|j,j-n\rangle\langle j,j-n|\left(\begin{array}{c}2j\\n\end{array}\right)\int_0^\infty\frac{x^n\,dx}{(1+x)^{2j+2}}$$

$$=\ (2j+1)|j,j-n\rangle\langle j,j-n|\left(\begin{array}{c}2j\\n\end{array}\right)\frac{n!(2j-n)!}{(2j+1)!}$$

$$=\ |j,j-n\rangle\langle j,j-n|\ . \tag{11.17.177}$$

Here we have used the integral given in the "hint". Thus, finally we have

$$\frac{2j+1}{\pi}\int|z\rangle\langle z|\frac{d^2z}{(1+z^*z')^2}\ =\ \sum_{n=0}^{2j}|j,j-n\rangle\langle j,j-n|$$

$$=\ \sum_{m=-j}^{j}|j,m\rangle\langle j,m|$$

$$=\ 1\ . \tag{11.17.178}$$

11.18 Useful Commutator

Prove that

$$[a^\dagger a, (a^\dagger)^n] = n(a^\dagger)^n .$$

Hint: Use induction.

Solution

We want to prove, using induction, that

$$[a^\dagger a, (a^\dagger)^n] = n(a^\dagger)^n . \qquad (11.18.179)$$

For $n = 1$ we have

$$[a^\dagger a, a^\dagger] = a^\dagger[a, a^\dagger] = a^\dagger . \qquad (11.18.180)$$

Now we assume that (11.18.179) holds for n and we consider $n + 1$. Then we have

$$
\begin{aligned}
[a^\dagger a, (a^\dagger)^{n+1}] &= [a^\dagger a, (a^\dagger)^n]a^\dagger + (a^\dagger)^n[a^\dagger a, a^\dagger] \\
&= n(a^\dagger)^n a^\dagger + (a^\dagger)^n a^\dagger \\
&= (n+1)(a^\dagger)^{n+1} . \qquad (11.18.181)
\end{aligned}
$$

11.19 Forced SHO

Given the Hamiltonian for a "forced" simple harmonic oscillator

$$H(t) = \frac{p^2}{2m} + \frac{1}{2}m\omega^2 x^2 - xF(t) - pG(t)$$

where both $F(t)$ and $G(t)$ vanish outside the interval $0 < t < T$, calculate the probability $P(n)$ that the system is in the state of n quanta of the Hamiltonian $H(t)$ ($t > T$) if it was originally in the ground state of the Hamiltonian $H(t)$ ($t < 0$).
Hint: Use the Heisenberg picture.

Solution

The Hamiltonian may be rewritten to read

$$H = \frac{1}{2}\hbar\omega(a^\dagger a + 1/2) - \hbar f(t)^* a^\dagger - \hbar f(t) a \qquad (11.19.182)$$

where we have defined

$$f(t) = \frac{1}{\sqrt{2m\hbar\omega}}F(t) - i\sqrt{\frac{m\omega}{2\hbar}}G(t) . \qquad (11.19.183)$$

The Heisenberg equations are

$$\frac{da}{dt} + i\omega\, a = if^*$$

$$\frac{da^\dagger}{dt} - i\omega\, a^\dagger = -if \ . \tag{11.19.184}$$

These equations are solved by using integrating factors $e^{i\omega t}$ and $e^{-i\omega t}$ respectively to give

$$a(t) = i\, e^{-i\omega t} \int_0^t e^{i\omega t'}\, f^*(t')\, dt' + a\, e^{-i\omega t}$$

$$a^\dagger(t) = i\, e^{i\omega t} \int_0^t e^{-i\omega t'}\, f(t')\, dt' + a^\dagger\, e^{i\omega t} \tag{11.19.185}$$

where

$$a = a(0) \quad , \quad a^\dagger = a^\dagger(0) \ . \tag{11.19.186}$$

We now define

$$
\begin{aligned}
b(t) &= a\, e^{-i\omega t} + i\, e^{-i\omega t} \int_0^t e^{i\omega t'}\, f^*(t')\, dt' \\
&= e^{-i\omega t} (a + c) \\
b^\dagger(t) &= e^{i\omega t} (a^\dagger + c^*) \ .
\end{aligned} \tag{11.19.187}
$$

In this case the state $|n;b\rangle$ of n quanta of the Hamiltonian $H(t > T)$ is given by

$$|n;b\rangle = \frac{(b^\dagger)^n}{\sqrt{n!}} |0;b\rangle \tag{11.19.188}$$

where

$$b|0;b\rangle = e^{-i\omega t} (a + c)|0;b\rangle = 0 \ . \tag{11.19.189}$$

This shows that the state $|0;b\rangle$ is a coherent state, namely an eigenstate of the annihilation operator a. Thus, we can write

$$|0;b\rangle = e^{-i\omega t}\, e^{-|c|^2/2}\, e^{-ca^\dagger} |0;a\rangle \tag{11.19.190}$$

where $|0,a\rangle$ is the state annihilated by a and gives the ground state of the Hamiltonian $H(t < 0)$.

We want the amplitude

$$
\begin{aligned}
\langle n;b|0;a\rangle &= e^{-in\omega t}\, e^{-|c|^2/2} \langle 0;a|e^{-c^*a} \frac{(a+c)^n}{\sqrt{n!}} |0;a\rangle \\
&= e^{-in\omega t}\, e^{-|c|^2/2} \frac{c^n}{\sqrt{n!}} \ .
\end{aligned} \tag{11.19.191}
$$

Therefore, the desired probability is given by

$$
\begin{aligned}
P(n) &= |\langle n;b|0;a\rangle|^2 \\
&= e^{-|c|^2} \frac{|c|^{2n}}{n!} \ .
\end{aligned} \tag{11.19.192}
$$

This is a Poisson distribution.

11.20 3-D Simple Harmonic Oscillator

A particle moves as a three-dimensional simple harmonic oscillator so that its Hamiltonian may be written

$$H = \frac{\vec{p}^2}{2m} + \frac{1}{2}m\omega^2\vec{r}^2 \ .$$

At time $t = 0$ the position of the particle is measured to be at $\vec{r} = \vec{r}_0$. Show that a measurement of the position of the particle at time $t = T/2$, where T is the period of the oscillator, is sure to yield the value $-\vec{r}_0$.
Hint: Use the Heisenberg equations of motion.

Solution

The Heisenberg equations of motion are formally identical with the classical equations. Furthermore, in the Heisenberg picture the state vector does not change with time. Thus, the Heisenberg equations are

$$\frac{d\vec{r}}{dt} = \frac{\vec{p}}{m}$$
$$\frac{d\vec{p}}{dt} = -m\omega^2\vec{r} \ . \tag{11.20.193}$$

The solutions of these equations are formally the same as the classical solutions.

$$\vec{r} = \hat{r}_0 \cos\omega t + \frac{\hat{p}_0}{m\omega}\sin\omega t \ . \tag{11.20.194}$$

Here we have already imposed the initial condition that at $t = 0$ we have $\vec{r} = \hat{r}_0$ where $\hat{r}_0$ is that operator whose eigenvalue at time $t = 0$ is $\vec{r}_0$.
Now, the period T of the oscillator is given by

$$T = \frac{2\pi}{\omega} \ .$$

Therefore, at $t = T/2$ we have that

$$\vec{r} = \hat{r}_0 \cos\pi + \frac{\hat{p}_0}{m\omega}\sin\pi = -\hat{r}_0 \ . \tag{11.20.195}$$

But, as stated, the state of the system is an eigenstate of $\hat{r}_0$. Therefore, we are certain to measure the value $-\vec{r}_0$ at time $t = T/2$.

11.21 Quadrupole Tensor

The quadrupole tensor of a system is defined by

$$Q_{ik} = \langle 3x_i x_k - \delta_{ik}|\vec{r}|^2\rangle \ ,$$

where $\vec{r}$ is the position vector and x_i are the components of $\vec{r}$. For a system in the state of definite angular momentum

$$\psi(r,\,\theta,\,\varphi) = R(r)\,Y_{lm}(\theta,\,\varphi)$$

evaluate the components of Q_{ik}.
Hint: Use the relation

$$\cos\theta\,Y_{lm} = \sqrt{\frac{(l+1)^2 - m^2}{(2l+1)(2l+3)}}Y_{l+1,m} + \sqrt{\frac{l^2 - m^2}{(2l-1)(2l+1)}}Y_{l-1,m}\ .$$

Solution

We begin by writing the components x_i in spherical coordinates and use the fact that

$$\int |R(r)|^2\,r^2\,dr = 1 \tag{11.21.196}$$

to write

$$
\begin{aligned}
Q_{xx} &= \langle r^2 \rangle \int (3\sin^2\theta\cos^2\varphi - 1)Y_{lm}^*\,Y_{lm}\,d\Omega \\[4pt]
Q_{xy} &= Q_{yx} = \langle r^2 \rangle \int 3\sin^2\theta\cos^2\varphi\,Y_{lm}^*\,Y_{lm}\,d\Omega \\[4pt]
Q_{yy} &= \langle r^2 \rangle \int (3\sin^2\theta\sin^2\varphi - 1)Y_{lm}^*\,Y_{lm}\,d\Omega \\[4pt]
Q_{xz} &= Q_{zx} = \langle r^2 \rangle \int 3\sin\theta\cos\varphi\cos\theta\,Y_{lm}^*\,Y_{lm}\,d\Omega \\[4pt]
Q_{yz} &= Q_{zy} = \langle r^2 \rangle \int 3\sin\theta\sin\varphi\cos\theta\,Y_{lm}^*\,Y_{lm}\,d\Omega \\[4pt]
Q_{zz} &= \langle r^2 \rangle \int (3\cos^2\theta - 1)Y_{lm}^*\,Y_{lm}\,d\Omega\ .
\end{aligned}
\tag{11.21.197}
$$

Next we recall that

$$Y_{lm}(\theta,\,\varphi) \propto P_l^m(\cos\theta)\,e^{im\varphi}\ . \tag{11.21.198}$$

In this case it follows that

$$
\begin{aligned}
&\int_0^{2\pi} \sin\varphi\,Y_{lm}^*(\theta,\,\varphi)Y_{lm}(\theta,\,\varphi)\,d\varphi \\[4pt]
&= \int_0^{2\pi} \cos\varphi\,Y_{lm}^*(\theta,\,\varphi)Y_{lm}(\theta,\,\varphi)\,d\varphi \\[4pt]
&= 0
\end{aligned}
\tag{11.21.199}
$$

and hence that

$$Q_{xy} = Q_{yx} = Q_{xz} = Q_{zx} = Q_{yz} = Q_{zy} = 0\ . \tag{11.21.200}$$

By exactly the same argument and using that

$$\sin^2 \varphi = \frac{1}{2}[1 - \cos 2\varphi]$$

$$\cos^2 \varphi = \frac{1}{2}[1 + \cos 2\varphi]$$

we see that

$$
\begin{aligned}
Q_{xx} &= Q_{yy} = \langle r^2 \rangle \int (\frac{3}{2}\sin^2 \theta - 1)Y_{lm}^* Y_{lm} \, d\Omega \\
&= \frac{1}{2}\langle r^2 \rangle \int (1 - 3\cos^2 \theta)Y_{lm}^* Y_{lm} \, d\Omega \\
&= -\frac{1}{2}Q_{zz} \ .
\end{aligned}
\tag{11.21.201}
$$

Now using the hint twice we have that

$$
\begin{aligned}
\cos^2 \theta \, Y_{lm} &= AY_{l+2,m} + BY_{l-2,m} \\
&+ \left[\frac{(l+1)^2 - m^2}{(2l+1)(2l+3)} + \frac{l^2 - m^2}{(2l+1)(2l-1)} \right] Y_{l,m} \ ,
\end{aligned}
\tag{11.21.202}
$$

where A and B are irrelevant constants. It now follows that

$$
Q_{zz} = \left\{ 3 \left[\frac{(l+1)^2 - m^2}{(2l+1)(2l+3)} + \frac{l^2 - m^2}{(2l+1)(2l-1)} \right] - 1 \right\} \langle r^2 \rangle \ .
\tag{11.21.203}
$$

After some simplification this reads

$$
Q_{zz} = \frac{2l(2l+1) - 6m^2}{(2l-1)(2l+3)} \langle r^2 \rangle \ .
\tag{11.21.204}
$$

As a direct check we also see immediately from the definitions for Q_{xx}, Q_{yy} and Q_{zz} that

$$
Q_{xx} + Q_{yy} + Q_{zz} = 0 \ .
\tag{11.21.205}
$$

11.22   Eigenfunction of J^2, L^2, and J_z

A spin 1/2 particle is in a central potential. Find the explicit form of the wavefunction which is simultaneously an eigenfunction of

$$J^2, \ L^2 \text{ and } J_z = L_z + S_z \ .$$

Solution

In terms of the Pauli spinors we have that

$$\vec{J} = \vec{L}\mathbf{1} + \frac{1}{2}\hbar\vec{\sigma} \ . \tag{11.22.206}$$

So, in matrix form we have

$$J_z = \hbar \begin{pmatrix} -i\frac{\partial}{\partial\varphi} + \frac{1}{2} & 0 \\ 0 & -i\frac{\partial}{\partial\varphi} - \frac{1}{2} \end{pmatrix} \ . \tag{11.22.207}$$

The eigenfunctions of J_z corresponding to the eigenvalue $m\hbar$ are

$$\Psi = \begin{pmatrix} R_1(r)\, F_1(\theta)\, e^{i(m-1/2)\varphi} \\ R_2(r)\, F_2(\theta)\, e^{i(m+1/2)\varphi} \end{pmatrix} \ . \tag{11.22.208}$$

Since we also want these to be eigenfunctions of L^2 we require that

$$L^2\Psi = l(l+1)\hbar^2\Psi \tag{11.22.209}$$

or, written out, that

$$\begin{aligned} L^2 R_1(r)\, F_1(\theta)\, e^{i(m-1/2)\varphi} &= l(l+1)\hbar^2 R_1(r)\, F_1(\theta)\, e^{i(m-1/2)\varphi} \\ L^2 R_2(r)\, F_2(\theta)\, e^{i(m+1/2)\varphi} &= l(l+1)\hbar^2 R_2(r)\, F_2(\theta)\, e^{i(m+1/2)\varphi}. \end{aligned} \tag{11.22.210}$$

This means that

$$\begin{aligned} F_1(\theta)\, e^{i(m-1/2)\varphi} &= Y_{l,m-1/2}(\theta,\varphi) \\ F_2(\theta)\, e^{i(m+1/2)\varphi} &= Y_{l,m+1/2}(\theta,\varphi) \ . \end{aligned} \tag{11.22.211}$$

Finally, we require that Ψ also be an eigenfunction of J^2. To achieve this we write out J^2 explicitly as a matrix using the fact that, acting on Y_{lm}

$$J^2 = l(l+1)\hbar^2 + \frac{3}{4}\hbar^2 + \hbar\vec{\sigma}\cdot\vec{L} \ . \tag{11.22.212}$$

Then,

$$J^2 = \hbar^2 \begin{pmatrix} l(l+1) + \frac{3}{4} + \frac{1}{\hbar}L_z & \frac{1}{\hbar}L_- \\ \frac{1}{\hbar}L_+ & l(l+1) + \frac{3}{4} - \frac{1}{\hbar}L_z \end{pmatrix} \tag{11.22.213}$$

where

$$L_\pm Y_{l,m} = (L_x \pm iL_y)Y_{l,m} = \sqrt{l(l+1) - m(m\pm 1)}\,\hbar Y_{l,m\pm 1} \ . \tag{11.22.214}$$

So, writing out the pair of equations resulting from the eigenvalue equation

$$J^2\,\Psi = j(j+1)\hbar^2\,\Psi \tag{11.22.215}$$

we get

$$[(l+1/2)^2 - j(j+1) + m]R_1 + \sqrt{(l+1/2)^2 - m^2}\,R_2 \;=\; 0$$
$$\sqrt{(l+1/2)^2 - m^2}\,R_1 + [(l+1/2)^2 - j(j+1) - m]R_2 \;=\; 0 \;.(11.22.216)$$

For a nontrivial solution we require that the determinant of the coefficients of R_1 and R_2 must vanish. This yields

$$[(l+1/2)^2 - j(j+1)]^2 - m^2 - (l+1/2)^2 + m^2 = 0 \;. \tag{11.22.217}$$

The two roots are

$$j = l + 1/2 \quad \text{and} \quad j = l - 1/2 \;. \tag{11.22.218}$$

The corresponding (normalized) eigenfunctions are
$j = l + 1/2, \quad l = 0, 1, 2, \cdots$

$$\Psi_{l,j=l+1/2,m} = \frac{R_{j-1/2}(r)}{\sqrt{2l+1}} \begin{pmatrix} \sqrt{l+m+1/2}\,Y_{l,m-1/2} \\ \sqrt{l-m+1/2}\,Y_{l,m+1/2} \end{pmatrix} \tag{11.22.219}$$

$j = l - 1/2, \quad l = 1, 2, 3, \cdots$

$$\Psi_{l,j=l-1/2,m} = \frac{R_{j+1/2}(r)}{\sqrt{2l+1}} \begin{pmatrix} \sqrt{l-m+1/2}\,Y_{l,m-1/2} \\ -\sqrt{l+m+1/2}\,Y_{l,m+1/2} \end{pmatrix} \;. \tag{11.22.220}$$

11.23   SHO: A Time-independent Operator

Show that for a simple harmonic oscillator, the operator

$$A(t) = m\omega x(t)\cos\omega t - p(t)\sin\omega t$$

is independent of the time t.

Can this operator be simultaneously diagonalized with the Hamiltonian?

Solution

The equation of motion (Heisenberg equation) for a time-dependent operator is

$$\frac{dA}{dt} = \frac{1}{i\hbar}[A, H] + \frac{\partial A}{\partial t} \;. \tag{11.23.221}$$

Here, the Hamiltonian is

$$H = \frac{p^2}{2m} + \frac{1}{2}m\omega^2 x^2 \;. \tag{11.23.222}$$

Thus, we get

$$\begin{aligned} \frac{dA}{dt} &= \frac{m\omega}{2m}2p\cos\omega t + \frac{1}{2}m\omega^2 2x\sin\omega t \\ &- m\omega^2 x\sin\omega t - \omega p\cos\omega t \\ &= 0 \;. \end{aligned} \tag{11.23.223}$$

This shows that the operator A is independent of t. We can go even further and obtain an explicit form for A by using the solutions for $x(t)$ and $p(t)$.

$$
\begin{aligned}
x &= x_0 \cos \omega t + \frac{p_0}{m\omega} \sin \omega t \\
p &= p_0 \cos \omega t - m\omega x_0 \sin \omega t \ .
\end{aligned}
\tag{11.23.224}
$$

Here, x_0 and p_0 are the initial conditions at $t = 0$ for x and p respectively. They are nothing other than the Schrödinger picture operators for x and p. Then, we see that

$$
\begin{aligned}
A &= m\omega \left[x_0 \cos \omega t + \frac{p_0}{m\omega} \sin \omega t \right] \cos \omega t \\
  &\quad - \left[p_0 \cos \omega t - m\omega x_0 \sin \omega t \right] \sin \omega t \\
  &= m\omega x_0 \ .
\end{aligned}
\tag{11.23.225}
$$

Thus, it is obvious that A is indeed a constant operator, independent of t.
    It is also clear that

$$
[A, H] \neq 0 \ .
$$

Therefore, it is not possible to simultaneously diagonalize A and H.

11.24 SHO with Time-Dependent Spring

Consider a simple harmonic oscillator whose Hamiltonian is

$$
H(t) = \frac{p^2}{2m} + \frac{1}{2} m\omega^2(t) x^2 \ .
$$

a) Define the time-dependent annihilation and creation operators

$$
\begin{aligned}
a(t) &= \sqrt{\frac{m\omega(t)}{2\hbar}} x + i \frac{1}{\sqrt{2m\hbar\omega(t)}} p \\
a^\dagger(t) &= \sqrt{\frac{m\omega(t)}{2\hbar}} x - i \frac{1}{\sqrt{2m\hbar\omega(t)}} p \ .
\end{aligned}
\tag{11.24.226}
$$

Use the Heisenberg equation, either directly or by first finding the equations of motion for x and p, to obtain the equations of motion for $a(t)$ and $a^\dagger(t)$.
b) Now rewrite $H(t)$ in terms of $a(t)$ and $a^\dagger(t)$. Next use the fact that the operators $a(t)$ and $a^\dagger(t)$ are related to the operators $a(0)$ and $a^\dagger(0)$ by a Bogoliubov transformation to show that the expectation value of $H(t)$ for a state $|n\rangle$, that is an eigenstate of $H(0)$,

$$
H(0)|n\rangle = (n + 1/2)\hbar\omega(0)|n\rangle
\tag{11.24.227}
$$

is of the form

$$
\langle n|H(t)|n\rangle = (n + 1/2)\hbar\omega(t)f(t)
\tag{11.24.228}
$$

where $f(t)$ is some function of t.

c) Find expressions, in terms of the parameters of the Bogoliubov transformation, for the expectation values at time t of x, p, as well as x^2 and p^2 and hence also for Δx and Δp at time t. Use the results obtained to conclude that it is possible to alternately squeeze x and p.

Hint: Use the Heisenberg picture. Also do not attempt to obtain explicit solutions for the time dependence. [11.3]

Solution

The Heisenberg equations are obtained in the standard manner by using

$$i\hbar\frac{dA}{dt} = [A, H] + i\hbar\frac{\partial A}{\partial t} \ . \tag{11.24.229}$$

As a consequence we get

$$\frac{dx}{dt} = \frac{p}{m} \quad , \quad \frac{dp}{dt} = -\omega^2(t)x \ . \tag{11.24.230}$$

Next, we introduce the explicitly time-dependent annihilation and creation operators

$$a(t) \;=\; \sqrt{\frac{m\omega(t)}{2\hbar}}x + i\frac{1}{\sqrt{2m\hbar\omega(t)}}p$$

$$a^\dagger(t) \;=\; \sqrt{\frac{m\omega(t)}{2\hbar}}x - i\frac{1}{\sqrt{2m\hbar\omega(t)}}p \ . \tag{11.24.231}$$

A straightforward computation now shows that these satisfy the commutation relations for creation and annihilation operators.

$$[a(t), a^\dagger(t)] = -\frac{i}{2\hbar}[x, p] + \frac{i}{2\hbar}[p, x] = 1 \ . \tag{11.24.232}$$

b) Also, by straightforward computation we see that

$$\frac{1}{2}\hbar\omega(t)[a(t)a^\dagger(t) + a^\dagger(t)a(t)] \;=\; \frac{p^2}{2m} + \frac{1}{2}m\omega^2(t)x^2$$

$$=\; H(t) \ . \tag{11.24.233}$$

Thus, using the commutation relation (11.24.232) we find that

$$H(t) = \hbar\omega(t)[a^\dagger(t)a(t) + 1/2] \ . \tag{11.24.234}$$

The Heisenberg equations for $a(t)$ are

$$i\hbar\frac{da}{dt} \;=\; [a(t), H(t)] + i\hbar\frac{\partial a}{\partial t}$$

$$=\; \hbar\omega(t)a(t) + i\hbar\frac{\partial a}{\partial t} \ . \tag{11.24.235}$$

But, using the explicit form for $a(t)$ and denoting differentiation with respect to t by a dot the last term on the right hand side of (11.24.235) yields

$$\frac{\partial a}{\partial t} = \frac{\dot{\omega}}{2\omega}[\sqrt{\frac{m\omega(t)}{2\hbar}}x - i\frac{1}{\sqrt{2m\hbar\omega(t)}}p] = \frac{\dot{\omega}}{2\omega}a^\dagger \quad. \tag{11.24.236}$$

Thus,

$$\frac{da}{dt} = -i\omega(t)a + \frac{\dot{\omega}}{2\omega}a^\dagger \quad. \tag{11.24.237}$$

In a similar manner or by taking the dagger of (11.24.237) we find that

$$\frac{da^\dagger}{dt} = i\omega(t)a^\dagger + \frac{\dot{\omega}}{2\omega}a \quad. \tag{11.24.238}$$

One can check that these equations are correct by writing out the Heisenberg equations for x and p

$$\frac{dx}{dt} = \frac{p}{m} \quad, \quad \frac{dp}{dt} = -m\omega^2(t)x \quad. \tag{11.24.239}$$

Now inverting equations (11.24.231) and substituting $a(t)$ and $a^\dagger(t)$ for x and p we again get (11.24.237) and (11.24.238).

The operators $a(t)$ and $a^\dagger(t)$ must be related to the operators $a(0)$ and $a^\dagger(0)$ by a Bogoliubov transformation. We parametrize the Bogoliubov transformation with the three real functions $\alpha(t)$, $\beta(t)$, and $\gamma(t)$ to get:

$$\begin{aligned} a(t) &= e^{-i\alpha(t)}a(0)\cosh\beta(t) + e^{i\gamma(t)}a^\dagger(0)\sinh\beta(t) \\ a^\dagger(t) &= e^{-i\gamma(t)}a(0)\sinh\beta(t) + e^{i\alpha(t)}a^\dagger(0)\cosh\beta(t) \quad. \end{aligned} \tag{11.24.240}$$

Now consider a state $|n\rangle$ such that

$$H(0)|n\rangle = (n + 1/2)\hbar\omega(0)|n\rangle \quad. \tag{11.24.241}$$

In other words we also have that

$$\begin{aligned} a(0)|n\rangle &= \sqrt{n}|n-1\rangle \\ a^\dagger(0)|n\rangle &= \sqrt{n+1}|n+1\rangle \quad. \end{aligned} \tag{11.24.242}$$

Using (11.24.234) we see that

$$\begin{aligned} \langle n|H(t)|n\rangle &= \langle n|\hbar\omega(t)[a^\dagger(t)a(t) + 1/2]|n\rangle \\ &= \langle n|\hbar\omega(t)[\sinh^2\beta a(0)a^\dagger(0) + \cosh^2\beta a^\dagger(0)a(0) + 1/2]|n\rangle \\ &= \hbar\omega(t)[(n+1)\sinh^2\beta + n\cosh^2\beta + 1/2] \\ &= (n+1/2)\hbar\omega(t)\cosh(2\beta) \quad. \end{aligned} \tag{11.24.243}$$

c) The expectation values of x and p vanish since both of these operators are linear in $a(0)$ and $a^\dagger(0)$. Thus,

$$\langle n|x(t)|n\rangle = \langle n|p(t)|n\rangle = 0 \quad. \tag{11.24.244}$$

On the other hand,

$$\langle n|x^2(t)|n\rangle = \frac{\hbar}{2m\omega(t)}\langle n|[a^\dagger(0) + a(0)]^2|n\rangle$$

$$= \frac{\hbar}{2m\omega(t)}\langle n|\cosh\beta\sinh\beta(e^{-i(\alpha-\gamma)} + e^{i(\alpha-\gamma)})[a(0)a^\dagger(0) + a^\dagger(0)a(0)]$$

$$+ \quad (\cosh^2\beta + \sinh^2\beta)[a(0)a^\dagger(0) + a^\dagger(0)a(0)]|n\rangle$$

$$= \frac{\hbar}{2m\omega(t)}(2n+1)[\cosh(2\beta) + \cos(\alpha - \gamma)\sinh(2\beta)] \ . \qquad (11.24.245)$$

Similarly,

$$\langle n|p^2(t)|n\rangle \quad = \quad -\frac{m\hbar\omega(t)}{2}\langle n|[a^\dagger(0) - a(0)]^2|n\rangle$$

$$= \quad \frac{m\hbar\omega(t)}{2}(2n+1) \times$$

$$\times \quad [\cosh(2\beta) - \cos(\alpha - \gamma)\sinh(2\beta)] \ . \qquad (11.24.246)$$

Since,

$$\langle n|x(t)|n\rangle = \langle n|p(t)|n\rangle = 0$$

it follows that

$$(\Delta x)^2 \quad = \quad \langle n|x^2(t)|n\rangle$$

$$= \quad \frac{\hbar}{2m\omega(t)}(2n+1)[\cosh(2\beta) + \cos(\alpha - \gamma)\sinh(2\beta)] \quad (11.24.247)$$

and

$$(\Delta p)^2 \quad = \quad \langle n|p^2(t)|n\rangle$$

$$= \quad \frac{m\hbar\omega(t)}{2}(2n+1)[\cosh(2\beta) - \cos(\alpha - \gamma)\sinh(2\beta)] \ . \ (11.24.248)$$

Notice that if

$$\cos(\alpha - \gamma) = 1 \qquad (11.24.249)$$

then

$$\langle n|p^2(t)|n\rangle \quad = \quad m\hbar\omega(t)(n + 1/2)\, e^{-2\beta}$$

$$\langle n|x^2(t)|n\rangle \quad = \quad \frac{\hbar}{m\omega(t)}(n + 1/2)\, e^{2\beta} \ . \qquad (11.24.250)$$

Thus, if $\beta > 0$ the variable p is squeezed. Similarly, by taking

$$\cos(\alpha - \gamma) = -1 \qquad (11.24.251)$$

we get

$$\langle n|p^2(t)|n\rangle \quad = \quad m\hbar\omega(t)(n + 1/2)\, e^{2\beta}$$

$$\langle n|x^2(t)|n\rangle \quad = \quad \frac{\hbar}{m\omega(t)}(n + 1/2)\, e^{-2\beta} \qquad (11.24.252)$$

and for positive β the variable x is squeezed. Of course, in both cases, if β is negative the situtation is reversed.

For completeness we write out the equations that determine the functions $\alpha(t)$, $\beta(t)$, and $\gamma(t)$. Substituting (11.24.240) into the equations of motion for $a(t)$ or else $a^\dagger(t)$ we get

$$
\begin{aligned}
\frac{da}{dt} &= \left(-i\dot{\alpha}\cosh\beta + \dot{\beta}\sinh\beta\right)e^{-i\alpha}a(0) \\
&\quad + \left(i\dot{\gamma}\sinh\beta + \dot{\beta}\cosh\beta\right)e^{i\gamma}a^\dagger(0) \\
&= -i\omega\left(e^{-i\alpha}\cosh\beta a(0) + e^{i\gamma}\sinh\beta a^\dagger(0)\right) \\
&\quad + \frac{\dot{\omega}}{\omega}\left(e^{-i\gamma}\sinh\beta a(0) + e^{i\alpha}\cosh\beta a^\dagger(0)\right) \ .
\end{aligned}
\tag{11.24.253}
$$

Now using the algebra

$$
[a(0), a^\dagger(0)] = 1
\tag{11.24.254}
$$

we get

$$
\begin{aligned}
\left(-i\dot{\alpha}\cosh\beta + \dot{\beta}\sinh\beta\right)e^{-i\alpha} &= -i\omega e^{-i\alpha}\cosh\beta + \frac{\dot{\omega}}{2\omega}e^{-i\gamma}\sinh\beta \\
\left(i\dot{\gamma}\sinh\beta + \dot{\beta}\cosh\beta\right)e^{i\gamma} &= -i\omega e^{i\gamma}\sinh\beta + \frac{\dot{\omega}}{2\omega}e^{i\alpha}\cosh\beta \ .
\end{aligned}
\tag{11.24.255}
$$

This simplifies to

$$
\begin{aligned}
(-i\dot{\alpha} + i\omega)\cosh\beta + (\dot{\beta} - \frac{\dot{\omega}}{2\omega}e^{i(\alpha-\gamma)})\sinh\beta &= 0 \\
(i\dot{\gamma} + i\omega)\sinh\beta + (\dot{\beta} - \frac{\dot{\omega}}{2\omega}e^{i(\alpha-\gamma)})\cosh\beta &= 0 \ .
\end{aligned}
\tag{11.24.256}
$$

We next separate real and imaginary parts and find

$$
\begin{aligned}
(\dot{\alpha} - \omega)\cosh\beta + \frac{\dot{\omega}}{2\omega}\sin(\alpha-\gamma)\sinh\beta &= 0 \\
(\dot{\gamma} + \omega)\sinh\beta - \frac{\dot{\omega}}{2\omega}\sin(\alpha-\gamma)\cosh\beta &= 0 \\
\dot{\beta} - \frac{\dot{\omega}}{2\omega}\cos(\alpha-\gamma) &= 0 \ .
\end{aligned}
\tag{11.24.257}
$$

The initial conditions are

$$
\alpha(0) = \beta(0) = \gamma(0) = 0 \ .
$$

We would have to solve these three equations, subject to these initial conditions, if we wanted the explicit time dependence.

Bibliography

[11.1] A.Z. Capri, *Nonrelativistic Quantum Mechanics* 3rd edition, World Scientific Publishing Co. Pte. Ltd., chapter 7, (2002) .

[11.2] C. Aragone, E. Chalbaud, and S. Salamó, J . Math. Phys. **17**, 1963, (1976).

[11.3] H.R.Lewis Jr. and W.B.Riesenfeld, J. Math. Phys. **10** ,1458, (1969). see also
H.R.Lewis Jr., J. Math. Phys. **9**,1976, (1968).

Chapter 12

Non-degenerate Perturbation Theory

12.1   Expansion of $1/|\vec{r}_1 - \vec{r}_2|$

Verify the formula

$$\frac{1}{|\vec{r}_1 - \vec{r}_2|} = \sum_{l=0}^{\infty} \frac{r_<^l}{r_>^{l+1}} \frac{4\pi}{2l+1} \sum_{m=-l}^{l} Y_{l,m}^*(\hat{r}_1) Y_{l,m}(\hat{r}_2) ,$$

where

$$\hat{r}_1 = (\theta, \varphi) \quad , \quad \hat{r}_2 = (\theta', \varphi') .$$

Hint: Solve the problem

$$\nabla^2 \psi(\vec{r}) = \delta(\vec{r} - \vec{r}')$$

by
a) expanding in spherical harmonics,
b) realizing that $\psi(\vec{r})$ is the potential for a unit charge located at $\vec{r}'$, and comparing the two solutions.

Solution

We begin with Poisson's equation for the potential ψ due to a unit point charge

$$\nabla^2 \psi = \delta(\vec{r} - \vec{r}') . \tag{12.1.1}$$

The solution of this equation is just

$$\psi(\vec{r}) = -\frac{1}{4\pi} \frac{1}{|\vec{r} - \vec{r}'|} . \tag{12.1.2}$$

Now, we also solve this differential equation in spherical coordinates. For this purpose we write

$$\psi(\vec{r}) = \sum_{l,m} a_{l,m}(r) Y_{l,m}(\theta, \varphi) \ . \tag{12.1.3}$$

In spherical coordinates we can use the completeness of the spherical harmonics to write

$$\delta(\vec{r} - \vec{r}') = \frac{1}{r^2} \delta(r - r') \sum_{l,m} Y_{l,m}^*(\hat{r}') Y_{l,m}(\hat{r}) \tag{12.1.4}$$

where

$$\hat{r} = (\theta, \varphi) \ , \quad \hat{r}' = (\theta', \varphi') \ . \tag{12.1.5}$$

Substituting all this in the Poisson equation we get

$$\frac{1}{r} \frac{d^2}{dr^2} [r \, a_{l,m}(r)] - \frac{l(l+1)}{r^2} a_{l,m}(r) = \frac{1}{r^2} \delta(r - r') Y_{l,m}^*(\hat{r}') \ . \tag{12.1.6}$$

Integrating this equation with rdr about r' from $r' - \epsilon$ to $r' + \epsilon$ we get in the limit as $\epsilon \to 0$

$$\int_{r'-\epsilon}^{r'+\epsilon} \frac{d^2}{dr^2} [r \, a_{l,m}(r)] \, dr = \frac{1}{r'} Y_{l,m}^*(\hat{r}') \ . \tag{12.1.7}$$

This shows that there is a discontinuity in the first derivative of $a_{l,m}(r)$.

$$\left. \frac{da_{l,m}}{dr} \right|_{r'+\epsilon} - \left. \frac{da_{l,m}}{dr} \right|_{r'-\epsilon} = \frac{1}{r'^2} Y_{l,m}^*(\hat{r}') \ . \tag{12.1.8}$$

For $r \neq r'$ the radial equation has the solutions

$$a_{l,m}(r) = A_{l,m} \, r^n \ \text{ with } \ n = l \ \text{ or } \ n = -(l+1) \tag{12.1.9}$$

so we have that

$$a_{l,m}(r) = \begin{cases} A_{l,m} \, r^l & r < r' \\ B_{l,m} \, r^{-(l+1)} & r > r' \end{cases} \ . \tag{12.1.10}$$

Here we have already imposed the boundary conditions that the solution has to be finite at $r = 0$ and has to vanish for $r \to \infty$. The solution also has to be continuous at $r = r'$ and to satisfy the condition of discontinuity in the first derivative. Imposing these two conditions we get

$$A_{l,m} \, r'^l = B_{l,m} \, r'^{-(l+1)} \tag{12.1.11}$$

or

$$B_{l,m} = A_{l,m} \, r'^{(2l+1)} \tag{12.1.12}$$

as well as

$$A_{l,m}\left[-(l+1)r'^{2l+1}r'^{-(l+2)} - lr'^{(l-1)}\right] = \frac{1}{r'^2}Y^*_{l,m}(\hat{r}') .$$ (12.1.13)

Thus,

$$A_{l,m} = -\frac{1}{2l+1}\frac{1}{r'^{(l+1)}}Y^*_{l,m}(\hat{r}') .$$ (12.1.14)

So finally we get:

$$\frac{1}{|\vec{r}-\vec{r}'|} = 4\pi \begin{cases} \sum_{l,m}\frac{1}{2l+1}\frac{r^l}{r'^{l+1}}Y^*_{l,m}(\hat{r}')Y_{l,m}(\hat{r}) & r < r' \\ \sum_{l,m}\frac{1}{2l+1}\frac{r'^l}{r^{l+1}}Y^*_{l,m}(\hat{r}')Y_{l,m}(\hat{r}) & r > r' \end{cases} .$$ (12.1.15)

But, this is just the desired result.

12.2   Second Order Correction to State

In the equation for the rth order correction to the state $|m\rangle$

$$^{(0)}\langle m|n\rangle^{(r)} = \frac{1}{E^{(0)}_n - E^{(0)}_m}\left[^{(0)}\langle m|H' - E^1_n|n\rangle^{(r-1)} - E^{(2)}_n \,^{(0)}\langle m|n\rangle^{(r-2)}\right.$$

$$\left. - E^{(3)}_n \,^{(0)}\langle m|n\rangle^{(r-3)} - \ldots - E^{(r-1)}_n \,^{(0)}\langle m|n\rangle^{(1)}\right] \quad m \neq n$$ (12.2.16)

set $r = 2$ and derive the equation

$$|n\rangle^{(2)} = \sum_{m,r\neq n}\frac{|m\rangle^{(0)}\,^{(0)}\langle m|H'|r\rangle^{(0)}\,^{(0)}\langle r|H'|n\rangle^{(0)}}{\left(E^{(0)}_n - E^{(0)}_m\right)\left(E^{(0)}_n - E^{(0)}_r\right)}$$

$$- \sum_{m\neq n}\frac{|m\rangle^{(0)}\,^{(0)}\langle m|H'|n\rangle^{(0)}\,^{(0)}\langle n|H'|n\rangle^{(0)}}{\left(E^{(0)}_n - E^{(0)}_m\right)^2}$$

$$- \frac{1}{2}\sum_{m\neq n}\frac{|m\rangle^{(0)}\left|\,^{(0)}\langle m|H'|n\rangle^{(0)}\right|^2}{\left(E^{(0)}_n - E^{(0)}_m\right)^2}$$ (12.2.17)

for the second order correction to the wave function.

Solution

We start with the equation (12.2.16) for the rth order correction to a state, namely

$$^{(0)}\langle m|n\rangle^{(r)} = \frac{1}{E^{(0)}_n - E^{(0)}_m}\left[^{(0)}\langle m|H' - E^1_n|n\rangle^{(r-1)} - E^{(2)}_n \,^{(0)}\langle m|n\rangle^{(r-2)}\right.$$

$$\left. - E^{(3)}_n \,^{(0)}\langle m|n\rangle^{(r-3)} - \ldots - E^{(r-1)}_n \,^{(0)}\langle m|n\rangle^{(1)}\right] \quad m \neq n$$ (12.2.18)

and we set $r = 2$. Thus, we get

$$^{(0)}\langle m|n\rangle^{(2)} = \frac{^{(0)}\langle m|H' - E_n^{(1)}|n\rangle^{(1)}}{E_n^{(0)} - E_m^{(0)}} .$$ (12.2.19)

Inserting, from equation (12.2.16) the expression for $^{(0)}\langle m|n\rangle^{(1)}$, namely

$$\begin{aligned}
^{(0)}\langle m|n\rangle^{(1)} &= \frac{1}{E_n^{(0)} - E_m^{(0)}} \, ^{(0)}\langle m|H' - E_n^1|n\rangle^{(0)} \\
&= \frac{1}{E_n^{(0)} - E_m^{(0)}} \, ^{(0)}\langle m|H'|n\rangle^{(0)} \quad m \neq n
\end{aligned}$$ (12.2.20)

we find, after writing everything out, the desired result, namely equation (12.2.17).

12.3   1/2 λx² Perturbation of SHO

Consider the Hamiltonian

$$H = \frac{p^2}{2m} + \frac{1}{2}kx^2 + \frac{1}{2}\lambda x^2 \quad k > 0$$

a) Find the exact energy of the nth state of this Hamiltonian and expand it to order λ^2 assuming $|\lambda| < k$.
b) Use perturbation theory, treating $(1/2)\lambda x^2$ as a perturbation, and find the energy of the nth state to order λ^2 .
c) Find a bound for the rth order correction and hence show that the perturbation series converges for $|\lambda| < k$.
Hint: For part c) find a simple diagonal bound for the perturbation Hamiltonian.

Solution

a) Here

$$H = \frac{p^2}{2m} + \frac{1}{2}kx^2 + \frac{1}{2}\lambda x^2 \quad k > |\lambda| .$$ (12.3.21)

We now define

$$\Omega^2 = \frac{k+\lambda}{m} \ , \quad \omega^2 = \frac{k}{m} \ , \quad \alpha^2 = \frac{\lambda}{m} .$$ (12.3.22)

The exact energy eigenvalues are given by

$$E_n = (n+1/2)\hbar\Omega = (n+1/2)\hbar\sqrt{\omega^2 + \alpha^2} .$$ (12.3.23)

If we now expand this in a binomial series in λ to get the corrections due to the perturbation we find

$$\begin{aligned}
E_n &= (n+1/2)\hbar\omega\sqrt{1 + \alpha^2/\omega^2} \\
&= (n+1/2)\hbar\omega\left[1 + \frac{\alpha^2}{2\omega^2} - \frac{\alpha^4}{8\omega^4} + \cdots\right] \\
&= (n+\frac{1}{2})\hbar\omega + (n+\frac{1}{2})\hbar\frac{\lambda}{2m\omega} - (n+\frac{1}{2})\hbar\frac{\lambda^2}{8m^2\omega^3} + \cdots .
\end{aligned}$$ (12.3.24)

b) Perturbation Theory

We have that

$$E_n^{(0)} = (n + 1/2)\hbar\omega . \qquad (12.3.25)$$

Then the first order correction is given by

$$
\begin{aligned}
E_n^{(1)} &= \lambda^{(0)}\langle n|H'|n\rangle^{(0)} \\
&= \frac{\lambda}{2}^{(0)}\langle n|x^2|n\rangle^{(0)} \\
&= \frac{\lambda\hbar}{4m\omega}^{(0)}\langle n|a^2 + a^{\dagger 2} + 2a^{\dagger}a + 1|n\rangle^{(0)} \\
&= \frac{\lambda\hbar}{2m\omega}(n + 1/2) .
\end{aligned}
\qquad (12.3.26)
$$

So this result agrees with the exact result to this order. Next we have

$$E_n^{(2)} = \lambda^2 \sum_{s \neq n} \frac{|^{(0)}\langle n|H'|s\rangle^{(0)}|^2}{E_n^{(0)} - E_s^{(0)}} . \qquad (12.3.27)$$

For $s \neq n$ we have

$$
\begin{aligned}
\lambda^{(0)}\langle n|H'|s\rangle^{(0)} &= \frac{\lambda\hbar}{2m\omega}^{(0)}\langle n|a^2 + a^{\dagger 2} + 2a^{\dagger}a + 1|s\rangle^{(0)} \\
&= \frac{\lambda\hbar}{2m\omega}\left[\sqrt{s(s-1)}\delta_{n,s-2} + \sqrt{(s+1)(s+2)}\delta_{n,s+2}\right] .
\end{aligned}
\qquad (12.3.28)
$$

So,

$$E_n^{(2)} = \frac{\hbar^2(\lambda^2/4)}{4m^2\omega^2} \sum_s \frac{s(s-1)\delta_{n,s-2} + (s+1)(s+2)\delta_{n,s+2}}{\hbar\omega(n + 1/2 - s - 1/2)} . \qquad (12.3.29)$$

Then,

$$
\begin{aligned}
E_n^{(2)} &= \frac{\hbar\lambda^2}{16m^2\omega^3}\left[\frac{(n+2)(n+1)}{-2} + \frac{n(n-1)}{2}\right] \\
&= -\frac{\hbar\lambda^2}{8m^2\omega^3}\left(n + \frac{1}{2}\right) .
\end{aligned}
\qquad (12.3.30)
$$

This again agrees with the exact result to this order.

c) The total Hamiltonian may be rewritten to read

$$H = H_0 + \lambda H' = \frac{p^2}{2m} + \frac{1}{2}kx^2 + \frac{\lambda}{k}[\frac{1}{2}kx^2] . \qquad (12.3.31)$$

This Hamiltonian is clearly bounded by

$$H = H_0 + \lambda H'' = \frac{p^2}{2m} + \frac{1}{2}kx^2 + \lambda\frac{1}{k}[\frac{p^2}{2m} + \frac{1}{2}kx^2] . \qquad (12.3.32)$$

It is now clear that

$$|\lambda|H'' < H_0 \quad \text{for} \quad |\lambda| < k . \qquad (12.3.33)$$

But, as stated, we also have

$$H' < H'' \ . \tag{12.3.34}$$

Therefore, the perturbation series is bounded by the perturbation series for H'' as long as $|\lambda| < k$. On the other hand, the rth order term for the perturbation series for H'' is

$$
\begin{aligned}
\lambda^r E_n^{(r)} &= (\lambda/k)^r \ {}^{(0)}\langle n| \frac{p^2}{2m} + \frac{1}{2}kx^2 |n\rangle^{(r-1)} \\
&= (\lambda/k)^r (n+1/2)\hbar\omega \ {}^{(0)}\langle n|n\rangle^{(r-1)} \\
&= (\lambda/k)^r (n+1/2)\hbar\omega \delta_{r,1} \ .
\end{aligned}
\tag{12.3.35}
$$

From this we see that we can bound the perturbation series for H' by a convergent result if $|\lambda| < k$. In fact,

$$
\begin{aligned}
|E_n - E_n^{(0)}| &< \sum_{r=1}^{\infty} (\lambda/k)^r (n+1/2)\hbar\omega\delta_{r,1} \\
&= (\lambda/k)(n+1/2)\hbar\omega \ .
\end{aligned}
\tag{12.3.36}
$$

12.4   1/4 λx⁴ **Perturbation of SHO**

a) Find the approximate ground state energy to second order for the Hamiltonian

$$H = \frac{p^2}{2m} + \frac{1}{2}kx^2 + \frac{1}{4}\lambda x^4 \quad k > 0$$

using the Rayleigh-Schrödinger perturbation theory.
b) Find the ground state correct to order λ.

Solution

a) The Hamiltonian is

$$H = H_0 + \lambda H' = \frac{p^2}{2m} + \frac{1}{2}kx^2 + \frac{1}{4}\lambda x^4 \ . \tag{12.4.37}$$

The unperturbed ground state energy is

$$E_0^{(0)} = \frac{1}{2}\hbar\omega \ . \tag{12.4.38}$$

The first order perturbation gives

$$
\begin{aligned}
\lambda E_0^{(1)} &= {}^{(0)}\langle 0| \frac{1}{4}\lambda x^4 |0\rangle^{(0)} \\
&= \frac{1}{4}\lambda \left(\frac{\hbar}{2m\omega} \right)^2 {}^{(0)}\langle 0|(a^\dagger + a)^4 |0\rangle^{(0)}
\end{aligned}
$$

$$= \frac{1}{4}\lambda\left(\frac{\hbar}{2m\omega}\right)^2 {}^{(0)}\langle 0|a^2a^{\dagger 2} + aa^{\dagger}aa^{\dagger}|0\rangle^{(0)}$$

$$= \frac{3}{4}\lambda\left(\frac{\hbar}{2m\omega}\right)^2 . \tag{12.4.39}$$

The second order perturbation correction is given by

$$\lambda^2 E_0^{(2)} = \frac{\lambda^2}{16}\sum_{n\neq 0}\frac{|{}^{(0)}\langle 0|x^4|n\rangle^{(0)}|^2}{\hbar\omega(1/2 - n - 1/2)}$$

$$= -\frac{\lambda^2}{16}\left(\frac{\hbar}{2m\omega}\right)^4 \sum_{n\neq 0}\frac{|{}^{(0)}\langle 0|(a^{\dagger} + a)^4|n\rangle^{(0)}|^2}{n\hbar\omega}$$

$$= -\frac{1}{16\hbar\omega}\left(\frac{\hbar\lambda}{2m\omega}\right)^4 \sum_{n}\frac{(\sqrt{4!}\delta_{n,4} + 6\sqrt{2}\delta_{n,2})^2}{n}$$

$$= -\frac{21}{8}\frac{1}{\hbar\omega}\left(\frac{\hbar\lambda}{2m\omega}\right)^4 . \tag{12.4.40}$$

b) The wavefunction correct to order λ is given by

$$|0\rangle = |0\rangle^{(0)} + \lambda\sum_{m\neq n}\frac{|m\rangle^{(0)}\,{}^{(0)}\langle m|H'|0\rangle^{(0)}}{E_0^{(0)} - E_m^{(0)}} . \tag{12.4.41}$$

But, for $m \neq 0$ we have

$$^{(0)}\langle m|H'|0\rangle^{(0)} = \frac{1}{4}\left(\frac{\hbar}{2m\omega}\right)^2 {}^{(0)}\langle m|(a^{\dagger} + a)^4|0\rangle^{(0)}$$

$$= \frac{1}{4}\left(\frac{\hbar}{2m\omega}\right)^2 {}^{(0)}\langle m|a^{\dagger 4} + a^{\dagger 2}aa^{\dagger} + a^{\dagger}aa^{\dagger 2} + aa^{\dagger 3}|0\rangle^{(0)}$$

$$= \frac{1}{4}\left(\frac{\hbar}{2m\omega}\right)^2 \left[\sqrt{24}\,\delta_{m,4} + 6\sqrt{2}\,\delta_{m,2}\right] . \tag{12.4.42}$$

Therefore,

$$|0\rangle = |0\rangle^{(0)} - \frac{\lambda}{4}\left(\frac{\hbar}{2m\omega}\right)^2 \left[\frac{6\sqrt{2}}{2\hbar\omega}|2\rangle^{(0)} + \frac{2\sqrt{6}}{4\hbar\omega}|4\rangle^{(0)}\right]$$

$$= |0\rangle^{(0)} - \frac{\lambda}{16}\frac{\hbar}{4m^2\omega^3}\left[12\sqrt{2}\,|2\rangle^{(0)} + 2\sqrt{6}\,|4\rangle^{(0)}\right] . \tag{12.4.43}$$

12.5 $1/4\,\lambda x^4$ - Brillouin-Wigner Perturbation

Repeat part a of the problem above (problem 12.4) using Brillouin-Wigner perturbation theory.

Solution

To first order the calculation is the same as for Rayleigh-Schrödinger perturbation theory. In second order the appropriate formula is

$$\lambda^2 E_0^{(2)} = \frac{\lambda^2}{4} \sum_{n \neq 0} \frac{^{(0)}\langle 0|x^4|0\rangle^{(0)}}{E_0 - E_n^{(0)}} . \tag{12.5.44}$$

Substituting the results from problem 12.4 this becomes

$$\lambda^2 E_0^{(2)} = \frac{\lambda^2}{4} \left(\frac{\hbar}{2m\omega}\right)^2 \sum_{n \neq 0} \frac{^{(0)}\langle 0|(a^\dagger + a)^4|0\rangle^{(0)}}{E_0 - E_n^{(0)}}$$

$$= \frac{\lambda^2}{4} \left(\frac{\hbar}{2m\omega}\right)^2 \frac{\sqrt{4!}\delta_{n,4} + n\sqrt{2}\delta_{n,2} + \sqrt{2}\delta_{n,2}}{E_0 - E_n^{(0)}}$$

$$= \frac{\lambda^2}{4} \left(\frac{\hbar}{2m\omega}\right)^2 \left[\frac{\sqrt{4!}}{E_0 - (4 + 1/2)\hbar\omega} + \frac{3\sqrt{2}}{E_0 - (2 + 1/2)\hbar\omega}\right] . \tag{12.5.45}$$

To this order we have that

$$E_0 = \frac{1}{2}\hbar\omega + \frac{3}{4}\lambda \left(\frac{\hbar}{2m\omega}\right)^2 + \lambda^2 E_0^{(2)} . \tag{12.5.46}$$

After substituting this in the equation above and rearranging we get the following equation for E_0.

$$E_0 - \frac{1}{2}\hbar\omega - \frac{3}{4}\lambda \left(\frac{\hbar}{2m\omega}\right)^2$$

$$= \lambda^2 \left(\frac{\hbar}{2m\omega}\right)^2 \left[\frac{\sqrt{4!}}{E_0 - (4 + 1/2)\hbar\omega} + \frac{3\sqrt{2}}{E_0 - (2 + 1/2)\hbar\omega}\right] . \tag{12.5.47}$$

Solving this equation to order λ^2 we get the same result as in 12.4 .

12.6   Two-level System

Consider the Hamiltonian

$$H = H_0 + \lambda H'$$

where

$$H_0 = \begin{pmatrix} E_1 & 0 \\ 0 & E_2 \end{pmatrix} \qquad H' = \begin{pmatrix} 0 & ia \\ -ia & 0 \end{pmatrix}$$

a) solve for the exact eigenvalues and eigenfunctions.
b) Solve for both eigenvalues and eigenfunctions to second order using Rayleigh-Schrödinger perturbation theory.

Solution

The Hamiltonian is as always

$$H = H_0 + \lambda H' \tag{12.6.48}$$

where

$$H_0 = \begin{pmatrix} E_1 & 0 \\ 0 & E_2 \end{pmatrix} , \quad H' = \begin{pmatrix} 0 & i\lambda a \\ -i\lambda a & 0 \end{pmatrix} \tag{12.6.49}$$

Exact solutions
The eigenvalues are given by

$$\det \begin{pmatrix} E_1 - E & i\lambda a \\ -i\lambda a & E_2 - E \end{pmatrix} = 0 . \tag{12.6.50}$$

We assume that $E_2 > E_1$ so that $|E_2 - E_1| = E_2 - E_1$. Then we have to solve the quadratic equation

$$E^2 - (E_1 + E_2)E + E_1 E_2 - \lambda^2 a^2 = 0 . \tag{12.6.51}$$

The solutions are

$$E_\pm = \frac{1}{2}(E_1 + E_2) \pm \frac{1}{2}\sqrt{(E_2 - E_1)^2 + 4\lambda^2 a^2} . \tag{12.6.52}$$

Expanding in powers of λ we get

$$\begin{aligned} E_+ &= \frac{1}{2}(E_1 + E_2) + \frac{1}{2}(E_2 - E_1) + \frac{\lambda^2 a^2}{E_2 - E_1} \\ &= E_2 + \frac{\lambda^2 a^2}{E_2 - E_1} . \\ E_- &= \frac{1}{2}(E_1 + E_2) - \frac{1}{2}(E_2 - E_1) - \frac{\lambda^2 a^2}{E_2 - E_1} \\ &= E_1 - \frac{\lambda^2 a^2}{E_2 - E_1} . \end{aligned} \tag{12.6.53}$$

This result shows that

$$E_+ - E_- = E_2 - E_1 + \frac{2\lambda^2 a^2}{E_2 - E_1} \tag{12.6.54}$$

so that the two energy levels are "repelled" by the perturbation. This result is quite general.

The eigenvectors are of the form

$$\psi = \begin{pmatrix} \alpha \\ \beta \end{pmatrix} \tag{12.6.55}$$

where

$$(E_1 - E_\pm)\alpha_\pm + i\lambda a \beta_\pm = 0 \tag{12.6.56}$$

so that

$$\alpha_\pm = \frac{-i\lambda a}{E_1 - E_\pm} \beta_\pm \ . \tag{12.6.57}$$

We can now write

$$\psi_\pm = A_\pm \begin{pmatrix} -i\lambda a \\ E_1 - E_\pm \end{pmatrix} \tag{12.6.58}$$

where, with an arbitrary choice of phase

$$|A_\pm| = \frac{1}{\sqrt{(E_1 - E_\pm)^2 + \lambda^2 a^2}} \ . \tag{12.6.59}$$

Now, to later compare with the perturbation theory we let $\lambda \to 0$ and find that

$$E_1 - E_+ \to \quad -(E_2 - E_1) - \frac{\lambda^2 a^2}{E_2 - E_1}$$

$$E_1 - E_- \to \quad \frac{\lambda^2 a^2}{E_2 - E_1} \tag{12.6.60}$$

$$A_+ \to \quad (E_2 - E_1) + \frac{\lambda^2 a^2}{2(E_2 - E_1)}$$

$$A_- \to \quad \lambda a \ . \tag{12.6.61}$$

So, with an arbitrary choice of phase we have

$$\psi_+ \to \begin{pmatrix} \frac{i\lambda a}{E_2 - E_1} \\ 1 \end{pmatrix} \quad \psi_- \to \begin{pmatrix} 1 \\ \frac{i\lambda a}{E_2 - E_1} \end{pmatrix} \ . \tag{12.6.62}$$

Perturbation Theory

The 0th order eigenvalues and eigenvectors are

$$E_1{}^{(0)} = E_1 \ , \quad E_2{}^{(0)} = E_2 \tag{12.6.63}$$

and

$$\psi_1^{(0)} = \begin{pmatrix} 1 \\ 0 \end{pmatrix} \quad \psi_2^{(0)} = \begin{pmatrix} 0 \\ 1 \end{pmatrix} \ . \tag{12.6.64}$$

The first order correction to the eigenvalues is zero since

$$(\psi_1^{(0)}, H'\psi_1^{(0)}) = (\psi_2^{(0)}, H'\psi_2^{(0)}) = 0 \ . \tag{12.6.65}$$

The first order correction to the wavefunctions is

$$\psi_1^{(1)} = \frac{\psi_2^{(0)}(\psi_2^{(0)}, H'\psi_1^{(0)})}{E_1{}^{(0)} - E_2{}^{(0)}}$$

$$= \frac{-ia}{E_1 - E_2} \begin{pmatrix} 0 \\ 1 \end{pmatrix}$$

$$\psi_2^{(1)} = \frac{\psi_1^{(0)}(\psi_1^{(0)}, H'\psi_2^{(0)})}{E_2{}^{(0)} - E_1{}^{(0)}}$$

$$= \frac{ia}{E_1 - E_2} \begin{pmatrix} 1 \\ 0 \end{pmatrix} \ . \tag{12.6.66}$$

So, to first order we have

$$
\begin{aligned}
\psi_1 &= \begin{pmatrix} 1 \\ 0 \end{pmatrix} + \frac{-i\lambda a}{E_1 - E_2} \begin{pmatrix} 0 \\ 1 \end{pmatrix} \\
&= \begin{pmatrix} 1 \\ \frac{i\lambda a}{E_2 - E_1} \end{pmatrix} \\
\psi_2 &= \begin{pmatrix} 0 \\ 1 \end{pmatrix} + \frac{i\lambda a}{E_2 - E_1} \begin{pmatrix} 1 \\ 0 \end{pmatrix} \\
&= \begin{pmatrix} \frac{i\lambda a}{E_2 - E_1} \\ 1 \end{pmatrix} .
\end{aligned}
\tag{12.6.67}
$$

To order λ we now have agreement with the exact solutions.

The energy to second order is

$$
\begin{aligned}
E_1^{(2)} &= \sum_{n \neq 1} \frac{\left| (\psi_2^{(0)}, H'\psi_1^{(0)}) \right|^2}{E_1^{(0)} - E_2^{(0)}} = \frac{a^2}{E_1 - E_2} \\
E_2^{(2)} &= \sum_{n \neq 2} \frac{\left| (\psi_1^{(0)}, H'\psi_2^{(0)}) \right|^2}{E_2^{(0)} - E_1^{(0)}} = \frac{a^2}{E_2 - E_1} .
\end{aligned}
\tag{12.6.68}
$$

Therefore, the two energies to order λ^2 are

$$
\begin{aligned}
E_- &= E_1 - \frac{\lambda^2 a^2}{E_2 - E_1} \\
E_+ &= E_2 + \frac{\lambda^2 a^2}{E_2 - E_1} .
\end{aligned}
\tag{12.6.69}
$$

These results again coincide with those obtained by expanding the exact solutions in powers of λ.

12.7   Approximate SHO

A particle of mass m moves in a potential

$$
V = \frac{1}{2} k |x|^{2+\epsilon} \qquad |\epsilon| < 1 .
$$

Estimate the energy of the ground state.

Hint:

$$
\frac{1}{2} k |x|^{2+\epsilon} = \frac{1}{2} k x^2 - \frac{1}{2} k (x^2 - |x|^{2+\epsilon}) \approx \frac{1}{2} k x^2 + \frac{\epsilon}{2} k x^2 \ln |x| .
$$

Also,

$$
\int_0^\infty e^{-\alpha x^2} x^2 \ln x \, dx = \frac{\sqrt{\pi}}{4} a^{-3/2} \left[1 - \frac{1}{2} (c + \ln 4a) \right]
$$

where

$$
c = 0.577216 \ldots = \text{Euler's constant} .
$$

Solution

Using the first hint and writing, as suggested,

$$\frac{1}{2}k|x|^{2+\epsilon} = \frac{1}{2}kx^2 + \frac{1}{2}k\epsilon x^2 \ln|x| \quad \epsilon \ll 1 \tag{12.7.70}$$

we immediately find that the ground state energy may be approximated by

$$E_0 = \frac{1}{2}\hbar\omega + \frac{1}{2}k\epsilon\langle 0|x^2 \ln|x||0\rangle \tag{12.7.71}$$

where

$$\omega^2 = k/m \tag{12.7.72}$$

and $|0\rangle$ is the corresponding simple harmonic oscillator ground state. Thus, we need to evaluate

$$\langle 0|x^2 \ln|x||0\rangle = \sqrt{\frac{m\omega}{2\pi\hbar}} \int_{-\infty}^{\infty} x^2 \ln|x| e^{-(m\omega/\hbar)x^2} \, dx \ . \tag{12.7.73}$$

Now using the second hint

$$\int_{0}^{\infty} a^{-\alpha x^2} x^2 \ln|x| \, dx = \frac{\sqrt{\pi}4^{-3/2}}{\alpha}[1 - 1/2(c + \ln(4\alpha))] \tag{12.7.74}$$

where $c = 0.577\,216\ldots = $ Euler's constant, we get

$$E_0 = \frac{1}{2}\hbar\omega \left[1 + \frac{\epsilon}{2\sqrt{2}}\left[1 - 1/2\left(c + \ln(4\frac{m\omega}{\hbar})\right)\right]\right] \ . \tag{12.7.75}$$

12.8 Two-dimensional SHO

For a particle of mass m moving in the potential

$$V = \frac{1}{2}k_1x^2 + \frac{1}{2}k_2x^2 + \lambda xy \quad |k_1 - k_2| > 2\lambda$$

a) find the exact energy levels.
b) Use perturbation theory to find to order λ^2 the energy of all the levels and compare with the exact solution to this order.

Solution

The Hamiltonian H is

$$H = \frac{\vec{p}^2}{2m} + V(x, y) = \frac{p_x{}^2}{2m} + \frac{1}{2}k_1x^2 + \frac{p_y{}^2}{2m} + \frac{1}{2}k_2y^2 + \lambda xy \ . \tag{12.8.76}$$

This can be rewritten as

$$H = \frac{p_x{}^2}{2m} + \frac{1}{2}m\omega_1{}^2x^2 + \frac{p_y{}^2}{2m} + \frac{1}{2}m\omega_2{}^2y^2 + m\alpha^2 xy \ , \tag{12.8.77}$$

where

$$\omega_1{}^2 = k_1/m \quad , \quad \omega_2{}^2 = k_2/m \quad , \quad \alpha^2 = \lambda/m \ . \tag{12.8.78}$$

a) Exact Solution
We now rotate the axes to remove the cross term.

$$
\begin{aligned}
x &= X\cos\theta + Y\sin\theta \\
y &= -X\sin\theta + Y\cos\theta
\end{aligned}
\tag{12.8.79}
$$

Then,

$$
\begin{aligned}
\frac{2V}{m} ={}& \omega_1^2 \left[X^2 \cos^2\theta + Y^2 \sin^2\theta + XY \sin 2\theta \right] \\
&+ \omega_2^2 \left[X^2 \sin^2\theta + Y^2 \cos^2\theta - XY \sin 2\theta \right] \\
&+ \alpha^2 \left[-X^2 \sin 2\theta + Y^2 \sin 2\theta + 2XY \cos 2\theta \right] \ .
\end{aligned}
\tag{12.8.80}
$$

So, to remove the cross term, we choose the angle θ to be given by

$$(\omega_1^2 - \omega_2^2)\sin 2\theta + 2\alpha^2 \cos 2\theta = 0 \ . \tag{12.8.81}$$

Also, since this is a rotation we have

$$\frac{p_x^2}{2m} + \frac{p_y^2}{2m} = \frac{p_X^2}{2m} + \frac{p_Y^2}{2m} \ . \tag{12.8.82}$$

So,

$$H = \frac{p_X^2}{2m} + \frac{p_Y^2}{2m} + \frac{1}{2}m\,\Omega_1^2 X^2 + \frac{1}{2}m\,\Omega_2^2 Y^2 \tag{12.8.83}$$

where

$$
\begin{aligned}
\Omega_1^2 &= \omega_1^2 \cos^2\theta + \omega_2^2 \sin^2\theta - \alpha^2 \sin 2\theta \\
&= \frac{1}{2}(\omega_1^2 + \omega_2^2) - \frac{1}{2}\sqrt{(\omega_2^2 - \omega_1^2)^2 + 4\alpha^4} \\
&\to \omega_1^2 + \frac{\alpha^4}{\omega_1^2 - \omega_2^2} \quad \text{as} \quad \alpha \to 0 \ . \\
\Omega_2^2 &= \omega_1^2 \cos^2\theta + \omega_2^2 \sin^2\theta + \alpha^2 \sin 2\theta \\
&= \frac{1}{2}(\omega_1^2 + \omega_2^2) + \frac{1}{2}\sqrt{(\omega_2^2 - \omega_1^2)^2 + 4\alpha^4} \\
&\to \omega_2^2 - \frac{\alpha^4}{\omega_1^2 - \omega_2^2} \quad \text{as} \quad \alpha \to 0 \ .
\end{aligned}
\tag{12.8.84}
$$

The (exact) energy eigenvalues are

$$E_{n_1,n_2} = (n_1 + 1/2)\hbar\Omega_1 + (n_2 + 1/2)\hbar\Omega_2 \ . \tag{12.8.85}$$

So, to lowest order in λ, for later comparison with perturbation theory, we have

$$
\begin{aligned}
E_{n_1,n_2} ={}& (n_1 + 1/2)\hbar\omega_1 + (n_2 + 1/2)\hbar\omega_2 \\
&+ \frac{\hbar\lambda^2}{2m^2}\frac{(n_2 + 1/2)\omega_1 - (n_1 + 1/2)\omega_2}{\omega_1\omega_2(\omega_2^2 - \omega_1^2)} \ .
\end{aligned}
\tag{12.8.86}
$$

b) Perturbation Theory

The Hamiltonian H_0 represents two uncoupled SHO's. The energy eigenvalues are

$$E^{(0)}_{n_1,n_2} = (n_1 + 1/2)\hbar\omega_1 + (n_2 + 1/2)\hbar\omega_2 . \tag{12.8.87}$$

Also, the eigenkets may be written

$$|n_1, n_2\rangle^{(0)} = \frac{(a_1^\dagger)^{n_1}(a_2^\dagger)^{n_2}}{\sqrt{n_1!\, n_2!}}|0, 0\rangle^{(0)} . \tag{12.8.88}$$

The first order correction to the energy is zero since

$$^{(0)}\langle n_1, n_2|xy|n_1, n_2\rangle^{(0)} = 0 . \tag{12.8.89}$$

In second order we get

$$E^{(2)}_{n_1,n_2} = \sum_{m_1 \neq n_1, m_2 \neq n_2} \frac{\left|\, ^{(0)}\langle n_1, n_2|xy|m_1, m_2\rangle^{(0)}\right|^2}{E^0_{n_1,n_2} - E^0_{m_1,m_2}} . \tag{12.8.90}$$

Now,

$$^{(0)}\langle n_1, n_2|xy|m_1, m_2\rangle^{(0)}$$

$$= \ ^{(0)}\langle n_1, n_2|\frac{\hbar}{2m}\frac{1}{\sqrt{\omega_1\omega_2}}(a_1^\dagger + a_1)(a_2^\dagger + a_2)|m_1, m_2\rangle^{(0)}$$

$$= \frac{\hbar}{2m}\frac{1}{\sqrt{\omega_1\omega_2}}\Big[\ ^{(0)}\langle n_1, n_2|\sqrt{(m_1 + 1)(m_2 + 1)}|m_1 + 1, m_2 + 1\rangle^{(0)}$$

$$+ \ ^{(0)}\langle n_1, n_2|\sqrt{(m_1 + 1)m_2}|m_1 + 1, m_2 - 1\rangle^{(0)}$$

$$+ \ ^{(0)}\langle n_1, n_2|\sqrt{m_1(m_2 + 1)}|m_1 - 1, m_2 + 1\rangle^{(0)}$$

$$+ \ ^{(0)}\langle n_1, n_2|\sqrt{m_1 m_2}|m_1 - 1, m_2 - 1\rangle^{(0)}\Big]$$

$$= \frac{\hbar}{2m}\frac{1}{\sqrt{\omega_1\omega_2}}\Big[\sqrt{(m_1 + 1)(m_2 + 1)}\,\delta_{n_1,m_1+1}\,\delta_{n_2,m_2+1}$$

$$+ \ \sqrt{(m_1 + 1)m_2}\,\delta_{n_1,m_1+1}\,\delta_{n_2,m_2-1}$$

$$+ \ \sqrt{m_1(m_2 + 1)}\,\delta_{n_1,m_1-1}\,\delta_{n_2,m_2+1}$$

$$+ \ \sqrt{m_1 m_2}\,\delta_{n_1,m_1-1}\,\delta_{n_2,m_2-1}\Big] . \tag{12.8.91}$$

So, after collecting terms we get

$$E^{(2)}_{n_1,n_2} = \frac{\hbar}{2m^2}\frac{(n_2 + 1/2)\omega_1 - (n_1 + 1/2)\omega_2}{\omega_1\omega_2(\omega_2^2 - \omega_1^2)} . \tag{12.8.92}$$

This is the same result that we obtained by expanding the exact solution.

12.9   Kuhn-Thomas-Reiche Sum Rule

Classically the polarizability α of an atom is defined as the induced electric dipole moment $e|\vec{r}|$ divided by the strength of the inducing electric field $\vec{E}$. So,

$$\alpha = \frac{e|\vec{r}|}{|\vec{E}|}$$

and for harmonically bound electrons takes the form

$$\alpha = \frac{e^2}{4\pi^2 m} \sum_j \frac{f_j}{\nu_j^2 - \nu^2} \ .$$

Here, f_j are dimensionless constants called the "oscillator strengths". In quantum mechanics these are defined by

$$
\begin{aligned}
f_j &= \frac{4\pi m}{3\hbar e^2} \nu_{j0} |\vec{m}_{j0}|^2 \\
&= \frac{4\pi m}{3\hbar e^2} \frac{E_j - E_0}{h} |e\vec{r}_{j0}|^2 \\
&= \frac{2m}{3\hbar^2} (E_j - E_0) |\vec{r}_{j0}|^2 \ .
\end{aligned}
$$

For N uncoupled electrons one then has the Kuhn-Thomas-Reiche sum rule

$$\sum_j f_j = N \ .$$

This polarizability can be used to describe the absorption of light which carries an electron from its ground state $|0\rangle$ to an excited state $|n\rangle$ in an atom. If, $\vec{r}_{n0} = \langle n|\vec{r}|0\rangle$ and the Hamiltonian for the bound electron is

$$H = \frac{\vec{p}^2}{2m} + V(\vec{r}) \ .$$

a) Show that

$$(E_n - E_0)\vec{r}_{n0} = -\frac{i\hbar}{m} \langle n|\vec{p}|0\rangle \ .$$

Hint: Use the commutator $[H, \vec{r}]$ and work component by component.

b) Use the commutators

$$[x, p_x] = [y, p_y] = [z, p_z] = i\hbar$$

together with the results of part a) to prove that for a single electron

$$\sum_n \frac{2m(E_n - E_0)}{3\hbar^2} [|x_{n0}|^2 + |y_{n0}|^2 + |z_{n0}|^2] = 1$$

and hence deduce the Kuhn-Thomas-Reiche sum rule.

For this problem and all subsequent problems dealing with sum rules it may be useful to consult [12.1].

Solution

We start with the Hamiltonian

$$H = \frac{\vec{p}^2}{2m} + V(\vec{r}) \ . \tag{12.9.93}$$

Then,

$$[x, H] = \frac{1}{2m}[x, \vec{p}^2] = \frac{i\hbar}{m} p_x \tag{12.9.94}$$

and similarly for $[y, H]$ and $[z, H]$. Therefore,

$$\begin{aligned}
\frac{i\hbar}{m}\langle n|p_x|k\rangle &= \langle n|[x, H]|k\rangle \\
&= (E_k - E_n)\langle n|x|k\rangle \ .
\end{aligned} \tag{12.9.95}$$

Hence, we have

$$\frac{i\hbar}{m}\langle 0|p_x|k\rangle = (E_k - E_0)\langle 0|x|k\rangle \tag{12.9.96}$$

and by complex conjugation

$$-\frac{i\hbar}{m}\langle k|p_x|0\rangle = (E_k - E_0)\langle k|x|0\rangle \ . \tag{12.9.97}$$

It then follows that

$$\begin{aligned}
\sum_k (E_k - E_0)|x_{k0}|^2 &= \sum_k (E_k - E_0)\langle 0|x|k\rangle\langle k|x|0\rangle \\
&= \sum_k \frac{i\hbar}{m}\langle 0|p_x|k\rangle\langle k|x|0\rangle \\
&= \frac{i\hbar}{m}\langle 0|p_x x|0\rangle \ .
\end{aligned} \tag{12.9.98}$$

Similarly,

$$\begin{aligned}
\sum_k (E_k - E_0)|x_{k0}|^2 &= \sum_k (E_k - E_0)\langle 0|x|k\rangle\langle k|x|0\rangle \\
&= \sum_k \frac{-i\hbar}{m}\langle 0|x|k\rangle\langle k|p_x|0\rangle \\
&= -\frac{i\hbar}{m}\langle 0|x p_x|0\rangle \ .
\end{aligned} \tag{12.9.99}$$

Therefore,

$$\begin{aligned}
\sum_k (E_k - E_0)|x_{k0}|^2 &= \frac{i\hbar}{2m}\langle 0|p_x x - x p_x|0\rangle \\
&= \frac{\hbar^2}{2m} \ .
\end{aligned} \tag{12.9.100}$$

This means that for a single electron

$$\sum_k \frac{2m}{\hbar^2}(E_k - E_0)|x_{k0}|^2 = 1 \ . \tag{12.9.101}$$

Hence, since all three directions x, $y\,z$ are equivalent we have for a single electron that

$$\sum_k \frac{2m}{3\hbar^2}(E_k - E_0)[|x_{k0}|^2 + |y_{k0}|^2 + |z_{k0}|^2] = 1 \ . \tag{12.9.102}$$

For N independent (uncoupled) electrons we therefore find the Kuhn-Thomas-Reiche sum rule

$$\sum_k \frac{2m}{3\hbar^2}(E_k - E_0)[|x_{k0}|^2 + |y_{k0}|^2 + |z_{k0}|^2] = \sum_k f_k = N \ . \tag{12.9.103}$$

12.10   Electron in Box Perturbed by Electric Field

An electron confined to a one-dimensional box $0 \le x \le L$ is acted on by an electric field $\mathcal{E}$ acting in the x-direction. Assuming that $e\mathcal{E}L <<$ energy of the lowest unperturbed level, find the lowest energy in the presence of the electric field $\mathcal{E}$.

Solution

The unperturbed energy for the ground state is

$$E_1^{(0)} = \frac{\hbar^2\pi^2}{2mL^2} \ . \tag{12.10.104}$$

The corresponding ground state wavefunction is

$$\psi_0 = \sqrt{\frac{2}{L}}\sin(\pi x/L) \tag{12.10.105}$$

The perturbation Hamiltonian is

$$V = e\mathcal{E}x \ . \tag{12.10.106}$$

Therefore, the energy correction is

$$\begin{aligned} E_1^{(1)} &= (\psi_0, V\psi_0) \\ &= \frac{2}{L}e\mathcal{E}\int_0^L \sin^2(\pi x/L)\,x\,dx \\ &= \frac{1}{4}e\mathcal{E}L \ . \end{aligned} \tag{12.10.107}$$

Thus, to this accuracy, the lowest energy is

$$E_1 = \frac{\hbar^2\pi^2}{2mL^2} + \frac{1}{4}e\mathcal{E}L \ . \tag{12.10.108}$$

12.11   Positronium

Positronium is a hydrogen-like system consisting of a bound state of an electron and a positron (positive electron). The ground state consists of a singlet and three triplet substates. The singlet state is the most stable lying about 8.2×10^{-4} eV below the triplet levels which are degenerate. Field theoretic calculations show that this is due to a spin-spin interaction of the form

$$H_0 = -\frac{A}{\hbar^2} \vec{s}_1 \cdot \vec{s}_2 \ .$$

a) Determine the value of the constant A.
b) Using the fact that the positron has a charge and magnetic moment opposite to that of the electron calculate the effect of a magnetic field on these levels.

Solution

a) The Hamiltonian for these four levels, in the absence of a magnetic field, is as given

$$H_0 = -\frac{A}{\hbar^2} \vec{s}_1 \cdot \vec{s}_2 \ . \tag{12.11.109}$$

If we choose the representation of good total spin $|s, m\rangle$ where

$$(s, m) = (1, 1); (1, 0); (1, -1) \ \text{and} \ (0, 0)$$

then, in terms of the spin states for the electron and positron, respectively, we have

$$\begin{pmatrix} |1,1\rangle \\ |1,0\rangle \\ |1,-1\rangle \\ |0,0\rangle \end{pmatrix} = \begin{pmatrix} |\uparrow\uparrow\rangle \\ \frac{1}{\sqrt{2}}[|\uparrow\downarrow\rangle + |\downarrow\uparrow\rangle] \\ |\downarrow\downarrow\rangle \\ \frac{1}{\sqrt{2}}[|\uparrow\downarrow\rangle - |\downarrow\uparrow\rangle] \end{pmatrix} \ . \tag{12.11.110}$$

Also, since

$$\vec{s}_1 \cdot \vec{s}_2 = \frac{1}{2}\left[(\vec{s}_1 + \vec{s}_2)^2 - \vec{s}_1^2 - \vec{s}_2^2\right] \tag{12.11.111}$$

we see that H_0 is diagonal in this representation.

$$H_0 = \frac{A}{2} \begin{pmatrix} 1 & 0 & 0 & 0 \\ 0 & 1 & 0 & 0 \\ 0 & 0 & 1 & 0 \\ 0 & 0 & 0 & -3 \end{pmatrix} \tag{12.11.112}$$

Therefore, the splitting of these levels is

$$\Delta E = \frac{A}{2}[1 - (-3)] = 2A = 8.2 \times 10^{-4} \ \text{eV} \ . \tag{12.11.113}$$

So, we have

$$A = 4.1 \times 10^{-4} \text{ eV} \ . \tag{12.11.114}$$

b) If we introduce a magnetic field $\vec{B}$ pointing in the z-direction then the Hamiltonian becomes

$$H = H_0 - \vec{\mu} \cdot \vec{B} = -\frac{A}{\hbar^2} \vec{s}_1 \cdot \vec{s}_2 + \frac{e\hbar}{2mc} \vec{B} \cdot (\vec{\sigma}_1 - \vec{\sigma}_2) \tag{12.11.115}$$

where $\vec{\sigma}_i$ are the Pauli matrices and the index 1 refers to the electron and the index 2 refers to the positron. Calling

$$\frac{eB}{2mc} = \omega$$

we find that in the representation already used we get that

$$-\vec{\mu} \cdot \vec{B} = \hbar\omega \begin{pmatrix} 0 & 0 & 0 & 0 \\ 0 & 0 & 2 & 0 \\ 0 & 0 & 0 & 0 \\ 0 & 2 & 0 & 0 \end{pmatrix} \ . \tag{12.11.116}$$

The eigenvalues of the total Hamiltonian are therefore (listed in the order in which the states are listed in (12.11.110))
$A/2$, $A/2 + 2\hbar\omega$, $A/2$, $A/2 - 2\hbar\omega$.

12.12   Rigid Rotator in Electric Field

Consider a three-dimensional rigid rotator with moment of inertia I and electric dipole moment $\vec{P}$ parallel to the axis of the rotator. This rotator is placed in an uniform electric field $\vec{E}$. Compute, to lowest non-vanishing order in E, the strength of the electric field, the ground state energy of the rotator.

Solution

The unperturbed Hamiltonian is

$$H_0 = \frac{L^2}{2I} \ . \tag{12.12.117}$$

The energy eigenvalues are

$$E_l = \frac{l(l+1)\hbar^2}{2I} \ . \tag{12.12.118}$$

The interaction Hamiltonian, if we take the axis of the rotator parallel to the z-axis, is

$$H' = -\vec{P} \cdot \vec{E} = -PE \cos\theta \ . \tag{12.12.119}$$

The effect of this perturbation is to produce (in lowest order) a change in energy

$$E_l^{(1)} = -PE\langle lm| \cos\theta|lm\rangle \ . \tag{12.12.120}$$

This yields no change. So we have to go to second order. In this case, the ground state is shifted by

$$E_0^{(2)} = (PE)^2 \sum_{lm} \frac{\langle 00| \cos\theta|lm\rangle\langle lm| \cos\theta|00\rangle}{0 - l(l+1)\hbar^2} \ . \tag{12.12.121}$$

The only non-zero matrix elements in this sum are

$$\langle 00| \cos\theta|10\rangle = \langle 10| \cos\theta|00\rangle \ . \tag{12.12.122}$$

This illustrates the selection rules

$$\Delta l = 1 \ , \quad \Delta m = 0$$

that apply to electric dipole transitions. Using the fact that

$$\cos\theta = \sqrt{\frac{4\pi}{3}}Y_{10} \ \text{and} \ Y_{00} = \frac{1}{\sqrt{4\pi}}$$

we get

$$\langle 00| \cos\theta|10\rangle = \langle 10| \cos\theta|00\rangle = \frac{1}{\sqrt{3}} \ . \tag{12.12.123}$$

So, finally

$$E_0^{(2)} = -\frac{(PE)^2}{6} \ . \tag{12.12.124}$$

12.13   Electric Dipole Moment Sum Rule

Show that for a system of N particles with charges q_j and masses M_j, $j = 1 \ldots N$ confined to a finite region of space we have the following sum rule [12.1] for the electric dipole moment.

$$\frac{2}{\hbar^2} \sum_m^{\infty} (E_m - E_n)|\vec{d}_{nm}|^2 = \sum_{j=1}^{N} \frac{q_j^2}{M_j} \ .$$

The sum here extends over a complete set of energy eigenstates.

Solution

For any operator A the matrix element of the time derivative of the operator is given by

$$\langle n|\dot{A}|m\rangle = -\frac{i}{\hbar}\langle n|[A, H]|m\rangle \tag{12.13.125}$$

where H is the total Hamiltonian. Applying this to the operator $\vec{d}$ we find

$$
\begin{aligned}
\dot{\vec{d}}_{nm} &= -\frac{i}{\hbar}\langle n|[\vec{d}, H]|m\rangle \\
&= -\frac{i}{\hbar}\left[\langle n|\vec{d}H|m\rangle - \langle n|H\vec{d}|m\rangle\right] \\
&= -\frac{i}{\hbar}(E_m - E_n)\vec{d}_{nm} \, .
\end{aligned}
\tag{12.13.126}
$$

From this it follows that

$$
\begin{aligned}
\sum_m^\infty (E_m - E_n)|\vec{d}_{nm}|^2 &= \sum_m^\infty (E_m - E_n)\vec{d}_{nm} \cdot \vec{d}_{mn} \\
&= \frac{i\hbar}{2}\sum \left[\dot{\vec{d}}_{nm} \cdot \vec{d}_{mn} - \vec{d}_{nm} \cdot \dot{\vec{d}}_{mn}\right] \\
&= \frac{i\hbar}{2}|\dot{\vec{d}} \cdot \vec{d} - \vec{d} \cdot \dot{\vec{d}}\,|_{nn} \, .
\end{aligned}
\tag{12.13.127}
$$

Now, using (12.13.125) again on the operator $\vec{d}$ we get

$$
\begin{aligned}
i\hbar\langle n|\dot{\vec{d}}|m\rangle &= \langle n|[\vec{d}, H]|m\rangle \\
&= \sum_j q_j \langle n|[\vec{x}_j, H]|m\rangle \\
&= i\hbar \sum_j q_j \langle n|\frac{p_j}{M_j}|m\rangle \, .
\end{aligned}
\tag{12.13.128}
$$

So,

$$
\begin{aligned}
\sum_m^\infty (E_m - E_n)|\vec{d}_{nm}|^2 &= \frac{i\hbar}{2}\sum_m^\infty \sum_j \frac{q_j}{M_j}\left[(\vec{p}_j)_{nm} \cdot (\vec{d})_{mn} - (\vec{d})_{nm} \cdot (\vec{p}_j)_{mn}\right] \\
&= \frac{i\hbar}{2}\sum_j \frac{q_j}{M_j}[\vec{p}_j, \vec{d}]_{nn} \\
&= \frac{i\hbar}{2}\sum_j \frac{q_j}{M_j}(-i\hbar q_j) \\
&= \hbar^2 \sum_j \frac{q_j^2}{M_j}
\end{aligned}
\tag{12.13.129}
$$

as required.

12.14 Another Sum Rule

Use the double commutator

$$\left[\left[H,\sum_j e^{i\vec{q}\cdot\vec{r}_j}\right],\sum_k e^{-i\vec{q}\cdot\vec{r}_k}\right]$$

to derive the sum rule (see [12.1]

$$\sum_m (E_m - E_n)\left|\langle n|\sum_j e^{i\vec{q}\cdot\vec{r}_j}|m\rangle\right|^2 = N\frac{(\hbar q)^2}{2M}$$

for a system of N interacting particles. Here $|n\rangle$ represents an energy eigenket of the Hamiltonian H.

Solution

If we first write out the double commutator explicitly we find

$$\left[\left[H,\sum_j e^{i\vec{q}\cdot\vec{r}_j}\right],\sum_k e^{-i\vec{q}\cdot\vec{r}_k}\right]$$

$$= \left[H\sum_j e^{i\vec{q}\cdot\vec{r}_j} - \sum_j e^{i\vec{q}\cdot\vec{r}_j}H, \sum_k e^{-i\vec{q}\cdot\vec{r}_k}\right]$$

$$= H\sum_j e^{i\vec{q}\cdot\vec{r}_j}\sum_k e^{-i\vec{q}\cdot\vec{r}_k} - \sum_j e^{i\vec{q}\cdot\vec{r}_j}H\sum_k e^{-i\vec{q}\cdot\vec{r}_k}$$

$$- \sum_k e^{-i\vec{q}\cdot\vec{r}_k}H\sum_j e^{i\vec{q}\cdot\vec{r}_j} + \sum_k e^{-i\vec{q}\cdot\vec{r}_k}\sum_j e^{i\vec{q}\cdot\vec{r}_j}H \ . \qquad (12.14.130)$$

Next, we take the diagonal matrix elements (in the energy representation) of this expression and insert a complete set of intermediate states.

$$\langle n|\left[\left[H,\sum_j e^{i\vec{q}\cdot\vec{r}_j}\right],\sum_k e^{-i\vec{q}\cdot\vec{r}_k}\right]|n\rangle$$

$$= E_n\langle n|\sum_j e^{i\vec{q}\cdot\vec{r}_j}\sum_k e^{-i\vec{q}\cdot\vec{r}_k} + \sum_k e^{-i\vec{q}\cdot\vec{r}_k}\sum_j e^{i\vec{q}\cdot\vec{r}_j}|n\rangle$$

$$- \sum_m E_m\left\{\langle n|\sum_j e^{i\vec{q}\cdot\vec{r}_j}|m\rangle\langle m|\sum_k e^{-i\vec{q}\cdot\vec{r}_k}|n\rangle\right.$$

$$+ \left.\langle n|\sum_k e^{-i\vec{q}\cdot\vec{r}_k}|m\rangle\langle m|\langle n|\sum_j e^{i\vec{q}\cdot\vec{r}_j}|n\rangle\right\}$$

$$= 2\sum_m (E_n - E_m) \left| \langle n | \sum_j e^{i\vec{q}\cdot\vec{r}_j} | m \rangle \right|^2 . \tag{12.14.131}$$

On the other hand, if we evaluate the double commutator we find

$$\left[\left[H, \sum_j e^{i\vec{q}\cdot\vec{r}_j} \right], \sum_k e^{-i\vec{q}\cdot\vec{r}_k} \right]$$

$$= \frac{\hbar\vec{q}}{2M} \cdot \sum_{jk} \left[\vec{p}_j\, e^{i\vec{q}\cdot\vec{r}_j} + e^{i\vec{q}\cdot\vec{r}_j}\, \vec{p}_j\, , \, e^{-i\vec{q}\cdot\vec{r}_k} \right]$$

$$= -2\frac{(\hbar\vec{q})^2}{2M} \sum_{jk} e^{i\vec{q}\cdot\vec{r}_j}\, e^{-i\vec{q}\cdot\vec{r}_k}\, \delta_{jk}$$

$$= -2\frac{(\hbar\vec{q})^2}{2M} N . \tag{12.14.132}$$

This proves the desired result.

It is worth noting that if we make the dipole approximation by putting

$$e^{i\vec{q}\cdot\vec{r}_j} \approx 1 + i\vec{q} \cdot \vec{r}_j \tag{12.14.133}$$

we get the result derived in the previous problem.

12.15   Gaussian Perturbation of SHO Bosons

Two identical bosons move in the one-dimensional simple harmonic oscillator potential

$$V = \frac{1}{2}m\omega^2 \left(x_1^2 + x_2^2 \right)$$

and also interact with each other via the potential

$$v_{int} = V_0\, e^{-\alpha(x_1 - x_2)^2} \quad \alpha > 0 .$$

Find the ground state energy correct to first order in V_0.

Solution

The unperturbed ground state of the two bosons is

$$\psi(x_1, x_2) = \phi_0(x_1)\phi_0(x_2) \tag{12.15.134}$$

with energy

$$E_0 = 2 \times \frac{1}{2}\hbar\omega = \hbar\omega .$$

Here

$$\phi_0(x) = \left(\frac{m\omega}{2\pi\hbar}\right)^{1/4} \exp\left(-\frac{m\omega}{2\hbar}x^2\right) \ . \tag{12.15.135}$$

Also, the wavefunction $\psi(x_1, x_2)$ is already properly symmetrized. The energy shift ΔE due to the perturbation is given to lowest order by

$$\begin{aligned} \Delta E &= (\psi, v_{int}\psi) \\ &= V_0 \left(\frac{m\omega}{2\pi\hbar}\right)^{1/2} \int_{-\infty}^{\infty} e^{-\frac{m\omega}{\hbar}(x_1^2+x_2^2)} e^{-\alpha(x_1-x_2)^2} \, dx_1 dx_2 \ . \end{aligned} \tag{12.15.136}$$

We now change variables to

$$\begin{aligned} R &= \frac{x_1 + x_2}{2} \\ r &= x_1 - x_2 \ . \end{aligned} \tag{12.15.137}$$

The Jacobian of this transformation is 1. Therefore,

$$\begin{aligned} \Delta E &= V_0 \left(\frac{m\omega}{2\pi\hbar}\right)^{1/2} \int_{-\infty}^{\infty} e^{-\alpha r^2} \exp\left(-\frac{m\omega}{\hbar}(2R^2 + r^2/2)\right) \, dR\, dr \\ &= V_0 \left(\frac{m\omega}{2\pi\hbar}\right)^{1/2} \left(\frac{2\pi\hbar}{m\omega + 2\alpha\hbar}\right)^{1/2} \left(\frac{\pi\hbar}{2m\omega}\right)^{1/2} \\ &= \frac{V_0}{2} \left(\frac{2\pi\hbar}{m\omega + 2\alpha\hbar}\right)^{1/2} \ . \end{aligned} \tag{12.15.138}$$

12.16   Gaussian Perturbation of SHO Fermions

Two identical spin $1/2$ fermions move in the one-dimensional simple harmonic oscillator potential

$$V = \frac{1}{2}m\omega^2 \left(x_1^2 + x_2^2\right)$$

and also interact with each other via the potential

$$v_{int} = V_0\, e^{-\alpha(x_1-x_2)^2} \qquad \alpha > 0 \ .$$

a) Find the ground state energy correct to first order in V_0 for the case of the singlet spin state.
b) Find the ground state energy correct to first order in V_0 for the case of the triplet spin state.

Solution

a) In the singlet case, the spin wavefunction is antisymmetric and the space wavefunction is therefore symmetric in the interchange of the two coordinates. Thus, this case is identical to the case of two bosons discussed in problem 12.15.

b) In the triplet case the spin wavefunction is symmetric so the space wavefunction must be antisymmetric. This forces one of the particles to be in the first excited state. Therefore, the unperturbed spatial wavefunction for the ground state of the two fermions is

$$\psi(x_1, x_2) = \frac{1}{\sqrt{2}} [\phi_0(x_1)\phi_1(x_2) - \phi_1(x_1)\phi_0(x_2)] \qquad (12.16.139)$$

with energy

$$E_0 = \frac{1}{2}\hbar\omega + \frac{3}{2}\hbar\omega = 2\hbar\omega \ .$$

Here

$$\phi_0(x) = \left(\frac{m\omega}{2\pi\hbar}\right)^{1/4} \exp\left(-\frac{m\omega}{2\hbar}x^2\right)$$

$$\phi_1(x) = \sqrt{\frac{2m\omega}{\hbar}} x\phi_0(x) \ . \qquad (12.16.140)$$

The energy shift ΔE due to the perturbation is now given to lowest order by

$$\Delta E = \frac{V_0}{2} \int_{-\infty}^{\infty} e^{-\alpha(x_1-x_2)^2} [\phi_0^2(x_1)\phi_1^2(x_2) + \phi_1^2(x_1)\phi_0^2(x_2)$$
$$- 2\phi_0(x_1)\phi_1(x_2)\phi_1(x_1)\phi_0(x_2)] \ dx_1 dx_2 \ . \qquad (12.16.141)$$

We now again change variables to

$$R = \frac{x_1 + x_2}{2}$$

$$r = x_1 - x_2 \ . \qquad (12.16.142)$$

The Jacobian of this transformation is again 1. Therefore,

$$\Delta E = \frac{V_0}{2} \left(\frac{m\omega}{2\pi\hbar}\right)^{1/2} \frac{2m\omega}{\hbar} \int_{-\infty}^{\infty} e^{-\alpha r^2} \exp\left(-\frac{m\omega}{\hbar}(2R^2 + r^2/2)\right) r^2 \ dR \ dr$$

$$= V_0 2\pi \left(\frac{m\omega}{2\pi\hbar}\right)^{3/2} \left(\frac{\pi\hbar}{2m\omega}\right)^{1/2} \frac{1}{2\pi} \left(\frac{2\pi\hbar}{m\omega + 2\alpha\hbar}\right)^{3/2}$$

$$= V_0 \sqrt{\frac{\pi\hbar}{2m\omega}} \left(\frac{m\omega}{m\omega + 2\alpha\hbar}\right)^{3/2} \ . \qquad (12.16.143)$$

12.17 Polarizability: Particle in a Box

For a particle of mass m and charge e in its ground state when confined to a box $-a \le x \le a$, $-b \le y \le b$, $-c \le z \le c$ find the electric polarizability. Hint: The polarizability α is obtained from the shift in energy

$$\Delta E = -\frac{1}{2}\alpha\mathcal{E}^2$$

when the particle is placed in an electric field $\mathcal{E}$.

Solution

If we take the electric field pointing in the x-direction then the perturbation is

$$V = -e\mathcal{E}x \ .$$ (12.17.144)

The problem is now, for all practical purposes, one-dimensional. The ground state wavefunction is

$$\psi_{0,0,0}^{+} = (abc)^{-1/2}\cos(\pi x/2a)\cos(\pi y/2b)\cos(\pi x/2c) \ .$$ (12.17.145)

Here the superscript $+$ refers to "positive parity". The corresponding ground state energy is

$$E_{0,0,0}^{+} = \frac{\hbar^2\pi^2}{8m}\left(\frac{1}{a^2} + \frac{1}{b^2} + \frac{1}{c^2}\right) \ .$$ (12.17.146)

To lowest order the perturbation has no effect since

$$(\psi_{0,0,0}^{+}, V\psi_{0,0,0}^{+}) = 0 \ .$$ (12.17.147)

Therefore we require second order perturbation theory. This gives

$$\Delta E = \sum_{n\neq 0} \frac{(\psi_{0,0,0}^{+}, V\psi_{n,0,0}^{-})(\psi_{n,0,0}^{-}, V\psi_{0,0,0}^{+})}{E_{0,0,0}^{+} - E_{n,0,0}^{-}}$$ (12.17.148)

where we have included only the non-vanishing matrix elements and the superscript "$-$" refers to parity $= -1$.

$$\psi_{n,0,0}^{-} = (abc)^{-1/2}\sin(n\pi x/2a)\cos(\pi y/2b)\cos(\pi x/2c)$$ (12.17.149)

as well as

$$E_{n,0,0}^{-} = \frac{\hbar^2\pi^2}{8m}\left(\frac{4n^2}{a^2} + \frac{1}{b^2} + \frac{1}{c^2}\right) \ .$$ (12.17.150)

Thus, we need the matrix element

$$\begin{aligned}
M &= \frac{1}{a}\int_a^a x\cos(\pi x/2a)\sin(n\pi x/a)\,dx \\
&= \frac{1}{2a}\int_a^a x\sin[(n+1/2)\pi x/a]\,dx + \frac{1}{2a}\int_a^a x\sin[(n-1/2)\pi x/a]\,dx \\
&= (-1)^{n+1}\frac{32a}{\pi^2}\frac{n}{(4n^2-1)^2} \ .
\end{aligned}$$ (12.17.151)

Also,

$$E_{0,0,0}^{+} - E_{n,0,0}^{-} = \frac{\hbar^2\pi^2}{8ma^2}\left(1 - 4n^2\right) \ .$$ (12.17.152)

Combining these results we obtain

$$\Delta E = -\frac{8192\,e^2\mathcal{E}^2 ma^4}{\pi^6}\frac{1}{\hbar^2}\sum_{n=1}^{\infty}\frac{n^2}{(4n^2-1)^5} \ .$$ (12.17.153)

After equating this to the expression for the energy shift in terms of the polarizability we find that the polarizability is given by

$$\alpha = \frac{4096}{\pi^6} \frac{e^2 m a^4}{\hbar^2} \sum_{n=1}^{\infty} \frac{n^2}{(4n^2 - 1)^5} \cdot$$ (12.17.154)

12.18   Atomic Isotope Effect

Every nucleus has a finite radius $R = r_0 A^{1/3}$ where

$$r_0 = 1.2 \times 10^{-13} \quad \text{cm}$$

and A is the atomic number of the nucleus. Thus, the potential energy experienced by an electron near a nucleus is not simply

$$V(r) = -\frac{Ze^2}{r} \quad .$$

If we assume that the charge density in the nucleus is constant then we have instead the potential energy

$$V(r) = \begin{cases} \frac{Ze^2}{R} \left[\frac{r^2}{2R^2} - \frac{3}{2} \right] & r \leq R \\ -\frac{Ze^2}{r} & r \geq R \end{cases} \quad .$$ (12.18.155)

a) Use perturbation theory to calculate the isotope shift, that is the dependence on A of the K-electron (1s state) for an atom with Z protons and atomic number A.

b) Use this result to compute the energy splitting for the K-electron between the heaviest lead $(Z = 82)$ isotope $A = 214$ and the lightest $A = 195$. Neglect the presence of the other electrons.

Solution

a) The unperturbed Hamiltonian is

$$H_0 = \frac{\vec{p}^2}{2m} - \frac{Ze^2}{r} \quad .$$ (12.18.156)

The perturbation is

$$\begin{aligned} H' &= V(r) - \left(-\frac{Ze^2}{r} \right) \\ &= \begin{cases} \frac{Ze^2}{R} \left[\frac{r^2}{2R^2} - \frac{3}{2} + \frac{R}{r} \right] & r \leq R \\ 0 & r \geq R \end{cases} \quad . \end{aligned}$$ (12.18.157)

The unperturbed ground state energy of the K-electron is

$$E_0^{(0)} = -\frac{1}{2} \frac{Ze^2}{a/Z}$$ (12.18.158)

where $a = 5.292 \times 10^{-9}$ cm is the Bohr radius. The corresponding wavefunction is

$$\psi_0^{(0)}(r) = \frac{1}{\sqrt{8\pi}} \left(\frac{2Z}{a} \right)^{3/2} e^{-Zr/a} \ . \tag{12.18.159}$$

The first order correction to $E_0^{(0)}$ is given by

$$E_0^{(1)} = (\psi_0^{(0)}, \, H' \, \psi_0^{(0)}) \ . \tag{12.18.160}$$

Thus,

$$E_0^{(1)} = \frac{1}{2} \left(\frac{2Z}{a} \right)^3 \frac{Ze^2}{R} \int_0^R e^{-2Zr/a} \left[\frac{r^2}{2R^2} - \frac{3}{2} + \frac{R}{r} \right] r^2 \, dr \ . \tag{12.18.161}$$

We now let

$$\alpha = \frac{2ZR}{a} \qquad x = \frac{2Zr}{a} \ . \tag{12.18.162}$$

Then,

$$\begin{aligned} E_0^{(1)} &= \frac{1}{2} \frac{Ze^2}{R} \int_0^\alpha e^{-x} \left[\frac{x^2}{2\alpha^2} - \frac{3}{2} + \frac{\alpha}{x} \right] x^2 \, dx \\ &= \frac{1}{2} \frac{Ze^2}{a/Z} \frac{2}{\alpha} \left[\frac{12}{\alpha^2} - 3 + \alpha - e^{-\alpha} \left(\frac{12}{\alpha^2} + \frac{12}{\alpha} + 3 \right) \right] \ . \end{aligned} \tag{12.18.163}$$

If we now make the dependence on the atomic number A explicit by writing

$$\alpha = \frac{2Zr_0}{a} A^{1/3} = \gamma A^{1/3} \tag{12.18.164}$$

we have the desired dependence on A.

$$\begin{aligned} E_0^{(1)} &= \frac{1}{2} \frac{Ze^2}{a/Z} \frac{2}{\gamma} A^{-1/3} \times \left[\frac{12}{\gamma^2} A^{-2/3} - 3 + \gamma A^{1/3} \right. \\ &\quad \left. - e^{-\gamma A^{1/3}} \left(\frac{12}{\gamma^2} A^{-2/3} + \frac{12}{\gamma} A^{-1/3} + 3 \right) \right] \ . \end{aligned} \tag{12.18.165}$$

b) If we take $Z = 82$ and $A = 195$ we get that $\alpha = 0.238$. Substituting these values we find that

$$E_0^{(1)}(A = 195) = \frac{1}{2} \frac{Ze^2}{a/Z} \times 9.91 \times 10^{-3} \ . \tag{12.18.166}$$

Similarly, for $Z = 82$ and $A = 214$ we get that $\alpha = 0.245$. Thus, repeating the calculation we find that in this case

$$E_0^{(1)}(A = 214) = \frac{1}{2} \frac{Ze^2}{a/Z} \times 1.08 \times 10^{-2} \ . \tag{12.18.167}$$

Thus, recalling that

$$\frac{1}{2} \frac{e^2}{a} = 13.6 \text{ eV}$$

the energy difference in energy of the K-electrons between the two isotopes is

$$\Delta E = \frac{1}{2} \frac{Ze^2}{a/Z} \times 9.4 \times 10^{-4} = 86 \text{ eV} \ . \tag{12.18.168}$$

12.19   Relativistic Correction to H atom

The kinetic energy for a relativistic particle is

$$T = \sqrt{c^2 \vec{p}^2 + m^2 c^4} - mc^2 \approx \frac{\vec{p}^2}{2m} - \frac{1}{2mc^2}\left(\frac{\vec{p}^2}{2m}\right)^2 . \qquad (12.19.169)$$

Use the last term as a perturbation to calculate the first order correction to the energy levels of a hydrogenic atom.

Hint: The following expectation values for a hydrogenic atom may be useful

$$\langle 1/r \rangle = \frac{1}{n^2} \frac{1}{a/Z}$$

$$\langle 1/r^2 \rangle = \frac{1}{n^3(l + 1/2)} \frac{1}{(a/Z)^2} . \qquad (12.19.170)$$

Here, a is the Bohr radius.

Solution

To save labour we use the fact that

$$\frac{\vec{p}^2}{2m} = E - V = -\frac{1}{2}\frac{Ze^2}{(a/Z)}\frac{1}{n^2} + \frac{Ze^2}{r} . \qquad (12.19.171)$$

Thus, the first order correction is given by

$$
\begin{aligned}
E_{n,l}^{(1)} &= -\frac{1}{2mc^2}\left\langle \left(\frac{(\vec{p})^2}{2m}\right)^2 \right\rangle \\
&= -\frac{1}{2mc^2}\left\langle \left(\frac{1}{2}\frac{Ze^2}{(a/Z)}\frac{1}{n^2} - \frac{Ze^2}{r}\right)^2 \right\rangle \\
&= -\frac{1}{2mc^2}\left[\frac{1}{4}\frac{Z^2 e^4}{(a/Z)^2}\frac{1}{n^4} - \frac{Z^2 e^4}{(a/Z)}\frac{1}{n^2}\langle 1/r \rangle + Z^2 e^2 \langle 1/r^2 \rangle\right] \\
&= -\frac{Z^2 e^4}{2mc^2}\left[\frac{1}{4}\left(\frac{Z}{a}\right)^2\frac{1}{n^4} - \left(\frac{Z}{a}\right)^2\frac{1}{n^4} + \left(\frac{Z}{a}\right)^2\frac{1}{n^3(l + 1/2)}\right] \\
&= \frac{Z^2 e^4}{mc^2}\left(\frac{Z}{a}\right)^2\left[\frac{3}{8n^4} - \frac{1}{n^3(2l + 1)}\right] . \qquad (12.19.172)
\end{aligned}
$$

12.20   van der Waals' Interaction

Two widely separated hydrogen atoms interact via a dipole-dipole interaction whose potential, known as the *van der Waals potential*, is given by

$$V = \frac{e^2}{R^3}[\vec{r_1} \cdot \vec{r_2} - z_1 z_2]$$

where R is the separation of the centres of the two hydrogen atoms and $\vec{r_1} = (x_1, y_1, z_1)$ and $\vec{r_2} = (x_2, y_2, z_2)$ are respectively the coordinates of the electrons associated with atom 1 and 2. Using perturbation theory calculate the interaction energy of two widely separated hydrogen atoms. As an approximation for the unperturbed energies use

$$E_1 - E_n \approx E_1 \; .$$

Solution

The total Hamiltonian for the two hydrogen atoms is

$$H = H_{01} + H_{02} + V \tag{12.20.173}$$

where

$$H_{0i} = \frac{\vec{p_i}^2}{2m} - \frac{Ze^2}{r_i} \quad i = 1, 2 \; . \tag{12.20.174}$$

The unperturbed energies for H_{0i} are

$$E_n = -\frac{e^2}{a_0}\frac{1}{n^2} \tag{12.20.175}$$

with the corresponding eigenstates $|n_i, l_i, m_i\rangle$. We are interested in the energy shift of the ground state

$$|\psi(1,2)\rangle = |1_1, 0_1, 0_1\rangle|1_2, 0_2, 0_2\rangle = |1\rangle|2\rangle \; . \tag{12.20.176}$$

To lowest order this energy shift is

$$
\begin{aligned}
E^{(1)} &= \langle\psi(1,2)|V|\psi(1,2)\rangle \\
&= \frac{e^2}{R^3}[\langle 1|x_1|1\rangle\langle 2|x_2|2\rangle + \langle 1|y_1|1\rangle\langle 2|y_2|2\rangle + \langle 1|z_1|1\rangle\langle 2|z_2|2\rangle] \\
&= 0 \; .
\end{aligned}
\tag{12.20.177}
$$

So we have to go to second order. In this case we find

$$
\begin{aligned}
E^{(2)} &= \sum_{n_1, n_2 \neq 1} \frac{|\langle\psi(1,2)|V|n_1, l_1, m_1\rangle|n_2, l_2, m_2\rangle|^2}{2E_1 - E_{n_1} - E_{n_2}} \\
&\approx \frac{1}{2E_1} \sum_{n_1, n_2 \neq 1} \langle\psi(1,2)|V|n_1, l_1, m_1\rangle|n_2, l_2, m_2\rangle \\
&\quad \times \langle n_1, l_1, m_1|\langle n_2, l_2, m_2|V|\psi(1,2)\rangle \\
&= \frac{1}{2E_1}\langle\psi(1,2)|V^2|\psi(1,2)\rangle \; .
\end{aligned}
\tag{12.20.178}
$$

Here we have used the indicated approximation to go to the second line and the completeness relation to obtain the last line. These matrix elements are now easy to evaluate using the symmetry of the ground state wavefunction. In fact the only non-zero matrix elements are

$$\langle i|x_i^2|i\rangle = \langle i|y_i^2|i\rangle = \langle i|z_i^2|i\rangle = \frac{1}{3}\langle i|r_i^2|i\rangle \ . \tag{12.20.179}$$

But,

$$\begin{aligned}
\langle i|r_i^2|i\rangle &= \frac{4}{a_0^3}\int_0^\infty e^{-2r/a_0}\, r^4\, dr \\
&= 3a_0^2 \ .
\end{aligned} \tag{12.20.180}$$

Thus, we finally have the desired result

$$E^{(2)} = \frac{3e^4 a_0^4}{2E_1 R^6} \ . \tag{12.20.181}$$

Bibliography

[12.1] R.Jackiw, Phys.Rev. **157**, 1220, (1967) .

[12.2] A.Z. Capri, *Nonrelativistic Quantum Mechanics* 3rd edition, World Scientific Publishing Co. Pte. Ltd., chapter 12, (2002) .

Chapter 13

Degenerate Perturbation Theory

13.1 Stark Effect for $n = 2$ Level in H

Find the shift in the energy of the $n = 2$ levels of a hydrogen atom, to first order due to a constant electric field (linear Stark effect). The potential is

$$V' = -e\mathbf{E} \cdot \mathbf{r} = -e\mathcal{E}z \ .$$

Solution

We first write out the $n = 2$ levels for the hydrogen atom in units such that length is scaled by the Bohr radius.

$$\psi_{2,0,0} = -\frac{1}{\sqrt{8\pi}}(1 - r/2)\, e^{-r/2}$$

$$\psi_{2,1,0} = -\frac{1}{\sqrt{32\pi}} r\, e^{-r/2} \cos\theta$$

$$\psi_{2,1,1} = \frac{1}{\sqrt{64\pi}} r\, e^{-r/2} \sin\theta\, e^{i\phi}$$

$$\psi_{2,1,-1} = -\frac{1}{\sqrt{64\pi}} r\, e^{-r/2} \sin\theta\, e^{-i\phi} \tag{13.1.1}$$

The matrix elements of the perturbation in this degenerate subspace are

$$\langle \psi_{2,l,m} | e E\, r \cos\theta | \psi_{2,l',m'} \rangle \ .$$

The only non-zero elements are between $\psi_{2,0,0}$ and $\psi_{2,1,0}$. Thus, the matrix to be diagonalized is

$$e\mathcal{E}a \begin{pmatrix} 0 & -3 & 0 & 0 \\ -3 & 0 & 0 & 0 \\ 0 & 0 & 0 & 0 \\ 0 & 0 & 0 & 0 \end{pmatrix} .$$

Here we have restored the usual units so that the Bohr radius a is no longer of unit length. The eigenvalues of this matrix are $+e\mathcal{E}a$, $-e\mathcal{E}a$, 0, 0. Therefore the degeneracy of two of the states is lifted in first order and the new energies are $E_2^{(0)} + e\mathcal{E}a$, $E_2^{(0)} - e\mathcal{E}a$, $E_2^{(0)}$, and $E_2^{(0)}$ where

$$E_2^{(0)} = -\frac{e^2}{2a} . \tag{13.1.2}$$

13.2   Perturbation of Particle in a Box

A particle is in a two-dimensional box of sides a. If a perturbation

$$V' = \lambda xy$$

is applied, find the change in the energy of the first excited state to first non-trivial order.

Solution

The eigenfunctions for the particle in the box are given by

$$\psi_{n,k} = \frac{2}{a}\sin\frac{n\pi x}{a}\sin\frac{k\pi y}{a} \quad 0 \le x, y \le a . \tag{13.2.3}$$

The corresponding energies are:

$$E_{n,k} = \frac{\hbar^2\pi^2}{2ma^2}(n^2 + k^2) . \tag{13.2.4}$$

This shows that the ground state with $n = k = 1$ is the only nondegenerate state. All other states are clearly degenerate. The first excited state is two-fold degenerate. The two degenerate states are $|1, 2\rangle^{(0)}$ and $|2, 1\rangle^{(0)}$. The perturbation Hamiltonian in the degenerate subspace has matrix elements $^{(0)}\langle i, j|xy|i, j\rangle^{(0)}$. When this is written out we get

$$H' = \begin{pmatrix} \frac{a}{2} & \frac{16a}{9\pi^2} \\ \frac{16a}{9\pi^2} & \frac{a}{2} \end{pmatrix} . \tag{13.2.5}$$

The corresponding eigenvalues and eigenvectors are respectively

$$a\left[\frac{1}{2} + \frac{16}{9\pi^2}\right] \quad , \quad a\left[\frac{1}{2} - \frac{16}{9\pi^2}\right] \tag{13.2.6}$$

and

$$\frac{1}{\sqrt{2}}\begin{pmatrix} 1 \\ 1 \end{pmatrix} \quad , \quad \frac{1}{\sqrt{2}}\begin{pmatrix} 1 \\ -1 \end{pmatrix} . \tag{13.2.7}$$

Thus, instead of $|1, 2\rangle^{(0)}$ and $|2, 1\rangle^{(0)}$ we begin with

$$|+\rangle = \frac{1}{\sqrt{2}}\left[|1, 2\rangle^{(0)} + |2, 1\rangle^{(0)}\right] \tag{13.2.8}$$

and

$$|-\rangle = \frac{1}{\sqrt{2}} \left[|1,2\rangle^{(0)} - |2,1\rangle^{(0)} \right] \tag{13.2.9}$$

in our perturbation calculation. The first order correction is given by λ times the eigenvalues above. The second order terms are given by

$$E_\pm^{(2)} = \sum_{n+k \neq 3} \lambda^2 \frac{|\langle \pm |xy| n,k \rangle^{(0)}|^2}{(\hbar^2 \pi^2)/(2ma^2)(5 - n^2 - k^2)} . \tag{13.2.10}$$

Now,

$$\langle \pm |xy| n,k \rangle^{(0)} = \frac{2\sqrt{2}}{a^2} \int_0^a dx\, dy \left[\sin \frac{\pi x}{a} \sin \frac{2\pi y}{a} \right.$$
$$\left. \pm \sin \frac{2\pi x}{a} \sin \frac{\pi y}{a} \right] xy \sin \frac{n\pi x}{a} \sin \frac{k\pi y}{a} . \tag{13.2.11}$$

But,

$$\frac{2}{a^2} \int_0^a x\, dx \sin \frac{n\pi x}{a} \sin \frac{m\pi x}{a}$$
$$= \frac{1}{a^2} \int_0^a x\, dx \left[\cos \frac{(n-m)\pi x}{a} - \cos \frac{(n+m)\pi x}{a} \right]$$
$$= \frac{1}{a^2} \int_0^a x \left[\frac{a}{(n-m)\pi} d\sin \frac{(n-m)\pi x}{a} \right.$$
$$- \frac{a}{(n+m)\pi} d\sin \frac{(n+m)\pi x}{a} \right]$$
$$= \frac{1}{(n-m)\pi a} \int_0^a \sin \frac{(n-m)\pi x}{a} dx$$
$$- \frac{1}{(n+m)\pi a} \int_0^a \sin \frac{(n+m)\pi x}{a} dx$$
$$= \frac{1}{(n-m)^2 \pi^2} [(-1)^{n-m} - 1] - \frac{1}{(n+m)^2 \pi^2} [(-1)^{n+m} - 1]$$
$$= [(-1)^{n+m} - 1] \frac{1}{\pi^2} \frac{4nm}{(n^2 - m^2)^2} . \tag{13.2.12}$$

With this result we find

$$\langle \pm |xy| n,k \rangle^{(0)} = \frac{16\sqrt{2}}{\pi^4} \left\{ \frac{[(-1)^{n+1} - 1]}{(n^2 - 1)^2} \frac{[(-1)^{k+2} - 1]}{(k^2 - 4)^2} \right.$$
$$\left. \pm \frac{[(-1)^{n+2} - 1]}{(n^2 - 4)^2} \frac{[(-1)^{k+1} - 1]}{(k^2 - 1)^2} \right\} . \tag{13.2.13}$$

Substituting this in the sum for the perturbation series we have the desired result.

13.3 Perturbation of Isotropic Two-dimensional SHO

For the two-dimensional simple harmonic oscillator with

$$H_0 = \hbar\omega(a_1^\dagger a_1 + a_2^\dagger a_2)$$

calculate the effect, to second order, of the perturbation

$$H' = \lambda(a_1^\dagger a_1^\dagger a_1 a_1 + a_2^\dagger a_2^\dagger a_2 a_2)$$

on the second excited states and to first order on the third excited states. What are the effects on the ground state and first excited states?

Solution

In this case we have

$$H_0 = \hbar\omega \left(a_1^\dagger a_1 + a_2^\dagger a_2 \right) \tag{13.3.14}$$

$$H' = \lambda \left(a_1^\dagger a_1^\dagger a_2 a_2 + a_2^\dagger a_2^\dagger a_1 a_1 \right) \tag{13.3.15}$$

and

$$H_0|n_1, n_2\rangle^{(0)} = (n_1 + n_2)\hbar\omega|n_1, n_2\rangle^{(0)} \tag{13.3.16}$$

This shows that all states except the ground state are degenerate. For the second excited state we have three degenerate states:

$$|2, 0\rangle^{(0)}, \ |1, 1\rangle^{(0)}, \ |0, 2\rangle^{(0)} \ .$$

All have the same unperturbed energy $2\hbar\omega$. In this degenerate subspace, the perturbation Hamiltonian is

$$H' = \begin{pmatrix} 0 & 0 & 2\lambda \\ 0 & 0 & 0 \\ 2\lambda & 0 & 0 \end{pmatrix} . \tag{13.3.17}$$

The eigenvalues are

$$2\lambda \, , 0 \, , \, -2\lambda \ .$$

The corresponding eigenvectors are respectively

$$\frac{1}{\sqrt{2}} \begin{pmatrix} 1 \\ 0 \\ 1 \end{pmatrix} , \ \begin{pmatrix} 0 \\ 1 \\ 0 \end{pmatrix} , \ \frac{1}{\sqrt{2}} \begin{pmatrix} 1 \\ 0 \\ -1 \end{pmatrix} . \tag{13.3.18}$$

Thus, we have as new basis states in this degenerate subspace

$$\begin{aligned}
|\psi_{2,+}\rangle &= \frac{1}{\sqrt{2}}\left[|2,0\rangle^{(0)} + |0,2\rangle^{(0)}\right] \\
|\psi_{2,0}\rangle &= |1,1\rangle^{(0)} \\
|\psi_{2,-}\rangle &= \frac{1}{\sqrt{2}}\left[|2,0\rangle^{(0)} - |0,2\rangle^{(0)}\right] .
\end{aligned}$$
(13.3.19)

Therefore, writing $r = +1, 0, -1$ we find

$$\begin{aligned}
E_{2,r} &= 2\hbar\omega + r2\lambda + \sum_{n_1+n_2\neq 2} \frac{\left|^{(0)}\langle n_1, n_2|H'|\psi_{2,r}\rangle\right|^2}{2\hbar\omega - (n_1 + n_2)\hbar\omega} \\
&= 2\hbar\omega + r2\lambda + 0 \quad \text{to second order in } \lambda .
\end{aligned}$$
(13.3.20)

The third excited states are
$|3,0\rangle^{(0)}$, $|2,1\rangle^{(0)}$, $|1,2\rangle^{(0)}$, $|0,3\rangle^{(0)}$.
The perturbation Hamiltonian in this degenerate subspace is

$$H' = \begin{pmatrix} 0 & 0 & \sqrt{12}\lambda & 0 \\ 0 & 0 & 0 & \sqrt{12}\lambda \\ \sqrt{12}\lambda & 0 & 0 & 0 \\ 0 & \sqrt{12}\lambda & 0 & 0 \end{pmatrix}$$
(13.3.21)

The eigenvalues are:

$$0, ,0, \sqrt{12}\lambda, \text{ and, } -\sqrt{12}\lambda .$$

Not all the degeneracies are lifted. For this reason we only compute to first order in λ. To this order the eigenvalues are

$$E_{3,r} = 3\hbar\omega + r\sqrt{12}\lambda$$
(13.3.22)

where $r = 0, 0, \pm 1$.

13.4   Two-dimensional SHO with Off-diagonal Term

a) Repeat the previous problem with

$$H' = \lambda[(a_1^\dagger a_2 + a_2^\dagger a_1)] .$$

b) This problem can also be solved exactly by introducing operators

$$A_1 = a_1 \cos\theta + a_2 \sin\theta$$

$$A_2 = -a_1 \sin\theta + a_2 \cos\theta$$

and choosing θ appropriately. Do this and compare with the perturbation result.

Solution

a) Again we have

$$H_0|n_1, n_2\rangle^{(0)} = (n_1 + n_2)\hbar\omega|n_1, n_2\rangle^{(0)} . \tag{13.4.23}$$

We first solve for the eigenvalues to second order in λ, then we solve the problem exactly and compare. For $n_1 + n_2 = 2$ (second excited state) we have the three degenerate states:
$|2, 0\rangle^{(0)}$, $|1, 1\rangle^{(0)}$, and $|0, 2\rangle^{(0)}$.
The perturbation Hamiltonian in this degenerate subspace is

$$H' = \begin{pmatrix} 0 & \sqrt{2}\lambda & 0 \\ \sqrt{2}\lambda & 0 & \sqrt{2}\lambda \\ 0 & \sqrt{2}\lambda & 0 \end{pmatrix} . \tag{13.4.24}$$

The eigenvalues are: 2λ , 0 , -2λ. The corresponding eigenvectors are respectively

$$\begin{pmatrix} 1/2 \\ 1/\sqrt{2} \\ 1/2 \end{pmatrix} , \quad \begin{pmatrix} 1/\sqrt{2} \\ 0 \\ -1/\sqrt{2} \end{pmatrix} , \quad \begin{pmatrix} 1/2 \\ -1/\sqrt{2} \\ 1/2 \end{pmatrix} . \tag{13.4.25}$$

Thus, our new basis states are

$$
\begin{aligned}
|\psi_{2,+}\rangle &= \frac{1}{2}|2, 0\rangle^{(0)} + \frac{1}{\sqrt{2}}|1, 1\rangle^{(0)} + \frac{1}{2}|0, 2\rangle^{(0)} \\
|\psi_{2,0}\rangle &= \frac{1}{\sqrt{2}}|2, 0\rangle^{(0)} - \frac{1}{\sqrt{2}}|0, 2\rangle^{(0)} \\
|\psi_{2,-}\rangle &= \frac{1}{2}|2, 0\rangle^{(0)} - \frac{1}{\sqrt{2}}|1, 1\rangle^{(0)} + \frac{1}{2}|0, 2\rangle^{(0)} .
\end{aligned}
\tag{13.4.26}
$$

This yields:

$$
\begin{aligned}
E_{2,r} &= 2\hbar\omega + r2\lambda + \sum_{n_1+n_2\neq 2} \frac{\left|{}^{(0)}\langle n_1, n_2|H'|\psi_{2,r}\rangle\right|^2}{2\hbar\omega - (n_1 + n_2)\hbar\omega} \\
&= 2\hbar\omega + r2\lambda .
\end{aligned}
\tag{13.4.27}
$$

Since H' does not change the sum over $n_1 + n_2$ there is no second order contribution. The third excited states are:

$$|3, 0\rangle^{(0)}, \ |2, 1\rangle^{(0)}, \ |1, 2\rangle^{(0)}, \ \text{and} |0, 3\rangle^{(0)} .$$

The perturbation Hamiltonian in this degenerate subspace is

$$H' = \begin{pmatrix} 0 & \sqrt{3}\lambda & 0 & 0 \\ \sqrt{3}\lambda & 0 & 2\lambda & 0 \\ 0 & 2\lambda & 0 & \sqrt{3}\lambda \\ 0 & 0 & \sqrt{3}\lambda & 0 \end{pmatrix} . \tag{13.4.28}$$

The eigenvalues for this matrix are: 3λ , -3λ , λ , $-\lambda$. The corresponding eigenvectors are

$$
\frac{1}{2}\begin{pmatrix} 1/\sqrt{2} \\ \sqrt{3}/2 \\ \sqrt{3}/2 \\ 1/\sqrt{2} \end{pmatrix} \quad , \quad \frac{1}{2}\begin{pmatrix} 1/\sqrt{2} \\ -\sqrt{3}/2 \\ \sqrt{3}/2 \\ -1/\sqrt{2} \end{pmatrix}
$$

$$
\frac{1}{2}\begin{pmatrix} \sqrt{3}/2 \\ 1/\sqrt{2} \\ -1/\sqrt{2} \\ -\sqrt{3}/2 \end{pmatrix} \quad , \quad \frac{1}{2}\begin{pmatrix} \sqrt{3}/2 \\ -1/\sqrt{2} \\ -1/\sqrt{2} \\ \sqrt{3}/2 \end{pmatrix} . \tag{13.4.29}
$$

We could now again compute the new basis vectors and go on to second order, but because $[H_0, H'] = 0$ there is no second order contribution. Thus, correct to second order the energies are

$$
E_2 = 3\hbar\omega \pm 3\lambda \quad , \quad 3\hbar\omega \pm \lambda . \tag{13.4.30}
$$

b) We next examine the exact solution. For this purpose we make the principal axis transformation

$$
\begin{aligned}
A &= a_1 \cos\theta + a_2 \sin\theta \\
B &= -a_1 \sin\theta + a_2 \cos\theta .
\end{aligned} \tag{13.4.31}
$$

Then, A , $A^\dagger$, B , $B^\dagger$ satisfy the same commutation relations as a_1 , $a_1^\dagger$, a_2 , $a_2^\dagger$. Rewriting the Hamiltonian we find

$$
\begin{aligned}
H_0 &= \hbar\omega(a_1^\dagger a_1 + a_2^\dagger a_2) \\
&= \hbar\omega(A^\dagger A + B^\dagger B)
\end{aligned} \tag{13.4.32}
$$

and

$$
\begin{aligned}
H' &= \lambda\left[(A^\dagger \cos\theta - B^\dagger \sin\theta)(A \sin\theta + B \cos\theta)\right. \\
&+ \left. (A^\dagger \sin\theta + B^\dagger \cos\theta)(A \cos\theta - B \sin\theta)\right] \\
&= \lambda\left[A^\dagger A \sin 2\theta - B^\dagger B \sin 2\theta + (A^\dagger B + B^\dagger A)\cos 2\theta\right] . \tag{13.4.33}
\end{aligned}
$$

Therefore, choosing $\cos 2\theta = 0$ so that $\sin 2\theta = 1$ we get

$$
H = (\hbar\omega + \lambda)A^\dagger A + (\hbar\omega - \lambda)B^\dagger B . \tag{13.4.34}
$$

The eigenstates may be labelled $|N_1, N_2\rangle$ and the eigenvalues are

$$
H = (\hbar\omega + \lambda)N_1 + (\hbar\omega - \lambda)N_2 = (N_1 + N_2)\hbar\omega + (N_1 - N_2)\lambda . \tag{13.4.35}
$$

For $N_1 + N_2 = 2$ we get $2\hbar\omega + 2r\lambda$ where $r = 1, 0, -1$.

For $N_1 + N_2 = 3$ we get $3\hbar\omega \pm 3\lambda$, $3\hbar\omega \pm \lambda$.

These results are exactly the same as those obtained from perturbation theory.

13.5 Non-diagonal Two-dimensional SHO

For a particle of mass m moving in the potential

$$V = \frac{1}{2}k_1 x^2 + \frac{1}{2}k_2 y^2 + \lambda xy \qquad |k_1 - k_2| < 2\lambda .$$

Find to order λ^2 the energy of the ground state and first excited state. Compare your answer with the exact solution obtained in problem 12.8c.

Solution

In this case $k_1 \approx k_2$ and all levels except the ground state are almost degenerate. Thus, for the first two excited states we need to use degenerate perturbation theory.

Ground state: $n_1 = n_2 = 0$

$$E_{0,0}^{(0)} = \frac{1}{2}\hbar(\omega_1 + \omega_2) \tag{13.5.36}$$

and

$$E_{0,0}^{(1)} = 0 \tag{13.5.37}$$

while

$$E_{0,0}^{(2)} = \lambda^2 \sum_{m_1+m_2 \neq 0} \frac{|\langle 0|x|m_1\rangle\langle 0|y|m_2\rangle|^2}{-\hbar(\omega_1 m_1 + \omega_2 m_2)} \tag{13.5.38}$$

where we have

$$\langle 0|x|m_1\rangle\langle 0|y|m_2\rangle = \frac{\hbar}{2m}\frac{1}{\sqrt{\omega_1\omega_2}}[\delta_{0,m_1-1}\delta_{0,m_2-1}] \tag{13.5.39}$$

Thus, we get

$$E_{0,0}^{(2)} = \frac{\lambda^2\hbar^2}{4m^2}\frac{1}{\omega_1\omega_2}\frac{1}{\hbar(\omega_1 + \omega_2)} . \tag{13.5.40}$$

Notice, that this expression does not involve

$$\omega_1^2 - \omega_2^2 = \frac{k_1 - k_2}{m} . \tag{13.5.41}$$

First two excited states: $n_1 = 0$, $n_2 = 1$; $n_1 = 1$, $n_2 = 0$
The Hamiltonian is

$$H = \frac{p_x^2 + p_y^2}{2m} + \frac{1}{2}k_1 x^2 + \frac{1}{2}k_2 y^2 + \lambda xy . \tag{13.5.42}$$

Our perturbation is

$$H' = \lambda xy = \frac{\hbar\lambda}{2m\omega}(a_1 + a_1^\dagger)(a_2 + a_2^\dagger) . \tag{13.5.43}$$

To do degenerate perturbation theory we diagonalize this perturbation in the degenerate subspace. That is, we diagonalize the matrix

$$\langle n_1, n_2 | H' | n_1', n_2' \rangle = \frac{\hbar \lambda}{2m\sqrt{\omega_1 \omega_2}} \begin{pmatrix} 0 & 1 \\ 1 & 0 \end{pmatrix} . \tag{13.5.44}$$

The eigenvalues are

$$E_\pm^{(1)} = \pm \frac{\hbar \lambda}{2m\sqrt{\omega_1 \omega_2}} \tag{13.5.45}$$

and the eigenvectors are

$$\psi_\pm = \frac{1}{\sqrt{2}} \begin{pmatrix} 1 \\ \pm 1 \end{pmatrix} . \tag{13.5.46}$$

The energies are therefore

$$E_\pm = 2\hbar\omega \pm \frac{\hbar \lambda}{2m\sqrt{\omega_1 \omega_2}} . \tag{13.5.47}$$

To compare with the exact solution obtained in problem 12.8a we must use the fact that $|k_1 - k_2| < 2\lambda$ and put $\omega_1 = \omega_2$. It is then easy to see that the results agree to this order.

We can now do higher order perturbation theory using that linear combination of $|0, 1\rangle$ and $|1, 0\rangle$ which diagonalizes H'. In other words, to find the perturbation to second order of the levels $|0, 1\rangle$ and $|1, 0\rangle$ we use as a basis the set:

$|0, 0\rangle$, $|\psi_+\rangle$, $|\psi_-\rangle$, $|n_1, n_2\rangle$ where $n_1 + n_2 \geq 2$.

Thus, we get

$$E_{n_1', n_2'}^{(2)} = \sum_{n_1 + n_2 \neq 1} \frac{|\langle \psi_{n_1', n_2'} | H' | n_1, n_2 \rangle|^2}{E_{n_1', n_2'} - E_{n_1, n_2}} . \tag{13.5.48}$$

But,

$$\langle \psi_\pm | H' | n_1, n_2 \rangle = \frac{\lambda \hbar}{2m\sqrt{\omega_1 \omega_2}} \frac{1}{\sqrt{2}} \Big[\langle 0, 1 | (a_1^\dagger + a_1)(a_2^\dagger + a_2) | n_1, n_2 \rangle$$

$$\pm \langle 1, 0 | (a_1^\dagger + a_1)(a_2^\dagger + a_2) | n_1, n_2 \rangle \Big] . \tag{13.5.49}$$

Here, we have already introduced the explicit form

$$\psi_\pm = \frac{1}{\sqrt{2}} [|0, 1\rangle \pm |1, 0\rangle] . \tag{13.5.50}$$

So, for $n_1 + n_2 \neq 1$ we then find that

$$\langle \psi_\pm | H' | n_1, n_2 \rangle = \frac{\lambda \hbar}{2m\sqrt{\omega_1 \omega_2}} [\delta_{n_1, 1} \delta_{n_2, 2} \pm \delta_{n_1, 2} \delta_{n_2, 1}] . \tag{13.5.51}$$

Combining these results we get

$$E_{0,1}^{(2)} = -\frac{\lambda^2\hbar}{4m^2\omega_1\omega_2}\left[\frac{1}{\omega_1+\omega_2} + \frac{1}{2\omega_1}\right]$$

(13.5.52)

and

$$E_{1,0}^{(2)} = -\frac{\lambda^2\hbar}{4m^2\omega_1\omega_2}\left[\frac{1}{2\omega_2} + \frac{1}{\omega_1+\omega_2}\right] \ .$$

(13.5.53)

Again, if we set $\omega_1 = \omega_2$, we find that the results agree to order λ^2 with the exact results obtained in problem 12.8a.

13.6 Particle in a Box Perturbed by Electric Field

A particle of mass m and a charge q is placed in a box of sides (a, a, b) where $b < a$. A weak electric field

$$\vec{E} = \mathcal{E}(y/a\,,x/a\,,0)$$

is applied to this particle. Find the energy of the ground state and first excited states correct to order $|\mathcal{E}|$.

Solution

For a box with sides (a, a, b) the energy eigenstates for a particle in this box are

$$\psi_{n,m,r}^{(0)} = \frac{2\sqrt{2}}{a\sqrt{b}}\sin\left(\frac{n\pi x}{a}\right)\sin\left(\frac{m\pi y}{a}\right)\sin\left(\frac{r\pi z}{b}\right)$$

(13.6.54)

where $n, m, r = 1, 2, 3,\ldots.$. The corresponding energies are:

$$E_{n,m,r}^{(0)} = \frac{\hbar^2}{2m}\left[\frac{(n^2+m^2)\pi^2}{a^2} + \frac{r^2\pi^2}{b^2}\right] \ .$$

(13.6.55)

Since $b \ll a$ the ground state energy is given by $E_{1,1,1}^{(0)}$ and the first excited states are degenerate and their energy is given by $E_{1,2,1}^{(0)} = E_{2,1,1}^{(0)}$. Thus, the ground state perturbation is given by taking the expectation value of the perturbing potential which is given by

$$-e\phi = \frac{e\mathcal{E}}{a}xy \ .$$

(13.6.56)

The result is

$$\begin{aligned}
E_{1,1,1}^{(0)} &= \frac{e\mathcal{E}}{a}\langle 1, 1, 1|xy|1, 1, 1\rangle \\
&= \frac{e\mathcal{E}}{a}\frac{4}{a^2}\int_0^a \sin^2\left(\frac{\pi x}{a}\right)x\,dx \int_0^a \sin^2\left(\frac{2\pi y}{a}\right)y\,dy \\
&= \frac{e\mathcal{E}a}{4} \ .
\end{aligned}$$

(13.6.57)

For the first excited states we need to diagonalize the perturbation in the degenerate subspace. Thus, we have to evaluate

$$\frac{e\mathcal{E}}{a}\langle 1,2,1|xy|1,2,1\rangle = \frac{e\mathcal{E}}{a}\langle 2,1,1|xy|2,1,1\rangle$$

$$= \frac{e\mathcal{E}}{a}\frac{4}{a^2}\int_0^a \sin^2\left(\frac{\pi x}{a}\right) x\, dx \int_0^a \sin^2\left(\frac{2\pi y}{a}\right) y\, dy$$

$$= \frac{e\mathcal{E}a}{4} \tag{13.6.58}$$

as well as

$$\frac{e\mathcal{E}}{a}\langle 1,2,1|xy|2,1,1\rangle = \frac{e\mathcal{E}}{a}\langle 2,1,1|xy|1,2,1\rangle$$

$$= \frac{e\mathcal{E}}{a}\frac{4}{a^2}\int_0^a \sin\left(\frac{\pi x}{a}\right)\sin\left(\frac{2\pi x}{a}\right) x\, dx \int_0^a \sin\left(\frac{\pi y}{a}\right)\sin\left(\frac{2\pi y}{a}\right) y\, dy$$

$$= e\mathcal{E}a\frac{256}{81\pi^4} . \tag{13.6.59}$$

Next we have to diagonalize the matrix

$$e\mathcal{E}a\begin{pmatrix} \frac{1}{4} & \frac{256}{81\pi^4} \\ \frac{256}{81\pi^4} & \frac{1}{4} \end{pmatrix} . \tag{13.6.60}$$

The corresponding eigenvalues are:

$$e\mathcal{E}a\left[\frac{1}{4} \pm \frac{256}{81\pi^4}\right] . \tag{13.6.61}$$

Thus, the degeneracy is lifted.

13.7 Unusual Particle on Interval

Consider the Hamiltonian

$$H_0 = \frac{p^2}{2m} \quad \text{on} \quad -a \le x \le a$$

with the domain for p given by

$$D_p = \{f(x) \in C^2 \mid f(-a) = -f(a)\} .$$

Find the shift in energy of all the energy levels to first nontrivial order due to a perturbation

$$\lambda H' = \lambda x .$$

Hint: The eigensolutions for H_0 are given in [13.1] and are

$$f_n(x) = \frac{1}{\sqrt{2a}} e^{i\pi(n+1/2)x/a} \quad n = 0, \pm 1, \pm 2, \ldots \tag{13.7.62}$$

with the corresponding energy eigenvalues

$$E_n = E_{-(n+1)} = \frac{\pi^2\hbar^2}{2ma^2}(n+1/2)^2 . \tag{13.7.63}$$

Solution

The eigenfunctions of the unperturbed Hamiltonian are, as in [13.1], given by

$$f_n(x) = \frac{1}{\sqrt{2a}} e^{i\pi(n+1/2)x/a} \quad n = 0, \pm 1, \pm 2, \dots \tag{13.7.64}$$

The energy eigenvalues are

$$E_n = E_{-(n+1)} = \frac{\pi^2 \hbar^2}{2ma^2} (n + 1/2)^2 . \tag{13.7.65}$$

This shows that every eigenvalue is two-fold degenerate. We thus need to diagonalize the perturbation Hamiltonian in the degenerate subspaces. This means we have to evaluate the matrix elements

$$(f_n, x f_n) = (f_{-(n+1)}, x f_{-(n+1)}) = 0 \tag{13.7.66}$$

as well as

$$(f_n, x f_{-(n+1)}) = (f_{-(n+1)}, x f_n)^* = \frac{a}{\pi(2n+1)} . \tag{13.7.67}$$

The energy corrections are now given by the eigenvalues of the matrix

$$\frac{\lambda a}{(2n+1)\pi} \begin{pmatrix} 0 & 1 \\ 1 & 0 \end{pmatrix} . \tag{13.7.68}$$

The eigenvalues are

$$\pm \frac{\lambda a}{(2n+1)\pi} . \tag{13.7.69}$$

Thus, all the degeneracies are lifted and the energies corresponding to E_n and $E_{-(n+1)}$ are

$$\frac{\pi^2 \hbar^2}{2ma^2} (n + 1/2)^2 \pm \frac{\lambda a}{(2n+1)\pi} . \tag{13.7.70}$$

13.8   Rigid Rotator in Magnetic Field

A system with moment of inertia I has the Hamiltonian

$$H_0 = \frac{L^2}{2I} .$$

a) What are the energies of the lowest and first excited states?
b) A perturbation

$$H' = g\frac{eB}{Mc} L_x$$

is applied. Find the splitting of the first excited states.

Solution

We have

$$H_0 = \frac{L^2}{2I} \ , \quad H' = g\frac{eB}{Mc} L_x \ . \tag{13.8.71}$$

The eigenfunctions of L^2 are $Y_{l,m}$ with eigenvalue $l(l+1)\hbar^2$. Therefore, we have

$$H_0 Y_{l,m} = \frac{l(l+1)\hbar^2}{2I} Y_{l,m} \ . \tag{13.8.72}$$

Every eigenvalue is $(2l+1)$-fold degenerate. Now, we can also write

$$L_x = \frac{1}{2}(L_+ + L_-) \ . \tag{13.8.73}$$

Thus, the first order corrections to the energy are given by

$$g\frac{eB}{Mc} \left(Y_{l,m}, \frac{1}{2}(L_+ + L_-)Y_{l,m} \right) = 0 \tag{13.8.74}$$

since

$$L_\pm Y_{l,m} = \sqrt{l(l+1) - m(m\pm 1)}\,\hbar Y_{l,m\pm 1} \ . \tag{13.8.75}$$

For the first excited states we have to diagonalize the matrix with matrix elements

$$(Y_{1,m'}, L_x Y_{1,m})$$
$$= \frac{\hbar}{2} \left[\sqrt{2 - m(m+1)}\delta_{m',m+1} + \sqrt{2 - m(m-1)}\delta_{m',m-1} \right] \ . \tag{13.8.76}$$

Written out as a matrix this looks like

$$\frac{1}{\sqrt{2}} \begin{pmatrix} 0 & 1 & 0 \\ 1 & 0 & 1 \\ 0 & 1 & 0 \end{pmatrix} \ . \tag{13.8.77}$$

The eigenvalues are: 0 , ± 1. Thus, the energies correct to second order are given by

$$\frac{\hbar^2}{I} + g\frac{eB}{Mc} \ , \quad \frac{\hbar^2}{I} \ , \quad \frac{\hbar^2}{I} - g\frac{eB}{Mc} \ . \tag{13.8.78}$$

Incidentally, since L_x commutes with L^2 and is known to have the eigenvalues $m\hbar$ with $m = -l, -(l-1), \ldots, (l-1), l$ we can write down the exact eigenvalues of the total Hamiltonian. These are

$$E_{l,m} = \frac{\hbar^2}{I}l(l+1) + g\frac{eB}{Mc}m\hbar \tag{13.8.79}$$

in complete agreement with the perturbation result.

13.9 $\lambda x p_y$ Perturbation of SHO

Find the energy correct to order λ^2 for the second excited state of the Hamiltonian

$$H = H_0 + \lambda H'$$

where

$$H_0 = \frac{1}{2m}(p_x^2 + p_y^2) + \frac{1}{2}m\omega^2(x^2 + y^2)$$

and

$$H' = x p_y \ .$$

Can this problem be solved exactly?

Solution

The energy levels of the second excited state are $|2,0\rangle$, $|1,1\rangle$, $|0,2\rangle$. They all have the same energy $3\hbar\omega$. We now diagonalize the perturbation part of the Hamiltonian in the degenerate subspace. To do this we use

$$x = \sqrt{\frac{\hbar}{2m\omega}}(a_1^\dagger + a_1) \quad , \quad p_y = i\sqrt{\frac{m\hbar\omega}{2}}(a_2^\dagger - a_2) \quad . \tag{13.9.80}$$

Then,

$$\lambda H' = i\lambda\frac{\hbar}{2}[a_1 a_2^\dagger + a_1^\dagger a_2^\dagger - a_1^\dagger a_2 - a_1 a_2] \ . \tag{13.9.81}$$

In the degenerate subspace the perturbation matrix to be diagonalized is now easily computed to be

$$i\frac{\lambda\hbar}{2}\begin{pmatrix} 0 & 1 & 0 \\ -1 & 0 & 1 \\ 0 & -1 & 0 \end{pmatrix} \ . \tag{13.9.82}$$

The eigenvalues and corresponding eigenvectors are $\lambda\hbar/2$, 0 , $-\lambda\hbar/2$ and

$$\frac{1}{\sqrt{6}}\begin{pmatrix} 1 \\ 2i \\ -1 \end{pmatrix} \ , \quad \frac{1}{\sqrt{2}}\begin{pmatrix} 1 \\ 0 \\ 1 \end{pmatrix} \ , \quad \frac{1}{\sqrt{6}}\begin{pmatrix} 1 \\ -2i \\ -1 \end{pmatrix} \ . \tag{13.9.83}$$

Thus, the new basis states we use for starting the perturbation computation are

$$|\psi_+\rangle = \frac{1}{\sqrt{6}}(|2,0\rangle + 2i|1,1\rangle - |0,2\rangle)$$

$$|\psi_0\rangle = \frac{1}{\sqrt{2}}((|2,0\rangle + |0,2\rangle))$$

$$|\psi_-\rangle = \frac{1}{\sqrt{6}}(|2,0\rangle - 2i|1,1\rangle - |0,2\rangle) \ .$$

$$\tag{13.9.84}$$

We now have to compute the terms $H'|\psi_+\rangle$, $H'|\psi_0\rangle$, and $H'|\psi_-\rangle$ which do not contain terms $|n, m\rangle$ with $n + m = 2$. Then we have (after omitting the terms with $n + m = 2$)

$$\lambda\langle n, m|H'|\pm\rangle$$
$$= -i\frac{\lambda\hbar}{\sqrt{6}}\left[\sqrt{2}(\delta_{n,3}\delta_{m,1} - \delta_{n,1}\delta_{m,3}) \pm 2i(2\delta_{n,2}\delta_{m,2} - \delta_{n,0}\delta_{m,0})\right] \quad (13.9.85)$$

and

$$\lambda\langle n, m|H'|0\rangle = i\lambda\hbar\left[\delta_{n,3}\delta_{m,1} + \delta_{n,1}\delta_{m,3}\right] \quad (13.9.86)$$

so that

$$E_{2,\pm}^{(2)} = \frac{(\lambda\hbar)^2}{16\hbar\omega}\left[\frac{2}{(3-5)} + \frac{2}{(3-5} + \frac{16}{3-5} + \frac{4}{3-1}\right]$$
$$= -\frac{4\lambda^2\hbar}{3\omega} . \quad (13.9.87)$$

Similarly we find

$$E_{2,0}^{(2)} = -\frac{\lambda^2\hbar}{\omega} . \quad (13.9.88)$$

Yes, this problem can be solved exactly since the total Hamiltonian is quadratic in the annihilation and creation operators.

13.10 Paschen-Back Effect

Calculate (to first order) the splitting of the n, l levels for $l = 0, 1$ of the energy levels of the valence electron of an alkali atom in a strong magnetic field. For weak fields this is the Zeeman effect. For magnetic fields so strong that their effect is greater than the spin-orbit splitting the result is known as the Paschen-Back effect.

Hint: For the Paschen-Back effect it may be convenient to use some of the results of problem 17.7 .

Solution

We take the magnetic field to point in the z-direction. In that case the Hamiltonian for the valence electron is

$$H = \frac{\vec{p}^2}{2m} + V(r) + \frac{1}{2mc^2}\frac{1}{r}\frac{dV}{dr}\vec{L}\cdot\vec{S} - \frac{eB}{2mc}(L_z + 2S_z) . \quad (13.10.89)$$

For a given solution of the unperturbed Hamiltonian with quantum numbers n, l and energy eigenvalue E_{nl} and

$$\langle n, l|\frac{1}{2mc^2}\frac{1}{r}\frac{dV}{dr}\vec{L}\cdot\vec{s}|n, l\rangle > \frac{eB}{2mc}\langle n, l|(L_z + 2S_z|n, l\rangle \quad (13.10.90)$$

we have the weak field or Zeeman effect. In that case it is convenient to choose states of good j, m_j and this situation is solved in general in problem 17.10.

On the other hand if

$$\langle n,l|\frac{1}{2mc^2}\frac{1}{r}\frac{dV}{dr}\vec{L}\cdot\vec{s}|n,l\rangle < \frac{eB}{2mc}\langle n,l|L_z+2S_z|n,l\rangle \tag{13.10.91}$$

we have the Paschen-Back effect. In this case we want states of good m_l, m_s. For this we use the results of problem 17.7 where we find

$$
\begin{aligned}
|3/2,3/2\rangle &= |1,1\rangle|1/2,1/2\rangle \\[4pt]
|3/2,1/2\rangle &= \sqrt{\frac{2}{3}}|1,0\rangle|1/2,1/2\rangle + \frac{1}{\sqrt{3}}|1,1\rangle|1/2,-1/2\rangle \\[4pt]
|3/2,-1/2\rangle &= \frac{1}{\sqrt{3}}|1,-1\rangle|1/2,1/2\rangle + \sqrt{\frac{2}{3}}|1,0\rangle|1/2,-1/2\rangle \\[4pt]
|3/2,-3/2\rangle &= |1,-1\rangle|1/2,-1/2\rangle
\end{aligned}
$$

$$
\begin{aligned}
|1/2,1/2\rangle &= \frac{1}{\sqrt{3}}|1,0\rangle|1/2,1/2\rangle - \sqrt{\frac{2}{3}}|1,1\rangle|1/2,-1/2\rangle \\[4pt]
|1/2,-1/2\rangle &= \sqrt{\frac{2}{3}}|1,-1\rangle|1/2,1/2\rangle - \frac{1}{\sqrt{3}}|1,0\rangle|1/2,-1/2\rangle \;. \tag{13.10.92}
\end{aligned}
$$

The inverse of these equations is

$$
\begin{aligned}
|1,1\rangle|1/2,1/2\rangle &= |3/2,3/2\rangle \\[4pt]
|1,0\rangle|1/2,1/2\rangle &= \sqrt{\frac{2}{3}}|3/2,1/2\rangle + \frac{1}{\sqrt{3}}|1/2,1/2\rangle \\[4pt]
|1,1\rangle|1/2,-1/2\rangle &= \frac{1}{\sqrt{3}}|3/2,1/2\rangle - \sqrt{\frac{2}{3}}|1/2,1/2\rangle \\[4pt]
|1,-1\rangle|1/2,1/2\rangle &= \sqrt{\frac{2}{3}}|3/2,-1/2\rangle + \frac{1}{\sqrt{3}}|1/2,-1/2\rangle \\[4pt]
|1,0\rangle|1/2,-1/2\rangle &= \frac{1}{\sqrt{3}}|3/2,-1/2\rangle - \sqrt{\frac{2}{3}}|1/2,-1/2\rangle \\[4pt]
|1,-1\rangle|1/2,-1/2\rangle &= |3/2,-3/2\rangle \;. \tag{13.10.93}
\end{aligned}
$$

For $l=0$ the only possibilities are

$$
\begin{aligned}
|0,0\rangle|1/2,m_s\rangle &= |1/2,m_j\rangle \\[4pt]
|0,0\rangle|1/2,1/2\rangle &= |1/2,1/2\rangle \\[4pt]
|0,0\rangle|1/2,-1/2\rangle &= |1/2,-1/2\rangle \;. \tag{13.10.94}
\end{aligned}
$$

Thus, we can evaluate the magnetic part of the energy. If we call

$$
\begin{aligned}
\hbar^2\langle n,l|\frac{1}{2mc^2}\frac{1}{r}\frac{dV}{dr}|n,l\rangle &= \epsilon_{n,l} \\[4pt]
\frac{eB}{2mc} &= \omega \tag{13.10.95}
\end{aligned}
$$

the result is

$$
\begin{array}{ll}
|l, m_l\rangle|1/2, m_s\rangle & E \\
|1, 1\rangle|1/2, 1/2\rangle & \epsilon_{n,1} + 2\hbar\omega \\
|1, 0\rangle|1/2, 1/2\rangle & \epsilon_{n,1} + \hbar\omega \\
|1, 1\rangle|1/2, -1/2\rangle & 0 \\
|1, -1\rangle|1/2, 1/2\rangle & 0 \\
|1, 0\rangle|1/2, -1/2\rangle & \epsilon_{n,1} - \hbar\omega \\
|1, -1\rangle|1/2, -1/2\rangle & \epsilon_{n,1} - 2\hbar\omega \\
|0, 0\rangle|1/2, 1/2\rangle & \epsilon_{n,1} + \hbar\omega \\
|0, 0\rangle|1/2, -1/2\rangle & \epsilon_{n,1} + \hbar\omega \ .
\end{array}
\tag{13.10.96}
$$

The spin-orbit term is now simply a perturbation. Since,

$$
\vec{L} \cdot \vec{S} = \frac{1}{2}\left[\vec{J}^2 - \vec{L}^2 - \vec{S}^2 \right] \ .
\tag{13.10.97}
$$

We immediately get that for the states $|j = 3/2, \pm 3/2\rangle$

$$
\begin{aligned}
&\langle n, l, j = 3/2, m_j = \pm 3/2 | \vec{L} \cdot \vec{S} | n, l, j = 3/2, m_j = \pm 3/2 \rangle \\
&= \frac{\hbar^2}{2}[3/2(3/2 + 1) - 1(1 + 1) - 3/4] \\
&= \frac{\hbar^2}{2} \ .
\end{aligned}
\tag{13.10.98}
$$

Similarly, for the states $|j = 1/2, \pm 1/2\rangle$ corresponding to $l = 0$ we immediately get that

$$
\begin{aligned}
&\langle n, l, j = 1/2, m_j = \pm 1/2 | \vec{L} \cdot \vec{S} | n, l, j = 1/2, m_j = \pm 1/2 \rangle \\
&= \frac{\hbar^2}{2}[1/2(1/2 + 1) - 0 - 3/4] = 0 \ .
\end{aligned}
\tag{13.10.99}
$$

This leaves only four states to consider. These split into two pairs: those with $m_j = 1/2$ and those with $m_j = -1/2$. Thus, we can consider them separately. However, $\vec{L} \cdot \vec{S}$ is not diagonal in this representation. In fact, we find that

$$
\vec{L} \cdot \vec{S} = \frac{\hbar}{2}\begin{pmatrix} L_z & L_- \\ L_+ & -L_z \end{pmatrix} \ .
\tag{13.10.100}
$$

For $m_j = 1/2$, when evaluated with the corresponding states, namely

$$
|1, 0\rangle|1/2, 1/2\rangle \quad \text{and} \quad |1, 1\rangle|1/2, -1/2\rangle
$$

we get

$$
\frac{\epsilon_{nl}}{2}\begin{pmatrix} 0 & \sqrt{2} \\ \sqrt{2} & -1 \end{pmatrix} \ .
\tag{13.10.101}
$$

The energy corrections are just the eigenvalues of this matrix. These are easily found to be

$$\frac{1}{2}\epsilon_{nl} \quad \text{and} \quad -\epsilon_{nl} \ .$$

Thus, the energy shifts are

$$\frac{1}{2}\epsilon_{nl} \quad \text{and} \quad -\epsilon_{nl} \ . \tag{13.10.102}$$

Similarly for $m_j = -1/2$ we have the states

$$|1,0\rangle|1/2,-1/2\rangle \quad \text{and} \quad |1,-1\rangle|1/2,1/2\rangle \ .$$

This leads to the energy matrix

$$\frac{\epsilon_{nl}}{2}\begin{pmatrix} 0 & \sqrt{2} \\ \sqrt{2} & 1 \end{pmatrix} \tag{13.10.103}$$

with the eigenvalues

$$-\frac{1}{2}\epsilon_{nl} \quad \text{and} \quad \epsilon_{nl} \ .$$

Thus, in this case, the energy shifts are

$$-\frac{1}{2}\epsilon_{nl} \quad \text{and} \quad \epsilon_{nl} \ . \tag{13.10.104}$$

To identify which energy level belongs to which state it is convenient to include the magnetic energy in these considerations. Since the magnetic energy is diagonal this causes only a slight complication. Thus, for $m_j = 1/2$ we get the energy matrix

$$\begin{pmatrix} \hbar\omega & \epsilon_{nl}/\sqrt{2} \\ \epsilon_{nl}/\sqrt{2} & \epsilon_{nl}/2 \end{pmatrix} \ . \tag{13.10.105}$$

The eigenvalues are

$$\frac{1}{2}\left(\hbar\omega - \epsilon_{nl}/2 \pm \sqrt{\hbar^2\omega^2 + \epsilon_{nl}\hbar\omega + 9/4\,\epsilon_{nl}^2} \right) \ .$$

If we now let

$$\epsilon_{nl}/(\hbar\omega) \to 0$$

we recover the strong field (Paschen-Back) case. The energy shifts then approach

$$\frac{1}{2}\left(\hbar\omega - \epsilon_{nl}/2 + \sqrt{\hbar^2\omega^2 + \epsilon_{nl}\hbar\omega + 9/4\,\epsilon_{nl}^2} \right) \quad \to \quad \hbar\omega$$

$$\frac{1}{2}\left(\hbar\omega - \epsilon_{nl}/2 - \sqrt{\hbar^2\omega^2 + \epsilon_{nl}\hbar\omega + 9/4\,\epsilon_{nl}^2} \right) \quad \to \quad -\frac{\epsilon_{nl}}{2} \ . \tag{13.10.106}$$

This shows that the term with the $+$ sign corresponds to the case $m_l = 0$, $m_s = 1/2$ and the term with the $-$ sign corresponds to $m_l = 1$, $m_s = -1/2$.

Similarly, for $m_j = -1/2$ we find the eigenvalues

$$\frac{1}{2}\left(\hbar\omega + \epsilon_{nl}/2 \pm \sqrt{\hbar^2\omega^2 + 3\epsilon_{nl}\hbar\omega + 9/4\,\epsilon_{nl}^2}\right) .$$

Again, letting

$$\epsilon_{nl}/(\hbar\omega) \to 0$$

we recover the strong field (Paschen-Back) case. We then get

$$\frac{1}{2}\left(\hbar\omega + \epsilon_{nl}/2 + \sqrt{\hbar^2\omega^2 + 3\epsilon_{nl}\hbar\omega + 9/4\,\epsilon_{nl}^2}\right) \to \hbar\omega + \epsilon_{nl}$$

$$\frac{1}{2}\left(\hbar\omega + \epsilon_{nl}/2 - \sqrt{\hbar^2\omega^2 + 3\epsilon_{nl}\hbar\omega + 9/4\,\epsilon_{nl}^2}\right) \to -\frac{\epsilon_{nl}}{2} . \qquad (13.10.107)$$

So, the $+$ sign corresponds to $m_l = 0$, $m_s = -1/2$ while the $-$ sign corresponds to $m_l = -1$, $m_s = 1/2$.

13.11 H Atom: Weak Field Stark Effect

Using the results of problem 10.11 and 11.22 calculate the energy shift in the hydrogen atom energy levels (fine structure) due to a weak electric field.

Solution

We have the weak perturbation

$$V(\vec{r}) = e\mathcal{E}z \qquad (13.11.108)$$

where we have chosen the electric field to point in the z-direction. Also, from problem 11.22 we have that the unperturbed normalized hydrogen atom wavefunctions are

$$\Psi_{n,l,j=l+1/2,m} = \frac{R_{n,j-1/2}(r)}{\sqrt{2l+1}}\left(\begin{array}{c}\sqrt{l+m+1/2}\,Y_{l,m-1/2} \\ \\ \sqrt{l-m+1/2}\,Y_{l,m+1/2}\end{array}\right)$$

$$(13.11.109)$$

$$\Psi_{n,l,j=l-1/2,m} = \frac{R_{n,j+1/2}(r)}{\sqrt{2l+1}}\left(\begin{array}{c}\sqrt{l-m+1/2}\,Y_{l,m-1/2} \\ \\ -\sqrt{l+m+1/2}\,Y_{l,m+1/2}\end{array}\right) .$$

$$(13.11.110)$$

To simplify the writing we call

$$\Psi_{n,l,j=l+1/2,m} = \psi_1$$
$$\Psi_{n,l,j=l-1/2,m} = \psi_2 . \qquad (13.11.111)$$

The energy levels in the unperturbed hydrogen atom are labelled by n, j. Under the perturbation the orbital angular momentum no longer commutes with the Hamiltonian and thus, l is no longer a good quantum number. However, the z-component of the total angular momentum J^2, namely m, continues to be a good quantum number. Thus, the matrix elements of the perturbation Hamiltonian V between states with different m vanish. The diagonal matrix elements of V also vanish

$$\left(\Psi_{n,l,j=l+1/2,m},\ V\Psi_{n,l,j=l+1/2,m} \right) \ =\ (\psi_1, V\psi_1) = 0$$
$$\left(\Psi_{n,l,j=l-1/2,m},\ V\Psi_{n,l,j=l-1/2,m} \right) \ =\ (\psi_2, V\psi_2) = 0 \ . \qquad (13.11.112)$$

This means we must diagonalize V in the degenerate subspace corresponding to the states $\Psi_{n,l=j+1/2,m}$ and $\Psi_{n,l=j-1/2,m}$. We call these matrix elements

$$V_{12} \ =\ V_{21} = \left(\Psi_{n,j-1/2,m},\ V\Psi_{n,j+1/2,m} \right)$$

$$=\ \frac{e\mathcal{E}}{2\sqrt{j(j+1)}} \int_0^\infty r^3\, R_{n,j-1/2}(r) R_{n,j+1/2}(r)\, dr$$

$$\times\left[\sqrt{(j+m)(j-m+1)} \int Y^*_{j-1/2,m-1/2} Y_{j+1/2,m-1/2}\, \cos\theta\, d\Omega \right.$$

$$\left. -\sqrt{(j-m)(j+m+1)} \int Y^*_{j-1/2,m+1/2} Y_{j-1/2,m+1/2}\, \cos\theta\, d\Omega \right]\ .$$

$$(13.11.113)$$

To perform the angular integration we use that

$$\cos\theta\, Y_{l,m}(\theta,\varphi) \ =\ \sqrt{\frac{(l+m+1)(l-m+1)}{(2l+1)(2l+3)}} Y_{l+1,m}(\theta,\varphi)$$

$$+\ \sqrt{\frac{(l+m)(l-m)}{(2l-1)(2l+1)}} Y_{l-1,m}(\theta,\varphi)\ . \qquad (13.11.114)$$

Our matrix elements now reduce to

$$V_{12} \ =\ V_{21} = \frac{e\mathcal{E}}{2\sqrt{j(j+1)}} \int_0^\infty r^3\, R_{n,j-1/2}(r) R_{n,j+1/2}(r)\, dr$$

$$\times \frac{m}{\sqrt{j(j+1)}}\ . \qquad (13.11.115)$$

The remaining radial integral is

$$\int_0^\infty r^3\, R_{n,j-1/2}(r) R_{n,j+1/2}(r)\, dr \qquad (13.11.116)$$

and may be evaluated using the generating function as in problem 10.11 to yield

$$-\frac{3}{2} n\sqrt{n^2 - (j+1/2)^2}\ .$$

Combining these results we have

$$V_{12} = V_{21} = -\frac{3}{4}\frac{n\sqrt{n^2 - (j+1/2)^2}m}{j(j+1)}e\mathcal{E} \ . \tag{13.11.117}$$

The perurbation matrix to be diagonalized is now

$$V = \begin{pmatrix} 0 & V_{12} \\ V_{21} & 0 \end{pmatrix} \ . \tag{13.11.118}$$

The resulting energy shifts are

$$\Delta E = \pm V_{12} = \pm\frac{3}{4}\frac{n\sqrt{n^2 - (j+1/2)^2}m}{j(j+1)}e\mathcal{E} \ . \tag{13.11.119}$$

Finally we see that for a fixed value of n all terms of the fine structure, except the term with $j = n - 1/2$, are split into $2j + 1$ equidistant levels corresponding to $m = -j, \ldots, j$. The term with $j = n - 1/2$ is not split at all since l has a fixed value $l = j - 1/2 = n - 1$ and is not degenerate as regards the quantum number l.

Bibliography

[13.1] A.Z. Capri, *Nonrelativistic Quantum Mechanics* 3rd edition, World Scientific Publishing Co. Pte. Ltd., section 6.8, (2002) .

Chapter 14

Further Approximation Methods

14.1 Variational Ground State of SHO

Use the trial wavefunction

$$\psi(x) = A\,e^{-\alpha|x|}$$

to estimate the ground state energy of a simple harmonic oscillator.

Solution

The trial wavefunction is

$$\psi = A\,e^{-\alpha|x|} \ . \tag{14.1.1}$$

Normalization yields,

$$A = \sqrt{\alpha} \ . \tag{14.1.2}$$

Since this wavefunction has a discontinuity in its first derivative, something that is rather unphysical, we expect that it will yield a rather poor estimate of the energy.

The Hamiltonian is

$$H = \frac{p^2}{2m} + \frac{1}{2}m\omega^2 x^2 \ . \tag{14.1.3}$$

Applying the kinetic energy operator to ψ we get

$$
\begin{aligned}
\frac{p^2}{2m}\psi &= -\sqrt{\alpha}\,\frac{\hbar^2}{2m}\frac{d^2}{dx^2}e^{-\alpha|x|} \\
&= -\sqrt{\alpha}\,\frac{\hbar^2}{2m}\left[\alpha^2 - 2\alpha\delta(x)\right]e^{-\alpha|x|} \ .
\end{aligned}
\tag{14.1.4}
$$

Thus,

$$\langle T \rangle = \frac{\hbar^2 \alpha^2}{2m} \; . \tag{14.1.5}$$

Similarly,

$$\langle V \rangle = \frac{m\omega^2}{4\alpha^2} \; . \tag{14.1.6}$$

So, the expectation value of the Hamiltonian is

$$\langle H \rangle = \frac{\hbar^2 \alpha^2}{2m} + \frac{m\omega^2}{4\alpha^2} \; . \tag{14.1.7}$$

The extremum of this function is obtained by differentiating to get

$$\frac{\hbar^2 \alpha}{m} - \frac{m\omega^2}{2\alpha^3} = 0 \tag{14.1.8}$$

The solution of this equation yields

$$\alpha^2 = \pm \frac{m\omega}{\sqrt{2}\hbar} \; . \tag{14.1.9}$$

Only the + sign is acceptable (since ψ has to be square integrable) and it yields a minimum for the energy expectation value given by

$$E_0 \approx \langle H \rangle = \frac{1}{\sqrt{2}} \hbar\omega . \tag{14.1.10}$$

As expected this is larger than the exact ground state energy of $1/2 \; \hbar\omega$ by about 41 %.

14.2   Variational Ground State of H$_2$ Molecule

Use the trial wavefunction

$$\psi(\vec{r}_1, \vec{r}_2) = \frac{Z^3}{\pi a_0^3} e^{-Z(r_1 + r_2)/a_0}$$

to evaluate the expectation value of the Hamiltonian

$$H = \frac{\vec{p}_1^2}{2m} + \frac{\vec{p}_2^2}{2m} - 2e^2 \left(\frac{1}{\vec{r}_1} + \frac{1}{\vec{r}_2} \right) + \frac{e^2}{\vec{r}_{12}} \; .$$

Solution

The Hamiltonian is

$$H = \frac{\vec{p}_1^2}{2m} - \frac{2e^2}{r_1} + \frac{\vec{p}_2^2}{2m} - \frac{2e^2}{r_2} + \frac{e^2}{r_{12}} = H_1 + H_2 + \frac{e^2}{r_{12}} , \qquad (14.2.11)$$

where

$$H_i = \frac{\vec{p}_i^2}{2m} - \frac{2e^2}{r_i} \quad i = 1, 2 \qquad (14.2.12)$$

are hydrogenic Hamiltonians.

As an approximate wavefunction we choose a product of "screened" hydrogenic wavefunctions since we expect the electrons to screen the nucleus from each other.

$$\psi(\vec{r}_1, \vec{r}_2) = \frac{Z^3}{\pi a^3} E^{-Z(r_1+r_2)/a} = \phi(\vec{r}_1)\phi(\vec{r}_2) \qquad (14.2.13)$$

where

$$a = \frac{\hbar^2}{me^2} \qquad (14.2.14)$$

is the Bohr radius. Now,

$$
\begin{aligned}
T_1 \phi(\vec{r}_1) &= \frac{\vec{p}_1^2}{2m} \phi(\vec{r}_1) \\
&= -\frac{\hbar^2}{2m} \frac{1}{r} \frac{\partial^2}{\partial r^2} (r\phi(\vec{r}_1)) \\
&= -\frac{\hbar^2}{2m} \left[\frac{Z^2}{a^2} - \frac{2Z}{ar} \right] \phi(\vec{r}_1) .
\end{aligned}
\qquad (14.2.15)
$$

The same result holds for $T_2\phi(\vec{r}_2)$. Next using that $\hbar^2/m = e^2a$ we find that

$$
\begin{aligned}
\langle T_1 \rangle &= \langle T_2 \rangle \\
&= -\frac{e^2}{2a} \frac{Z^3}{\pi a^3} 4\pi \int_0^\infty \left[Z^2 r^2 - 2Zra \right] e^{-2Zr/a} \, dr \\
&= \frac{Z^2 e^2}{2a} .
\end{aligned}
\qquad (14.2.16)
$$

Also,

$$
\begin{aligned}
\langle -\frac{2e^2}{r_1} \rangle &= \langle -\frac{2e^2}{r_2} \rangle \\
&= -2e^2 \frac{Z^3}{\pi a^3} 4\pi \int_0^\infty e^{-2Zr/a} \, r\,dr \\
&= -\frac{2Ze^2}{a} .
\end{aligned}
\qquad (14.2.17)
$$

Thus,

$$\langle H_1 \rangle + \langle H_1 \rangle = 2\frac{Z^2 e^2}{2a} - 2\frac{2Z e^2}{a} = (Z^2 - 4Z)\frac{e^2}{a} . \qquad (14.2.18)$$

Next we evaluate $\langle e^2/r_{12} \rangle$ by expanding in spherical harmonics, just as we did in problem 12.1. Thus, we get

$$\langle \frac{e^2}{r_{12}} \rangle = \frac{5Z e^2}{8a} . \qquad (14.2.19)$$

Collecting all these terms we get

$$\langle H \rangle = \frac{e^2}{a} \left[Z^2 - \frac{27}{8} Z \right] . \qquad (14.2.20)$$

To complete the computation we now minimize the energy with respect to Z. Thus, we compute

$$\frac{\partial}{\partial z} \langle H \rangle = 0 = 2Z - \frac{27}{8} . \qquad (14.2.21)$$

Then,

$$Z = \frac{27}{16} . \qquad (14.2.22)$$

So, clearly $Z < 2$ and we see that the electrons do indeed screen each other. The best estimate for the ground state or "ionization" energy is now

$$E = - \left(\frac{27}{16} \right)^2 \frac{e^2}{a} = -2.85\frac{e^2}{a} . \qquad (14.2.23)$$

This is clearly lower than the perturbation result

$$-\frac{1}{2}\frac{4e^2}{a} = -2.00\frac{e^2}{a} . \qquad (14.2.24)$$

14.3 Square Barrier: WKB Approximation

a) Use the WKB approximation to solve the tunneling problem for a square barrier using the fact that in this case you have no need to use the connection formulae. This result agrees with the exact result in [14.1].
b) Repeat part a) by using the connection formulae.

Solution

a) In this case we have the potential

$$V(x) = \begin{cases} 0 & x < 0 \\ V_0 & 0 < x < a \\ 0 & x > a \end{cases} . \qquad (14.3.25)$$

Also, in this case we have special boundaries so we do not have to use the connection formulae. Calling

$$k^2 = \frac{2mE}{\hbar^2} \quad , \quad K^2 = \frac{2m(V_0 - E)}{\hbar^2} \tag{14.3.26}$$

we can write our solutions as follows for a wave incoming from the left

$$\psi(x) = \begin{cases} \psi_I(x) &= \frac{1}{\sqrt{k}} \left[e^{i(kx - \pi/4)} + R' e^{-i(kx - \pi/4)} \right] & x < 0 \\ \psi_{II}(x) &= \frac{1}{\sqrt{K}} \left[A' e^{Kx} + B' e^{-Kx} \right] & 0 < x < a \\ \psi_{III}(x) &= \frac{1}{\sqrt{k}} T\, e^{i[k(x-a) - \pi/4]} & x > a \end{cases} \tag{14.3.27}$$

If we define new constants

$$A' = A \frac{\sqrt{K}}{\sqrt{k}} \quad , \quad B' = B \frac{\sqrt{K}}{\sqrt{k}} \quad , \quad R' = R e^{-i\pi/2} \tag{14.3.28}$$

and use the fact that the solutions are valid right up to the turning points, we can impose the boundary conditions that both ψ and $d\psi/dx$ are continuous at $x = 0$ and $x = a$. Thus, we get

$$\begin{aligned} e^{-i\pi/4} [1 + R] &= A + B \\ ik\, e^{-i\pi/4} [1 - R] &= K[A - B] \\ e^{-i\pi/4} T &= A\, e^{Ka} + B\, e^{-Ka} \\ ik\, e^{-i\pi/4} T &= K[A\, e^{Ka} - B\, e^{-Ka}] \end{aligned} \tag{14.3.29}$$

From the first two of these equations we get

$$\begin{aligned} A &= \frac{1}{2} e^{-i\pi/4} \left[(1 + ik/K) + R(1 - ik/K) \right] \\ B &= \frac{1}{2} e^{-i\pi/4} \left[(1 - ik/K) + R(1 + ik/K) \right] \end{aligned} \tag{14.3.30}$$

Substituting this result in the second pair of equations we get

$$\begin{aligned} T &= \frac{1}{2} \left[(1 + ik/K)\, e^{Ka} + (1 - ik/K)\, e^{-Ka} \right] \\ &+ \frac{1}{2} R \left[(1 - ik/K)\, e^{Ka} + (1 + ik/K)\, e^{-Ka} \right] \\ T &= -\frac{iK}{2k} \left[(1 + ik/K)\, e^{Ka} - (1 - ik/K)\, e^{-Ka} \right] \\ &- \frac{iK}{2k} R \left[(1 - ik/K)\, e^{Ka} - (1 + ik/K)\, e^{-Ka} \right] \end{aligned} \tag{14.3.31}$$

Solving for R we find

$$R = \frac{(k^2 + K^2) \sinh Ka}{(k^2 - K^2) \sinh Ka + 2ikK \cosh Ka} . \tag{14.3.32}$$

Next, we use that

$$T = e^{i\pi/4} [A\, e^{Ka} + B\, e^{-Ka}] \tag{14.3.33}$$

and substitute for A and B to get, after a little algebra, the result that

$$T = \frac{2ikK}{(k^2 - K^2)\sinh Ka + 2ikK\cosh Ka} \quad . \tag{14.3.34}$$

Except for an irrelevant phase factor these results agree with the exact results obtained in [14.1].
b) The solutions are exactly those given by (14.3.27). However, this time we have to use the connection formulae. To do so we first rewrite ψ_I and ψ_{III} as follows.

$$\psi_I \quad = \quad \frac{1}{\sqrt{k}}\{(1+R)\cos(kx - \pi/4) + i(1-R)\sin(kx - \pi/4)\}$$

$$\psi_{III} \quad = \quad \frac{1}{\sqrt{k}}T\{\cos[k(x-a)\pi/4] + i\sin[k(x-a) - \pi/4]\} \quad . \tag{14.3.35}$$

Now using the connection formulae at the turning points $x = 0$ and $x = a$ and defining

$$S = e^{-Ka} \tag{14.3.36}$$

we get

$$A \quad = \quad \frac{1}{2}(1+R)$$

$$B \quad = \quad i(1-R) \tag{14.3.37}$$

as well as

$$T \quad = \quad 2BS^{-1} = 2i(1-R)S^{-1}$$

$$T \quad = \quad iAS = \frac{i}{2}(1+R)S \quad . \tag{14.3.38}$$

We solve these equations for T and R and get

$$T \quad = \quad \frac{iS}{1+S^2/4} \approx ie^{-Ka}$$

$$R \quad = \quad \frac{1-S^2/4}{1+S^2/4} \approx 1 - \frac{e^{-2Ka}}{2} \quad . \tag{14.3.39}$$

Clearly, the results from part a) and part b) are very different. On the other hand, to the approximation used, we still have conservation of probability since

$$|T|^2 + |R|^2 = e^{-2Ka} + 1 - e^{-2Ka} + \frac{e^{-4Ka}}{4} \approx 1 \quad . \tag{14.3.40}$$

14.4   Variational Ground State in Gaussian Potential

a) Use a variational approach to find the ground state energy for a particle in the potential

$$V(r) = -V_0 e^{-\alpha r^2}$$

if the particle involved is an electron and $\alpha = 5.29 \times 10^{13}$ cm^{-2} , $V_0 = 20$ eV.
A numerical answer is required.
b) For comparison approximate the Gaussian potential by its first two terms of
a Taylor expansion and estimate the energy that way.

Solution

a) The potential

$$V(r) = -V_0\, e^{-\alpha r^2} \tag{14.4.41}$$

looks like a simple harmonic oscillator near the bottom of the well. So a reasonable choice of wavefunction for the ground state seems to be

$$\psi(r) = A\, e^{-\beta r^2/2} \tag{14.4.42}$$

with the normalization

$$A = 2\left(\beta^3/\pi\right)^{1/4} . \tag{14.4.43}$$

Since this is an $l = 0$ state we have

$$\begin{aligned}
H\psi &= -\frac{\hbar^2}{2m}\frac{1}{r}\frac{d^2}{dr^2}(r\psi) - V_0\, e^{-\alpha r^2}\,\psi \\
&= A\left[-\frac{\hbar^2}{2m}(\beta^2 r^2 - 3\beta) - V_0\, e^{-\alpha r^2}\right]e^{-\beta r^2/2} .
\end{aligned} \tag{14.4.44}$$

Therefore,

$$\begin{aligned}
E(\beta) &= (\psi, H\psi) = |A|^2\left[-\frac{\hbar^2}{2m}\int_0^\infty r^2 dr(\beta^2 r^2 - 3\beta)e^{-\beta r^2}\right. \\
&\quad \left. - V_0\int_0^\infty r^2 dr\, e^{-(\alpha+\beta)r^2}\right] .
\end{aligned} \tag{14.4.45}$$

So,

$$E(\beta) = \frac{\hbar^2}{2m}\left[\frac{3\beta}{2} - \frac{2mV_0}{\hbar^2}\frac{\beta^{3/2}}{(\alpha+\beta)^{3/2}}\right] . \tag{14.4.46}$$

We now minimize $E(\beta)$. To do this we take the derivative with respect to β
and set the result equal to zero. Furthermore, we call

$$K = \frac{2mV_0}{\hbar^2\alpha} \tag{14.4.47}$$

and get

$$\frac{3}{2}\left[1 - \frac{K\alpha\beta^{1/2}}{(\beta+\alpha)^{3/2}} + \frac{K\alpha\beta^{3/2}}{(\beta+\alpha)^{5/2}}\right] = 0 . \tag{14.4.48}$$

After setting $\beta = \alpha x$, to make everything dimensionless, we find

$$(1+x)^{5/2} = Kx^{1/2} . \tag{14.4.49}$$

Now, substituting the numerical values we get $K = 10^3$. Thus, we see that x must be large compared to 1. In fact we see that

$$x \approx \sqrt{K} \approx 30 .$$

To find x we therefore make a table of values for the left side and the right side. Thus, we conclude that, to the accuracy of the data given, $x = 30.4$. Putting

x	$(1+x)^{5/2}$	$Kx^{1/2}$
30.0	5.35×10^3	5.48×10^3
30.2	5.44×10^3	5.50×10^3
30.4	5.52×10^3	5.51×10^3
30.5	5.57×10^3	5.52×10^3

this back into the expression for the energy we find $E \approx -15.4$ eV. Thus, the ground state energy is about $E \approx -15.4$ eV .
b) If we Taylor expand the potential we get

$$V(r) \approx -V_0 + V_0 \alpha r^2 . \tag{14.4.50}$$

This is just a simple harmonic oscillator potential and the ground state energy is given by

$$E \approx -V_0 + \frac{3}{2} \hbar \omega \tag{14.4.51}$$

where

$$\omega = \sqrt{\frac{2V_0 \alpha}{m}} \tag{14.4.52}$$

so that

$$E = -V_0 + \frac{3}{\sqrt{K}} V_0 . \tag{14.4.53}$$

Substituting the numerical values we get

$$E \approx -18.1 \text{ eV} . \tag{14.4.54}$$

This shows that the two calculations give similar results and agree to within about 20% . Although the variational calculation gives a larger result it should be considered the more reliable since the approximate potential is wider in the region where the wavefunction is appreciable and thus gives a lower kinetic energy.

14.5   Variational Ground State: Quartic Potential

An electron is in the spherically symmetric potential

$$V(r) = A r^2 (r^2 - a^2)$$

where

$$a = 2.00 \times 10^{-1} \text{ nm}$$

$$a^4 A = 1.90 \text{ eV}.$$

Use Rayleigh Ritz to estimate the ground state energy. A numerical answer is required.

Solution

We have

$$H = \frac{\vec{p}^2}{2m} + V(r) \tag{14.5.55}$$

where

$$V(r) = A r^2 (r^2 - a^2) . \tag{14.5.56}$$

Since we are looking for the ground state we have $l = 0$ and the kinetic energy is

$$T\psi = -\frac{\hbar^2}{2m} \frac{1}{r} \frac{d^2}{dr^2} (r\psi) . \tag{14.5.57}$$

We now scale in terms of dimensionless variables so that $x = r/a$. Then we find that

$$H\psi = \epsilon \left[-\frac{1}{x} \frac{d^2}{dx^2} (x\psi) + \frac{a^4 A}{\epsilon} x^2 (x^2 - 1)\psi \right] \tag{14.5.58}$$

with

$$\epsilon = \frac{\hbar^2}{2ma^2} = 0.95 \text{ eV} \tag{14.5.59}$$

and

$$\frac{a^4 A}{\epsilon} = 2.00 . \tag{14.5.60}$$

As possible trial wavefunctions we choose

$$\psi = x^n e^{-bx^2/2} \tag{14.5.61}$$

since near the bottom of the potential well we have something that looks like a simple harmonic oscillator and we want as few nodes as possible. To proceed we need the following integrals.

$$(\psi, \psi) = \int_0^\infty x^{2n+2} e^{-bx^2} \, dx = \frac{(2n+1)!!}{2(2b)^{n+1}} \sqrt{\frac{\pi}{b}} \, . \tag{14.5.62}$$

Also,

$$-\frac{1}{x} \frac{d^2}{dx^2}(x\psi) = -\left[n(n+1)x^{n-2} - b(2n+3)x^n + b^2 x^{n+2} \right] e^{-bx^2/2} . \tag{14.5.63}$$

Combining these results we find that the expectation value of the Hamiltonian is given by

$$
\begin{aligned}
E(n, b) &= \frac{(\psi, H\psi)}{(\psi, \psi)} \\
&= \epsilon \left[\frac{2n^2 + 6n + 3}{4n + 2} + \frac{4n^2 + 16n + 15}{2b^2} - \frac{2n + 3}{b} \right] .
\end{aligned} \tag{14.5.64}
$$

Differentiating with respect to b to find the minimum energy we get after some rearranging that b is given by

$$\frac{2n^2 + 6n + 3}{4n + 2} b^3 + (2n + 3)b - (4n^2 + 16n + 15) = 0 . \tag{14.5.65}$$

We now look at this for a couple of different values of n.

<u>n = 0</u>

$$\frac{3}{2} b^3 + 3b - 15 = 0 . \tag{14.5.66}$$

A few numerical attempts show that $b = 1.847$ to the accuracy warranted by the data. Substituting this value back into the expression for the energy we find

$$E = 3.345 \, \epsilon = 3.18 \text{ eV} . \tag{14.5.67}$$

<u>n=1</u>

$$\frac{11}{6} b^3 + 5b - 35 = 0 . \tag{14.5.68}$$

Again a few simple computations show that $b = 2.335$. This yields

$$E = 4.921 \, \epsilon = 4.667 \text{ eV} . \tag{14.5.69}$$

Since this result is greater than the previous one we conclude that the ground state energy for this electron is $E \leq 3.18$ eV.

14.6   WKB: Ball Bouncing on a Floor

A particle is in a potential

$$V = \begin{cases} mgz & z > 0 \\ \infty & z = 0 \end{cases} .$$

This corresponds to a perfectly elastic ball bouncing on a floor.
a) Find the WKB solution for all the energy levels.
b) Find the WKB solution for the ground state wavefunction of this particle.

Solution

a) We have the potential

$$V = \begin{cases} mgz & z > 0 \\ \infty & z \leq 0 \end{cases} . \tag{14.6.70}$$

This means that the point $z = 0$ is a special boundary and the wavefunction must vanish at $z = 0$. The other turning point is given by

$$E = mgz_0 \quad \text{or} \quad z_0 = \frac{E}{mg} . \tag{14.6.71}$$

Thus, the regions are as follows

I $0 < z \leq z_0$ is classically allowed.

II $z > z_0$ is classically forbidden.

In the classically forbidden region we need an exponentially damped solution. Writing,

$$K(z) = \frac{1}{\hbar}\sqrt{2m^2g(z - z_0)} \quad z > z_0 \tag{14.6.72}$$

$$k(z) = \frac{1}{\hbar}\sqrt{2m^2g(z_0 - z)} \quad 0 < z < z_0 \tag{14.6.73}$$

we have in the classically forbidden region

$$\psi_{II} = B\frac{1}{\sqrt{K(z)}} \exp\left[-\int_{z_0}^{z} K(z')\, dz'\right] \quad z > z_0 . \tag{14.6.74}$$

This connects onto the solution ψ_I in the classically allowed region if ψ_I is given by

$$\psi_I = 2B\frac{1}{\sqrt{k(z)}} \cos\left[\int_{z}^{z_0} k(z')\, dz' - \frac{\pi}{4}\right] \quad 0 < z < z_0 . \tag{14.6.75}$$

But, $\psi_I(0) = 0$. This means that

$$\cos\left[\int_{0}^{z_0} k(z')\, dz' - \frac{\pi}{4}\right] = 0 . \tag{14.6.76}$$

Therefore,

$$\int_0^{z_0} k(z')\,dz' - \frac{\pi}{4} = (n+1/2)\pi \quad n = 0, 1, 2, \ldots . \tag{14.6.77}$$

So we find,

$$\sqrt{2gm}\int_0^{z_0} \sqrt{z_0 - z}\,dz = (n + 3/4)\pi\hbar . \tag{14.6.78}$$

Integrating this expression and substituting for z_0 in terms of the energy E_n we get

$$E_n = \left[\frac{(3n+9)^2\pi^2}{128}\right]^{1/3}(mg^2\hbar^2)^{1/3} . \tag{14.6.79}$$

In particular the ground state energy, which corresponds to the classical situation of the ball at rest on the floor, is given by

$$E_0 = \left[\frac{81\pi^2}{128}\right]^{1/3}(mg^2\hbar^2)^{1/3} \approx 1.84(mg^2\hbar^2)^{1/3} . \tag{14.6.80}$$

b) The ground state wavefunction is given (up to normalization) by $n = 0$ and is

$$
\begin{aligned}
\psi &= 2B\frac{\hbar^{1/2}}{[2m^2g(z_0 - z)]^{1/4}}\cos\left[\frac{2m}{3\hbar}\sqrt{2g(z_0 - z)} - \frac{\pi}{4}\right] \quad 0 < z < z_0 \\
\psi &= 2B\frac{\hbar^{1/2}}{[2m^2g(z - z_0)]^{1/4}}\exp\left[-\frac{2m}{3\hbar}\sqrt{2g(z - z_0)}\right] \quad z > z_0 . \text{(14.6.81)}
\end{aligned}
$$

Here,

$$z_0 = \frac{E_0}{mg} = \left[\frac{81\pi^2}{128}\right]^{1/3}\left[\frac{\hbar^2}{m^2g}\right]^{1/3} . \tag{14.6.82}$$

This solution corresponds to the classical case of a particle sitting on an impenetrable floor.

This solution should also be compared with the variational calculation (problem 14.16) which yields the value

$$E_0 = 1.36\left(mg^2\hbar^2\right)^{1/3} . \tag{14.6.83}$$

Clearly the variational calculation gives a better result.

14.7 Ground State of H$_2^+$

Estimate the ground state energy of H$_2^+$, an ionized hydrogen molecule.

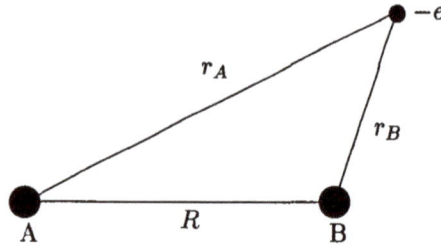

Figure 14.1: An ionized hydrogen molecule.

Solution

In this case the Hamiltonian is (see figure 14.1)

$$H = \frac{\vec{p}^2}{2m} - \frac{e^2}{r_A} - \frac{e^2}{r_B} + \frac{e^2}{R} \tag{14.7.84}$$

where r_A and r_B are the distances between the electron and the protons at A and B respectively and R is the distance between the two protons. As a trial wavefunction we choose a superposition of a wavefunction centered at A and at B. Thus, we use

$$\psi_A = \frac{1}{\sqrt{\pi a^3}} e^{-r_A/a} \quad , \quad \psi_B = \frac{1}{\sqrt{\pi a^3}} e^{-r_B/a} \tag{14.7.85}$$

where a is the Bohr radius, R is a variational parameter, and N is a normalization constant. We now find

$$N = (\psi_A + \psi_B, \psi_A + \psi_B) = 2[1 + S] \tag{14.7.86}$$

where S is the overlap integral

$$S = (\psi_A, \psi_B) = \frac{1}{\pi a^3} \int e^{-(r_A + r_B)/a} \, d^3 r_A \tag{14.7.87}$$

and is evaluated by using elliptical coordinates.

$$\xi = \frac{r_A + r_B}{R} \quad , \quad \eta = \frac{r_A - r_B}{R} \quad , \quad \varphi \tag{14.7.88}$$

where φ is the angle of rotation about the line joining the two protons. The volume element in these coordinates is [14.2]

$$d^3 r_A = \frac{R^3}{8} (\xi^2 - \eta^2) d\xi d\eta d\varphi \tag{14.7.89}$$

with the range of integration

$$1 \le \xi < \infty , \quad -1 \le \eta \le 1 , \quad 0 \le \varphi \le 2\pi . \tag{14.7.90}$$

Thus,

$$
\begin{aligned}
S &= \frac{1}{8\pi}\left(\frac{R}{a}\right)^3 \int_1^\infty e^{-(R\xi)/a}\, d\xi \int_{-1}^1 (\xi^2 - \eta^2)\, d\eta \int_0^{2\pi} d\varphi \\
&= \left[1 + \frac{R}{a} + \frac{1}{3} + \left(\frac{R}{a}\right)^2\right] e^{-R/a}\,.
\end{aligned}
\tag{14.7.91}
$$

The energy is given by

$$
E(R) = \frac{(\psi, H\psi)}{(\psi, \psi)} = \frac{H_{AA} + H_{AB} + H_{BA} + H_{BB}}{2[1 + S]}
\tag{14.7.92}
$$

where

$$
H_{AA} = H_{BB} = (\psi_A, H\psi_A)
\tag{14.7.93}
$$

$$
H_{AB} = H_{BA} = (\psi_A, H\psi_B)\,.
\tag{14.7.94}
$$

Thus,

$$
E(R) = \frac{H_{AA} + H_{AB}}{1 + S}
\tag{14.7.95}
$$

and we are left with three integrals to evaluate. Now,

$$
\begin{aligned}
H_{AA} &= \left(\psi_A, \left[\frac{\vec{p}^2}{2m} - \frac{e^2}{r_A}\right]\psi_A\right) - \left(\psi_A, \frac{e^2}{r_B}\psi_A\right) + \left(\psi_A, \frac{e^2}{R}\psi_A\right) \\
&= -\frac{e^2}{2a} + \frac{e^2}{R} - \left(\psi_A, \frac{e^2}{r_B}\psi_A\right)\,.
\end{aligned}
\tag{14.7.96}
$$

The last term is again evaluated by using elliptical coordinates and yields

$$
\begin{aligned}
\left(\psi_A, \frac{e^2}{r_B}\psi_A\right) &= \int |\psi_A|^2 \frac{e^2}{r_B}\, d^3r \\
&= \frac{e^2 R^2}{2a^3}\left[\int_1^\infty \xi\, e^{-R\xi/a}\, d\xi \int_{-1}^1 e^{-R\eta/a}\, d\eta \right. \\
&\qquad \left. + \int_1^\infty e^{-R\xi/a}\, d\xi \int_{-1}^1 \eta\, e^{-R\eta/a}\, d\eta\right] \\
&= \frac{e^2}{R}\left[1 - e^{-2R/a}\left(1 + \frac{R}{a}\right)\right]\,.
\end{aligned}
\tag{14.7.97}
$$

The term H_{AB} breaks up into

$$
\begin{aligned}
H_{AB} &= \left(\psi_A, \left[\frac{\vec{p}^2}{2m} - \frac{e^2}{r_A}\right]\psi_B\right) - \left(\psi_A, \frac{e^2}{r_A}\psi_B\right) + \left(\psi_A, \frac{e^2}{R}\psi_B\right) \\
&= \left[-\frac{e^2}{2a} + \frac{e^2}{R}\right]S - \left(\psi_A, \frac{e^2}{r_A}\psi_B\right)\,.
\end{aligned}
\tag{14.7.98}
$$

The final integral is also evaluated using elliptical coordinates and yields

$$\left(\psi_A, \frac{e^2}{r_A}\psi_B\right) = \frac{R^3}{4\pi a^3}\int_1^\infty e^{-R\mu/a}\,d\mu\int_1^1 \frac{\mu^2 - \nu^2}{(R/2a)(\mu - \nu)}\,d\nu\int_0^{2\pi} d\varphi$$

$$= \left[1 + \frac{R}{a}\right]e^{-R/a}\,. \tag{14.7.99}$$

Combining all these results we find with

$$x = \frac{R}{a} \tag{14.7.100}$$

that

$$E(x) = e^2\left[-\frac{1}{2} + \frac{1}{x} + \frac{(1+x)e^{-2x} - x(1+x)e^{-x} - 1}{x(1 + x + x^2/3)e^{-x} + x}\right]\,. \tag{14.7.101}$$

A graph of this function shows that $E(x)$ has a minimum at $x \approx 2.16$. This yields a value of about 1.76 eV below the energy for $x = \infty$, i.e. when one of the protons is removed to infinity.

14.8 Variational Solution: Particle in a Box

Use a variational technique to find the ground state energy of a particle of mass m in the potential

$$V = \begin{cases} \infty & x < -a \\ V_0 & -a \le x \le a \\ \infty & x > a \end{cases}\,.$$

Hint: Pay close attention to the boundary conditions that the trial wavefunction has to satisfy.

Solution

Since the wavefunction must vanish at $x = \pm a$ and must be a positive parity solution with no nodes we choose

$$\psi = A\left[|a|^\alpha - |x|^\alpha\right]\,. \tag{14.8.102}$$

Here, α is the variational parameter. Normalization yields

$$|A|^2 = \frac{(2\alpha + 1)(\alpha + 1)}{4\alpha^2 a^{2\alpha+1}}\,. \tag{14.8.103}$$

Differentiation of this wavefunction does not produce any singularities at $x = \pm a$ since $\psi(\pm a) = 0$. The kinetic energy operator T acting on ψ yields

$$T\psi = -\frac{\hbar^2}{2m}A[-\alpha(\alpha - 1)x^{\alpha-2}]\,. \tag{14.8.104}$$

Hence we find

$$\begin{aligned}
\langle T \rangle &= 2|A|^2 \int_0^\infty (|a|^\alpha - |x|^\alpha) \frac{\hbar^2}{2m} \alpha(\alpha - 1) x^{\alpha-2} \\
&= \frac{\hbar^2}{4ma^2} \frac{(2\alpha + 1)(\alpha + 1)}{(2\alpha - 1)} .
\end{aligned}$$

(14.8.105)

Also

$$\langle V \rangle = V_0 .$$

(14.8.106)

Therefore,

$$\langle H \rangle = \frac{\hbar^2}{4ma^2} \frac{(2\alpha + 1)(\alpha + 1)}{2\alpha - 1)} + V_0 .$$

(14.8.107)

To minimize this we differentiate with respect to α to get

$$\frac{d}{d\alpha} \langle H \rangle = 0 = (2\alpha - 1)(4\alpha + 3) - 2(2\alpha^2 + 3\alpha + 1) .$$

(14.8.108)

The solution yields

$$\alpha = \frac{1 + \sqrt{6}}{2} \approx 1.7247 .$$

(14.8.109)

Substituting this value back into $\langle H \rangle$ we obtain

$$\langle H \rangle = E_0 = \langle H \rangle \approx 1.003 \times \frac{\pi^2 \hbar^2}{8ma^2} + V_0 .$$

(14.8.110)

This agrees with the exact result to within better than 0.3 % since we had a very good choice of trial function.

14.9 Hydrogen Atom: Variational Technique

Use the ground state wavefunction for a three-dimensional simple harmonic oscillator, namely

$$\psi = \left(\frac{2a}{\pi}\right)^{3/4} e^{-ar^2}$$

as a trial wavefunction to calculate an approximate value for the hydrogen atom ground state energy.

Solution

The Hamiltonian for the hydrogen atom is

$$H = \frac{\vec{p}^2}{2M} - \frac{e^2}{r} \, . \tag{14.9.111}$$

Using the trial wavefunction we get

$$E(a) = (\psi, H\psi) \, . \tag{14.9.112}$$

This expectation value breaks up into two parts, the kinetic energy T plus the potential energy V. For the kinetic energy we get

$$
\begin{aligned}
\langle T \rangle &= -\frac{\hbar^2}{2M} 4\pi \left(\frac{2a}{\pi}\right)^{3/2} \int_0^\infty e^{-ar^2} \frac{1}{r} \frac{d^2}{dr^2} \left[r e^{-ar^2} \right] r^2 dr \\
&= -\frac{\hbar^2}{2M} 4\pi \left(\frac{2a}{\pi}\right)^{3/2} \int_0^\infty 2a(2ar^2 - 3) \, e^{-2ar^2} r^2 dr \\
&= \frac{\hbar^2}{2M} 4\pi \left(\frac{2a}{\pi}\right)^{3/2} \frac{3}{8} \sqrt{\frac{\pi}{2a}} \\
&= 3\frac{\hbar^2 a}{2M} \, .
\end{aligned}
\tag{14.9.113}
$$

For the potential energy we find

$$
\begin{aligned}
\langle V \rangle &= -e^2 4\pi \left(\frac{2a}{\pi}\right)^{3/2} \int_0^\infty e^{-ar^2} \frac{1}{r} e^{-ar^2} r^2 dr \\
&= -e^2 4\pi \left(\frac{2a}{\pi}\right)^{3/2} \int_0^\infty e^{-2ar^2} r \, dr \\
&= -e^2 4\pi \left(\frac{2a}{\pi}\right)^{3/2} \frac{1}{4a} \\
&= -2e^2 \sqrt{\frac{2a}{\pi}} \, .
\end{aligned}
\tag{14.9.114}
$$

Therefore,

$$E(a) = 3\frac{\hbar^2 a}{2M} - 2e^2 \sqrt{\frac{2a}{\pi}} \, . \tag{14.9.115}$$

The variational principle states that the best approximation is obtained by minimizing this expectation value with respect to the variational parameter a. Thus we compute dE/da and set it equal to zero. Therefore,

$$\frac{dE(a)}{da} = 3\frac{\hbar^2}{2M} - e^2 \sqrt{\frac{2}{\pi a}} = 0 \, . \tag{14.9.116}$$

Solving for a we get

$$a = \frac{8}{9\pi} \frac{M^2 e^4}{\hbar^4} \, . \tag{14.9.117}$$

After substituting this into the expression for $E(a)$ we get the best estimate for the ground state energy

$$E = -\frac{8}{3\pi}\frac{1}{2}\frac{Me^4}{\hbar^2} \quad . \tag{14.9.118}$$

This is to be compared with the exact ground state energy

$$E_0 = -\frac{1}{2}\frac{Me^4}{\hbar^2} \quad . \tag{14.9.119}$$

It is also to be noticed that the exact energy is lower than the estimate. This is in keeping with the variational principle.

14.10   Bound State in One Dimension

Use the variational principle to prove that, in one dimension, a potential that is everywhere attractive always has at least one bound state.

Solution

What we need to show is that for a potential $V(x)$ that is everywhere attractive

$$V(x) \leq 0 \quad \text{for all} \quad x$$

we can always find a solution such that the energy E is less than zero. To do this we assume a normalized trial function

$$\psi = \left(\frac{2\alpha}{\pi}\right)^{1/4} e^{-\alpha x^2} \tag{14.10.120}$$

together with the Hamiltonian

$$H = -\frac{\hbar^2}{2m}\frac{d^2}{dx^2} + V(x) \quad . \tag{14.10.121}$$

Using this wavefunction the expectation value of the energy is, by the variational principle such that

$$E(\alpha) = (\psi, H\psi) \geq E \quad \text{for all} \quad \alpha \tag{14.10.122}$$

where E is the energy of the possible bound state. Evaluating $E(\alpha)$ we get

$$E(\alpha) = \frac{\hbar^2\alpha}{2m} + \sqrt{\frac{2\alpha}{\pi}}\int_{-\infty}^{\infty} e^{-2\alpha x^2} V(x)\, dx \quad . \tag{14.10.123}$$

To obtain the best estimate for the energy we now minimize $E(\alpha)$ with respect to the variational parameter α.

$$\begin{aligned}
\frac{dE(\alpha)}{d\alpha} &= \frac{\hbar^2}{2m} + \sqrt{\frac{1}{2\alpha\pi}}\int_{-\infty}^{\infty} e^{-2\alpha x^2} V(x)\, dx \\
&\quad - \sqrt{\frac{2\alpha}{\pi}}\int_{-\infty}^{\infty} 2x^2 e^{-2\alpha x^2} V(x)\, dx \quad .
\end{aligned} \tag{14.10.124}$$

Solving for $\frac{\hbar^2 \alpha}{2m}$ by setting

$$\frac{dE(\alpha)}{d\alpha} = 0$$

we find

$$\frac{\hbar^2 \alpha}{2m} = \sqrt{\frac{2\alpha}{\pi}} \int_{-\infty}^{\infty} 2\alpha x^2 e^{-2\alpha x^2} V(x)\, dx$$

$$- \frac{1}{2}\sqrt{\frac{2\alpha}{\pi}} \int_{-\infty}^{\infty} e^{-2\alpha x^2} V(x)\, dx \ . \tag{14.10.125}$$

Substituting this value back into (14.10.123) we get that

$$E \leq \sqrt{\frac{2\alpha}{\pi}} \int_{-\infty}^{\infty} (1/2 + 2\alpha x^2)\, e^{-2\alpha x^2} V(x)\, dx < 0\ . \tag{14.10.126}$$

Thus, we have proven the desired result.

14.11   Field Emission: WKB Approximation

In field emission the electron is literally torn from inside the surface of a metal. A model for this process is as follows. The surface of the metal, located at $x = 0$, is subject to an electric field $\mathcal{E}$ outside the metal, i.e. for $x > 0$ so that the potential experienced by the electron is

$$V(x) = \begin{cases} -U & x < 0 \\ -e\mathcal{E}x & x > 0 \end{cases} \ .$$

The electron is bound by an amount ϕ, the so-called "work function". Thus, the energy of the electron is $E = -\phi$.

Use the WKB approximation to show that the transmission probability for an electron incident on the surface of the metal is well approximated by

$$T(E) = \exp\left\{ -\frac{\mathcal{E}_C}{\mathcal{E}} \right\}$$

where

$$\mathcal{E}_C = \frac{4\sqrt{2m}|E|^{3/2}}{3e\hbar} \ .$$

Solution

The transmission probability is given by $|S|^2$ where, in WKB approximation,

$$S = \exp\left[-\int_{x_1}^{x_2} \kappa(x')\, dx' \right] \tag{14.11.127}$$

and the integral runs over the classically forbidden region. Also,

$$\kappa(x) = \frac{1}{\hbar}\sqrt{2m(V-E)} \ . \tag{14.11.128}$$

In our case we have

$$x_1 = 0 \quad \text{and} \quad x_2 = -\frac{E}{e\mathcal{E}}$$

so that

$$\int_{x_1}^{x_2} \kappa(x)\,dx = \int_0^{x_2}\sqrt{x_2 - x}\,dx = \frac{2}{3}\sqrt{e\mathcal{E}}\left(\frac{|E|}{e\mathcal{E}}\right)^{3/2} \ . \tag{14.11.129}$$

Therefore,

$$T(E) = |S|^2 = \exp\left\{-\frac{\mathcal{E}_C}{\mathcal{E}}\right\} \tag{14.11.130}$$

as desired.

14.12 Deuteron: Variational Principle

Assume that a deuteron is an $l = 0$ bound state of a neutron and a proton in the potential

$$V(r) = -V_0\,e^{-r/a}$$

where

$$V_0 = 32 \text{ MeV} \quad \text{and} \quad a = 2.2 \text{ fm.}$$

Use the trial wavefunction

$$\psi(\alpha) = A\,e^{-\alpha r/a}$$

in a variational calculation to determine the binding energy of the deuteron.

Solution

The Hamiltonian of this system in the center of mass is

$$H = \frac{\vec{p}^2}{2m} + V(r) \tag{14.12.131}$$

where m is the reduced mass of the proton-neutron system. The energy is estimated from

$$E(\alpha) = \frac{(\psi(\alpha), H\psi(\alpha))}{(\psi(\alpha), \psi(\alpha))} \ . \tag{14.12.132}$$

Evaluating the denominator (normalization) we get

$$(\psi(\alpha), \psi(\alpha)) = |A|^2 \int_0^\infty e^{-2\alpha r/a} r^2 dr$$

$$= |A|^2 \frac{a^3}{4\alpha^3} . \tag{14.12.133}$$

The kinetic energy yields

$$T = (\psi(\alpha), \frac{\vec{p}^2}{2m}\psi(\alpha))$$

$$= -\frac{\hbar^2|A|^2}{2m} \int_0^\infty e^{-\alpha r/a} \left(\frac{1}{r}\frac{d^2}{dr^2} r e^{-\alpha r/a}\right) r^2 dr$$

$$= \frac{\hbar^2|A|^2}{2m} \int_0^\infty e^{-2\alpha r/a} \left(\frac{\alpha^2 r^2}{a^2} - \frac{2\alpha r}{a}\right) dr$$

$$= \frac{\hbar^2|A|^2}{2m} \frac{a}{4\alpha} . \tag{14.12.134}$$

The potential energy contributes

$$= (\psi(\alpha), V(r)\psi(\alpha))$$

$$= -|A|^2 V_0 \int_0^\infty e^{-2\alpha r/a} e^{-r/a} r^2 dr$$

$$= -|A|^2 V_0 \frac{2a^3}{(2\alpha + 1)^3} . \tag{14.12.135}$$

Therefore,

$$E(\alpha) = \frac{\hbar^2}{2ma^2}\alpha^2 - V_0 \left(\frac{\alpha}{\alpha + 1/2}\right)^3 . \tag{14.12.136}$$

To use the variational principle we must minimize the energy with respect to the parameter α. So we compute α from

$$\frac{dE(\alpha)}{d\alpha} = 0 = \frac{\hbar^2}{ma^2}\alpha - V_0\frac{3}{2}\frac{\alpha^2}{(\alpha + 1/2)^4} . \tag{14.12.137}$$

After inserting the appropriate numbers we get the following equation for α.

$$\frac{(\alpha + 1/2)^4}{\alpha} = \frac{3V_0 ma^2}{2\hbar^2} = 2.80 \tag{14.12.138}$$

A few straightforward numerical attempts yield the value

$$\alpha = 0.67 . \tag{14.12.139}$$

After substituting this value back into the equation for the energy we find

$$E = 8.52\,\alpha^2 - 32 \left(\frac{\alpha}{\alpha + 1/2}\right)^3 \text{ MeV} = -2.2 \text{ MeV} . \tag{14.12.140}$$

14.13 Bouncing Ball: Variational Calculation

A perfectly elastic ball, mass m is bouncing on a recoilless surface. Use a variational calculation to find the energy of its lowest eigenstate. Choose your own trial function.

Solution

We choose the potential energy to be zero when the ball is in contact with the surface at say $x = 0$. Furthermore, since the ball cannot penetrate into the region below the surface we must have that the wavefunction must vanish at $x = 0$. Thus, we choose as a trial function

$$\psi = A\,x\,e^{-ax} \ . \tag{14.13.141}$$

The normalization constant A, with an arbitrary choice of phase, is

$$A = 2a^{3/2} \ . \tag{14.13.142}$$

The Hamiltonian for this problem is

$$H = \frac{p^2}{2m} + mgx \ . \tag{14.13.143}$$

Then we have, using this trial function, that the expectation value for the energy is

$$
\begin{aligned}
E(a) \ &= \ (\psi,\, H\psi) = 4a^3 \int_0^\infty x\, e^{-ax} \left[\frac{-\hbar^2}{2m} \frac{d^2}{dx^2} + mgx \right] x\, e^{-ax}\, dx \\
&= \ 4a^3 \int_0^\infty \left[\frac{\hbar^2 a}{m} x - \frac{\hbar^2 a^2}{2m} x^2 + mgx^3 \right] e^{-2ax}\, dx \\
&= \ \frac{\hbar^2 a^2}{2m} + \frac{3mg}{2a} \ .
\end{aligned}
\tag{14.13.144}
$$

The best value is obtained by minimizing $E(a)$ with respect to the variational parameter a. Thus, we compute a from

$$\frac{dE}{da} = \frac{\hbar^2 a}{m} - \frac{3mg}{2a^2} = 0 \ . \tag{14.13.145}$$

This yields the value

$$a = \left(\frac{3m^2 g}{2\hbar^2} \right)^{1/3} \ . \tag{14.13.146}$$

Substituting this back into the expression for E we get

$$
\begin{aligned}
E \ &= \ \left(\frac{81}{32} \right)^{1/3} \left(mg^2 \hbar^2 \right)^{1/3} \\
&= \ 1.36 \left(mg^2 \hbar^2 \right)^{1/3} \ .
\end{aligned}
\tag{14.13.147}
$$

This result should be compared with the result obtained from the WKB approximation (problem 14.6) which yielded

$$E = 1.84 \left(mg^2\hbar^2\right)^{1/3} \ .$$
(14.13.148)

It is clear that the variational calculation gives a better result since the energy obtained is lower than the value obtained from the WKB approximation.

14.14   Beta Decay of Tritium

A tritium atom (H^3) in its ground state β-decays to form a singly ionized helium atom (He^3). What is the probability that this new atom will be found in its ground state? Assume that both nuclei have infinite mass and that there is no interaction between the beta-decay electron and the rest of the system.

Solution

With the stated assumptions we can neglect recoil and need only compute the overlap between the initial and final states. For the ground state of a hydrogenic atom with nuclear charge Z we have

$$\psi = \frac{1}{\sqrt{4\pi}} \left(\frac{2Z}{a_0}\right)^{3/2} e^{-Zr/a_0}$$
(14.14.149)

where a_0 is the Bohr radius

$$a_0 = \frac{\hbar^2}{me^2} \ .$$

For the initial state we have $Z = 1$ so that

$$\psi_i = \frac{1}{\sqrt{4\pi}} \left(\frac{2}{a_0}\right)^{3/2} e^{-r/a_0} \ .$$
(14.14.150)

For the final state we have $Z = 2$ so that

$$\psi_f = \frac{1}{\sqrt{4\pi}} \left(\frac{4}{a_0}\right)^{3/2} e^{-2r/a_0} \ .$$
(14.14.151)

The probability amplitude for finding the atom after the decay in its ground state is therefore given by

$$
\begin{aligned}
(\psi_f, \psi_i) &= \frac{1}{4\pi} \left(\frac{8}{a_0^2}\right)^{3/2} \int e^{-3r/a_0} r^2 dr \, d\Omega \\
&= 8^{3/2} \int_0^\infty e^{-x} x^2 dx \\
&= \frac{8^{3/2}}{3^4} \approx 0.559 \ .
\end{aligned}
$$
(14.14.152)

Thus, the desired probability is

$$P = |(\psi_f, \psi_i)|^2 \approx 0.312 \ .$$
(14.14.153)

14.15 Anharmonic Oscillator

For a potential of the form

$$V = \lambda x^4$$

use the variational method to find the energies of
a) the lowest energy level and
b) the first excited state.
The corresponding energies as found by numerical integration are respectively:

$$1.060 \left(\frac{\hbar^2}{2m}\right)^{2/3} \lambda^{1/3} \quad \text{and} \quad 3.800 \left(\frac{\hbar^2}{2m}\right)^{2/3} \lambda^{1/3} .$$

Solution

a) For the ground state we choose a trial wavefunction similar to the ground state wavefunction for the harmonic oscillator.

$$\psi_0(x) = \left(\frac{\alpha^2}{\pi}\right)^{1/4} e^{-\alpha^2 x^2/2} . \tag{14.15.154}$$

Then,

$$
\begin{aligned}
\langle H \rangle &= E_0(\alpha) = \frac{\alpha}{\sqrt{\pi}} \int_{-\infty}^{\infty} e^{-\alpha^2 x^2/2} \left[-\frac{\hbar^2}{2m} \frac{d^2}{dx^2} + \lambda x^4 \right] e^{-\alpha^2 x^2/2} \\
&= \frac{\alpha}{\sqrt{\pi}} \int_{-\infty}^{\infty} \left[-\frac{\hbar^2}{2m} \left(\alpha^4 x^2 - \alpha^2 \right) + \lambda x^4 \right] e^{-\alpha^2 x^2} \, dx \\
&= \frac{\hbar^2}{2m} \frac{\alpha^2}{2} + \frac{3}{4} \frac{\lambda}{\alpha^4} . \tag{14.15.155}
\end{aligned}
$$

The best value for α is obtained from

$$\frac{dE_0}{d\alpha} = 0 \quad \Rightarrow \quad \frac{\hbar^2}{2m}\alpha - \frac{3\lambda}{\alpha^5} = 0 . \tag{14.15.156}$$

So we find

$$\alpha^6 = \frac{6m\lambda}{\hbar^2} \quad \text{or} \quad \alpha = \left(\frac{6m\lambda}{\hbar^2}\right)^{1/6} . \tag{14.15.157}$$

Substituting this value for α in the expression for $E_0(\alpha)$ we get

$$
\begin{aligned}
E_0 &= \frac{\hbar^2}{4m} \left(\frac{6m\lambda}{\hbar^2}\right)^{1/3} + \frac{3\lambda}{4} \left(\frac{\hbar^2}{6m\lambda}\right)^{2/3} \\
&= \frac{3^{4/3}}{4} \left(\frac{\hbar^2}{2m}\right)^{2/3} \lambda^{1/3} \approx 1.082 \times \left(\frac{\hbar^2}{2m}\right)^{2/3} \lambda^{1/3} . \tag{14.15.158}
\end{aligned}
$$

This is within about 2 % of the value obtained by numerical integration.
b) In this case we need a trial wavefunction orthogonal to the wavefunction

for the ground state. For this reason we choose a wavefunction similar to the wavefunction for the first excited state of the simple harmonic oscillator.

$$\psi_1(x) = \left(\frac{4\alpha^6}{\pi}\right)^{1/4} x\, e^{-\alpha^2 x^2/2} \quad . \tag{14.15.159}$$

Then,

$$
\begin{aligned}
E_1(\alpha) &= \frac{2\alpha^3}{\sqrt{\pi}} \int_{-\infty}^{\infty} x\, e^{-\alpha^2 x^2/2} \left[-\frac{\hbar^2}{2m}\frac{d^2}{dx^2} + \lambda x^4\right] x\, e^{-\alpha^2 x^2/2} \\
&= \frac{2\alpha^3}{\sqrt{\pi}} \int_{-\infty}^{\infty} \left[-\frac{\hbar^2}{2m}\left(\alpha^4 x^4 - 3\alpha^2 x^2\right) + \lambda x^6\right] e^{-\alpha^2 x^2}\, dx \\
&= \frac{\hbar^2}{2m}\frac{3\alpha^2}{2} + \frac{15}{2}\frac{\lambda}{\alpha^4} \quad .
\end{aligned}
\tag{14.15.160}
$$

Again the best value for α is obtained from

$$\frac{dE_1}{d\alpha} = 0 \quad \Rightarrow \quad \frac{\hbar^2}{2m}\frac{3}{2}\alpha - \frac{15\lambda}{2\alpha^5} \quad . \tag{14.15.161}$$

Thus, we find

$$\alpha^6 = \frac{10m\lambda}{\hbar^2} \quad \text{or} \quad \alpha = \left(\frac{10m\lambda}{\hbar^2}\right)^{1/6} \quad . \tag{14.15.162}$$

Hence, our best estimate for the energy of the first excited state is

$$E_1 = \frac{9 \times 5^{1/3}}{4}\left(\frac{\hbar^2}{2m}\right)^{2/3} \lambda^{1/3} \approx 3.847 \times \left(\frac{\hbar^2}{2m}\right)^{2/3} \lambda^{1/3} \quad . \tag{14.15.163}$$

This is within better than 1.5% of the value obtained by numerical integration.

14.16 Nonlinear SHO: Variational Calculation

Use the trial wavefunction

$$|\psi\rangle = \cos\theta|0\rangle + \sin\theta|2\rangle \quad ,$$

where $|0\rangle$ and $|2\rangle$ are the ground state and second excited state of the unperturbed simple harmonic oscillator respectively corresponding to the Hamiltonian

$$H_0 = \frac{p^2}{2m} + \frac{1}{2}m\omega^2 x^2$$

and θ is a variational parameter, to calculate the ground state energy of the Hamiltonian

$$H = H_0 + \frac{1}{4}\lambda x^4 \quad .$$

Compare the result with that obtained by first order perturbation theory. (Problem 12.4 a).

Solution

The trial wavefunction is already normalized. So we need only compute

$$E(\theta) = \langle \psi | H | \psi \rangle \tag{14.16.164}$$

and minimize this result. To this end we need the matrix elements $\langle 0|H_0|0\rangle$, $\langle 2|H_0|2\rangle$, and $\langle 0|H_0|2\rangle = \langle 2|H_0|0\rangle$, as well as $\langle 0|x^4|0\rangle$, $\langle 2|x^4|2\rangle$, and $\langle 0|x^4|2\rangle = \langle 2|x^4|0\rangle$. Now, we have

$$\langle 0|H_0|0\rangle \;=\; \frac{1}{2}\hbar\omega$$

$$\langle 2|H_0|2\rangle \;=\; \frac{5}{2}\hbar\omega$$

$$\langle 2|H_0|0\rangle \;=\; \langle 0|H_0|2\rangle = 0 \;. \tag{14.16.165}$$

Also, in terms of creation and annihilation operators,

$$\begin{aligned}
x^4 \;=\;& \left(\frac{\hbar}{2m\omega}\right)^2 (a^\dagger + a)^4 \\[2mm]
\;=\;& \left(\frac{\hbar}{2m\omega}\right)^2 (a^4 + a^3 a^\dagger + a^2 a^\dagger a + a^2 {a^\dagger}^2 + aa^\dagger a^2 \\
& + aa^\dagger aa^\dagger + a{a^\dagger}^2 a + a{a^\dagger}^3 + a^\dagger a^3 + a^\dagger a^2 a^\dagger + a^\dagger aa^\dagger a \\
& + a^\dagger a{a^\dagger}^2 + {a^\dagger}^2 a^2 + {a^\dagger}^2 aa^\dagger + {a^\dagger}^3 a + {a^\dagger}^4) \;. \tag{14.16.166}
\end{aligned}$$

Next, using that

$$\begin{aligned}
a|n\rangle \;&=\; \sqrt{n}|n-1\rangle \\
a^\dagger|n\rangle \;&=\; \sqrt{n+1}|n+1\rangle \tag{14.16.167}
\end{aligned}$$

we get

$$\begin{aligned}
\left(\frac{\hbar}{2m\omega}\right)^{-2} \langle 0|x^4|0\rangle \;&=\; \langle 0|a^2 {a^\dagger}^2 + aa^\dagger aa^\dagger|0\rangle \\
&=\; 3 \tag{14.16.168}
\end{aligned}$$

as well as

$$\begin{aligned}
\left(\frac{\hbar}{2m\omega}\right)^{-2} \langle 2|x^4|0\rangle \;&=\; \left(\frac{\hbar}{2m\omega}\right)^{-2} \langle 0|x^4|2\rangle \\
&=\; \langle 2|aa^{\dagger 3} + a^\dagger aa^{\dagger 2} + a^{\dagger 2} aa^\dagger|0\rangle \\
&=\; 6\sqrt{2} \;. \tag{14.16.169}
\end{aligned}$$

Finally,

$$\begin{aligned}
\left(\frac{\hbar}{2m\omega}\right)^{-2} \langle 2|x^4|2\rangle \;&=\; \langle 2|a^2 {a^\dagger}^2 + aa^\dagger aa^\dagger + aa^{\dagger 2}a + a^\dagger aa^\dagger a + {a^\dagger}^2 a^2|2\rangle \\
&=\; 39 \;. \tag{14.16.170}
\end{aligned}$$

Combining all these results we get

$$
\begin{aligned}
E(\theta) \;=\;& \cos^2\theta \left[\frac{1}{2}\hbar\omega + 3\frac{\lambda}{4}\left(\frac{\hbar}{2m\omega}\right)^2\right] \\
& + \; \sin^2\theta \left[\frac{5}{2}\hbar\omega + 39\frac{\lambda}{4}\left(\frac{\hbar}{2m\omega}\right)^2\right] \\
& + \; 2\sin\theta\cos\theta\, 6\sqrt{2}\frac{\lambda}{4}\left(\frac{\hbar}{2m\omega}\right)^2 .
\end{aligned}
\tag{14.16.171}
$$

Using that

$$
\begin{aligned}
\cos^2\theta \;=\;& \frac{1}{2}[1+\cos(2\theta)] \\
\sin^2\theta \;=\;& \frac{1}{2}[1-\cos(2\theta)]
\end{aligned}
\tag{14.16.172}
$$

we find

$$
\begin{aligned}
E(\theta) \;=\;& -\cos(2\theta)\left[\hbar\omega + \frac{9}{2}\lambda\left(\frac{\hbar}{2m\omega}\right)^2\right] + \sin(2\theta)\frac{3\sqrt{2}}{2}\lambda\left(\frac{\hbar}{2m\omega}\right)^2 \\
& + \; \frac{3}{2}\hbar\omega + \frac{21\lambda}{4}\left(\frac{\hbar}{2m\omega}\right)^2 .
\end{aligned}
\tag{14.16.173}
$$

To minimize this energy we differentiate with respect to θ and set the result equal to 0 to get

$$
\sin(2\theta)\left[\hbar\omega + \frac{9}{2}\lambda\left(\frac{\hbar}{2m\omega}\right)^2\right] = -\cos(2\theta)\frac{3\sqrt{2}}{2}\lambda\left(\frac{\hbar}{2m\omega}\right)^2 .
\tag{14.16.174}
$$

Therefore,

$$
\tan(2\theta) = -\frac{3\sqrt{2}}{9+8(\hbar\omega/\lambda)(m\omega/\hbar)^2} ,
\tag{14.16.175}
$$

if λ is sufficiently small so that we can compare this result with perturbation theory we get

$$
\theta \approx -\frac{3\sqrt{2}\,\lambda}{16\hbar\omega}\left(\frac{\hbar}{m\omega}\right)^2 .
\tag{14.16.176}
$$

If we now substitute this result into the expression (14.16.171) for the energy we get the approximate ground state energy. To simplify the writing we introduce the parameter

$$
\gamma = \frac{\lambda}{4}\frac{\hbar^2}{m^2\omega^2} .
\tag{14.16.177}
$$

Then,

$$E_0 \approx \frac{1}{2}\hbar\omega + \frac{3}{4}\gamma - \frac{9}{16}\frac{\gamma^2}{\hbar\omega} + \frac{45}{16}\frac{\gamma^2}{\hbar\omega} - \frac{9}{2}\frac{\gamma^2}{\hbar\omega}$$

$$= \frac{1}{2}\hbar\omega + \frac{3}{4}\gamma - \frac{9}{4}\frac{\gamma^2}{\hbar\omega} \ . \tag{14.16.178}$$

A comparison with the perturbation result (Problem 12.4 a) shows that these results agree to order λ and differ by about 14 % in the λ^2 term.

14.17 WKB Solution and Parity

Show that for a parity invariant potential

$$V(x) = V(-x)$$

and an energy E such that there are four turning points, say at

$$x = \pm a \ , \qquad x = \pm b \quad b > a > 0$$

it is not possible to write WKB solutions that are simultaneously eigenstates of the parity operator.

Solution

We begin by writing the solutions in the four "external" regions

$$x < -b \ , \quad -b < x < -a \quad \text{and} \quad a < x < b \ , \quad x > b \ .$$

We do not need to consider the "central" region $-a < x < a$. Thus, with

$$\kappa(x) = \sqrt{\frac{2m}{\hbar^2}[V(x) - E]} \quad E < V \ , \quad x < -b \ \text{ or } \ x > b$$

$$k(x) = \sqrt{\frac{2m}{\hbar^2}[E - V(x)]} \quad E > V \ , \quad -b < x < -a \ \text{ or } \ a < x < b$$

$$\tag{14.17.179}$$

we have

$$\psi(x) = \begin{cases} \dfrac{A}{\sqrt{\kappa(x)}} \exp\left[-\int_{-b}^{x} \kappa(x')\,dx'\right] & x < -b \\[2.5ex] \dfrac{2A}{\sqrt{k(x)}} \cos\left[\int_{-b}^{x} k(x')\,dx' - \frac{\pi}{4}\right] & -b < x < -a \\[2.5ex] \dfrac{2B}{\sqrt{k(x)}} \cos\left[\int_{b}^{x} k(x')\,dx' - \frac{\pi}{4}\right] & a < x < b \\[2.5ex] \dfrac{B}{\sqrt{\kappa(x)}} \exp\left[-\int_{b}^{x} \kappa(x')\,dx'\right] & x > b \end{cases} \tag{14.17.180}$$

Here, we have already imposed the conditions that the solutions must be exponentially damped for $|x| \to \infty$ as well as the matching conditions at $x = -b$ and $x = b$. Letting $x \to -x$ and remembering that both $\kappa(x)$ and $k(x)$ are even functions of x the solutions become

$$\psi(-x) = \begin{cases} \frac{A}{\sqrt{\kappa(x)}} \exp\left[-\int_{-b}^{-x} \kappa(x')\, dx'\right] & x > b \\[2mm] \frac{2A}{\sqrt{k(x)}} \cos\left[\int_{-b}^{-x} k(x')\, dx' - \frac{\pi}{4}\right] & a < x < b \\[2mm] \frac{2B}{\sqrt{k(x)}} \cos\left[\int_{b}^{-x} k(x')\, dx' - \frac{\pi}{4}\right] & -b < x < -a \\[2mm] \frac{B}{\sqrt{\kappa(x)}} \exp\left[-\int_{b}^{-x} \kappa(x')\, dx'\right] & x < -b \end{cases} \qquad (14.17.181)$$

We now impose the conditions that these solutions should be eigenstates of the parity operator. Thus, we want

$$\psi(-x) = \pm\psi(x) \ . \tag{14.17.182}$$

Hence, we get the conditions

$$A \exp\left[-\int_{-b}^{x} \kappa(x')\, dx'\right] = \pm B \exp\left[-\int_{b}^{-x} \kappa(x')\, dx'\right] \tag{14.17.183}$$

as well as

$$A \cos\left[\int_{-b}^{x} k(x')\, dx' - \frac{\pi}{4}\right] = \pm B \cos\left[\int_{b}^{-x} k(x')\, dx' - \frac{\pi}{4}\right] \ . \tag{14.17.184}$$

From (14.17.183) we get by changing the integration variable on the right hand side from x to $-x$ that

$$B = \pm A \exp\left[-2\int_{-b}^{x} \kappa(x')\, dx'\right] \ . \tag{14.17.185}$$

Similarly by changing the integration variable on the right hand side of (14.17.184) we get

$$A \cos\left[\int_{-b}^{x} k(x')\, dx' - \frac{\pi}{4}\right] = \pm B \cos\left[\int_{-b}^{x} k(x')\, dx' + \frac{\pi}{4}\right] \ . \tag{14.17.186}$$

But, both A and B have to be constants independent of x. So, the only possibility is that

$$A = B = 0 \ . \tag{14.17.187}$$

In this case we have no solution at all. Thus, these solutions can never be parity eigenstates.

Bibliography

14.1 A.Z. Capri, *Nonrelativistic Quantum Mechanics* 3rd edition, World Scientific Publishing Co. Pte. Ltd., section 4.6, (2002) .
ibid, section 14.7.
ibid, section 14.8.

14.2 P.M. Morse and H. Feshbach, *Methods of Theoretical Physics* , Vol.1 - McGraw-Hill Book Co., Inc., New York (1953).

Chapter 15

Time-Dependent Perturbation Theory

15.1 Transition Probability: Bound State to Free

A particle is in the ground state of the Hamiltonian

$$H = \frac{p^2}{2m} + V$$

where

$$V = \begin{cases} 0 & x < -a \;,\; x > a \\ -V_0 & -a < x < a \end{cases}.$$

Find the transition probability per unit time to a state of energy $E_k > 0$, due to a perturbation

$$H'(t) = v\, e^{-x^2/\alpha^2} \sin \omega t$$

where v is a constant and $\alpha \ll a$.
Hint: The normalized bound state solution is given in [15.1] and the normalized continuum solution is given in problem 8.3.

Solution

We need both the normalized ground state solution and the normalized continuum solution. The normalized ground state solution is the even parity solution given in section 4.5 of [15.1].
If we define

$$\kappa_0^2 = -\frac{2m|E|}{\hbar^2}$$

$$\kappa^2 = \frac{2m(|E| + V_0)}{\hbar^2} \qquad\qquad (15.1.1)$$

then the ground state solution is

$$\psi_+(x) = \begin{cases} A\,e^{\kappa_0 x} & x < -a \\ B\,\cos\kappa x & |x| < a \\ A\,e^{-\kappa_0 x} & x > a \end{cases} \quad . \tag{15.1.2}$$

The equation for the binding energy is

$$\tan\kappa a = \frac{\kappa_0}{\kappa} \quad . \tag{15.1.3}$$

The normalization follows from

$$A = B\,e^{\kappa_0 a}\cos\kappa a \tag{15.1.4}$$

and

$$\begin{aligned} 1 &= |A|^2 \left[2\int_a^\infty e^{-2\kappa_0 x}\,dx + \frac{e^{-2\kappa_0 a}}{\cos^2\kappa a}\int_a^a \cos^2\kappa x\,dx \right] \\ &= |A|^2\,e^{-2\kappa_0 a}\left[\frac{1}{\kappa_0} + \frac{a}{\cos^2\kappa a} + \frac{\tan\kappa a}{\kappa}\right] \\ &= |A|^2\,e^{-2\kappa_0 a}\,\frac{\kappa^2 + \kappa_0^2}{\kappa^2}\left[a + \frac{1}{\kappa_0}\right] \quad . \end{aligned} \tag{15.1.5}$$

Therefore, we get,

$$A = \sqrt{\frac{\kappa_0}{1 + \kappa_0 a}}\,e^{\kappa_0 a}\cos\kappa a \quad , \quad B = \sqrt{\frac{\kappa_0}{1 + \kappa_0 a}} \quad . \tag{15.1.6}$$

The normalized continuum solution is obtained from the solution given in problem 8.1, as well as

$$A = \frac{\cos(ka + \delta_+)}{\cos Ka} \quad . \tag{15.1.7}$$

The overall normalization is found in problem 8.1 to be $1/\sqrt{\pi}$.

We only need the positive parity solution since the perturbation potential is an even function of x and the ground state is even parity. Now we are ready to use Fermi's golden rule. To do this we need the density of final states. This follows from putting the continuum solution in a box. Fortunately the perturbation is extremely short range and we only need the solution inside the well. So if the box is of length $L \gg 2a$ we get that

$$K = \frac{2\pi n}{L} \quad , \quad E + \frac{\hbar^2 k^2}{2m} = \frac{\hbar^2 K^2}{2m} - U \quad . \tag{15.1.8}$$

So,

$$dE = \frac{\hbar^2 K}{m}\,dK \quad \text{and} \quad dK = \frac{2\pi}{L}\,dn \quad . \tag{15.1.9}$$

Therefore, the density of final states is

$$\rho(E) = \frac{m}{2\pi\hbar^2 K} \quad . \tag{15.1.10}$$

We next need the matrix element. The perturbation is

$$H' = \frac{v}{2}\exp[-x^2/\alpha^2]\,2\sin\omega t \; . \tag{15.1.11}$$

Since $\alpha << a$ we need only consider the wavefunctions near $x = 0$, but we can extend the integrals to all of x. Therefore the relevant matrix element is

$$
\begin{aligned}
M &= \frac{v}{2}\sqrt{\frac{\kappa_0}{1+\kappa_0 a}}\frac{1}{\sqrt{\pi}}\int_{-\infty}^{\infty}\cos Kx\,\cos\kappa x\,\exp[-x^2/\alpha^2]\,dx \\
&= \frac{v}{2}\sqrt{\frac{\kappa_0}{1+\kappa_0 a}}\frac{1}{\sqrt{\pi}}\frac{\alpha\sqrt{\pi}}{2}\left[\exp[-\alpha(K+\kappa)^2/2]\right. \\
&+ \left.\exp[-\alpha(K-\kappa)^2/2]\right] \; .
\end{aligned}
\tag{15.1.12}
$$

Putting all this together we finally get w, the transition probability per unit time.

$$w = \frac{2\pi}{\hbar}|M|^2\rho(E) \tag{15.1.13}$$

or more explicitly

$$
\begin{aligned}
w &= \frac{mv^2a^2}{32\hbar^3}\frac{\kappa_0}{K}\frac{\alpha}{1+\kappa_0 a} \\
&\times \left[\exp[-\alpha(K+\kappa)^2/2]+\exp[-\alpha(K-\kappa)^2/2]\right] \; .
\end{aligned}
\tag{15.1.14}
$$

15.2   Photo-disintegration of Deuteron

The deuteron is an s-wave ($l = 0$) bound state of a proton and neutron with a binding energy of 2.226 MeV. It is well approximated as a bound state in a square well of depth $V_0 = 36.2$ MeV and width $a = 2.02 \times 10^{-13}$ cm . Using these data, compute the probability for photo-disintegration of the deuteron. Assume the incident photon can be approximated by a perturbation

$$V = \begin{cases} e\vec{A}\cdot\vec{r}\sin\omega t & t > 0 \\ 0 & t < 0 \end{cases}$$

where $\vec{A}$ is a constant vector of magnitude about 1×10^3 V/cm. Use whatever other approximations seem reasonable.

Solution

Since, V_0 is the depth of the well, we define

$$k^2 = -\frac{2mE}{\hbar^2} > 0 \; , \quad K^2 = \frac{2m(V_0-E)}{\hbar^2} > 0 \; . \tag{15.2.15}$$

Then the radial equation for $l = 0$ reads

$$
\begin{aligned}
\frac{1}{r}\frac{d^2}{dr^2}(rR) + K^2 R &= 0 \quad \text{for} \quad r < a \\
\frac{1}{r}\frac{d^2}{dr^2}(rR) - k^2 R &= 0 \quad \text{for} \quad r > a \; .
\end{aligned}
\tag{15.2.16}
$$

The solutions are

$$R = A \frac{\sin Kr}{r} \quad \text{for} \quad r < a$$

$$R = B \frac{e^{-kr}}{r} \quad \text{for} \quad r > a \ . \tag{15.2.17}$$

At $r = a$ both R and dR/dr are continuous. Therefore,

$$B = A e^{-ka} \sin Ka \ . \tag{15.2.18}$$

Equating the logarithmic derivatives now yields

$$Ka \cot Ka = -ka \ . \tag{15.2.19}$$

The solution of this transcendental equation yields the bound state energy which for the values given is stated to be 2.226 MeV. So the wavefunction is (up to normalization)

$$R(r) = \begin{cases} A \frac{\sin Kr}{r} & r < a \\ A \sin Ka \frac{e^{-k(r-a)}}{r} & r > a \end{cases} \ . \tag{15.2.20}$$

Throughout, m is the reduced mass $\approx 1/2\, m_{\text{proton}}$. Thus,

$$K \approx 9.66 \times 10^{12} \ \text{cm}^{-1} \ . \tag{15.2.21}$$

This means that we can neglect the part of the wavefunction that is inside the well and simply write

$$R(r) \approx B \frac{e^{-k(r-a)}}{r} = B' \frac{e^{-kr}}{r} \ . \tag{15.2.22}$$

The normalization of this wavefunction yields

$$|B'|^2 = \frac{k}{2\pi} \tag{15.2.23}$$

so that

$$\psi_{\text{in}} = \sqrt{\frac{k}{2\pi}} \frac{e^{-kr}}{r} \ . \tag{15.2.24}$$

This specifies the initial state with

$$k = \sqrt{\frac{2m|E_B|}{\hbar^2}} \approx 2.32 \times 10^{12} \ \text{cm}^{-1} \ . \tag{15.2.25}$$

For the outgoing state we take a plane wave

$$\psi_{\text{out}} = \frac{1}{(2\pi)^{3/2}} e^{-i\vec{q}\cdot\vec{r}} \ . \tag{15.2.26}$$

Next we compute the density of final states. To do this we take the outgoing plane wave and discretize the energy levels (to be able to count them) by placing the particle in a cube of sides L. In this case the permitted wavenumbers are

$$\vec{q} = \frac{2\pi}{L}(n_x, n_y, n_z) \tag{15.2.27}$$

where n_x, n_y, n_z are integers. Now we can count the levels. The number of modes lying between q_x and $q_x + dq_x$, q_y and $q_y + dq_y$, as well as q_z and $q_z + dq_z$ is

$$
\begin{aligned}
dN &= \Delta n_x \Delta n_y \Delta n_z = \frac{L}{2\pi} dq_x \frac{L}{2\pi} dq_y \frac{L}{2\pi} dq_z \\
&= \left(\frac{L}{2\pi}\right)^3 d^3 q = \left(\frac{L}{2\pi}\right)^3 q^2 \, dq \, \sin\theta \, d\theta \, d\varphi \ .
\end{aligned}
\tag{15.2.28}
$$

We use spherical coordinates in momentum space since energy conservation fixes the magnitude $|\vec{q}| = q$ according to

$$
E_q = E_B + \hbar\omega \ .
\tag{15.2.29}
$$

Now using (15.2.28) we get the density of final states

$$
\rho(q) = \frac{m}{(2\pi)^3 \hbar^2} q \, d\Omega \ .
\tag{15.2.30}
$$

We now have to evaluate the matrix element

$$
\begin{aligned}
\left(\psi_{out}, \frac{e}{2}\vec{A}\cdot\vec{r}\psi_{in}\right) &= \frac{e}{2}\frac{\sqrt{k}}{(2\pi)^2}\vec{A}\cdot\int \vec{r}e^{i\vec{q}\cdot\vec{r}}\frac{e^{-kr}}{r}r^2 \, dr \, \sin\theta \, d\theta \, d\varphi \\
&= \frac{e}{2}\frac{\sqrt{k}}{2\pi}\vec{A}\cdot\vec{I}
\end{aligned}
\tag{15.2.31}
$$

where

$$
\vec{I} = \int_0^\infty r \, dr \, e^{-kr} \int_0^\pi \vec{r}e^{iqr\cos\theta} \sin\theta \, d\theta \ .
\tag{15.2.32}
$$

By symmetry, this integral points in a direction parallel to $\vec{q}$ so

$$
\vec{I} = \vec{q}I \quad \text{or} \quad I = \frac{1}{q^2}\vec{q}\cdot\vec{I} \ .
\tag{15.2.33}
$$

Thus,

$$
\begin{aligned}
I &= \frac{1}{q^2}\int_0^\infty r \, dr \, e^{-kr} \int_{-1}^1 qru \, e^{iqru} \, du \\
&= \frac{4ik^2}{q^2(q^2 + k^2)^2} \ .
\end{aligned}
\tag{15.2.34}
$$

So,

$$
\vec{I} = \vec{q}\frac{4ik^2}{q^4(q^2 + k^2)^2}
\tag{15.2.35}
$$

and the matrix element squared is

$$
|M|^2 = \frac{e^2}{4}\frac{k}{(2\pi)^2}\frac{(\vec{q}\cdot\vec{A})^2 16k^4}{q^8(k^2 + q^2)^4} \ .
\tag{15.2.36}
$$

We now put all these results into Fermi's golden rule

$$dw = \frac{2\pi}{\hbar} \times \text{square of the matrix element} \times \text{density of final states}$$

$$= \frac{e^2}{8\pi\hbar} \frac{16k^5 q^2 A^2 \cos^2\theta}{q^8(k^2+q^2)^4} \frac{m}{(2\pi)^3\hbar^2} q\, d\Omega$$

$$= \frac{4e^2}{(2\pi)^4} \frac{1}{\hbar^3} \frac{mk^5 A^2 \cos^2\theta}{q^5(k^2+q^2)^4} d\Omega . \qquad (15.2.37)$$

15.3   Excitation of SHO

An atom is initially in the ground state of a simple harmonic oscillator

$$H = \hbar\omega\, a^\dagger a .$$

At $t = 0$ a perturbation

$$V' = \hbar\Omega(a^\dagger + a)$$

is turned on. Find the transition probability to any excited state of the system for $t > 0$. What is the probability that the atom remains in its ground state for $t > 0$?

Solution

Here we have an example of the sudden approximation. So for $t < 0$ we have

$$|\Psi(t)\rangle = |0, a\rangle e^{-i\omega t/2} \quad t < 0 \qquad (15.3.38)$$

where

$$a|0, a\rangle = 0 . \qquad (15.3.39)$$

For $t > 0$ we have to solve for the eigenstates of the Hamiltonian

$$H = \hbar\omega\, a^\dagger a + \hbar\Omega(a^\dagger + a) . \qquad (15.3.40)$$

Writing $b = a + c$ where c is just a c-number we find that

$$H = \hbar\omega(b^\dagger - c^*)(b - c) + \hbar\Omega(b^\dagger + b - c^* - c)$$

$$= H_b = \hbar\omega\, b^\dagger b - \hbar\frac{\Omega^2}{\omega} . \qquad (15.3.41)$$

Here we have chosen $c = \Omega/\omega$. Thus, in terms of the operators b, $b^\dagger$ the Hamiltonian H_b is diagonal. The energy eigenvalues are

$$E_n = (n_b + 1/2)\hbar\omega - \hbar\frac{\Omega^2}{\omega} . \qquad (15.3.42)$$

We have to re-express $|\Psi(t)\rangle$ in terms of the eigenstates of this Hamiltonian. For this purpose we find the unitary operator U such that

$$b = U\, a\, U^\dagger . \qquad (15.3.43)$$

This is easily done if we realize that for any two operators A, B such that $[A, B] = $ c-number we have

$$A e^B = e^B A + [A, B] e^B . \tag{15.3.44}$$

So we try

$$U = e^{ca - c^* a^\dagger} . \tag{15.3.45}$$

Then,

$$U^\dagger = e^{-(ca - c^* a^\dagger)} . \tag{15.3.46}$$

So,

$$U a U^\dagger = UU^\dagger a + U[a, -(ca - c^* a^\dagger)]U^\dagger = a + c^* . \tag{15.3.47}$$

Therefore, choosing

$$c = -\frac{\Omega}{\omega} \tag{15.3.48}$$

yields the desired transformation. It then follows that

$$|0, a\rangle = U^\dagger |0, b\rangle \tag{15.3.49}$$

where also

$$|n, b\rangle = \frac{1}{\sqrt{n!}} (b^\dagger)^n |0, b\rangle \tag{15.3.50}$$

is the eigenstate of H_b corresponding to the eigenvalue

$$E_n = (n_b + 1/2)\hbar\omega - \hbar\frac{\Omega^2}{\omega} . \tag{15.3.51}$$

The probability to find the system at time t in the state

$$|n, b, t\rangle = |n, b\rangle e^{-iE_n t/\hbar} \tag{15.3.52}$$

is

$$
\begin{aligned}
P_n(t) &= |\langle n, b, t | \Psi(t)\rangle|^2 \\
&= \frac{1}{n!} |\langle 0, b | b^n U^\dagger |0, b\rangle|^2 \\
&= \frac{1}{n!} |\langle 0, b | b^n |0, a\rangle|^2 .
\end{aligned}
\tag{15.3.53}
$$

Therefore,

$$
\begin{aligned}
P_n(t) &= \frac{1}{n!} \left| \langle 0, b | b^n e^{-\Omega b^\dagger/\omega} e^{\Omega b/\omega} e^{-(\Omega^2/2\omega^2)} |0, b\rangle \right|^2 \\
&= \frac{1}{n!} e^{-(\Omega^2/\omega^2)} \left| \langle 0, b | e^{-\Omega b^\dagger/\omega} e^{\Omega b/\omega} (b - \Omega/\omega)^n |0, b\rangle \right|^2 \\
&= \frac{1}{n!} e^{-(\Omega^2/\omega^2)} \left(-\frac{\Omega}{\omega} \right)^{2n} .
\end{aligned}
\tag{15.3.54}
$$

So finally,

$$P_n(t) = \frac{1}{n!} \left(\frac{\Omega}{\omega}\right)^{2n} e^{-(\Omega^2/\omega^2)} \tag{15.3.55}$$

which is a Poisson distribution.

15.4 Excitation of SHO by Stiffer Spring

Repeat the previous problem with

$$V' = \hbar\Omega\, a^\dagger a \ .$$

This amounts to increasing the spring constant suddenly.

Solution

Here again, as in problem 15.3, we have an example of the sudden approximation. So for $t < 0$ we have

$$|\Psi(t)\rangle = |0, a\rangle\, e^{-i\omega t/2} \quad t < 0 \tag{15.4.56}$$

where

$$a|0, a\rangle = 0 \ . \tag{15.4.57}$$

For $t > 0$ we have to solve for the eigenstates of the Hamiltonian

$$H = \hbar\omega a^\dagger a + \hbar\Omega a^\dagger a = \hbar(\omega + \Omega) a^\dagger a \ . \tag{15.4.58}$$

These are given by

$$|n, t\rangle = |n\rangle\, e^{-i E_n t/\hbar} \tag{15.4.59}$$

where

$$E_n = \hbar(\omega + \Omega)(n + 1/2) \tag{15.4.60}$$

and

$$a|n\rangle = \sqrt{n}|n - 1\rangle \ . \tag{15.4.61}$$

Therefore, the probability that at some time $t > 0$ we find the system in the state $|n, t\rangle$ is given by

$$P_n(t) = |\langle n, t|\Psi(t)\rangle|^2 = \delta_{n,0} \ . \tag{15.4.62}$$

This simply means that the perturbation does not change the state, only the energy. The evolution of the state for $t > 0$ is, however, governed by the energy $\hbar(\omega + \Omega)$ not the energy $\hbar\omega$.

15.5   Periodic Perturbation

An atom has two energy levels $\pm\hbar\Omega$. A weak disturbance $V(t)$ connecting these two levels and varying periodically in time such that

$$\langle 1|V(t)|2\rangle = \hbar\Omega_1 \sin\omega t$$

is turned on at $t = 0$.
a) Find a model Hamiltonian for this system.
b) If the atom was originally in its ground state, estimate the probability $P(t)$ that it is in its excited state at time t.

Solution

a) The model Hamiltonian is $H = H_0 + H'$ where

$$H_0 = \hbar\Omega \begin{pmatrix} 1 & 0 \\ 0 & -1 \end{pmatrix} = \hbar\Omega\sigma_3 \ . \tag{15.5.63}$$

$$H' = \hbar\Omega_1 \sin\omega t \begin{pmatrix} 0 & 1 \\ 1 & 0 \end{pmatrix} = \hbar\Omega_1 \sin\omega t\,\sigma_1 \quad \text{for} \quad t > 0 \ . \tag{15.5.64}$$

b) Let

$$\Psi = \begin{pmatrix} a \\ b \end{pmatrix} \ . \tag{15.5.65}$$

The Schrödinger equation now reads for $t > 0$

$$
\begin{aligned}
i\hbar\frac{da}{dt} &= \hbar\Omega\,a + \hbar\Omega_1 \sin\omega t\,b \\
i\hbar\frac{db}{dt} &= -\hbar\Omega\,b + \hbar\Omega_1 \sin\omega t\,a \ .
\end{aligned}
\tag{15.5.66}
$$

After simplifying these read

$$
\begin{aligned}
\frac{da}{dt} &= -i\Omega\,a - i\Omega_1 \sin\omega t\,b \\
\frac{db}{dt} &= i\Omega\,b - i\Omega_1 \sin\omega t\,a \ .
\end{aligned}
\tag{15.5.67}
$$

At $t = 0$ we have $a = 0$, $b = 1$. So, we put

$$
\begin{aligned}
a &= e^{-i\Omega t}\,A(t) \\
b &= e^{i\Omega t}\,B(t) \ .
\end{aligned}
\tag{15.5.68}
$$

Then,

$$
\begin{aligned}
\frac{dA}{dt} &= -i\Omega_1 \sin\omega t\,e^{2i\Omega t}\,B \\
\frac{dB}{dt} &= -i\Omega_1 \sin\omega t\,e^{-2i\Omega t}\,A
\end{aligned}
\tag{15.5.69}
$$

where at $t = 0$ we have $A = 0$ and $B = 1$. Thus, to lowest order in Ω_1 we get

$$
\begin{aligned}
A(t) &\approx -i\Omega_1 \int_0^t \sin \omega t' \, e^{2i\Omega t'} \, dt' \\
&= \frac{i\Omega_1}{2} \left[\frac{e^{i(\omega + 2\Omega)t} - 1}{\omega + 2\Omega} + \frac{e^{i(\omega - 2\Omega)t} - 1}{\omega - 2\Omega} \right] \\
&= \frac{2i\Omega_1}{\omega^2 - \Omega^2} e^{2i\Omega t} \left[\omega(\cos \omega t - e^{-2i\Omega t}) - 2i\Omega \sin \omega t \right] \qquad (15.5.70)
\end{aligned}
$$

and

$$
B(t) \approx 1 . \qquad (15.5.71)
$$

The probability to find the particle in the excited state is given by

$$
P(t) = |A(t)|^2 . \qquad (15.5.72)
$$

When evaluated this yields

$$
\frac{\omega^2 \Omega_1^2}{(\omega^2 - \Omega^2)^2} \left[2 + 3\sin^2 \omega t - 2\cos \omega t \cos 2\Omega t + 4\frac{\Omega}{\omega} \sin \omega t \sin 2\Omega t \right] . \ (15.5.73)
$$

15.6   Excitation of H-atom

A hydrogen atom in an excited state $|n, l, m\rangle$ is perturbed by a uniform electric field. If the interaction can be written

$$
V(t) = \begin{cases} e\vec{E} \cdot \vec{r} \, 2\sin \omega t & 0 < t < T \\ 0 & t < 0 , \ t > T \end{cases} .
$$

Find an expression for the transition probabilities to a lower level $|n', l', m'\rangle$. Do not attempt to evaluate the radial integrals. This is how intensities of spectral lines can be computed. You may use the results of problem 9.54. See also [15.2].

Solution

The state $|n, l, m\rangle$ is described in configuration space by the wavefunction

$$
\psi_{n,l,m}(r, \theta, \varphi) = R_{n,l}(r) \, Y_{l,m}(\theta, \varphi) . \qquad (15.6.74)
$$

The perturbation is

$$
V(t) = \begin{cases} e\mathcal{E} \, r(\sin \theta \cos \varphi, \ \sin \theta \sin \varphi, \ \cos \theta) \sin \omega t & 0 < t < t_0 \\ 0 & t < 0 , \ t > t_0 \end{cases} . \qquad (15.6.75)
$$

The transition probability is given by

$$
\begin{aligned}
P &= \frac{1}{\hbar^2} \left| \int_0^{t_0} \langle n, l, m | V(t) | n', l', m' \rangle \, e^{i\omega_{nn'} t} \right|^2 \\
&= \frac{e^2 \mathcal{E}^2}{\hbar^2} |\langle n, l, m | V(t) | n', l', m' \rangle|^2 \left| \frac{e^{i(\omega + \omega_{nn'})t_0} - 1}{2(\omega + \omega_{nn'})} \right|^2 . \qquad (15.6.76)
\end{aligned}
$$

Here we have $E_n < E_{n'}$ and

$$\omega_{nn'} = \frac{E_n - E_{n'}}{\hbar} = \frac{e^2}{2\hbar a_0}\left(\frac{1}{n^2} - \frac{1}{n'^2}\right) . \tag{15.6.77}$$

Now, we have

$$\int d\Omega\, Y_{lm}^*(\theta,\varphi)\sin\theta e^{i\varphi}\, Y_{l'm'}(\theta,\varphi)$$

$$= \delta_{m,m'+1}\left[\sqrt{\frac{(l-m+1)(l+2-m)}{(2l+3)(2l+1)}}\delta_{l',l+1}\right.$$

$$\left.+ \sqrt{\frac{(l+m)(l+m-1)}{(2l-1)(2l+1)}}\delta_{l',l-1}\right] . \tag{15.6.78}$$

$$\int d\Omega\, Y_{lm}^*(\theta,\varphi)\sin\theta e^{-i\varphi}\, Y_{l'm'}(\theta,\varphi)$$

$$= \delta_{m,m'-1}\left[\sqrt{\frac{(l+m+2)(l+m+1)}{(2l+3)(2l+1)}}\delta_{l',l+1}\right.$$

$$\left.+ \sqrt{\frac{(l-m)(lm-1)}{(2l-1)(2l+1)}}\delta_{l',l-1}\right] . \tag{15.6.79}$$

$$\int d\Omega\, Y_{lm}^*(\theta,\varphi)\cos\theta\, Y_{l'm'}(\theta,\varphi)$$

$$= \delta_{mm'}\left[\frac{2(l-m-1)(l+m)!}{(2l+3)(2l+1)(l-m)!}\delta_{l',l+1}\right.$$

$$\left.+ \frac{2(l+m+2)(l+m)!}{(2l-1)(2l+1)(l-m)!}\delta_{l',l-1}\right] . \tag{15.6.80}$$

This explicitly displays the selection rules $\Delta l = \pm 1$, $\Delta m = 0, \pm 1$. We are now left with only the radial integrals.

$$I_{n,l}^{(\pm)} = \int_0^\infty R_{n,l}(r)\, r\, R_{n,l\pm1}(r)\, dr . \tag{15.6.81}$$

These integrals can be expressed in terms of hypergeometric functions. The results, however, are no more illuminating than the integrals themselves so we leave them in this form. Thus, we finally get

$$P = \frac{4e^2\mathcal{E}^2\sin^2[\frac{1}{2}(\omega_{n,n'}+\omega)t_0]}{\hbar^2}\frac{1}{(\omega_{n,n'}+\omega)^2}\left[\frac{(l+m)!}{(2l+1)(l-m)!}\right]^2$$

$$\times \left\{\left(\frac{l-m+2}{2l+3}\right)^2|I_{n,l}^+|^2 + \left(\frac{l+m-1}{2l-1}\right)^2|I_{n,l}^-|^2\right\} . \tag{15.6.82}$$

15.7 Expanding Box

A particle is in the ground state in a one-dimensional box of length L. Suddenly the box expands (symmetrically) to a length $2L$ leaving the wavefunction undisturbed. Calculate the probability that the particle will be found in the ground state of the expanded box.

Solution

For a box extending from $-L/2$ to $L/2$ the ground state wavefunction is given by

$$\psi_0(x) = \sqrt{\frac{2}{L}} \cos\left(\frac{\pi x}{L}\right) \qquad -\frac{L}{2} \le x \le \frac{L}{2} . \qquad (15.7.83)$$

Also for a box extending from $-L$ to L the ground state wavefunction is given by

$$\psi(x) = \sqrt{\frac{1}{L}} \cos\left(\frac{\pi x}{2L}\right) \qquad -L \le x \le L . \qquad (15.7.84)$$

At $t = 0$, where this is the time at which the box expands, the state is the same as that of the unexpanded box. Thereafter the state evolves according to the Hamiltonian described by the box of length $2L$. Thus, for $t > 0$ the state is described by

$$\Psi(t, x) = \sqrt{\frac{2}{L^2}} \left(\int_{-L/2}^{L/2} \cos\left(\frac{\pi x'}{2L}\right) \cos\left(\frac{\pi x'}{L}\right) dx' \right) e^{i\omega t} \sqrt{\frac{1}{L}} \cos\left(\frac{\pi x}{2L}\right)$$
$$+ \quad \text{terms orthogonal to } \cos\left(\frac{\pi x}{2L}\right) \qquad (15.7.85)$$

where

$$\omega = \frac{\hbar \pi^2}{8mL^2} . \qquad (15.7.86)$$

This is the sudden approximation. The probability for finding the particle in the gound state is now given by the overlap of this wavefunction with the ground state wavefunction, namely

$$\Psi_0(t, x) = \begin{cases} e^{i\omega t} \sqrt{\frac{1}{L}} \cos\left(\frac{\pi x}{L}\right) & -\frac{L}{2} \le x \le \frac{L}{2} \\ 0 & |x| > \frac{L}{2} \end{cases} . \qquad (15.7.87)$$

Thus, the result is

$$P = \left| \sqrt{\frac{2}{L^2}} \int_{-L/2}^{L/2} \cos\left(\frac{\pi x'}{2L}\right) \cos\left(\frac{\pi x'}{L}\right) dx' \right|^2 = \left(\frac{8}{3\pi}\right)^2 . \qquad (15.7.88)$$

15.8   Sudden Displacement of SHO

A particle is bound by a simple harmonic oscillator potential and is in the first excited state. If a perturbation

$$V' = \lambda x$$

is turned on at time $t = 0$, find the probability that the particle will be in the new ground state for times $t > 0$.

Solution

For $t < 0$ we have a simple harmonic oscillator with

$$H_a = \hbar\omega(a^\dagger a + 1/2) .$$ (15.8.89)

For $t > 0$ we have a displaced simple harmonic oscillator with

$$H = \hbar\omega(a^\dagger a + 1/2) + \lambda\sqrt{\frac{\hbar}{2m\omega}}(a^\dagger + a) .$$ (15.8.90)

If we now introduce new operators

$$b = a + c \quad , \quad b^\dagger = a^\dagger + c$$ (15.8.91)

where

$$c^2 = \frac{\lambda^2}{2m\hbar\omega^3}$$ (15.8.92)

then, the Hamiltonian for $t > 0$ is again diagonalized and becomes

$$H_b = \hbar\omega(b^\dagger b + 1/2) - \frac{\lambda^2}{2m\omega^2} .$$ (15.8.93)

If we call the eigenstates of H_b, $|n, b\rangle$ with corresponding energies

$$E_{n,b} = (n + 1/2)\hbar\omega - \frac{\lambda^2}{2m\omega^2} .$$ (15.8.94)

Then the ground state $|0, a\rangle$ annihilated by a is a coherent state $|c\rangle$ and is given by

$$|c\rangle = e^{-c^2/2} \sum_{n=0}^{\infty} \frac{c^n}{\sqrt{n!}}|n, b\rangle .$$ (15.8.95)

At $t = 0$ the state is given by the first excited state of H_a and is

$$a^\dagger|c\rangle = (b^\dagger + c)|c\rangle .$$ (15.8.96)

But,

$$(b^\dagger + c)|c\rangle = e^{-c^2/2} \sum_{n=0}^{\infty} \frac{c^n}{\sqrt{n!}}[c|n, b\rangle + \sqrt{n+1}|n+1, b\rangle] .$$ (15.8.97)

This means that for $t > 0$ the state is given by

$$|\Psi\rangle = \exp i \left(\frac{\lambda^2}{2m\omega^2} - \frac{\omega t}{2} \right) e^{-c^2/2}$$

$$\times \sum_{n=0}^{\infty} \frac{c^n}{\sqrt{n!}} [c|n, b\rangle e^{-in\omega t} + \sqrt{n+1}|n+1, b\rangle e^{-i(n+1)\omega t}] . \qquad (15.8.98)$$

Therefore the probability amplitude for finding the particle in the new ground state is $\langle 0, b|\Psi\rangle$ and the probability is given by

$$|\langle 0, b|\Psi\rangle|^2 = c^2 e^{-c^2} . \qquad (15.8.99)$$

15.9   Sudden Perturbation of Two-level Atom

An atom has two energy levels of energy $\pm E$.   So the Hamiltonian may be written

$$H = E\sigma_3$$

where

$$\sigma_3 = \begin{pmatrix} 1 & 0 \\ 0 & -1 \end{pmatrix} .$$

If this atom is in the ground state, and at time $t = 0$ a perturbation

$$H' = V \begin{pmatrix} 0 & 1 \\ 1 & 0 \end{pmatrix}$$

is turned on, find the probability that this atom is still in its ground state at some later time t.

Solution

If we make the replacements

$$E = \hbar\Omega_0 \quad , \quad V = \hbar\Omega_1 \qquad (15.9.100)$$

the Hamiltonian is

$$H = \hbar\Omega_0 \begin{pmatrix} 1 & 0 \\ 0 & -1 \end{pmatrix} = H_0 = \hbar\Omega_0\sigma_3 \quad t < 0 \qquad (15.9.101)$$

and

$$H = \hbar\Omega_0\sigma_3 + \hbar\Omega_1\sigma_1 = H_1 \quad t > 0 \qquad (15.9.102)$$

or

$$H = \hbar\Omega_0 \begin{pmatrix} 1 & 0 \\ 0 & -1 \end{pmatrix} + \hbar\Omega_1 \begin{pmatrix} 0 & 1 \\ 1 & 0 \end{pmatrix} = H_1 \quad t > 0 . \qquad (15.9.103)$$

The eigenstates of H_0 are, as before,

$$|+\rangle = \begin{pmatrix} 1 \\ 0 \end{pmatrix} \qquad |-\rangle = \begin{pmatrix} 0 \\ 1 \end{pmatrix} \tag{15.9.104}$$

with

$$E_\pm = \pm\hbar\Omega_0 \ . \tag{15.9.105}$$

If we call the eigenfunctions and eigenvalues of H, v_μ and ϵ_μ respectively, we have for

$$v_\mu = \begin{pmatrix} a \\ b \end{pmatrix} \tag{15.9.106}$$

the eigenvalue equation

$$\hbar \begin{pmatrix} \Omega_0 & \Omega_1 \\ \Omega_1 & -\Omega_0 \end{pmatrix} \begin{pmatrix} a \\ b \end{pmatrix} = \epsilon_\mu \begin{pmatrix} a \\ b \end{pmatrix} \ . \tag{15.9.107}$$

The eigenvalues are:

$$\epsilon_\pm = \pm\hbar\sqrt{\Omega_0^2 + \Omega_1^2} \ . \tag{15.9.108}$$

Calling

$$\Omega^2 = \Omega_0^2 + \Omega_1^2 \tag{15.9.109}$$

we have

$$\epsilon_\pm = \pm\hbar\Omega \ . \tag{15.9.110}$$

We then get:

$$v_\pm = \frac{1}{\sqrt{2\Omega(\Omega - \Omega_0)}} \begin{pmatrix} \Omega_1 \\ \pm\Omega - \Omega_0 \end{pmatrix} \ . \tag{15.9.111}$$

If at $t = 0$ the system is in the state $u_+ = |+\rangle$, then for $t > 0$ the system will be in the state

$$\Psi(t) = b_+ v_+ e^{-i\Omega t} + b_- v_- e^{i\Omega t} \tag{15.9.112}$$

where $b_\pm$ are determined by the initial condition

$$u_+ = b_+ v_+ + b_- v_- \tag{15.9.113}$$

or

$$b_\pm = (v_\pm, u_+) \ . \tag{15.9.114}$$

Thus,

$$b_\pm = \frac{\Omega_1}{\sqrt{2\Omega(\Omega \mp \Omega_0)}} \ . \tag{15.9.115}$$

Combining these results we get

$$\Psi(t) = \begin{pmatrix} \cos\Omega t - i\frac{\Omega_0}{\Omega}\sin\Omega t \\ -i\frac{\Omega_0}{\Omega}\sin\Omega t \end{pmatrix} \ . \tag{15.9.116}$$

This wavefunction Ψ is an *exact* solution of the Hamiltonian (15.9.101) and (15.9.102) with the initial condition $\Psi(0) = u_+$. The approximation was made in writing the Hamiltonian (15.9.101) and (15.9.102) in the first place.

15.10 Berry's Phase

Consider the Hamiltonian

$$H(t) = -\frac{\mu}{2}\vec{\sigma} \cdot \vec{B}(t)$$

where

$$\vec{B}(t) = B_0 \left[\sin\theta\cos\omega t\, \hat{e}_x + \sin\theta\sin\omega t\, \hat{e}_y + \cos\theta\, \hat{e}_z\right]$$

so that

$$\vec{B}(t + 2\pi/\omega) = \vec{B}(t) .$$

If $\omega \ll \mu B_0$ show that

$$\langle\Psi(t = 2\pi/\omega)|\Psi(t = 0)\rangle = \exp\left(i\frac{2\pi}{\omega}\mu B_0\right)\exp[-i\pi(1 - \cos\theta)]$$

where $\frac{2\pi}{\omega}\mu B_0$ is just the dynamical phase

$$-\int_0^{2\pi/\omega} E(t)\, dt$$

and

$$-\pi(1 - \cos\theta) = -\Delta\Omega/2$$

is the geometric phase (Berry's phase).

a) Use the formula

$$\gamma_n(C) = \oint_C \langle n, \vec{B}|\nabla_B|n, \vec{B}\rangle \cdot d\vec{B}$$

for the geometrical or Berry's phase.

b) Use the formula

$$\frac{d\gamma_\pm}{dt} = i\left(v_\pm, \frac{dv_\pm}{dt}\right)$$

where $v_\pm$ are the "instantaneous" eigenvectors of $H(t)$.

c) Finally solve the exact eigenvalue equation for $H(t)$ and compute

$$\langle\Psi(t = 2\pi/\omega)|\Psi(t = 0)\rangle .$$

Solution

We are given

$$H(t) = -\frac{\mu}{2}\vec{\sigma}\cdot\vec{B}(t) \tag{15.10.117}$$

where

$$\vec{B}(t) = B_0(\sin\theta\,\cos\omega t\,\hat{e}_x + \sin\theta\,\sin\omega t\,\hat{e}_y + \cos\theta\,\hat{e}_z) \tag{15.10.118}$$

so that

$$\vec{B}(t + 2\pi) = \vec{B}(t) . \tag{15.10.119}$$

Using $\hbar\omega \ll \mu B_0$ we want to show that

$$\langle\Psi(t = 2\pi/\omega)|\Psi(t = 0)\rangle = \exp\left(\pm i\pi\frac{\mu B_0}{\hbar\omega}\right) e^{-i\pi(1\pm\cos\omega t)} \tag{15.10.120}$$

where the dynamical phase is

$$\pm\pi\frac{\mu B_0}{\hbar\omega} = \frac{-i}{\hbar}\int_0^{2\pi/\omega} E(t')\,dt' \tag{15.10.121}$$

and

$$-\pi(1 \pm \cos\theta) = -\frac{\Delta\Omega}{2} \tag{15.10.122}$$

is the geometric phase (Berry's phase).

We begin by writing

$$|\Psi(t)\rangle = \exp\left[\frac{-i}{\hbar}\int_0^{2\pi/\omega} E(t')\,dt'\right] e^{i\gamma_n(t)}|n,\vec{B}(t)\rangle . \tag{15.10.123}$$

Now we apply the theory developed for Berry's phase (see [15.3]). To illustrate the theory we solve the problem by three different methods.
a) Thus,

$$\gamma_n(C) = \oint_C \langle n,\vec{B}|\nabla_B|n,\vec{B}\rangle\cdot d\vec{B} \tag{15.10.124}$$

where the contour C is due to the fact that

$$\vec{B}(0) = \vec{B}(2\pi/\omega) . \tag{15.10.125}$$

To compute the dynamical phase we need the "instantaneous" energies which are the eigenvalues of

$$H(t) = -\frac{\mu}{2}\vec{\sigma}\cdot\vec{B}(t) . \tag{15.10.126}$$

Writing out the eigenvalue equation

$$\det|E - H(t)| = 0 \tag{15.10.127}$$

we get the two eigenvalues

$$E_\pm = \pm \frac{\mu B_0}{2} \ . \tag{15.10.128}$$

The corresponding eigenvectors are

$$v_+ = \begin{pmatrix} -\sin(\theta/2) \\ \cos(\theta/2)\,e^{i\omega t} \end{pmatrix} \ , \quad v_- = \begin{pmatrix} \cos(\theta/2) \\ \sin(\theta/2)\,e^{i\omega t} \end{pmatrix} \ . \tag{15.10.129}$$

It now follows that the dynamical phase is given by

$$\frac{i}{\hbar} \int_0^{\frac{2\pi}{\omega}} E_\pm(t')\,dt' = \frac{i}{\hbar} \int_0^{\frac{2\pi}{\omega}} \pm \frac{\mu B_0}{2}\,dt'$$
$$= \pm i\pi \frac{\mu B_0}{\hbar\omega} \ . \tag{15.10.130}$$

The geometrical phase (Berry's) phase is given by

$$\gamma_n(C) = i \oint_C \langle n, \vec{B} | \nabla_B | n, \vec{B} \rangle \cdot d\vec{B} \tag{15.10.131}$$

where

$$\vec{B}(t) = B_0(\sin\theta \, \cos\omega t \, \hat{e}_x + \sin\theta \, \sin\omega t \, \hat{e}_y + \cos\theta \, \hat{e}_z) \ . \tag{15.10.132}$$

Thus, using "spherical coordinates" in parameter space we have

$$\nabla_B v_\pm = \left(\frac{\partial}{\partial B_0}, \frac{1}{B_0}\frac{\partial}{\partial\theta}, \frac{1}{B_0\sin\theta}\frac{\partial}{\partial\omega t} \right) v_\pm \ . \tag{15.10.133}$$

Carrying out these differentiations and taking inner products we get

$$(v_+, \nabla_B v_+) = (0, 0, \frac{i}{2B_0}\cot(\theta/2)) = \frac{i}{2B_0}\cot(\theta/2)\hat{e}_\varphi \tag{15.10.134}$$

and

$$(v_-, \nabla_B v_-) = (0, 0, \frac{i}{2B_0}\tan(\theta/2)) = \frac{i}{2B_0}\tan(\theta/2)\hat{e}_\varphi \ . \tag{15.10.135}$$

We still need to compute $d\vec{B}/dt$. This is straightforward and yields

$$\frac{d\vec{B}}{dt} = B_0\omega \sin\theta(-\sin\omega t \, \hat{e}_x + \cos\omega t \, \hat{e}_y)$$
$$= B_0\omega \sin\theta\hat{e}_\varphi \ . \tag{15.10.136}$$

Now, we are finally ready to compute the geometrical phase using (15.10.131)

$$\gamma_+(C) = i \int_0^{\frac{2\pi}{\omega}} \frac{i}{2B_0}\cot(\theta/2)B_0\omega \sin\theta \, dt$$
$$= -\pi \cot(\theta/2)\sin\theta$$
$$= -\pi(1 + \cos\theta) \tag{15.10.137}$$

$$\begin{aligned}
\gamma_-(C) &= i \int_0^{\frac{2\pi}{\omega}} \frac{i}{2B_0} \tan(\theta/2) B_0 \omega \sin\theta \, dt \\
&= -\pi \tan(\theta/2) \sin\theta \\
&= -\pi(1 - \cos\theta) .
\end{aligned} \tag{15.10.138}$$

b) Another more direct way to compute the geometrical phases is to use the equations

$$\frac{d\gamma_\pm}{dt} = i\left(v_\pm, \frac{dv_\pm}{dt}\right) . \tag{15.10.139}$$

In this case we immediately find

$$\frac{d\gamma_+}{dt} = -\omega \cos^2(\theta/2) . \tag{15.10.140}$$

Then,

$$\begin{aligned}
\gamma_+(C) &= \left. -\omega t \cos^2\left(\frac{\theta}{2}\right) \right|_0^{2\pi/\omega} \\
&= -2\pi \cos^2(\theta/2) \\
&= -\pi(1 + \cos\theta) .
\end{aligned} \tag{15.10.141}$$

Similarly,

$$\frac{d\gamma_-}{dt} = -\omega \sin^2(\theta/2) \tag{15.10.142}$$

$$\begin{aligned}
\gamma_-(C) &= \left. -\omega t \sin^2\left(\frac{\theta}{2}\right) \right|_0^{2\pi/\omega} \\
&= -2\pi \sin^2(\theta/2) \\
&= -\pi(1 - \cos\theta) .
\end{aligned} \tag{15.10.143}$$

c) Finally, we can also simply just solve the time-dependent Schrödinger equation exactly. To do this we look for a solution of the form

$$\Psi(t) = \begin{pmatrix} a\, e^{-i\omega t/2} \\ b\, e^{i\omega t/2} \end{pmatrix} e^{i\lambda t} \tag{15.10.144}$$

with

$$|a|^2 + |b|^2 = 1 .$$

The time-dependent Schrödinger equation

$$i\hbar \frac{\partial \Psi}{\partial t} = H(t)\Psi(t)$$

now reads

$$\begin{pmatrix} \hbar(\omega/2 - \lambda)a\, e^{-i(\omega/2-\lambda)t} \\ -\hbar(\omega/2 + \lambda)b\, e^{i(\omega/2+\lambda)t} \end{pmatrix}$$

$$= -\frac{\mu B_0}{2} \begin{pmatrix} \cos\theta & \sin\theta\, e^{-i\omega t} \\ \sin\theta\, e^{i\omega t} & -\cos\theta \end{pmatrix} \begin{pmatrix} a\, e^{-i(\omega/2-\lambda)t} \\ b\, e^{i(\omega/2+\lambda)t} \end{pmatrix} . \quad (15.10.145)$$

Thus, after some simplification, we get the eigenvalue equation

$$-\lambda a = \left(\frac{\mu B_0}{2\hbar} \cos\theta - \frac{\omega}{2} \right) a + \frac{\mu B_0}{2\hbar} \sin\theta\, b$$

$$-\lambda b = \frac{\mu B_0}{2\hbar} \sin\theta\, a - \left(\frac{\mu B_0}{2\hbar} \cos\theta - \frac{\omega}{2} \right) b \ . \quad (15.10.146)$$

Hence,

$$\lambda = \pm\sqrt{ \left(\frac{\mu B_0}{2\hbar} \right)^2 - 2\left(\frac{\mu B_0}{2\hbar} \right) \frac{\omega}{2} \cos\theta + \frac{\omega^2}{4} }$$

$$\approx \pm\left(\frac{\mu B_0}{2\hbar} - \frac{\omega}{2} \cos\theta \right) \ . \quad (15.10.147)$$

Here we have used the approximation that

$$\omega << \mu B_0 \ .$$

With the initial condition

$$\Psi(0) = \begin{pmatrix} a \\ b \end{pmatrix} \quad (15.10.148)$$

we again obtain the same result as before.

$$(\Psi(2\pi/\omega), \Psi(0)) = \exp\left[i\frac{2\pi}{\omega}\lambda - i\pi \right]$$

$$= \exp i\pi \left[\pm \frac{\mu B_0}{2\hbar\omega} - (1 \pm \cos\theta) \right] \ . \quad (15.10.149)$$

15.11   Neutron in Rotating Magnetic Field

Consider a neutron (charge = 0), magnetic moment

$$\vec{\mu} = -g\frac{e\hbar}{2m}\vec{\sigma}$$

in a magnetic field consisting of a uniform component B in the z-direction and a component $b(t)$ rotating in the $x - y$ plane. Assume that at $t = 0$ the neutron has its spin pointing in the negative z-direction. Find $P_+(t)$, the probability that at time t the neutron has spin $\hbar/2$ in the positive z-direction. Find the condition between B and b such that the amplitude of oscillation of the spin between $+\hbar/2$ and $-\hbar/2$ is a maximum.

Solution

In this case the Hamiltonian consists of

$$H = \hbar\Omega\sigma_z + \hbar\omega[\sigma_x \cos at + \sigma_y \sin at] \tag{15.11.150}$$

where a is the angular frequency of rotation of the magnetic field $b(t)$ and

$$\Omega = g\frac{eB}{2mc} \quad , \quad \omega = g\frac{eb}{2mc} \ . \tag{15.11.151}$$

The state at $t = 0$ is given by

$$\Psi(0) = \begin{pmatrix} 0 \\ 1 \end{pmatrix} \ . \tag{15.11.152}$$

The state for a time $t > 0$ is taken to be

$$\Psi(t) = \begin{pmatrix} a_+ e^{-i\Omega t} \\ a_- e^{i\Omega t} \end{pmatrix} \ . \tag{15.11.153}$$

To first order we then have

$$a_- = 1 \tag{15.11.154}$$

and

$$a_+(t) = -i\omega \int_0^t (1, 0)(\sigma_x \cos at + \sigma_y \sin at) \begin{pmatrix} 0 \\ 1 \end{pmatrix} e^{-2i\Omega t} \, dt \ . \tag{15.11.155}$$

Therefore,

$$a_+(t) = \frac{i\omega}{2\Omega + a} \left[e^{-i(2\Omega + a)t} - 1 \right] \ . \tag{15.11.156}$$

Then the probability

$$P_+ = |a_+(t)|^2 \tag{15.11.157}$$

is given by

$$P_+ = \frac{4\omega^2}{(2\Omega + a)^2} \sin^2(2\Omega + a)t \ . \tag{15.11.158}$$

For the oscillations to be as large as possible requires that

$$\frac{4\omega^2}{(2\Omega + a)^2}$$

be a maximum. This means that we want ω or b as large as possible.

15.12   Excitation of Electron by Electric Field

Compute to lowest non-trivial order the probability that an electron in the ground state of a simple harmonic oscillator at $t = -\infty$ will be found in an excited state at $t = \infty$ if perturbed by an electric field

$$\frac{\mathcal{E}}{\sqrt{\pi}\tau} e^{-(t/\tau)^2} \ .$$

The electron continues to be bound by the simple harmonic oscillator potential.

Solution

The perturbation potential is due a dipole interaction and is

$$V(t) = -ex\frac{\mathcal{E}}{\sqrt{\pi}\tau} e^{-(t/\tau)^2} \ . \tag{15.12.159}$$

The probability amplitude for a transition from the state $|0\rangle$ to the state $|m\rangle$ is

$$a_{n,0} = \delta_{n,0} + \frac{1}{i\hbar} \sum_m \int_{-\infty}^{\infty} dt\, \langle n|V(t)|m\rangle a_{m,0}\, e^{i\omega_{nm}t} \tag{15.12.160}$$

where

$$\omega_{nm} = (E_n - E_m)/\hbar = (n-m)\omega \ . \tag{15.12.161}$$

The matrix element $\langle n|V(t)|m\rangle$ is given by

$$
\begin{aligned}
\langle n|V(t)|m\rangle &= -e\frac{\mathcal{E}}{\sqrt{\pi}\tau} \sqrt{\frac{\hbar}{2m\omega}} \langle n|(a+a^\dagger)|m\rangle\, e^{-(t/\tau)^2} \\
&= \frac{-e\mathcal{E}}{\sqrt{\pi}\tau} \sqrt{\frac{\hbar}{2m\omega}} e^{-(t/\tau)^2} \left(\sqrt{m}\,\delta_{n,m-1} + \sqrt{m+1}\,\delta_{n,m+1}\right) \ . \tag{15.12.162}
\end{aligned}
$$

Therefore,

$$
\begin{aligned}
a_{n,0} &= \delta_{n,0} - \frac{1}{i\hbar}\frac{e\mathcal{E}}{\sqrt{\pi}\tau}\sqrt{\frac{\hbar}{2m\omega}} \sum_m \left(\sqrt{m}\,\delta_{n,m-1} + \sqrt{m+1}\,\delta_{n,m+1}\right) \\
&\quad \times a_{m,0} \int_{-\infty}^{\infty} dt\, e^{-(t/\tau)^2}\, e^{i\omega(n-m)t} \\
&= \delta_{n,0} - \frac{1}{i\hbar}\frac{e\mathcal{E}}{\sqrt{\pi}\tau}\sqrt{\frac{\hbar}{2m\omega}} \sum_m \left(\sqrt{m}\,\delta_{n,m-1} + \sqrt{m+1}\,\delta_{n,m+1}\right) \\
&\quad \times a_{m,0}\sqrt{\pi}\,\tau\, e^{-(n-m)^2\omega^2\tau^2/4} \\
&= \delta_{n,0} + \frac{ie\mathcal{E}}{\sqrt{2m\hbar\omega}} e^{-\omega^2\tau^2/4} \left(\sqrt{n+1}\, a_{n+1,0} + \sqrt{n}\, a_{n-1,0}\right) \ . \tag{15.12.163}
\end{aligned}
$$

This equation is still exact. To use perturbation theory we simply iterate this equation. Then,

$$
\begin{aligned}
a_{n,0} &= \delta_{n,0} + \frac{ie\mathcal{E}}{\sqrt{2m\hbar\omega}} e^{-\omega^2\tau^2/4}\left(\sqrt{n+1}\,a_{n+1,0} + \sqrt{n}\,a_{n-1,0}\right) \\
&\approx \delta_{n,0} + \frac{ie\mathcal{E}}{\sqrt{2m\hbar\omega}} e^{-\omega^2\tau^2/4}\delta_{n-1,0} \quad .
\end{aligned}
$$
(15.12.164)

This equation also shows that, to lowest order, a transition can only occur to the state with $n = 1$. The resultant probability is

$$
|a_{1,0}|^2 = \frac{e^2\mathcal{E}^2}{2m\hbar\omega} e^{-\omega^2\tau^2/2} \quad .
$$
(15.12.165)

15.13   Neutron Magnetic Moment

A neutron has a magnetic moment μ_n. A free neutron is placed in a magnetic field B pointing in the z-direction and at time $t = 0$ the x-component of the spin of the neutron is measured to be $\hbar/2$. If this measurement is repeated at some later time t, calculate the probability of obtaining the same result.

Solution

The Hamiltonian describing the evolution of the spin of the neutron is given by

$$
H = -\vec{\mu}_n \cdot \vec{B}
$$
(15.13.166)

and may be written

$$
H = \hbar\omega\sigma_3
$$
(15.13.167)

where, as always,

$$
\sigma_3 = \begin{pmatrix} 1 & 0 \\ 0 & -1 \end{pmatrix} \quad .
$$
(15.13.168)

At time $t = 0$ the state is described by

$$
\psi(t = 0) = \frac{1}{\sqrt{2}} \begin{pmatrix} 1 \\ 1 \end{pmatrix} \quad .
$$
(15.13.169)

The equation for the evolution of this state reads

$$
i\hbar\frac{\psi}{dt} = \hbar\omega\sigma_3\psi \quad .
$$
(15.13.170)

The general solution of this equation is

$$
\psi(t) = \begin{pmatrix} a\,e^{i\omega t} \\ b\,e^{-i\omega t} \end{pmatrix}
$$
(15.13.171)

where a, b are integration constants. Imposing the initial condition we find

$$\psi(t) = \frac{1}{\sqrt{2}} \begin{pmatrix} e^{i\omega t} \\ e^{-i\omega t} \end{pmatrix} \ . \tag{15.13.172}$$

Therefore, the probability of obtaining the value $\hbar/2$ for a measurement of spin along the x direction is given by

$$
\begin{aligned}
|(\psi(0), \psi(t))|^2 &= \frac{1}{2} |e^{i\omega t} + e^{-i\omega t}|^2 \\
&= \cos^2 \omega t \ .
\end{aligned}
\tag{15.13.173}
$$

15.14 Electron Passing Through Magnetic Field

An electron (magnetic moment μ) enters a uniform magnetic field $\vec{B}$ which acts at right angles to its velocity $\vec{v}$. If the path length through the field is L and the electron's spin is initally perpendicular to both $\vec{B}$ and $\vec{v}$ find the probability that the electron emerges from the field with its spin pointing in the same direction as when it entered.

Solution

If we take the magnetic field pointing in the z-direction and the velocity in the y-direction then we can take the spin pointing in the x-direction. In this case the problem is like problem 15.13 above and the probability of obtaining the spin to point in the same direction after a time t as it was pointing at time $t = 0$ is given by

$$\cos^2 \omega t$$

where

$$\omega = \frac{\mu B}{\hbar} \ .$$

In this case the time t is just the time that the electron spends in the magnetic field and is given by

$$t = L/v \ .$$

Therefore, the probability that the electron's spin points in the same direction after emerging from the magnetic field is

$$P = \cos^2(\omega L/v) \ . \tag{15.14.174}$$

15.15 SHO: Sudden Transition

A one-dimensional simple harmonic oscillator with spring constant k is in its ground state. At $t = 0$ the spring constant is suddenly changed to $4k$. Calculate the probability that the system will end up in the new ground state.

Solution

Before the spring constant changes the state of the system is described by the wavefunction

$$\psi(x) = \left(\frac{m\omega}{\pi\hbar}\right)^{1/4} \exp\left[-\frac{m\omega}{2\hbar}x^2\right] \tag{15.15.175}$$

where

$$\omega = \sqrt{k/m} \ . \tag{15.15.176}$$

At $t = 0$ the spring constant increases to $4k$. Therefore, ω changes to 2ω and the new ground state wavefunction is

$$\phi_0(x) = \left(\frac{2m\omega}{\pi\hbar}\right)^{1/4} \exp\left[-\frac{m\omega}{\hbar}x^2\right] \ . \tag{15.15.177}$$

For $t > 0$ the wavefunction evolves according to the new Hamiltonian and is given by

$$\Phi(x,t) = \sum_n a_n \, \phi_n(x) \, e^{-i\omega(2n+1)t} \tag{15.15.178}$$

where ϕ_n is the wavefunction for the nth excited state corresponding to the energy $(2n+1)\hbar\omega$ of the new Hamiltonian. At $t = 0$ this wavefunction must coincide with ψ. So,

$$\psi(x) = \sum_n a_n \, \phi_n(x) \ . \tag{15.15.179}$$

The probability that the system winds up in the new ground state is now given by

$$
\begin{aligned}
|a_0|^2 &= |(\phi_0, \psi)|^2 \\
&= \frac{\sqrt{2}m\omega}{\pi\hbar} \left| \int_{-\infty}^{\infty} \exp\left[-\frac{3m\omega}{2\hbar}x^2\right] dx \right|^2 \\
&= \frac{\sqrt{2}m\omega}{\pi\hbar} \frac{2\pi\hbar}{3m\omega} \\
&= \frac{2\sqrt{2}}{3} = 0.9428 \ .
\end{aligned}
\tag{15.15.180}
$$

15.16   Coulomb Excitation

A particle with charge Q and initial momentum

$$\vec{P} = M v \hat{e}_z$$

is incident from infinity on an atom with Z electrons and energy levels E_n such that the impact parameter is b.

a) Use first order perturbation theory to show that the probability amplitude
for an excitation from the level

$$|i\rangle = |n, l, m\rangle \quad \text{to the level} \quad |f\rangle = |n', l', m'\rangle$$

is given by

$$A_{fi} = -\frac{iQe}{\hbar vb} \int_{-\pi}^{\theta(b)} e^{i\omega_{fi}t} \left(x_{if} \sin\theta + z_{if} \cos\theta\right) d\theta \qquad (15.16.181)$$

where

$$\hbar\omega_{fi} = E_{n'} - E_n$$

and the following approximations have to be made.
1) The incident particle follows a classical trajectory $\vec{R}(t)$ that lies entirely in
the $x - z$ plane and is determined by the classical equation

$$\frac{d\theta}{dt} = \frac{bv}{R^2(t)} \quad . \qquad (15.16.182)$$

Here we have used the fact that

$$\vec{R}(t) = R(t)(\sin\theta(t), 0, \cos\theta(t)) \quad . \qquad (15.16.183)$$

2) The dimensions of the target atom are much smaller than $R(t) = |\vec{R}(t)|$.
3) A dipole approximation is used.
For details on the validity of these approximations and further discussions see
[15.4].

Solution

Let H_0 be the unperturbed Hamiltonian so that

$$H_0|n, l, m\rangle = E_n|n, l, m\rangle \quad . \qquad (15.16.184)$$

The perturbation is given by

$$V = \sum_{i=1}^{z} \frac{-Qe}{|\vec{R}(t) - \vec{r}_i|} \qquad (15.16.185)$$

where $\vec{r}_i$ are the coordinates of the electrons in the target atom. Since the
dimensions of the target atom are much smaller than $R(t)$ we can write

$$\frac{1}{|\vec{R}(t) - \vec{r}_i|} = \sum_{l=0}^{\infty} \frac{r_i^l}{R(t)^{l+1}} P_l(\hat{r}_i \cdot \hat{R}(t)) \quad . \qquad (15.16.186)$$

Since we are using a dipole approximation we may truncate this series after
$l = 1$. However, for $l = 0$ the perturbation is just a constant and since the

initial and final states are different and therefore orthogonal we see that the $l = 0$ term does not contribute. Thus, the perturbation reduces to

$$V \approx \sum_{i=1}^{z} \frac{-Qe}{R(t)^3} \vec{r}_i \cdot \hat{R}(t)$$

$$= \sum_{i=1}^{z} \frac{-Qe}{R(t)^2} [x_i \sin \theta_i(t) + z_i \cos \theta_i(t)] . \qquad (15.16.187)$$

Here we have used the fact that the trajectory of the incoming particle lies in the $x - z$ plane so that $\theta_i(t)$ is the instantaneous angle between the z-axis and $\vec{R}(t)$, the vector from electron i to the instantaneous position of the incoming particle. Again, since $|\vec{r}_i| << R(t)$ we see that all angles θ_i as well as coordinates x_i, z_i are the same and we may as well call them θ and x, z. Therefore,

$$V = \frac{-QZe}{R(t)^2} [x \sin \theta(t) + z \cos \theta(t)] . \qquad (15.16.188)$$

Thus, using first order perturbation theory we have that the transition amplitude A_{if} from $|i\rangle$ to $|f\rangle$ is given by

$$A_{fi} \approx \frac{1}{i\hbar} \int_{-\infty}^{\infty} e^{i\omega_{fi}t} \langle i|V|f \rangle \, dt$$

$$= \frac{iQZe}{\hbar} \int_{-\infty}^{\infty} e^{i\omega_{fi}t} \frac{1}{R(t)^2} \langle f|x \sin \theta(t) + z \cos \theta(t)|i \rangle \, dt \quad (15.16.189)$$

But, the classical trajectory is given by (15.16.182) so that

$$dt = \frac{R^2(t) \, d\theta}{bv} . \qquad (15.16.190)$$

Therefore, we can write

$$A_{fi} = \frac{iQZe}{\hbar} \frac{1}{bv} \int_{-\pi}^{\theta(b)} e^{i\omega_{fi}t} \langle f|x \sin \theta + z \cos \theta|i \rangle \, d\theta . \qquad (15.16.191)$$

Here, $\theta(b)$ is the asymptotic value of $\theta(t)$ for $t \to \infty$.

Bibliography

[15.1] A.Z. Capri, *Nonrelativistic Quantum Mechanics* 3rd edition, World Scientific Publishing Co. Pte. Ltd. (2002), section 15.9.1.

[15.2] The computation of transition probabilities for atomic systems is carried out in
E.U. Condon and G.H. Shortley, *The Theory of Atomic Spectra* - Cambridge University Press (1963).

[15.3] A.Z. Capri, *Nonrelativistic Quantum Mechanics* 3rd edition, World Scientific Publishing Co. Pte. Ltd. (2002), section 15.11 .

[15.4] N. Bohr, Kgl. Danske Videnskab Selskab, Mat.-fys. Medd, bf18, No.8, (1948).

Chapter 16

Particle in a Uniform Magnetic Field

16.1 Estimate of Magnetic Energies

The strongest static magnetic fields currently achieved in laboratories are of the order of 3×10^5 gauss. For fields of this strength estimate the magnitude of the terms

$$\frac{e}{2mc}\vec{B} \cdot \vec{L} \quad \text{and} \quad \frac{e^2}{2mc^2}\vec{A} \cdot \vec{A} \ .$$

Solution

We want to estimate the two terms

$$E_1 = \frac{e}{2mc}\vec{B} \cdot \vec{L} \ , \quad E_2 = \frac{e^2}{2mc^2}\vec{A}^2 \ . \tag{16.1.1}$$

For this purpose we take

$$\vec{B} \cdot \vec{L} = \hbar B \quad \text{and} \quad |\vec{A}| = \frac{1}{2}|\vec{r}||\vec{B}| \tag{16.1.2}$$

with $r = |\vec{r}| = $ typical atomic distance $= 10^{-9}$ m.
Then we find

$$
\begin{aligned}
E_1 &= \frac{e\hbar}{2mc}B = \frac{4.8 \times 10^{-10} \times 1.05 \times 10^{-27}}{8 \times 9.1 \times 10^{-28} \times 3 \times 10^{10}} \times 3 \times 10^5 \\
&= 6.92 \times 10^{-16} \ \text{erg} = 4.33 \times 10^{-4} \ \text{eV} \ . \tag{16.1.3} \\
E_2 &= \frac{e^2 r^2}{2mc^2}B^2 = \frac{(4.8 \times 10^{-10})^2 \times (10^{-7})^2}{8 \times 9.1 \times 10^{-28} \times (3 \times 10^{10})^2} \times (3.0 \times 10^5)^2 \\
&= 3.16 \times 10^{-17} \ \text{erg} = 2 \times 10^{-5} \ \text{eV} \ . \tag{16.1.4}
\end{aligned}
$$

16.2   Radii of Landau Levels

Solve the eigenvalue problem for the Hamiltonian

$$r_0^2 = x_0^2 + y_0^2$$

where x_0 and y_0 are given by

$$\pi_x + M\omega y = M\omega y_0 \qquad\qquad (16.2.5)$$

$$\pi_y - M\omega x = -M\omega x_0 \qquad\qquad (16.2.6)$$

and π_x and π_y are given by

$$\vec{\pi} = \vec{p} + \frac{e}{c}\vec{A} \ . \qquad\qquad (16.2.7)$$

Here, $\vec{A}$ is the vector potential for a uniform magnetic field and $\omega = eB/Mc$ is the Larmor frequency. Interpret the meaning of this result.
Hint: Use the symmetric gauge

$$\vec{A} = \frac{B}{2}(-y, x, 0) \ .$$

Solution

We want to solve the eigenvalue equation

$$r_0^2 f_l = a_l^2 f_l \qquad\qquad (16.2.8)$$

where

$$r_0^2 = x_0^2 + y_0^2 \ . \qquad\qquad (16.2.9)$$

Now consider the "transverse" Hamiltonian

$$H_t = \frac{1}{2M}(\pi_x^2 + \pi_y^2) = \frac{1}{2M}M^2\omega^2[(x - x_0)^2 + (y - y_0)^2] \ . \qquad (16.2.10)$$

When written out this reads

$$H_t = \frac{1}{2M}(p_x^2 + p_y^2) + \frac{1}{2}M(\omega/2)^2(x^2 + y^2) + \frac{1}{2}\omega L_z \ . \qquad (16.2.11)$$

On the other hand

$$
\begin{aligned}
L_z &= xp_y - yp_x \\
&= x\pi_y - y\pi_x - \frac{1}{2}M\omega(x^2 + y^2) \\
&= M\omega[x(x - x_0) + y(y - y_0) - \frac{1}{2}(x^2 + y^2)] \\
&= \frac{1}{2}M\omega[(x - x_0)^2 + (y - y_0)^2 - x_0^2 - y_0^2] \\
&= \frac{1}{\omega}H_t - \frac{1}{2}M\omega(x_0^2 + y_0^2) \ . \qquad (16.2.12)
\end{aligned}
$$

Therefore,

$$r_0^2 = \frac{2}{M\omega^2}[H_t - \omega L_z] \ .$$

(16.2.13)

Notice that, by definition, r_0^2 is a positive operator so that the eigenvalues of H_t are greater than the eigenvalues of ωL_z. Now,

$$[L_z, H_t] = 0 \ .$$

So we can diagonalize them simultaneously to get

$$H_t = (2n+1)\hbar\frac{\omega}{2} + \frac{1}{2}\hbar\omega$$

(16.2.14)

since $L_z = m\hbar$. Therefore,

$$r_0^2 = \frac{2\hbar}{M\omega}(n + 1/2 - m/2) = \frac{\hbar}{M\omega}(2n - m + 1) \ .$$

(16.2.15)

If we now introduce the "magnetic length"

$$\lambda^2 = \frac{\hbar c}{eB} = \frac{\hbar}{M\omega}$$

(16.2.16)

and let $2n - m = l$, we get

$$r_0^2 = \lambda^2(l+1) \ .$$

(16.2.17)

Furthermore, the eigenfunctions of r_0^2 are products of the eigenfunctions of H_t and L_z. This may be interpreted to mean that the various energy levels with fixed z-component of angular momentum correspond to circular orbits with definite radii. However, the orbits are not observable since the "centres of the orbits" satisfy the uncertainty relation

$$\Delta x_0 \Delta y_0 \geq \frac{\lambda^2}{2} \ .$$

(16.2.18)

16.3   Equation of Continuity

Show that in the presence of a time-dependent electromagnetic field $(\phi, \vec{A})$ the equation of continuity holds for a particle of charge $-e$ if the charge density is given by

$$\rho = -e\,\Psi^*\Psi$$

and the current density is given by

$$\vec{j} = -e\frac{\hbar}{2im}\left[\Psi^*(\nabla + ie/(\hbar c)\vec{A})\Psi - \Psi(\nabla - ie/(\hbar c)\vec{A})\Psi^*\right] \ .$$

Hint: Start with the time-dependent Schrödinger equation.

Solution

The proof is a straightforward use of the Schrödinger equation and its complex conjugate

$$
i\hbar\frac{\partial\Psi}{\partial t} = -\frac{\hbar^2}{2m}[\nabla + ie/(\hbar c)\vec{A}]^2\Psi - e\phi\Psi
$$

$$
-i\hbar\frac{\partial\Psi^*}{\partial t} = -\frac{\hbar^2}{2m}[\nabla - ie/(\hbar c)\vec{A}]^2\Psi^* - e\phi\Psi^* \ . \tag{16.3.19}
$$

If we multiply the first of these equations by Ψ^* and the second by Ψ and subtract the two equations from each other we get

$$
i\hbar\frac{\partial}{\partial t}(\Psi^*\Psi) = -\frac{\hbar^2}{2m}\Big\{\Psi^*\nabla^2\Psi - \Psi\nabla^2\Psi^* + \frac{ie}{\hbar c}[\Psi^*\vec{A}\cdot\nabla\Psi
$$

$$
+\Psi^*\nabla\cdot(\vec{A}\Psi) + \Psi\vec{A}\cdot\nabla\Psi^* + \Psi\nabla(\vec{A}\Psi^*)]\Big\}
$$

$$
= -\frac{\hbar^2}{2m}\nabla\cdot\Big\{\Psi^*[\nabla + ie/(\hbar c)\vec{A}]\Psi - \Psi[\nabla - ie/(\hbar c)\vec{A}]\Psi^*\Big\} \ . \tag{16.3.20}
$$

After dividing both sides by $i\hbar$ we obtain the desired equation of continuity for the probability current density.

$$
\frac{\partial\rho}{\partial t} = \nabla\cdot\vec{j} \ . \tag{16.3.21}
$$

After multiplying by the electric charge $-e$ we have the electric current and charge densities given by

$$
\vec{j} = -e\frac{\hbar}{2im}\Big\{\Psi^*[\nabla + ie/(\hbar c)\vec{A}]\Psi - \Psi[\nabla - ie/(\hbar c)\vec{A}]\Psi^*\Big\}
$$

$$
\rho = -e\Psi^*\Psi \ . \tag{16.3.22}
$$

16.4   Gauge Invariance

Show that the Schrödinger equation is form invariant under the time-dependent gauge transformations.

Hint: The wavefunction acquires a phase under the gauge transformation.

Solution

We again start with the time-dependent Schrödinger equation in the presence of an electromagnetic field $(\phi, \vec{A})$

$$
i\hbar\frac{\partial\Psi}{\partial t} = \frac{1}{2m}\left(\vec{p} - \frac{q}{c}\vec{A}\right)^2\Psi + q\phi\Psi \ . \tag{16.4.23}
$$

Now define the gauge transformations

$$
\Psi' = \exp\left[\frac{iq}{\hbar c}\Lambda\right]\Psi \tag{16.4.24}
$$

and

$$\begin{aligned} \vec{A}' &= \vec{A} + \nabla\Lambda \\ \phi' &= \phi - \frac{1}{c}\frac{\partial\Lambda}{\partial t} . \end{aligned} \qquad (16.4.25)$$

Then, inverting these definitions and substituting in the Schrödinger equation we find

$$i\hbar\frac{\partial}{\partial t}\left[\exp\left(\frac{-iq}{\hbar c}\Lambda\right)\Psi'\right] = \frac{1}{2m}\left[\vec{p} - \frac{q}{c}(\vec{A}' - \nabla\Lambda)\right]^2\left[\exp\left(\frac{-iq}{\hbar c}\Lambda\right)\Psi'\right]$$
$$+ \ q\left(\phi' + \frac{1}{c}\frac{\partial\Lambda}{\partial t}\right)\left[\exp\left(\frac{-iq}{\hbar c}\Lambda\right)\Psi'\right] . \qquad (16.4.26)$$

The left hand side now becomes

$$\exp\left(\frac{-iq}{\hbar c}\Lambda\right)i\hbar\frac{\partial\Psi'}{\partial t} + \frac{q}{c}\frac{\partial\Lambda}{\partial t}\exp\left(\frac{-iq}{\hbar c}\Lambda\right)\Psi' . \qquad (16.4.27)$$

Also, by straightforward differentiation, we have

$$\left[\vec{p} - \frac{q}{c}(\vec{A}' - \nabla\Lambda)\right]\exp\left(\frac{-iq}{\hbar c}\Lambda\right)\Psi' = \exp\left(\frac{-iq}{\hbar c}\Lambda\right)\left[\vec{p} - \frac{q}{c}\vec{A}'\right]\Psi' . \qquad (16.4.28)$$

So that the right hand side reduces to

$$\exp\left(\frac{-iq}{\hbar c}\Lambda\right)\frac{1}{2m}\left[\vec{p} - \frac{q}{c}\vec{A}'\right]^2\Psi' + q\phi\exp\left(\frac{-iq}{\hbar c}\Lambda\right)\Psi' . \qquad (16.4.29)$$

Combining all this, and cancelling the phase factor, we get

$$i\hbar\frac{\partial\Psi'}{\partial t} = \frac{1}{2m}\left(\vec{p} - \frac{q}{c}\vec{A}\right)^2\Psi' + q\phi\Psi' . \qquad (16.4.30)$$

Thus, the Schrödinger equation is form invariant under a gauge transformation.

16.5   Gauge Transformations and Observables

Under a gauge transformation we have that for static electromagnetic fields

$$\vec{A} \to \vec{A}^\Lambda = \vec{A} + \nabla\Lambda$$

$$\phi \to \phi^\Lambda = \phi$$

$$\Psi \to \Psi^\Lambda = e^{i(q/\hbar c)\Lambda}\Psi$$

so that the Schrödinger equation remains form invariant. We also require, however, that observables be gauge invariant in the sense that their matrix elements remain invariant under gauge transformations. This means that if under a gauge transformation an observable

$$O(\vec{p}, \vec{A}, \phi) \to O(\vec{p}, \vec{A}^\Lambda, \phi^\Lambda)$$

we require that

$$(\Psi, O(\vec{p}, \vec{A}, \phi)\Phi) = (\Psi^\Lambda, O(\vec{p}, \vec{A}^\Lambda, \phi^\Lambda)\Phi^\Lambda) \ .$$

The left side may be rewritten as

$$(\Psi, O(\vec{p}, \vec{A}, \phi)\Phi) = (\Psi^\Lambda, O^\Lambda(\vec{p}, \vec{A}, \phi)\Phi^\Lambda) \ ,$$

where

$$O^\Lambda(\vec{p}, \vec{A}, \phi) = e^{i(q/\hbar c)\Lambda}\, O(\vec{p}, \vec{A}, \phi)\, e^{-i(q/\hbar c)\Lambda}$$

is the unitary transform of the operator $O(\vec{p}, \vec{A}, \phi)$ that maps $O \to O^\Lambda$. Show that the resulting necessary and sufficient condition, namely

$$O^\Lambda(\vec{p}, \vec{A}, \phi) = O(\vec{p}, \vec{A}^\Lambda, \phi^\Lambda)$$

is satisfied if and only if

$$O(\vec{p}, \vec{A}, \phi) = O(\vec{p} + q/c\vec{A}, \phi) \ .$$

See also [16.1] for further discussion.

Solution

Clearly the condition

$$O(\vec{p}, \vec{A}, \phi) = O(\vec{p} + \frac{q}{c}\vec{A}, \phi) \tag{16.5.31}$$

is sufficient since under the gauge transformation

$$
\begin{aligned}
\vec{A} \to \vec{A}^\Lambda &= \vec{A} + \nabla\Lambda \\
\phi \to \phi^\Lambda &= \phi \\
\Psi \to \Psi^\Lambda &= e^{i(q/\hbar c)\Lambda}\,\Psi
\end{aligned}
\tag{16.5.32}
$$

we have that

$$(\vec{p} + \frac{q}{c}\vec{A}^\Lambda)\Psi^\Lambda = e^{i(q/\hbar c)\Lambda}(\vec{p} + \frac{q}{c}\vec{A})\Psi \ . \tag{16.5.33}$$

Therefore,

$$O(\vec{p}, \vec{A}^\Lambda, \phi^\Lambda)\Psi^\Lambda = e^{i(q/\hbar c)\Lambda}O(\vec{p} + \frac{q}{c}\vec{A}, \phi)\Psi \tag{16.5.34}$$

and hence, with

$$O^\lambda(\vec{p}, \vec{A}, \phi) = O(\vec{p}, \vec{A}^\lambda, \phi^\lambda) \tag{16.5.35}$$

we have

$$(\Psi, O(\vec{p}, \vec{A}, \phi)\Phi^\Lambda) = (\Psi^\Lambda, O(\vec{p}, \vec{A}^\Lambda, \phi^\Lambda)\Psi^\Lambda) \ . \tag{16.5.36}$$

To prove necessity we try to consider the canonical momentum as an observable. In this case we have

$$O(\vec{p}, \vec{A}, \phi) = \vec{p} \tag{16.5.37}$$

and we need

$$\vec{p}\Psi^\Lambda = e^{i(q/\hbar c)\Lambda}\vec{p}\Psi \ . \tag{16.5.38}$$

This condition cannot be satisfied. Thus, we need to replace $\vec{p}$ by $\vec{p} + q/c\vec{A}$. This means that, in the presence of a magnetic field, the canonical momentum is not an observable, but the mechanical momentum

$$m\vec{v} = \vec{p} + (q/c)\vec{A} \tag{16.5.39}$$

is.

16.6   Spin 1/2 Particle in Magnetic Field

A neutral spin 1/2 particle with a magnetic moment $\vec{\mu}$ is placed in a constant uniform magnetic field $\vec{B}$. Compute how the average values of the three spin components vary with time if
a) the initial spin is in the direction of the magnetic field.
b) The initial spin is normal to the magnetic field.

Solution

The Hamiltonian for this problem is

$$H = -\vec{\mu} \cdot \vec{B} = -\mu\vec{\sigma} \cdot \vec{B} \ . \tag{16.6.40}$$

a) In this case we take

$$\vec{B} = (0, 0, B) \ , \quad \vec{s} = \frac{\hbar}{2}\vec{\sigma} \ . \tag{16.6.41}$$

The initial state is

$$\psi(0) = \begin{pmatrix} 1 \\ 0 \end{pmatrix} \ . \tag{16.6.42}$$

The Schrödinger equation may then be written

$$i\hbar\frac{\partial\psi}{\partial t} = -\hbar\omega\sigma_z\psi \tag{16.6.43}$$

where

$$\omega = \frac{\mu B}{\hbar} \ . \tag{16.6.44}$$

The solution is

$$\psi(t) = \exp(i\omega t\sigma_z)\psi(0) = e^{i\omega t}\begin{pmatrix} 1 \\ 0 \end{pmatrix} . \tag{16.6.45}$$

The expectation values of the spin operators in this time-dependent state are

$$
\begin{aligned}
\langle s_x \rangle &= 0 \\
\langle s_y \rangle &= 0 \\
\langle s_z \rangle &= \frac{\hbar}{2} .
\end{aligned}
\tag{16.6.46}
$$

b) In this case we take

$$\vec{B} = (B,0,0) . \tag{16.6.47}$$

The inital state is as in case a) above, but the Schrödinger equation may now be written

$$i\hbar\frac{\partial\psi}{\partial t} = -\hbar\omega\sigma_x\psi . \tag{16.6.48}$$

The solution is

$$\psi(t) = \exp(i\omega t\sigma_x)\psi(0) = \begin{pmatrix} \cos\omega t \\ i\sin\omega t \end{pmatrix} . \tag{16.6.49}$$

In this case the expectation values for the spin operators are

$$
\begin{aligned}
\langle s_x \rangle &= 0 \\
\langle s_y \rangle &= \frac{\hbar}{2}\sin 2\omega t \\
\langle s_z \rangle &= \frac{\hbar}{2}\cos 2\omega t .
\end{aligned}
\tag{16.6.50}
$$

The average values of the spin clearly precess about the magnetic field.

16.7 Spin 1/2 in Magnetic Field: Heisenberg Equations

A neutral spin 1/2 particle with a magnetic moment $\vec{\mu}$ is placed in a constant uniform magnetic field $\vec{B}$. Using the Heisenberg equations, compute how the average values of the three spin components vary with time if
a) the initial spin is in the direction of the magnetic field.
b) The initial spin is normal to the magnetic field. This is the same problem as 16.6, except that we now compute in the Heisenberg picture.

Solution

The Hamiltonian for this problem is as before

$$H = -\vec{\mu} \cdot \vec{B} = -\mu\vec{\sigma} \cdot \vec{B} \ . \tag{16.7.51}$$

a) In this case we again take

$$\vec{B} = (0, 0, B) \ , \quad \vec{s} = \frac{\hbar}{2}\vec{\sigma} \ . \tag{16.7.52}$$

The state, for all times, is

$$\psi(0) = \begin{pmatrix} 1 \\ 0 \end{pmatrix} \ . \tag{16.7.53}$$

The Heisenberg equations for the spin operators read

$$i\hbar\frac{d\vec{s}}{dt} = [\vec{s}, H] \tag{16.7.54}$$

and when written out in detail become

$$\begin{aligned}
\frac{ds_x}{dt} &= 2\omega s_y \\
\frac{ds_y}{dt} &= -2\omega s_x \\
\frac{ds_z}{dt} &= 0
\end{aligned} \tag{16.7.55}$$

where, as before,

$$\omega = \frac{\mu B}{\hbar} \ . \tag{16.7.56}$$

The solutions are

$$\begin{aligned}
s_x(t) &= s_x(0)\cos 2\omega t + \frac{\dot{s}_x(0)}{2\omega}\sin 2\omega t \\
s_y(t) &= s_y(0)\cos 2\omega t + \frac{\dot{s}_y(0)}{2\omega}\sin 2\omega t \\
s_z(t) &= s_z(0) \ .
\end{aligned} \tag{16.7.57}$$

Here, $s_x(0)$, $s_y(0)$, and $s_z(0)$ are operators such that

$$\begin{aligned}
\langle s_x(0)\rangle &= 0 \\
\langle s_y(0)\rangle &= 0 \\
\langle s_z(0)\rangle &= \frac{\hbar}{2} \ .
\end{aligned} \tag{16.7.58}$$

The two operators $\dot{s}_x(0)$ and $\dot{s}_y(0)$ are determined by the initial conditions that may be read off the Heisenberg equations.

$$\begin{aligned}
\dot{s}_x(0) &= 2\omega s_y(0) \\
\dot{s}_y(0) &= -2\omega s_x(0) \ .
\end{aligned} \tag{16.7.59}$$

Thus,

$$\begin{aligned}
\langle \dot{s}_x(0) \rangle &= 2\omega \langle s_y(0) \rangle = 0 \\
\langle \dot{s}_y(0) \rangle &= -2\omega \langle s_x(0) \rangle = 0 \ .
\end{aligned} \tag{16.7.60}$$

Therefore, we obtain for the time-dependent expectation values

$$\begin{aligned}
\langle s_x(t) \rangle &= 0 \\
\langle s_y(t) \rangle &= 0 \\
\langle s_z(t) \rangle &= \frac{\hbar}{2} \ .
\end{aligned} \tag{16.7.61}$$

b) In this case we take

$$\vec{B} = (B, 0, 0) \ . \tag{16.7.62}$$

The Heisenberg equations now read

$$\begin{aligned}
\frac{ds_x}{dt} &= 0 \\
\frac{ds_y}{dt} &= 2\omega s_z \\
\frac{ds_z}{dt} &= -2\omega s_y \ .
\end{aligned} \tag{16.7.63}$$

The solutions are

$$\begin{aligned}
s_x(t) &= s_x(0) \\
s_y(t) &= s_y(0) \cos 2\omega t + \frac{\dot{s}_y(0)}{2\omega} \sin 2\omega t \\
s_z(t) &= s_z(0) \cos 2\omega t + \frac{\dot{s}_z(0)}{2\omega} \sin 2\omega t
\end{aligned} \tag{16.7.64}$$

where, as above, $s_x(0)$, $s_y(0)$, and $s_z(0)$ are operators such that

$$\begin{aligned}
\langle s_x(0) \rangle &= 0 \\
\langle s_y(0) \rangle &= 0 \\
\langle s_z(0) \rangle &= \frac{\hbar}{2} \ .
\end{aligned} \tag{16.7.65}$$

Also, this time we find, by the same method as in part a), that

$$\begin{aligned}
\langle \dot{s}_y(0) \rangle &= 2\omega \langle s_z(0) \rangle = \hbar \omega \\
\langle \dot{s}_z(0) \rangle &= -2\omega \langle s_y(0) \rangle = 0 \ .
\end{aligned} \tag{16.7.66}$$

Therefore, the expectation values for the time-dependent spin operators are

$$\begin{aligned}
\langle s_x \rangle &= 0 \\
\langle s_y \rangle &= \frac{\hbar}{2} \sin 2\omega t \\
\langle s_z \rangle &= \frac{\hbar}{2} \cos 2\omega t \ .
\end{aligned} \tag{16.7.67}$$

These are, of course, the same results as obtained by using the Schrödinger equation.

16.8   Separation of Spin and Space for Spin 1/2

Show that for a spin 1/2 particle in a uniform, but time-dependent, magnetic field the wavefunction separates into the product of a spin and space function.

Solution

The Hamiltonian is

$$H = H_0 - \mu \vec{\sigma} \cdot \vec{B} = \frac{(\vec{p} - e/c\vec{A})^2}{2m} + eV - \mu \vec{\sigma} \cdot \vec{B} \ . \tag{16.8.68}$$

We attempt to find a solution of the form

$$\Psi = \psi(t, \vec{r}) \left(\begin{array}{c} a(t) \\ b(t) \end{array} \right) = \psi(t, \vec{r}) \chi(t) \ . \tag{16.8.69}$$

If we substitute this into the time-dependent Schrödinger equation (which in this instance is also called the Pauli equation) we find that

$$i\hbar \frac{\partial \psi}{\partial t} \chi + i\hbar \psi \frac{\partial \chi}{\partial t} = (H_0 \psi)\chi - \psi(\mu \vec{\sigma} \cdot \vec{B})\chi \ . \tag{16.8.70}$$

Thus, if we have that

$$i\hbar \frac{\partial \psi}{\partial t} = H_0 \psi \tag{16.8.71}$$

and

$$i\hbar \frac{\partial \chi}{\partial t} = -\mu \vec{\sigma} \cdot \vec{B} \chi \tag{16.8.72}$$

we have a solution. So, we have shown that the equation separates.

16.9   Spin 1/2 in Time-dependent Magnetic Field

A spin 1/2 particle (magnetic moment μ) is placed in a uniform, but time-dependent, magnetic field $\vec{B}(t)$ that points along the z-axis. The wavefunction at time $t = 0$ is given by

$$\psi = \left(\begin{array}{c} \cos \alpha \, e^{-i\beta} \\ \sin \alpha \, e^{i\beta} \end{array} \right) \ .$$

a) Calculate the expectation values of the x and y-components of the spin.
b) Find the direction in space such that at any time t the z-component of the spin has a well-defined value along this direction.

Solution

If we write

$$\psi(t) = \begin{pmatrix} a(t) \\ b(t) \end{pmatrix}$$

(16.9.73)

the Pauli equation becomes

$$i\hbar \frac{da}{dt} = -\mu B(t)a$$
$$i\hbar \frac{db}{dt} = +\mu B(t)b \ .$$

(16.9.74)

If we write

$$\hbar \omega(t) = \mu B(t)$$

the solutions may be written

$$a(t) = a_0 \exp\left[i \int_0^t \omega(t') \, dt'\right]$$

$$b(t) = b_0 \exp\left[-i \int_0^t \omega(t') \, dt'\right] \ .$$

(16.9.75)

After imposing the initial condition we find that

$$a_0 = \cos \alpha \, e^{-i\beta}$$
$$b_0 = \sin \alpha \, e^{i\beta} \ .$$

(16.9.76)

This shows that except for acquiring a phase, the two z-components of the spin remain the same. In other words, the probability of measuring a z-component $\hbar/2$ remains at $\cos^2 \alpha$ and the probability of measuring a z-component $-\hbar/2$ remains at $\sin^2 \alpha$.

a) The expectation values of the x and y-components are

$$\langle s_x \rangle = \frac{\hbar}{2} \left\{ a_0^* b_0 \exp\left[-2i \int_0^t \omega(t') \, dt'\right] + ab^* \exp\left[2i \int_0^t \omega(t') \, dt'\right] \right\}$$

$$= \frac{\hbar}{2} \cos \alpha \sin \alpha 2 \cos\left[-2i \int_0^t \omega(t') \, dt' + \beta\right]$$

$$= \frac{\hbar}{2} \sin(2\alpha) \cos\left[-2i \int_0^t \omega(t') \, dt' + \beta\right] \ .$$

(16.9.77)

Similarly,

$$\langle s_y \rangle = -i\frac{\hbar}{2} \left\{ a_0^* b_0 \exp\left[-2i \int_0^t \omega(t') \, dt'\right] - ab^* \exp\left[2i \int_0^t \omega(t') \, dt'\right] \right\}$$

$$= \frac{\hbar}{2} \cos \alpha \sin \alpha 2 \sin\left[2i \int_0^t \omega(t') \, dt' - \beta\right]$$

$$= \frac{\hbar}{2} \sin(2\alpha) \sin\left[2i \int_0^t \omega(t') \, dt' - \beta\right] \ .$$

(16.9.78)

b) In order that the $\hat{n}$-component of the spin have a well-defined value $\hbar/2$ we need that $\psi(t)$ should be the eigenfunction of $\hat{n} \cdot \vec{s}$ corresponding to the eigenvalue $\hbar/2$. Writing this out for

$$\hat{n} = (\sin\theta \cos\varphi, \sin\theta \sin\varphi, \cos\varphi)$$

we get

$$\cos\theta\, a(t) + \sin\theta\, e^{-i\varphi}\, b(t) = a(t) \ . \tag{16.9.79}$$

Hence,

$$\frac{b}{a} = \tan(\theta/2)\, e^{i\varphi} = \tan\alpha\, e^{2i\beta} \exp\left[-2i \int_0^t w(t')\, dt'\right] \ . \tag{16.9.80}$$

Therefore, a solution is

$$\begin{aligned} \theta &= 2\alpha \\ \varphi &= 2\left[\beta - \int_0^t w(t')\, dt'\right] \ . \end{aligned} \tag{16.9.81}$$

16.10   Spin 1/2 in Rotating Magnetic Field

A spin 1/2 particle (magnetic moment μ) is placed in a magnetic field

$$\vec{B}(t) = B(\sin\theta \cos\omega t, \sin\theta \sin\omega t, \cos\theta)$$

rotating with frequency ω about an axis making an angle θ with the z-axis. If at $t = 0$ the component of the spin parallel to $\vec{B}(t)$ has the value $\hbar/2$ find the probability that at some later time t the component of the spin parallel to $\vec{B}(t)$ has the value $-\hbar/2$.

Solution

The Schrödinger equation for the wavefunction

$$\psi(t) = \begin{pmatrix} a(t) \\ b(t) \end{pmatrix} \tag{16.10.82}$$

is

$$i\hbar \frac{\partial}{\partial t} \begin{pmatrix} a(t) \\ b(t) \end{pmatrix} = -\mu B \begin{pmatrix} \cos\theta & \sin\theta\, e^{-i\omega t} \\ \sin\theta\, e^{i\omega t} & -\cos\theta \end{pmatrix} \begin{pmatrix} a(t) \\ b(t) \end{pmatrix} \ . \tag{16.10.83}$$

If we write

$$\hbar q = \mu B \cos\theta \ , \quad \hbar p = \mu B \sin\theta$$

the Schrödinger equation becomes

$$\begin{aligned} \frac{da(t)}{dt} &= iq\, a(t) + ip\, e^{-i\omega t}\, b(t) \\ \frac{db(t)}{dt} &= ip\, e^{i\omega t}\, a(t) - iq\, b(t) \ . \end{aligned} \tag{16.10.84}$$

If we differentiate the first equation we obtain

$$\frac{d^2a}{dt^2} = iq\frac{da}{dt} + ip\,e^{i\omega t}\left[-i\omega b + \frac{db}{dt}\right] \, .$$

(16.10.85)

After substituting for $ip(db/dt)$ and ipb from (16.10.84) we get

$$\frac{d^2a}{dt^2} + i\omega\frac{da}{dt} + (p^2 + q^2 + \omega q)a = 0 \, .$$

(16.10.86)

If we now call

$$\begin{aligned}
\Omega_1 &= -\frac{\omega}{2} + \sqrt{\omega^2/4 + p^2 + q^2 + \omega q} \\
\Omega_2 &= -\frac{\omega}{2} - \sqrt{\omega^2/4 + p^2 + q^2 + \omega q}
\end{aligned}$$

(16.10.87)

then the solution for $a(t)$ may be written

$$a(t) = a_1\,e^{i\Omega_1 t} + a_2\,e^{i\Omega_2 t} \, .$$

(16.10.88)

From the first of (16.10.84) we get

$$b(t) = -\frac{i}{p}\,e^{i\omega t}\left[\frac{da}{dt} - iqa\right] \, .$$

(16.10.89)

Hence, we find that

$$b(t) = \frac{\Omega_1 - q}{p}a_1\,e^{i(\Omega_1+\omega)t} + \frac{\Omega_2 - q}{p}a_2\,e^{i(\Omega_2+\omega)t} \, .$$

(16.10.90)

The initial conditions require that

$$\begin{aligned}
\psi(0) &= \begin{pmatrix} a(0) \\ b(0) \end{pmatrix} \\
&= \begin{pmatrix} a_1 + a_2 \\ \frac{\Omega_1-q}{p}a_1 + \frac{\Omega_2-q}{p}a_2 \end{pmatrix}
\end{aligned}$$

(16.10.91)

should be the eigenfunction corresponding to the eigenvalue $\hbar/2$ of the operator

$$(\sin\theta, 0, \cos\theta)\cdot\vec{s} = \frac{\hbar}{2}\begin{pmatrix} \cos\theta & \sin\theta \\ \sin\theta & -\cos\theta \end{pmatrix} \, .$$

(16.10.92)

After a short computation we get that, up to normalization

$$\begin{aligned}
a_1 &= A\left[-\tan\theta/2 + \frac{\Omega_2 - q}{p}\right] \\
a_2 &= A\left[\tan\theta/2 - \frac{\Omega_1 - q}{p}\right] \, .
\end{aligned}$$

(16.10.93)

The normalization yields that

$$|A|^2 = \frac{p^2\cos^2\theta/2}{(\Omega_2 - \Omega_1)^2} \, .$$

(16.10.94)

We also need the state $\phi(t)$ that at time t is the eigenstate of

$$(\sin\theta\cos\omega t,\ \sin\theta\sin\omega t,\ \cos\theta)\cdot\vec{s} = \frac{\hbar}{2}\begin{pmatrix} \cos\theta & \sin\theta\, e^{-i\omega t} \\ \sin\theta\, e^{i\omega t} & -\cos\theta \end{pmatrix} \quad (16.10.95)$$

corresponding to the eigenvalue $-\hbar/2$. Again, a short computation yields

$$\phi(t) = \begin{pmatrix} \sin\theta/2 \\ -\cos\theta/2\, e^{i\omega t} \end{pmatrix} . \quad (16.10.96)$$

Then, the probability that at time t the particle will be found to have the spin value $-\hbar/2$ aligned along the magnetic field is given by

$$P(t) = |(\phi(t), \psi(t))|^2 .$$

Thus,

$$
\begin{aligned}
P ={}& \frac{p^2\cos^2\theta/2}{(\Omega_2-\Omega_1)^2}\left|\sin\theta/2\left[-\tan\theta/2+\frac{\Omega_2-q}{p}\right]e^{i\Omega_1 t}\right.\\
&+\ \sin\theta/2\left[\tan\theta/2-\frac{\Omega_1-q}{p}\right]e^{i\Omega_2 t}\\
&-\ \cos\theta/2\frac{\Omega_1-q}{p}\left[-\tan\theta/2+\frac{\Omega_2-q}{p}\right]e^{i\Omega_1 t}\\
&\left.-\ \cos\theta/2\frac{\Omega_2-q}{p}\left[\tan\theta/2-\frac{\Omega_1-q}{p}\right]e^{i\Omega_2 t}\right|^2\\
={}& \frac{p^2}{(\Omega_2-\Omega_1)^2}\left(\sin\theta/2-\frac{\Omega_1-q}{p}\cos\theta/2\right)^2\\
&\times\ \left(\sin\theta/2-\frac{\Omega_2-q}{p}\cos\theta/2\right)^2[1-\cos(\Omega_1-\Omega_2)t] . \quad (16.10.97)
\end{aligned}
$$

Next, using the definitions of Ω_1 and Ω_2 and writing

$$\cos(\Omega_1-\Omega_2)t = 1 - 2\cos^2(\Omega_1-\Omega_2)t/2$$

we can simplify the result to read

$$
P = \frac{2p^2}{\omega^2+4(p^2+q^2+\omega q)}\left(\sin\theta/2-\frac{\Omega_1-q}{p}\cos\theta/2\right)^2\times
$$
$$
\left(\sin\theta/2-\frac{\Omega_2-q}{p}\cos\theta/2\right)^2\cos^2\left(\sqrt{\omega^2/4+p^2+q^2+\omega q}\ t\right) .
$$
$$(16.10.98)$$

Here,

$$q = \omega_L\cos\theta\ ,\quad p = \omega_L\sin\theta \quad (16.10.99)$$

and

$$\omega_L = \frac{\mu B}{\hbar} \quad (16.10.100)$$

is the Larmor frequency.

Bibliography

[16.1] D.H. Kobe, Am. J. Phys. **54**, 77, (1986).

Chapter 17

Angular Momentum, Etc.

17.1 Operator to Lower Total J

Show that the operator

$$T = A[j_2 J_{1-} - j_1 J_{2-}]$$

where A is a normalization constant, has the property that

$$T|j, j, j_1, j_2\rangle = |j-1, j-1, j_1, j_2\rangle$$

for $j = j_1 + j_2$.

Solution

We consider the "unnormalized" operator

$$T = (j_2 J_1 - j_1 J_2)/\hbar \, . \tag{17.1.1}$$

Then, we find that

$$
\begin{aligned}
T|j_1 + j_2, j_1 + j_2, j_1, j_2\rangle &= T|j_1, j_1\rangle|j_2, j_2\rangle \\
&= j_2\sqrt{2j_1}|j_1, j_1 - 1\rangle - j_1\sqrt{2j_2}|j_2, j_2 - 1\rangle \, .
\end{aligned}
\tag{17.1.2}
$$

Now we have that

$$J^2 = J_1^2 + J_2^2 + 2J_{1z}J_{2z} + J_{1+}J_{2-} + J_{1-}J_{2+} \, . \tag{17.1.3}$$

Also we see that

$$
\begin{aligned}
&(J_1^2 + J_2^2 + 2J_{1z}J_{2z})T|j_1 + j_2, j_1 + j_2, j_1, j_2\rangle \\
&= [j_1(j_1 + 1) + j_2(j_2 + 1)]\hbar^2 T|j_1 + j_2, j_1 + j_2, j_1, j_2\rangle \\
&+ j_2\sqrt{2j_1}2(j_1 - 1)j_2\hbar^2|j_1, j_1 - 1\rangle|j_2, j_2\rangle \\
&- j_1\sqrt{2j_2}2j_1(j_2 - 1)\hbar^2|j_1, j_1\rangle|j_2, j_2 - 1\rangle \, .
\end{aligned}
\tag{17.1.4}
$$

Furthermore,

$$(J_{1+}J_{2-} + J_{1-}J_{2+})T|j_1 + j_2, j_1 + j_2, j_1, j_2\rangle$$
$$= 2j_1j_2\sqrt{2j_2}\hbar^2|j_1, j_1\rangle|j_2, j_2 - 1\rangle$$
$$-j_1j_2\sqrt{2j_1}\hbar^2|j_1, j_1 - 1\rangle|j_2, j_2\rangle . \tag{17.1.5}$$

So adding all these results we find that

$$J^2T|j_1 + j_2, j_1 + j_2, j_1, j_2\rangle$$
$$= [j_1(j_1 + 1) + j_2(j_2 + 1) + 2j_1j_2$$
$$- 2(j_1 + j_2)]\hbar^2 T|j_1 + j_2, j_1 + j_2, j_1, j_2\rangle$$
$$= (j_1 + j_2 - 1)(j_1 + j_2)\hbar^2 T|j_1 + j_2, j_1 + j_2, j_1, j_2\rangle . \tag{17.1.6}$$

This shows that $T|j_1 + j_2, j_1 + j_2, j_1, j_2\rangle$ is proportional to a linear combination of $|j_1 + j_2 - 1, m, j_1, j_2\rangle$ for various m. The last step is to show that m is restricted to only one value, namely $j_1 + j_2 - 1$. But this follows immediately from applying $J_{1z} + J_{2z}$ to

$$T|j_1 + j_2, j_1 + j_2, j_1, j_2\rangle .$$

17.2 Energy Shift Due to a Magnetic Field

Evaluate the expectation value $\langle S_z \rangle$ for the states $|l\pm1/2, m, l\rangle$. These expectation values are used to obtain the shift in energy due to a uniform magnetostatic field.

Solution

The state $|l + 1/2, m, l\rangle$ is obtained by applying

$$\sqrt{\frac{(l + 1/2 + m)!}{(2l + 1)!(l + 1/2 - m)!}}\hbar^{-(l+1/2-m)} J_-^{(l+1/2-m)}$$

to the state $|l, l\rangle|1/2, 1/2\rangle$. Here

$$J_- = L_- + S_- . \tag{17.2.7}$$

So we find

$$\langle S_z \rangle = \frac{(l + 1/2 + m)!}{(2l + 1)!(l + 1/2 - m)!}\hbar^{-(2l+1-2m)} \times$$
$$\langle l, l|\langle\frac{1}{2}, \frac{1}{2}|J_+^{l-(m-1/2)} S_z J_-^{l-(m-1/2)}|l, l\rangle|\frac{1}{2}, \frac{1}{2}\rangle$$
$$= \frac{(l + 1/2 + m)!}{(2l + 1)!(l + 1/2 - m)!}\hbar^{-(2l+1-2m)} \times$$
$$\langle l, l|\langle\frac{1}{2}, \frac{1}{2}|(L_+ + S_+)^{l-(m-1/2)} S_z (L_- + S_-)^{l-(m-1/2)}|l\rangle|\frac{1}{2}, \frac{1}{2}\rangle$$

$$= \frac{(l+1/2+m)!}{(2l+1)!(l+1/2-m)!}\hbar^{-(2l+1-2m)} \times$$

$$\left[\frac{\hbar}{2} \times \langle l,l|L_+^{l-(m-1/2)} L_-^{l-(m-1/2)}|l,l\rangle \right.$$

$$\left. -(l-m+1/2)\frac{\hbar^3}{2}\langle l,l|L_+^{l-(m+1/2)} L_-^{l-(m+1/2)}|l,l\rangle \right] \qquad (17.2.8)$$

where we have used the fact that more than one application of S_- to the state $|1/2, 1/2\rangle$ annihilates it. Thus, what we have to evaluate are quantities like

$$\langle l,l|L_+^{l-(m-1/2)} L_-^{l-(m-1/2)}|l,l\rangle \quad \text{and} \quad \langle l,l|L_+^{l-(m+1/2)} L_-^{l-(m+1/2)}|l,l\rangle .$$

Now, using that

$$|l,m\rangle = \sqrt{\frac{(l+m)!}{(2l)!(l-m)!}}\hbar^{-(l-m)} L_-^{(l-m)}|l,l\rangle \qquad (17.2.9)$$

we imediately get that

$$\langle l,l|L_+^{l-(m-1/2)} L_-^{l-(m-1/2)}|l,l\rangle = \frac{(2l)!(l-m+1/2)!}{(l+m-1/2)!}\hbar^{2l-2m+1} \qquad (17.2.10)$$

and

$$\langle l,l|L_+^{l-(m+1/2)} L_-^{l-(m+1/2)}|l,l\rangle = \frac{(2l)!(l-m-1/2)!}{(l+m+1/2)!}\hbar^{2l-2m-1} . \qquad (17.2.11)$$

Hence we get that

$$\langle S_z \rangle = \hbar \frac{(l+1/2)^2 - m^2 + 1}{(2l+1)(2l+1-m)} . \qquad (17.2.12)$$

For the states $|l-1/2, m, l\rangle$ we proceed in exactly the same manner starting with the equation

$$|l-1/2, l-1/2\rangle$$

$$= \frac{1}{\sqrt{2l+1}}|l,l-1\rangle|1/2,1/2\rangle - \sqrt{\frac{2l}{2l+1}}|l,l\rangle|1/2,-1/2\rangle . \qquad (17.2.13)$$

Then,

$$\langle S_z \rangle = \frac{(l+m+1/2)!}{(2l+1)!(l-m+1/2)!}\left[\sqrt{\frac{1}{2l+1}}\langle l,l-1|\langle 1/2,1/2| \right.$$

$$\left. - \sqrt{\frac{2l}{2l+1}}\langle l,l|\langle 1/2,-1/2| \right] J_+^{l-m+1/2}$$

$$\times S_z J_-^{l-m+1/2}\left[\sqrt{\frac{1}{2l+1}}|l,l-1\rangle|1/2,1/2\rangle \right.$$

$$\left. - \sqrt{\frac{2l}{2l+1}}|l,l\rangle|1/2,-1/2\rangle \right] . \qquad (17.2.14)$$

This simplifies to

$$\langle S_z \rangle = -\hbar \frac{(l + m + 1/2)}{4l(2l + 1)^2} [3l(l + 1) + m(m - 2) + 7/4] . \qquad (17.2.15)$$

17.3 Coupling of Spin 1 to Spin 1/2

A particle of angular momentum 1/2 is coupled to a particle of angular momentum 1. List the states that are eigenstates of

$$J^2 = (\vec{J}_1 + \vec{J}_2)^2$$

and

$$J_z = J_{1z} + J_{2z}$$

and express them in terms of the eigenstates of

$$(J_1^2, J_{1z}) \quad \text{and} \quad (J_2^2, J_{2z}) .$$

Solution

The possible eigenstates of J^2 and J_z are

$$|3/2, 3/2\rangle \quad , \quad |3/2, 1/2\rangle \quad , \quad |3/2, -1/2\rangle \quad , \quad |3/2, -3/2\rangle$$

and

$$|1/2, 1/2\rangle \quad , \quad |1/2, -1/2\rangle .$$

These are obtained as follows. We begin with

$$|3/2, 3/2\rangle = |1, 1\rangle|1/2, 1/2\rangle . \qquad (17.3.16)$$

Now apply J_- to both sides to get

$$\sqrt{3}|3/2, 1/2\rangle = \sqrt{2}|1, 0\rangle|1/2, 1/2\rangle + |1, 1\rangle|1/2, -1/2\rangle \qquad (17.3.17)$$

so that

$$|3/2, 1/2\rangle = \frac{\sqrt{2}}{\sqrt{3}}|1, 0\rangle|1/2, 1/2\rangle + \frac{1}{\sqrt{3}}|1, 1\rangle|1/2, -1/2\rangle . \qquad (17.3.18)$$

Now apply J_- to both sides again to get

$$2|3/2, -1/2\rangle = \frac{2}{\sqrt{3}}|1, -1\rangle|1/2, 1/2\rangle + \frac{\sqrt{2}}{\sqrt{3}}|1, 0\rangle|1/2, -1/2\rangle$$

$$+ \frac{\sqrt{2}}{\sqrt{3}}|1, 0\rangle|1/2, -1/2\rangle \qquad (17.3.19)$$

so that

$$|3/2, -1/2\rangle = \frac{1}{\sqrt{3}}|1, -1\rangle|1/2, 1/2\rangle + \frac{\sqrt{2}}{\sqrt{3}}|1, 0\rangle|1/2, -1/2\rangle . \qquad (17.3.20)$$

Finally,

$$|3/2, 3/2\rangle = |1, -1\rangle|1/2, -1/2\rangle . \tag{17.3.21}$$

The remaining two states must be linear combinations of

$$|1, 0\rangle|1/2, 1/2\rangle \quad , \quad |1, 1\rangle|1/2, -1/2\rangle$$

and

$$|1, -1\rangle|1/2, 1/2\rangle \quad , \quad |1, 0\rangle|1/2, -1/2\rangle$$

and must be orthogonal to all the states above. Thus we find (up to an arbitrary phase factor)

$$|1/2, 1/2\rangle \;=\; \frac{1}{\sqrt{3}}|1, 0\rangle|1/2, 1/2\rangle - \frac{\sqrt{2}}{\sqrt{3}}|1, 1\rangle|1/2, -1/2\rangle \tag{17.3.22}$$

$$|1/2, -1/2\rangle \;=\; -\frac{\sqrt{2}}{\sqrt{3}}|1, -1\rangle|1/2, 1/2\rangle + \frac{1}{\sqrt{3}}|1, 0\rangle|1/2, -1/2\rangle . \tag{17.3.23}$$

Another way to proceed for these two states would be to use the operator T of problem 17.1. Then we find that, up to normalization,

$$
\begin{aligned}
|1/2, 1/2\rangle &= A[S_- - 1/2\, L_-]|1, 1\rangle|1/2, 1/2\rangle \\
&= A[|1, 1\rangle|1/2, -1/2\rangle - \frac{1}{2}\sqrt{2}|1, 0\rangle|1/2, 1/2\rangle] . \tag{17.3.24}
\end{aligned}
$$

Normalization now yields, up to an arbitrary phase factor,

$$A = -\sqrt{2/3} . \tag{17.3.25}$$

Thus,

$$|1/2, 1/2\rangle = \frac{1}{\sqrt{3}}|1, 0\rangle|1/2, 1/2\rangle - \frac{\sqrt{2}}{\sqrt{3}}|1, 1\rangle|1/2, -1/2\rangle \tag{17.3.26}$$

which agrees with our previous result. The state $|1/2, -1/2\rangle$ is obtained by applying J_- to this result.

17.4   Example of Wigner-Eckart

Consider a set of three operators T_m $m = 1, 0, -1$ such that

$$T_m^\dagger = T_m$$

$$[J_\pm, T_m] = \sqrt{2 - m(m \pm 1)}\hbar T_m$$

$$[J_z, T_m] = m\hbar T_m$$

where $\vec{J}$ are the total angular momentum operators. Evaluate the total m', m'' dependence of the matrix elements $\langle j, m'|T_m|j, m''\rangle$.

Hint: Express T_m in terms of 3×3 matrices. This is an example of the Wigner-Eckart Theorem.

Solution

We begin by taking

$$
T_0 = \begin{pmatrix} a & d & e \\ f & b & g \\ h & k & c \end{pmatrix} . \tag{17.4.27}
$$

Then using

$$
[J_z, T_0] = 0 \tag{17.4.28}
$$

we find

$$
T_0 = \begin{pmatrix} a & 0 & 0 \\ 0 & b & 0 \\ 0 & 0 & c \end{pmatrix} . \tag{17.4.29}
$$

Next we use

$$
[J_+, T_0] = \sqrt{2}\hbar T_1 \tag{17.4.30}
$$

and find

$$
T_1 = \begin{pmatrix} 0 & b-a & 0 \\ 0 & b & c-b \\ 0 & 0 & 0 \end{pmatrix} . \tag{17.4.31}
$$

Similarly we find using

$$
[J_-, T_0] = \sqrt{2}\hbar T_{-1} \tag{17.4.32}
$$

that

$$
T_{-1} = \begin{pmatrix} 0 & 0 & 0 \\ a-b & b & 0 \\ 0 & b-c & 0 \end{pmatrix} . \tag{17.4.33}
$$

The condition

$$
T_m^\dagger = (-1)^m T_{-m} \tag{17.4.34}
$$

simply shows that the three numbers a, b, c are all real. If we now use

$$
[J_-, T_{-1}] = [J_+, T_1] = 0 \tag{17.4.35}
$$

we find the condition that

$$
a - 2b + c = 0 . \tag{17.4.36}
$$

Thus, we get

$$
b = \frac{a+c}{2} . \tag{17.4.37}
$$

Therefore,

$$T_1 = \frac{c-a}{2\sqrt{2}}\frac{1}{\hbar}J_+$$

$$T_0 = -\frac{c-a}{2}\frac{1}{\hbar}J_z - \frac{c+a}{2}\mathbf{1}$$

$$T_{-1} = -\frac{c-a}{2\sqrt{2}}\frac{1}{\hbar}J_- \ . \tag{17.4.38}$$

17.5 Rotations for Spin 1/2 and Spin 1

Consider the unitary operator

$$R_n(\varphi) = e^{i(\vec{J}\cdot\hat{n})\varphi/\hbar}$$

where $\vec{J}$ is the angular momentum operator.
a) If $j = 1/2$ expand $R_n(\varphi)$ in a Taylor series to obtain a simpler expression and apply it to the states

$$\begin{pmatrix} 1 \\ 0 \end{pmatrix} \text{ and } \begin{pmatrix} 0 \\ 1 \end{pmatrix}.$$

b) If $j = 1$ repeat part a) but consider the states

$$\begin{pmatrix} 1 \\ 0 \\ 0 \end{pmatrix}, \begin{pmatrix} 0 \\ 1 \\ 0 \end{pmatrix} \text{ and } \begin{pmatrix} 0 \\ 0 \\ 1 \end{pmatrix}.$$

Solution

a) We start with

$$\hat{n} = (\sin\alpha\cos\beta, \sin\alpha\sin\beta, \cos\alpha) \ . \tag{17.5.39}$$

Then for $j = 1/2$

$$\frac{\vec{J}}{\hbar} = \frac{1}{2}(\sigma_x, \sigma_y, \sigma_z) \tag{17.5.40}$$

and

$$\frac{\hat{n}\cdot\vec{J}}{\hbar} = \frac{1}{2}\begin{pmatrix} \cos\alpha & \sin\alpha\, e^{-i\beta} \\ \sin\alpha\, e^{i\beta} & \cos\alpha \end{pmatrix} \tag{17.5.41}$$

$$\left(\frac{\hat{n}\cdot\vec{J}}{\hbar}\right)^2 = \frac{1}{4}\begin{pmatrix} 1 & 0 \\ 0 & 1 \end{pmatrix} = \frac{1}{2^2}\mathbf{1} \ . \tag{17.5.42}$$

Thus,

$$\left(\frac{\hat{n}\cdot\vec{J}}{\hbar}\right)^{2n} = \frac{1}{2^{2n}}\mathbf{1} \tag{17.5.43}$$

$$\left(\frac{\hat{n} \cdot \vec{J}}{\hbar}\right)^{2n+1} = \frac{1}{2^{2n+1}} \frac{\hat{n} \cdot \vec{J}}{\hbar} . \tag{17.5.44}$$

Hence we find by expanding the exponential that

$$R_n(\varphi) = \cos(\varphi/2)\mathbf{1} + i\sin(\varphi/2)\frac{\hat{n} \cdot \vec{J}}{\hbar} . \tag{17.5.45}$$

Applying this to the states indicated we find

$$R_n(\varphi) \begin{pmatrix} 1 \\ 0 \end{pmatrix} = \begin{pmatrix} \cos(\varphi/2) + i\sin(\varphi/2)\cos\alpha \\ i\sin(\varphi/2)\sin\alpha\, e^{i\beta} \end{pmatrix} \tag{17.5.46}$$

$$R_n(\varphi) \begin{pmatrix} 0 \\ 1 \end{pmatrix} = \begin{pmatrix} i\sin(\varphi/2)\sin\alpha\, e^{-i\beta} \\ \cos(\varphi/2) - i\sin(\varphi/2)\cos\alpha \end{pmatrix} . \tag{17.5.47}$$

b) Similarly we find, by explicitly writing everything out, that for $j = 1$ with

$$\frac{J_x}{\hbar} = \frac{1}{\sqrt{2}} \begin{pmatrix} 0 & 1 & 0 \\ 1 & 0 & 1 \\ 0 & 1 & 0 \end{pmatrix}$$

$$\frac{J_y}{\hbar} = \frac{1}{\sqrt{2}} \begin{pmatrix} 0 & -i & 0 \\ i & 0 & -i \\ 0 & i & 0 \end{pmatrix}$$

$$\frac{J_z}{\hbar} = \begin{pmatrix} 1 & 0 & 0 \\ 0 & 0 & 0 \\ 0 & 0 & -1 \end{pmatrix} \tag{17.5.48}$$

we have

$$\frac{\hat{n} \cdot \vec{J}}{\hbar} = \begin{pmatrix} \cos\alpha & \frac{1}{\sqrt{2}}\sin\alpha\, e^{-i\beta} & 0 \\ \frac{1}{\sqrt{2}}\sin\alpha\, e^{i\beta} & 0 & \cos\alpha \\ 0 & \frac{1}{\sqrt{2}}\sin\alpha\, e^{i\beta} & -\cos\alpha \end{pmatrix} \tag{17.5.49}$$

and by multiplication that

$$\left(\frac{\hat{n} \cdot \vec{J}}{\hbar}\right)^3 = \frac{\hat{n} \cdot \vec{J}}{\hbar} . \tag{17.5.50}$$

Thus,

$$\left(\frac{\hat{n} \cdot \vec{J}}{\hbar}\right)^{2n+1} = \frac{\hat{n} \cdot \vec{J}}{\hbar} , \qquad \left(\frac{\hat{n} \cdot \vec{J}}{\hbar}\right)^{2n} = \left(\frac{\hat{n} \cdot \vec{J}}{\hbar}\right)^2 \tag{17.5.51}$$

where

$$\left(\frac{\hat{n} \cdot \vec{J}}{\hbar}\right)^2$$

$$= \begin{pmatrix} \frac{1}{2}(1 + \cos\alpha)^2 & \frac{1}{\sqrt{2}}\sin\alpha\cos\alpha\, e^{-i\beta} & \frac{1}{2}\sin^2\alpha\, e^{-2i\beta} \\ \frac{1}{\sqrt{2}}\sin\alpha\cos\alpha\, e^{i\beta} & \sin^2\alpha & \frac{-1}{\sqrt{2}}\sin\alpha\cos\alpha\, e^{-i\beta} \\ \frac{1}{2}\sin^2\alpha\, e^{2i\beta} & \frac{-1}{\sqrt{2}}\sin\alpha\cos\alpha\, e^{i\beta} & \frac{1}{2}(1 + \cos\alpha)^2 \end{pmatrix} . \tag{17.5.52}$$

Expanding the exponential in a power series we get

$$R_n(\phi) = 1 - \left(\frac{\hat{n}\cdot\vec{J}}{\hbar}\right)^2 + \frac{\hat{n}\cdot\vec{J}}{\hbar}\sum_{n=0}^{\infty}\frac{(i\phi)^{2n+1}}{(2n+1)!} + \left(\frac{\hat{n}\cdot\vec{J}}{\hbar}\right)^2\sum_{n=0}^{\infty}\frac{(i\phi)^{2n}}{(2n)!}$$

$$= 1 - \left(\frac{\hat{n}\cdot\vec{J}}{\hbar}\right)^2 + i\frac{\hat{n}\cdot\vec{J}}{\hbar}\sin\phi + \left(\frac{\hat{n}\cdot\vec{J}}{\hbar}\right)^2\cos\phi . \qquad (17.5.53)$$

Applying this to the three states we get

$$R_n(\phi)\begin{pmatrix}1\\0\\0\end{pmatrix} = \begin{pmatrix}\frac{1}{2}\sin^2\alpha + i\cos\alpha\sin\phi + \frac{1}{2}(1+\cos\alpha)\cos\phi\\ \frac{-1}{\sqrt{2}}\sin\alpha\,\cos\alpha\,e^{i\beta}(1-\cos\phi) + \frac{i}{\sqrt{2}}\sin\alpha\sin\phi\,e^{i\beta}\\ -\frac{1}{2}\sin^2\alpha\,e^{2i\beta}(1-\cos\phi)\end{pmatrix}$$

$$R_n(\phi)\begin{pmatrix}0\\1\\0\end{pmatrix} = \begin{pmatrix}\frac{-1}{\sqrt{2}}\sin\alpha\,\cos\alpha\,e^{-i\beta}(1-\cos\phi) + \frac{i}{\sqrt{2}}\sin\alpha\sin\phi\,e^{-i\beta}\\ \cos^2\alpha + \sin^2\alpha\cos\phi\\ \frac{1}{\sqrt{2}}\sin\alpha\,\cos\alpha\,e^{i\beta}(1-\cos\phi) + \frac{i}{\sqrt{2}}\sin\alpha\sin\phi\,e^{i\beta}\end{pmatrix}$$

$$R_n(\phi)\begin{pmatrix}0\\0\\1\end{pmatrix} = \begin{pmatrix}-\frac{1}{2}\sin^2\alpha\,e^{-2i\beta}(1-\cos\phi)\\ \frac{1}{\sqrt{2}}\sin\alpha\,\cos\alpha\,e^{-i\beta}(1-\cos\phi) + \frac{i}{\sqrt{2}}\sin\alpha\sin\phi\,e^{-i\beta}\\ \frac{1}{2}\sin^2\alpha - i\cos\alpha\sin\phi + \frac{1}{2}(1+\cos\alpha)\cos\phi\end{pmatrix} .$$

$$(17.5.54)$$

17.6 Spin 1/2 Coupled to Spin 3/2

A particle of total angular momentum $j_1 = 1/2$ is coupled to another particle with total angular momentum $j_2 = 3/2$. What are the states of possible total j? Express all the states with the lowest possible j in terms of the states $|j_1, m_1\rangle$, $|j_2, m_2\rangle$.

Solution

The states of total possible angular momentum j are given by

$$j = 3/2 + 1/2 = 2 \quad , \quad 3/2 - 1/2 = 1 .$$

Thus, we only have two possibilities $j = 2$ and $j = 1$. The possible states of total $j = 1$ are

$$|1,1\rangle \quad , \quad |1,0\rangle \quad , \quad |1,-1\rangle .$$

To get the state $|1,1\rangle$ we use the result of problem 17.1 that, up to a normalization factor A we have

$$|1,1\rangle = AT|2,2\rangle = \frac{A}{\hbar}\left(\frac{1}{2}J_{3/2-} - \frac{3}{2}J_{1/2-}\right)|3/2,3/2\rangle|1/2,1/2\rangle$$

$$= \frac{A}{2}\left[\sqrt{\frac{3}{2}\frac{3}{2}(\frac{3}{2}+1)-\frac{3}{2}\frac{3}{2}(\frac{3}{2}-1)}|3/2,1/2\rangle|1/2,1/2\rangle\right.$$

$$\left. - \frac{3A}{2}\sqrt{\frac{1}{2}\frac{1}{2}(\frac{1}{2}+1)-\frac{1}{2}\frac{1}{2}(\frac{1}{2}-1)}|3/2,3/2\rangle|1/2,-1/2\rangle\right]$$

$$= \frac{A}{2}\left[\sqrt{3}|3/2,1/2\rangle|1/2,1/2\rangle - 3|3/2,3/2\rangle|1/2,-1/2\rangle\right] . (17.6.55)$$

Normalizing we find

$$|1,1\rangle = \frac{1}{2}|3/2,1/2\rangle|1/2,1/2\rangle - \frac{\sqrt{3}}{2}|3/2,3/2\rangle|1/2,-1/2\rangle . \qquad (17.6.56)$$

Then, again up to a normalization factor A,

$$|1,0\rangle = AJ_-|1,1\rangle$$
$$= A[|3/2,-1/2\rangle|1/2,1/2\rangle - |3/2,1/2\rangle|1/2,-1/2\rangle] \qquad (17.6.57)$$

so that

$$|1,0\rangle = \frac{1}{\sqrt{2}}[|3/2,-1/2\rangle|1/2,1/2\rangle - |3/2,1/2\rangle|1/2,-1/2\rangle] . \qquad (17.6.58)$$

Finally,

$$|1,-1\rangle = AJ_-|1,0\rangle$$
$$= \frac{A}{\sqrt{2}}[\sqrt{3}|3/2,-3/2\rangle|1/2,1/2\rangle - |3/2,-1/2\rangle|1/2,-1/2\rangle] .$$
$$(17.6.59)$$

So

$$|1,-1\rangle = \frac{\sqrt{3}}{2}|3/2,-3/2\rangle|1/2,1/2\rangle - \frac{1}{2}|3/2,-1/2\rangle|1/2,-1/2\rangle . \quad (17.6.60)$$

17.7   Coupling of Spin 1 or 0 and Spin 1/2

An electron (spin $= 1/2$) is in a state of either $l = 0$ or $l = 1$. Express all states of total angular momentum $|j,m_j\rangle$ in terms of the states $|l,m_l\rangle|1/2,m_s\rangle$ where $l = 0$ or $l = 1$.

Solution

Starting with the state of highest weight we have

$$|3/2,3/2\rangle = |1,1\rangle|1/2,1/2\rangle . \qquad (17.7.61)$$

Applying

$$J_- = L_- + S_- \qquad (17.7.62)$$

to this equation we get

$$\sqrt{3}|3/2, 1/2\rangle = \sqrt{2}|1, 0\rangle|1/2, 1/2\rangle + |1, 1\rangle|1/2, -1/2\rangle \ . \tag{17.7.63}$$

Thus,

$$|3/2, 1/2\rangle = \sqrt{2/3}|1, 0\rangle|1/2, 1/2\rangle + 1/\sqrt{3}|1, 1\rangle|1/2, -1/2\rangle \ . \tag{17.7.64}$$

Applying J_- once more to this equation we find

$$|3/2, -1/2\rangle = 1/\sqrt{3}|1, -1\rangle|1/2, 1/2\rangle + \sqrt{2/3}|1, 0\rangle|1/2, -1/2\rangle \ . \tag{17.7.65}$$

The state of lowest weight is simply

$$|3/2, -3/2\rangle = |1, -1\rangle|1/2, -1/2\rangle \ . \tag{17.7.66}$$

If $l = 0$ we only have two states of total angular momentum 1/2 and these are

$$\begin{aligned} |1/2, 1/2\rangle &= |0, 0\rangle|1/2, 1/2\rangle \\ |1/2, -1/2\rangle &= |0, 0\rangle|1/2, -1/2\rangle \ . \end{aligned} \tag{17.7.67}$$

17.8   Identity for Constant Magnetic Field

Show that if

$$\vec{A} = -\frac{1}{2}\vec{r} \times \vec{B}$$

where $\vec{B}$ is a constant vector, then

$$\vec{p} \cdot \vec{A} + \vec{A} \cdot \vec{p} = \vec{B} \cdot \vec{L} \ .$$

Solution

We are given that

$$\vec{A} = -\frac{1}{2}\vec{r} \times \vec{B} \tag{17.8.68}$$

where $\vec{B}$ is a constant vector. Then, using the Lorentz condition

$$\nabla \cdot \vec{A} = 0 \tag{17.8.69}$$

we see that

$$\vec{p} \cdot \vec{A} = \vec{A} \cdot \vec{p} \ . \tag{17.8.70}$$

Now, using this we can write

$$\begin{aligned} \vec{p} \cdot \vec{A} + \vec{A} \cdot \vec{p} &= -\frac{1}{2}\vec{p} \cdot (\vec{r} \times \vec{B}) - \frac{1}{2}(\vec{r} \times \vec{B}) \cdot \vec{p} \\ &= -(\vec{r} \times \vec{B}) \cdot \vec{p} \\ &= (\vec{B} \times \vec{r}) \cdot \vec{p} \ . \end{aligned} \tag{17.8.71}$$

Next, using the fact that the dot and cross product may be interchanged as long as the order of the vectors is maintained we get

$$\vec{p} \cdot \vec{A} + \vec{A} \cdot \vec{p} = \vec{B} \cdot (\vec{r} \times \vec{p}) = \vec{B} \cdot \vec{L} \ . \tag{17.8.72}$$

17.9 The State $|n, j, m, l\rangle$

Use induction to show that

$$|n, j, m, l\rangle = \sqrt{\frac{(j+m)!}{(2j)!(j-m)!}}\hbar^{m-j} J_-^{j-m}|n, j, j, l\rangle$$

Solution

We want to use induction to show that

$$|n, j, m, l\rangle = \sqrt{\frac{(j+m)!}{(2j)!(j-m)!}}\hbar^{m-j} J_-^{j-m}|n, j, j, l\rangle \qquad (17.9.73)$$

We start with $m = j$. In this case the (17.9.73) is clearly correct. For $m = j-1$ we have

$$J_-|n, j, j, l\rangle = \sqrt{j(j+1) - j(j-1)}\hbar|n, j, j-1, l\rangle . \qquad (17.9.74)$$

So solving for $|n, j, j-1, l\rangle$ we again see that (17.9.73) holds. Now assume that (17.9.73) holds for $m = j - M$. Then, using

$$J_-|n, j, j - M, l\rangle$$
$$= \sqrt{j(j+1) - (j-M)(j-M-1)}\hbar|n, j, j-M-1, l\rangle \qquad (17.9.75)$$

so that

$$|n, j, j - M - 1, l\rangle$$
$$= \frac{1}{\sqrt{j(j+1) - (j-M)(j-M-1)}}\hbar^{-1}J_-|n, j, j-M, l\rangle$$
$$= \sqrt{\frac{(2j-M)!}{(2j)!M!}}\frac{1}{\sqrt{(2j-M)(M+1)}}\hbar^{-(M+1)}J_-^{M+1}|n, j, j, l\rangle$$
$$= \sqrt{\frac{(2j-M-1)!}{(2j)!(M+1)!}}\hbar^{-(M+1)}J_-^{M+1}|n, j, j, l\rangle \qquad (17.9.76)$$

as required.

17.10 Landé g-factor

Show that if $\vec{A}$ is a vector operator such that

$$[J_x, A_x] = 0 \quad , \quad [J_x, A_y] = i\hbar A_z \quad , \quad [J_x, A_z] = -i\hbar A_y$$

and cyclic permutations. Then,
a)

$$[J^2, [J^2, \vec{A}]] = 2\hbar^2(J^2\vec{A} + \vec{A}J^2) - 4\hbar^2(\vec{A} \cdot \vec{J})\vec{J} .$$

b) Use this result to show that

$$\langle JM|S_z|JM\rangle = M\hbar \frac{\langle JM|\vec{S}\cdot\vec{J}|JM\rangle}{J(J+1)\hbar^2}$$

and

c) hence evaluate the matrix element for the weak-field Zeeman effect

$$\Delta E = \frac{\mu_B B}{\hbar}\{M\hbar + \langle nLSJM|S_z|nLSJM\rangle\} = \mu_B BMg$$

where

$$g = 1 + \frac{J(J+1)+S(S+1)-L(L+1)}{2J(J+1)}$$

is the *Landé g-factor*.

Solution

The first part is simply a tedious computation. We therefore first compute $[J^2,[J^2,A_x]]$. Then, by symmetry, the result also holds for A_y and A_z. As a first step we compute

$$
\begin{aligned}
[J^2, A_x] &= \vec{J}\cdot[\vec{J}, A_x] + [A_x, \vec{J}]\cdot\vec{J} \\
&= i\hbar[-J_y A_z + J_z A_y - A_z J_y + A_y J_z] .
\end{aligned}
\tag{17.10.77}
$$

Similarly we find

$$
\begin{aligned}
[J^2, A_y] &= i\hbar[J_x A_z - J_z A_x + A_z J_x - A_x J_z] \\
[J^2, A_z] &= i\hbar[-J_x A_y + J_y A_x - A_y J_x + A_x J_y] .
\end{aligned}
\tag{17.10.78}
$$

Then,

$$
\begin{aligned}
[J^2,[J^2, A_x]] &= i\hbar[-J_y[J^2, A_z] + J_z[J^2, A_y] \\
&\quad - [J^2, A_z]J_y + [J^2, A_y]J_z] \\
&= -\hbar^2[J_y J_x A_y - J_y^2 A_x + J_y A_y J_x - J_y A_x J_y \\
&\quad + J_y A_y J_y - J_y A_x J_y + A_y J_x J_y - A_x J_y^2 \\
&\quad + J_z J_x A_z - J_z^2 A_x + J_z A_z J_x - J_z A_x J_z \\
&\quad + J_x A_z J_z - J_z A_x J_z + A_x J_x J_z - A_x J_z^2] \\
&= \hbar^2[(J_y^2 + J_z^2)A_x + A_x(J_y^2 + J_z^2)] \\
&\quad - \hbar^2[J_y(2A_y J_x + i\hbar A_z) - (A_x J_y - i\hbar A_z)J_y \\
&\quad - J_y(J_y + i\hbar A_z) + (A_y J_x + i\hbar A_z)J_y + A_y(J_y J_x + i\hbar J_z) \\
&\quad + J_z(2A_z J_x - i\hbar A_y) - (A_x J)_z + i\hbar A_y)J_z - J_z(J_z A_x - i\hbar A_y) \\
&\quad + (A_z J_x - i\hbar A_y)J_z + A_z(J_z J_x - i\hbar J_y)] .
\end{aligned}
\tag{17.10.79}
$$

Here we have used

$$[J_x, A_x] = 0 .\tag{17.10.80}$$

Now collecting terms and using (17.10.80) again we find that

$$
\begin{aligned}
[J^2, [J^2, A_x]] &= \hbar^2[2(J_y^2 + J_z^2)A_x + A_x(J_y^2 + J_z^2) \\
&\quad - 4(A_y J_y + A_z J_z)J_x)] \\
&= \hbar^2[2(J_x^2 + J_y^2 + J_z^2)A_x + A_x(J_x^2 + J_y^2 + J_z^2) \\
&\quad - 4(A_x J_x + A_y J_y + A_z J_z)J_x)] \\
&= 2\hbar^2(J^2 A_x + A_x J^2) - 4\hbar^2(\vec{A} \cdot \vec{J})J_x \ .
\end{aligned}
\tag{17.10.81}
$$

Thus, applying identical computations for A_y and A_z we find the desired result

$$
[J^2, [J^2, \vec{A}]] = 2\hbar^2(J^2\vec{A} + \vec{A}J^2) - 4\hbar^2(\vec{A} \cdot \vec{J})\vec{J} \ .
\tag{17.10.82}
$$

b) We now choose $\vec{A} = \vec{S}$ and then we immediately get

$$
\begin{aligned}
&\langle JM | [J^2, [J^2, \vec{S}]] | JM \rangle \\
&= 2\hbar^2 \langle JM | (J^2\vec{S} + \vec{S}J^2) | JM \rangle - 4\hbar^2 \langle JM | (\vec{S} \cdot \vec{J})\vec{J} | JM \rangle \ .
\end{aligned}
\tag{17.10.83}
$$

But the left-hand side of this equation vanishes. Thus, evaluating this for the z-component we obtain

$$
0 = 2\hbar^2 \langle JM | (J^2 S_z + S_z J^2) | JM \rangle - 4\hbar^2 \langle JM | (\vec{S} \cdot \vec{J})J_z | JM \rangle
\tag{17.10.84}
$$

or

$$
0 = 4\hbar^4 J(J+1)\langle JM | S_z | JM \rangle - 4\hbar^3 M \langle JM | \vec{S} \cdot \vec{J} | JM \rangle \ .
\tag{17.10.85}
$$

Thus, we have the desired result

$$
\langle JM | S_z | JM \rangle = M\hbar \frac{\langle JM | \vec{S} \cdot J | JM \rangle}{J(J+1)\hbar^2} \ .
\tag{17.10.86}
$$

c) Here we have to evaluate

$$
\langle nLSJM | S_z | nLSJM \rangle = M\hbar \frac{\langle nLSJM | \vec{S} \cdot \vec{J} | nLSJM \rangle}{J(J+1)\hbar^2} \ .
\tag{17.10.87}
$$

But,

$$
L^2 = (\vec{J} - \vec{S})^2 = J^2 + S^2 - 2\vec{S} \cdot \vec{J} \ .
\tag{17.10.88}
$$

Therefore,

$$
\begin{aligned}
&M\hbar \frac{\langle nLSJM | \vec{S} \cdot \vec{J} | nLSJM \rangle}{J(J+1)\hbar^2} \\
&= M\hbar^3 \frac{J(J+1) + S(S+1) - L(L+1)}{2J(J+1)\hbar^2}
\end{aligned}
\tag{17.10.89}
$$

and the desired result follows that

$$
\Delta E = \mu_B B M g
\tag{17.10.90}
$$

where the Landé g-factor is given by

$$
g = 1 + \frac{J(J+1) + S(S+1) - L(L+1)}{2J(J+1)} \ .
\tag{17.10.91}
$$

17.11   Spin and Space Coordinates

A spin 1/2 particle has its spin components aligned with a vector

$$\hat{n} = (\sin\alpha\cos\beta,\ \sin\alpha\sin\beta,\ \cos\alpha)\ .$$

This means that the eigenvalue of $\vec{s}\cdot\hat{n}$ is $\hbar/2$. The angular momentum wavefunction of this particle is given by the wavefunction for total angular momentum j derived in section 17.7 of [17.1], namely

$$\psi_{j=l\pm1/2,m,l} = \frac{1}{\sqrt{2l+1}}\left(\begin{array}{c}\sqrt{l\pm m+1/2}\,Y_{l,m-1/2}(\theta,\varphi)\\ \pm\sqrt{l\mp m+1/2}\,Y_{l,m+1/2}(\theta,\varphi)\end{array}\right).$$

Find the relationship between the angles (α,β) and (θ,φ).

Solution

The eigenfunction of $\vec{s}\cdot\hat{n}$ corresponding to the eigenvalue $\hbar/2$ is given by

$$\frac{\hbar}{2}\left(\begin{array}{cc}\cos\alpha & \sin\alpha\,e^{-i\beta}\\ \sin\alpha\,e^{i\beta} & -\cos\alpha\end{array}\right)\left(\begin{array}{c}a\\ b\end{array}\right) = \frac{\hbar}{2}\left(\begin{array}{c}a\\ b\end{array}\right). \qquad (17.11.92)$$

From this we get that

$$\cos\alpha\,a + \sin\alpha\,e^{-i\beta}b = a\ . \qquad (17.11.93)$$

The second equation is an identity. Therefore, solving for a/b we get

$$a(1-1+2\sin^2\alpha/2) = 2\sin\alpha/2\ \cos\alpha/2\ e^{-i\beta}\,b \qquad (17.11.94)$$

and

$$\frac{a}{b} = \cot\alpha/2\,e^{-i\beta}\ . \qquad (17.11.95)$$

For the given wavefunction we have

$$\frac{a}{b} = \sqrt{\frac{l\pm m+1/2}{l\mp m+1/2}\frac{Y_{l,m-1/2}(\theta,\varphi)}{Y_{l,m+1/2}(\theta,\varphi)}}$$

$$= \sqrt{\frac{l\pm m+1/2}{l\mp m+1/2}\frac{P_l^{m-1/2}(\cos\theta)}{P_l^{m+1/2}(\cos\theta)}}\,e^{-i\varphi}\ . \qquad (17.11.96)$$

Comparing this with (17.11.95) we get

$$\beta = \varphi$$

$$\cot\alpha/2 = \sqrt{\frac{l\pm m+1/2}{l\mp m+1/2}\frac{P_l^{m-1/2}(\cos\theta)}{P_l^{m+1/2}(\cos\theta)}}\ . \qquad (17.11.97)$$

17.12   Clebsch-Gordon for $j= 3/2$

A spin 1/2 particle in a P-state ($l = 1$) has a total angular momentum $j = 3/2$ and z-component of the total angular momentum $m_j = 1/2$. What is the probability of finding the z-component of the spin of the particle to have the value $m_s = 1/2$?

Solution

By starting from

$$|3/2, 3/2\rangle = |1, 1\rangle|1/2, 1/2\rangle \qquad (17.12.98)$$

and applying

$$J_- = L_- + S_- \qquad (17.12.99)$$

to both sides of this equation, we find the angular momentum wavefunction for the particle

$$|3/2, 1/2\rangle = \frac{1}{\sqrt{3}} \left[\sqrt{2}|1, 0\rangle|1/2, 1/2\rangle + |1, 1\rangle|1/2, -1/2\rangle \right] . \qquad (17.12.100)$$

Therefore, the probability that the z-component of the spin is +1/2 is

$$\left| \frac{\sqrt{2}}{\sqrt{3}} \right|^2 = \frac{2}{3} .$$

17.13   Rigid Rotator in a Step Potential

A rigid plane rotator, moment of inertia I, rotates subject to a potential field

$$V(\varphi) = \begin{cases} 0 & 0 < \varphi < \pi \\ -V_0 & \pi < \varphi, 2\pi \end{cases}$$

where φ is the azimuthal angle of the rotator relative to some fixed axis and V_0 is a positive constant. Find the transcendental equation for the positive energy eigenvalues.

Solution

The Hamiltonian for this problem may be written

$$H = \frac{L_z^2}{2I} + V(\varphi) \qquad (17.13.101)$$

where we impose periodic boundary conditions on the eigenfunctions, namely

$$\psi(2\pi) = \psi(0) . \qquad (17.13.102)$$

If, using the fact that we restrict ourselves to positive energies, we write

$$k^2 = \frac{2IE}{\hbar^2} \quad , \quad \kappa^2 = \frac{2I(E + V_0)}{\hbar^2} \tag{17.13.103}$$

the eigenvalue equation for the energy becomes

$$\frac{d^2\psi}{d\varphi^2} + k^2\psi = 0 \quad 0 < \varphi < \pi$$

$$\frac{d^2\psi}{d\varphi^2} + \kappa^2\psi = 0 \quad \pi < \varphi < 2\pi \ . \tag{17.13.104}$$

The solutions are

$$\psi = A e^{ik\varphi} + B e^{-ik\varphi} \quad 0 < \varphi < \pi$$

$$\psi = C e^{i\kappa\varphi} + D e^{-i\kappa\varphi} \quad \pi < \varphi < 2\pi \ . \tag{17.13.105}$$

Imposing the condition that ψ and its first derivative are continuous at $\varphi = \pi$ we get

$$A e^{ik\pi} + B e^{-ik\pi} = C e^{i\kappa\pi} + D e^{-i\kappa\pi}$$

$$k\left[A e^{ik\pi} - B e^{-ik\pi}\right] = \kappa\left[C e^{i\kappa\pi} - D e^{-i\kappa\pi}\right] \ . \tag{17.13.106}$$

Finally, imposing the periodic boundary condition

$$\psi(2\pi) = \psi(0) \tag{17.13.107}$$

we get

$$A e^{ik2\pi} + B e^{-ik2\pi} = A + B$$

$$C e^{i\kappa2\pi} + D e^{-i\kappa2\pi} = C + D \ . \tag{17.13.108}$$

Writing all this out we find a set of four homogeneous equations in four unknowns. For a nontrivial solution to exist requires that the determinant of the coefficients must vanish. This is the desired transcendental equation. The required determinant is

$$D = \begin{vmatrix} e^{i\pi k} & e^{-i\pi k} & -e^{i\pi\kappa} & -e^{-i\pi\kappa} \\ k e^{i\pi k} & -k e^{-i\pi k} & -\kappa e^{i\pi\kappa} & \kappa e^{-i\pi\kappa} \\ [e^{i2\pi k} - 1] & [e^{-i2\pi k} - 1] & 0 & 0 \\ 0 & 0 & [e^{i2\pi\kappa} - 1] & [e^{-i2\pi\kappa} - 1] \end{vmatrix} \ . \tag{17.13.109}$$

This determinant when evaluated yields the value

$$D = -4i\kappa \sin \pi(\kappa + k) \ . \tag{17.13.110}$$

Therefore, the energy eigenvalues are given by the equation

$$\kappa + k = n \quad n = 1, 2, 3, \ldots \ . \tag{17.13.111}$$

Negative values of n are excluded since

$$\kappa = n - k > 0 \quad \text{and} \quad k > 0 \ . \tag{17.13.112}$$

17.14   Spin Dependent Operators for Two Particles

If we have two particles with spin operators

$$\vec{S}_1 = \frac{\hbar}{2}\vec{\sigma}_1 \quad , \quad \vec{S}_2 = \frac{\hbar}{2}\vec{\sigma}_2$$

separated by a distance $\vec{r} = \hat{n}r$ show that any positive integral power of either of the operators

$$A = \vec{\sigma}_1 \cdot \vec{\sigma}_2$$

or

$$S_{12} = 3(\vec{\sigma}_1 \cdot \hat{n})(\vec{\sigma}_2 \cdot \hat{n}) - \vec{\sigma}_1 \cdot \vec{\sigma}_2$$

as well as any product of such powers can again be written as a linear combination of A, S_{12} and the unit matrix.

Solution

We show the result step by step. Recall that for the Pauli matrices we have

$$\sigma_x^2 = \sigma_y^2 = \sigma_z^2 = 1 \qquad (17.14.113)$$

and

$$\sigma_x\sigma_y = i\sigma_z \quad , \quad \sigma_y\sigma_z = i\sigma_x \quad , \quad \sigma_z\sigma_x = i\sigma_y \ . \qquad (17.14.114)$$

We are now ready to compute

$$
\begin{aligned}
A^2 &= \vec{\sigma}_1 \cdot \vec{\sigma}_2 \vec{\sigma}_1 \cdot \vec{\sigma}_2 \\
&= (\sigma_{1x}\sigma_{2x} + \sigma_{1y}\sigma_{2y} + \sigma_{1z}\sigma_{2z})(\sigma_{1x}\sigma_{2x} + \sigma_{1y}\sigma_{2y} + \sigma_{1z}\sigma_{2z}) \\
&= 3 - 2(\sigma_{1x}\sigma_{2x} + \sigma_{1y}\sigma_{2y} + \sigma_{1z}\sigma_{2z}) \\
&= 3 - 2A \ .
\end{aligned}
\qquad (17.14.115)
$$

Next we compute

$$S_{12}^2 = 9(\vec{\sigma}_1 \cdot \hat{n})^2(\vec{\sigma}_2 \cdot \hat{n})^2 + A^2 - 6(\vec{\sigma}_1 \cdot \vec{\sigma}_2)(\vec{\sigma}_1 \cdot \hat{n})(\vec{\sigma}_2 \cdot \hat{n}) \ . \quad (17.14.116)$$

But, by using (17.14.113), we see that

$$(\vec{\sigma}_1 \cdot \hat{n})^2 = (\vec{\sigma}_2 \cdot \hat{n})^2 = 1 \ . \qquad (17.14.117)$$

Also, by simply multiplying out the matrices we see that

$$\vec{\sigma}(\vec{\sigma} \cdot \hat{n}) = \hat{n} + i(\vec{\sigma} \times \hat{n}) \ . \qquad (17.14.118)$$

Therefore,

$$
\begin{aligned}
(\vec{\sigma}_1 \cdot \vec{\sigma}_2)(\vec{\sigma}_1 \cdot \hat{n})(\vec{\sigma}_2 \cdot \hat{n}) &= [\hat{n} + i(\vec{\sigma}_1 \times \hat{n})][\hat{n} + i(\vec{\sigma}_2 \times \hat{n})] \\
&= 1 - (\vec{\sigma}_1 \times \hat{n})(\vec{\sigma}_2 \times \hat{n}) \\
&= 1 - \vec{\sigma}_1 \cdot \vec{\sigma}_2 + (\vec{\sigma}_1 \cdot \hat{n})(\vec{\sigma}_2 \cdot \hat{n}) \ .
\end{aligned}
\qquad (17.14.119)
$$

So, we find

$$
\begin{aligned}
S_{12}^2 &= 9 + A^2 - 6 + 6A - 6(\vec{\sigma}_1 \cdot \hat{n})(\vec{\sigma}_2 \cdot \hat{n}) \\
&= 6 + 4A - 2(S_{12} + A) \\
&= 6 + 2A - 2S_{12} \ .
\end{aligned}
\tag{17.14.120}
$$

Finally,

$$
\begin{aligned}
AS_{12} &= -A^2 + 3(\vec{\sigma}_1 \cdot \vec{\sigma}_2)(\vec{\sigma}_1 \cdot \hat{n})(\vec{\sigma}_2 \cdot \hat{n}) \\
&= -3 + 2A + 3 - 3A + 3(\vec{\sigma}_1 \cdot \hat{n})(\vec{\sigma}_2 \cdot \hat{n}) \\
&= S_{12} \ .
\end{aligned}
\tag{17.14.121}
$$

In exactly the same manner we find that

$$
S_{12}A = S_{12} \ .
\tag{17.14.122}
$$

This proves the desired result.

Bibliography

[17.1] A.Z. Capri, *Nonrelativistic Quantum Mechanics* 3rd edition, World Scientific Publishing Co. Pte. Ltd., chapter 17, (2002) .

Chapter 18

Scattering - Time Dependent

18.1 Cross-section from Experiment

A proton beam producing a current of 5×10^{-9} amps is incident on a target of copper. Assume the target thickness is such that the areal density is 0.2 mg/cm^2. The detector has an area of 0.5 cm^2, normal to the scattered beam, and is 20 cm from the target. If 10 protons are counted by the detector every second at a particular angle, calculate the differential cross-section for protons scattering off copper at that angle.

Solution

We use the equation

$$\Delta N = J N_0 \frac{d\sigma}{d\Omega} \Delta\Omega \ . \tag{18.1.1}$$

Now we have

$$
\begin{aligned}
I &= 5 \times 10^{-9} \text{A} = \text{incident proton current} \\
&= \frac{5 \times 10^{-9} \text{ A}}{1.6 \times 10^{-19} \text{ C/particle}} = 3.12 \times 10^{10} \text{ particles/s} \ . \tag{18.1.2}
\end{aligned}
$$

$$A = \text{cross-sectional area of the beam} \tag{18.1.3}$$

$$J = I/A \tag{18.1.4}$$

$$
\begin{aligned}
\rho &= 0.2 \text{ mg/cm}^2 = \text{areal density} \\
&= \frac{0.2 \text{mg/cm}^2}{1.09 \times 10^{-19} \text{ mg/atom}} = 1.84 \times 10^{18} \text{ atoms/cm}^2 \tag{18.1.5}
\end{aligned}
$$

$$\Delta N = 10 \text{protons/s} \tag{18.1.6}$$

$$\Delta S = 0.5 \text{ cm}^2 = \text{area of the detector} \tag{18.1.7}$$

$$R = 20 \text{ cm} = \text{distance of the target from the detector} \tag{18.1.8}$$

$$\Delta\Omega = \Delta S/R^2 = 1.25 \times 10^{-3} . \tag{18.1.9}$$

From our equation above we find

$$\frac{d\sigma}{d\Omega} = \frac{1}{JN_0}\frac{\Delta N}{\Delta\Omega} = \frac{A}{I}\frac{1}{A\rho}\frac{\Delta N}{\Delta\Omega} = \frac{1}{I\rho}\frac{\Delta N}{\Delta\Omega} . \tag{18.1.10}$$

Therefore we have

$$\frac{d\sigma}{d\Omega} = 1.34 \times 10^{-25} \text{ cm}^2/\text{atom/steradian} . \tag{18.1.11}$$

18.2   Green's Functions for Free Particle States

Use the expressions

$$G_0^+(t) = -\frac{i}{\hbar} e^{-iH_0 t/\hbar}\,\theta(t) \tag{18.2.12}$$

$$G_0^-(t) = +\frac{i}{\hbar} e^{-iH_0 t/\hbar}\,\theta(-t) \tag{18.2.13}$$

and evaluate the matrix elements $\langle\vec{p}|G_0^\pm(t)|\vec{k}\rangle$ where $|\vec{p}\rangle$, $|\vec{k}\rangle$ are free particle states of momentum $\vec{p}$ and $\vec{k}$ respectively.

Solution

We begin with

$$G_0^\pm(t) = \mp\frac{i}{\hbar}\exp\left(-i\frac{H_0 t}{\hbar}\right)\theta(\pm t) . \tag{18.2.14}$$

Now let $|\vec{p}\rangle$ be an eigenstate of H_0.

$$H_0|\vec{p}\rangle = \frac{\vec{p}^2}{2m}|\vec{p}\rangle . \tag{18.2.15}$$

Then,

$$\langle\vec{p}|G_0^\pm|\vec{k}\rangle = \mp\frac{i}{\hbar}\langle\vec{p}|\exp\left(-i\frac{H_0 t}{\hbar}\right)|\vec{k}\rangle\,\theta(\pm t)$$
$$= \mp\frac{i}{\hbar}\exp\left(-i\frac{\vec{k}^2 t}{2m\hbar}\right)\delta(\vec{p}-\vec{k})\,\theta(\pm t) . \tag{18.2.16}$$

18.3 Dispersion Relations

a) Let $f(z)$ be defined by

$$f(z) = \int dt \, e^{itz} \, F(t)$$

where $F(t) = 0$ for $t < 0$. Show that $f(z)$ is an analytic function for $\Im(z) > 0$.
b) Consider a function $f(z)$ which is analytic in the upper half of the complex z-plane as well as on the real axis and vanishes rapidly for $|z| \to \infty$. In that case we may write

$$f(z) = \frac{1}{2\pi i} \oint \frac{f(z')}{z' - z} dz' \qquad (18.3.17)$$

where the contour runs along the real axis from $-R$ to R and is closed by a semicircle in the upper half-plane. Let $R \to \infty$ and let $z \to$ the real axis from above and use this result to express $\Re(f)$ in terms of $\Im(f)$ and vice versa.
Hint: Recall that

$$\lim_{\epsilon \to 0+} \frac{1}{x \pm i\epsilon} = P\frac{1}{x} \mp i\pi\delta(x) \ . \qquad (18.3.18)$$

c) Verify the result obtained by applying it to the function

$$f(z) = \frac{1}{z + ia} \qquad a > 0 \ . \qquad (18.3.19)$$

Solution

a) The proof that $f(z)$ is an analytic function for $\Im(z) > 0$ is straightforward since by assumption the integral

$$f(z) = \int dt \, e^{itz} \, F(t) \qquad (18.3.20)$$

is assumed convergent. This convergence is only improved by letting

$$z = x + iy \ , \quad y > 0 \ .$$

In that case we may differentiate with respect to z under the integral sign and get that

$$\frac{df}{dz}(z) = i \int t \, dt \, e^{itz} \, F(t) \qquad (18.3.21)$$

is finite for all $\Im(z) > 0$. Thus, $f(z)$ is an analytic function for $\Im(z) > 0$.
b) Following the instructions we see that for $R \to \infty$ the contribution to the integral from the semicircle in (18.3.17) vanishes and by taking the limit as $z \to$ the real axis from above we get (writing $z = x + i\epsilon$)

$$f(x + i\epsilon) = \frac{1}{2\pi i} \int_{-\infty}^{\infty} \frac{f(z')}{z' - x - i\epsilon} dz'$$

$$= \frac{1}{2\pi i} P \int_{-\infty}^{\infty} \frac{f(z')}{z' - x} dz' + \frac{i\pi}{2\pi i} \int_{-\infty}^{\infty} f(z')\delta(z' - z) \, dz' \ . \qquad (18.3.22)$$

So carrying out the delta function integration and simplifying we find

$$f(x) = \frac{1}{i\pi} P \int_{-\infty}^{\infty} \frac{f(z')}{z' - x} \, dz' \; . \tag{18.3.23}$$

Next, we take the real and imaginary parts of this equation and find

$$\begin{aligned}
\Re[f(x)] &= \frac{1}{\pi} P \int_{-\infty}^{\infty} \frac{\Im[f(z')]}{z' - x} \, dz' \\
\Im[f(x)] &= -\frac{1}{\pi} P \int_{-\infty}^{\infty} \frac{\Re[f(z')]}{z' - x} \, dz' \; .
\end{aligned} \tag{18.3.24}$$

This is the required result. The pair of equations (18.3.24) are known as dispersion relations. These relations may be combined with a statement of causality, such as is assumed in part a), to yield important conditions for scattering amplitudes. In optics the resulting equations are known as "Kramers-Kronig relations".

c) If the given function is

$$f(z) = \frac{1}{z + ia} \quad a > 0 \tag{18.3.25}$$

then we have that

$$\begin{aligned}
\Re[f(x)] &= \frac{x}{x^2 + a^2} \\
\Im[f(x)] &= \frac{-a}{x^2 + a^2} \; .
\end{aligned} \tag{18.3.26}$$

To verify the dispersion relations for this example we simply write out the integrals

$$\begin{aligned}
\frac{1}{\pi} P \int_{-\infty}^{\infty} \frac{\Im[f(z')]}{z' - x} \, dz' &= \frac{-a}{\pi} P \int_{-\infty}^{\infty} \frac{dz'}{(z'^2 + a^2)(z' - x)} \\
&= \frac{x}{x^2 + a^2} = \Re[f(x)]
\end{aligned} \tag{18.3.27}$$

as required. Similarly,

$$\begin{aligned}
-\frac{1}{\pi} P \int_{-\infty}^{\infty} \frac{\Re[f(z')]}{z' - x} \, dz' &= -\frac{1}{\pi} P \int_{-\infty}^{\infty} \frac{z' \, dz'}{(z'^2 + a^2)(z' - x)} \\
&= -\frac{a}{x^2 + a^2} = \Im[f(x)] \; .
\end{aligned} \tag{18.3.28}$$

This is again the expected result.

18.4 Källén-Yang-Feldman Equations

Assume that V is independent of time and use the equation

$$G^{\pm}(t) = G_0^{\pm}(t) + \int_{-\infty}^{\infty} G_0^{\pm}(t - t') V G^{\pm}(t') \, dt' \tag{18.4.29}$$

to obtain an equation for the Fourier transform $G^{\pm}(\omega)$ of $G^{\pm}(t)$ in terms of the Fourier transforms of $G_0^{\pm}(t)$. Write a formal solution for $G^{\pm}(\omega)$.

Solution

We begin with the integral equation for $G^{\pm}(t)$.

$$G^{\pm}(t) = G_0^{\pm}(t) + \int_{-\infty}^{\infty} G_0^{\pm}(t - t') \, V \, G^{\pm}(t') \, dt' \, . \tag{18.4.30}$$

Now define

$$G_0^{\pm}(\omega) = \int_{-\infty}^{\infty} e^{i\omega t} \, G_0^{\pm}(t) \, dt \tag{18.4.31}$$

$$G^{\pm}(\omega) = \int_{-\infty}^{\infty} e^{i\omega t} \, G^{\pm}(t) \, dt \, . \tag{18.4.32}$$

Then,

$$G^{\pm}(\omega) = G_0^{\pm}(\omega) + \int_{-\infty}^{\infty} dt \int_{-\infty}^{\infty} e^{i\omega(t-t')} e^{i\omega t'} G_0^{\pm}(t-t') \, V \, G^{\pm}(t') \, dt' \, . \tag{18.4.33}$$

So, changing the variable of integration from t to $t-t'$ and since V is independent of t, we have:

$$G^{\pm}(\omega) = G_0^{\pm}(\omega) + G_0^{\pm}(\omega) \, V \, G^{\pm}(\omega) \, . \tag{18.4.34}$$

Therefore solving formally we get

$$[1 - G_0^{\pm}(\omega) \, V] \, G^{\pm}(\omega) = G_0^{\pm}(\omega) \tag{18.4.35}$$

and

$$G^{\pm}(\omega) = [1 - G_0^{\pm}(\omega) \, V]^{-1} G_0^{\pm}(\omega) \, . \tag{18.4.36}$$

18.5   Born Approximation

In the equation

$$T_{fi} = \langle \Psi^0(E_f) | V | \Psi^+(E_i) \rangle \quad \text{for} \quad E_f = E_i \tag{18.5.37}$$

approximate $|\Psi^+(E_i)\rangle$ by a free particle state. If V is a screened Coulomb potential

$$V = -V_0 \frac{e^{-\mu r}}{r}$$

calculate the scattering amplitude. The approximation used is known as the first Born approximation.

Solution

The T-matrix on shell is defined by

$$T_{fi} = \langle \Psi^0(E_f)|V|\Psi^+(E_i)\rangle \quad \text{for} \quad E_f = E_i . \tag{18.5.38}$$

We are using the approximation that

$$\langle \vec{r}|\Psi^+(E_i)\rangle = (2\pi)^{-3/2} e^{i\vec{k}\cdot\vec{r}} \tag{18.5.39}$$

where

$$E_i = \frac{\hbar^2 k^2}{2m} \quad , \quad E_f = \frac{\hbar^2 q^2}{2m} . \tag{18.5.40}$$

Thus,

$$T_{fi} = \frac{-1}{(2\pi)^3} \int d^3 r \, e^{-i\vec{q}\cdot\vec{r}} \, V_0 \frac{e^{-\mu r}}{r} \, e^{i\vec{k}\cdot\vec{r}} . \tag{18.5.41}$$

So,

$$\begin{aligned}
T_{fi} &= \frac{-V_0(2\pi)}{(2\pi)^3} \int_0^\infty r dr \, e^{-\mu r} \int_0^\pi e^{i|\vec{k}-\vec{q}|r \cos\theta} \sin\theta \, d\theta \\
&= \frac{V_0(2\pi)}{(2\pi)^3} \int_0^\infty r dr \, e^{-\mu r} \frac{2\sin(pr)}{pr}
\end{aligned} \tag{18.5.42}$$

where

$$p = |\vec{k} - \vec{q}| \tag{18.5.43}$$

is the magnitude of the change in momentum that the particle experiences. So finally,

$$T_{fi} = \frac{i}{2\pi^2} \frac{V_0}{|\vec{k} - \vec{q}|^2 + \mu^2} . \tag{18.5.44}$$

18.6   Scattering in CM and Laboratory Frame

Establish the connection between the centre of mass frame scattering amplitude $f(\theta, \varphi)$ and the laboratory frame scattering amplitude $f_L(\theta_L, \varphi_L)$ for particles of mass m_1 and m_2 scattering off each other.

Solution

Comment: The first thing to realize is that the calculations involved here are completely classical.

In the laboratory frame a particle of mass m_1, velocity $\vec{v}_1$ is incident along the z-axis on a particle of mass m_2 at rest so that its velocity is $v_2 = 0$. This gives rise to a differential cross section $d\sigma_L(\theta_L, \varphi_L)/d\Omega_L$ for the incident particle.

The number of scattered particles that are observed in a unit time in a cone $\Delta\Omega_L$ centered at θ_l, φ_L is then given by

$$N\frac{d\sigma_L(\theta_L, \varphi_L)}{d\Omega_L}\Delta\Omega_L$$

where we have assumed a single scattering centre and an incident flux of N particles per unit area per unit time.

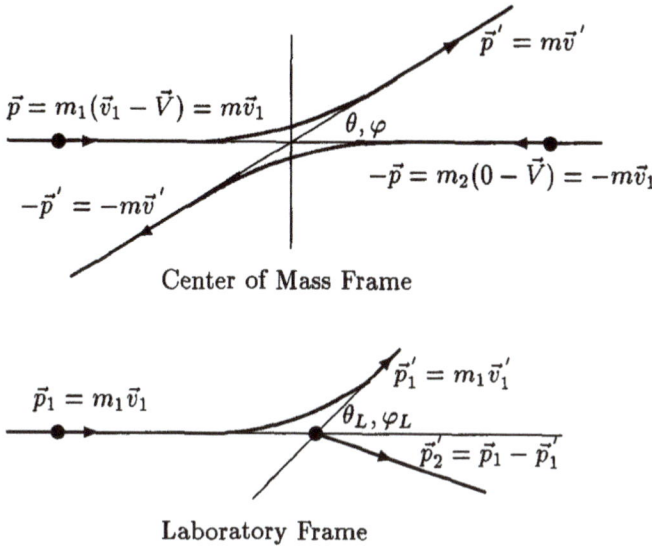

Center of Mass Frame

Laboratory Frame

Figure 18.1: Scattering in the center of mass and laboratory frames.

The situation in the two cases is as depicted in figure 18.1. In the centre of mass frame a "fictitious" particle with the reduced mass

$$m = \frac{m_1 m_2}{m_1 + m_2} = \frac{m_1 m_2}{M} \tag{18.6.45}$$

is incident on a fixed centre of force. The centre of mass moves with a constant speed

$$V = \frac{m_1}{M}v_1 . \tag{18.6.46}$$

We still need the relation between θ_L, φ_L and θ, φ. This is obtained from the relation between the velocity $\vec{v}_1'$ of the scattered particle in the laboratory frame where it has components

$$\vec{v}_1' = v_1(\sin\theta_L \cos\varphi_L, \sin\theta_L \sin\varphi_L, \cos\theta_L) \tag{18.6.47}$$

and the centre of mass velocity $\vec{V}$ and the velocity $\vec{v}'$ of the scattered particle in the centre of mass frame. The speed v' of the scattered particle in the centre of mass frame is given by

$$v' = \frac{m_2}{M}v_1 . \tag{18.6.48}$$

The components of the velocity of the scattered particle in the centre of mass frame are therefore

$$\vec{v}' = \frac{m_2}{M} v_1 (\sin\theta\cos\varphi, \ \sin\theta\sin\varphi, \ \cos\theta) \ . \tag{18.6.49}$$

The relationship between the two sets of angles follows from

$$\vec{v}_1' = \vec{v}' + \vec{V} \ . \tag{18.6.50}$$

Writing out the components, we get

$$\begin{aligned}
v_1 \sin\theta_L \cos\varphi_L &= \frac{m_2}{M} v_1 \sin\theta\cos\varphi \\
v_1 \sin\theta_L \sin\varphi_L &= \frac{m_2}{M} v_1 \sin\theta\sin\varphi \\
v_1 \cos\theta_L &= \frac{m_2}{M} v_1 \cos\theta + \frac{m_1}{M} v_1 \ .
\end{aligned} \tag{18.6.51}$$

Thus, we have the desired relation

$$\begin{aligned}
\tan\theta_L &= \frac{\sin\theta}{m_1/m_2 + \cos\theta} \\
\varphi_L &= \varphi \ .
\end{aligned} \tag{18.6.52}$$

Now, as stated at the beginning of this section, the number of scattered particles that are observed in a unit time in a direction $\Delta\Omega_L$ is given by

$$N \frac{d\sigma_L(\theta_L, \varphi_L)}{d\Omega_L} \Delta\Omega_L \ . \tag{18.6.53}$$

The same number of scattered particles must be observed in the centre of mass frame. Therefore, we have

$$\frac{d\sigma_L(\theta_L, \varphi_L)}{d\Omega_L} \sin\theta_L \, d\theta_L \, d\varphi_L = \frac{d\sigma(\theta, \varphi)}{d\Omega} \sin\theta \, d\theta \, d\varphi. \tag{18.6.54}$$

Hence,

$$\frac{d\sigma_L(\theta_L, \varphi_L)}{d\Omega_L} = \frac{d\sigma(\theta, \varphi)}{d\Omega} \frac{\sin\theta}{\sin\theta_L} \frac{d\theta}{d\theta_L} \frac{d\varphi}{d\varphi_L} \ . \tag{18.6.55}$$

Now, from (18.6.52) we get

$$\cos\theta_L = \frac{|m_1/m_2 + \cos\theta|}{\sqrt{(m_1/m_2)^2 + 2(m_1/m_2)\cos\theta + 1}} \ . \tag{18.6.56}$$

So, after differentiating and some rearranging we find

$$\frac{d\sigma_L(\theta_L, \varphi_L)}{d\Omega_L} = \frac{[(m_1/m_2)^2 + 2(m_1/m_2)\cos\theta + 1]^{3/2}}{|m_1/m_2 + \cos\theta|} \frac{d\sigma(\theta, \varphi)}{d\Omega} \ . \tag{18.6.57}$$

Therefore, in terms of the scattering amplitudes we obtain

$$|f_L(\theta_L, \varphi_L)|^2 = \frac{[(m_1/m_2)^2 + 2(m_1/m_2)\cos\theta + 1]^{3/2}}{|m_1/m_2 + \cos\theta|} |f(\theta, \varphi)|^2 \ . \tag{18.6.58}$$

18.7   Propagator for a Free Particle

The retarded propagator is defined as an operator by the differential equation

$$\left(i\hbar\frac{\partial}{\partial t} - H\right)G^+(t) = \delta(t)\mathbf{1} \qquad (18.7.59)$$

with the initial condition that

$$G^+(t) = 0 \quad \text{for} \quad t < 0 \ .$$

a) Show that if the complete set of eigenfunctions of H is

$$\{\phi_n(x)\} = \{\langle x|n\rangle\}$$

with corresponding eigenvalues E_n, then the configuration space representation of $G^+(t)$, namely $\langle x|G^+(t)|y\rangle$ may be written

$$\langle x|G^+(t)|y\rangle = -\frac{i}{\hbar}\theta(t)\sum_n e^{-iE_nt/\hbar}\,\phi_n(x)\phi_n^*(y) \ . \qquad (18.7.60)$$

Hint: Insert a complete set of intermediate states at the necessary places.
b) Use this result to obtain the propagator for a free particle.

Solution

a) Formally the differential equation (18.7.59) with its accompanying initial condition is solved by

$$G^+(t) = -\frac{i}{\hbar}\theta(t)\,e^{-iHt/\hbar} \qquad (18.7.61)$$

where $\theta(t)$ is defined by

$$\theta(t) = \begin{cases} 1 & t > 0 \\ 0 & t < 0 \end{cases} \ . \qquad (18.7.62)$$

Taking matrix elements of this equation we get

$$\begin{aligned} G^+(x,y;t,0) &\equiv \langle x|G^+(t)|y\rangle \\ &= -\frac{i}{\hbar}\theta(t)\sum_{n,n'}\langle x|n\rangle\langle n|e^{-iHt/\hbar}|n'\rangle\langle n'|y\rangle \\ &= -\frac{i}{\hbar}\theta(t)\sum_{n,n'}e^{-iE_{n'}t/\hbar}\langle x|n\rangle\langle n|n'\rangle\langle n'|y\rangle \\ &= -\frac{i}{\hbar}\theta(t)\sum_n e^{-iE_nt/\hbar}\phi_n(x)\phi_n^*(y) \ . \qquad (18.7.63) \end{aligned}$$

b) For a free particle the normalized complete set of eigenfunctions are

$$\phi_{\vec{k}}(\vec{x}) = \frac{1}{(2\pi)^{3/2}}\,e^{i\vec{k}\cdot\vec{x}} \qquad (18.7.64)$$

with corresponding energy

$$E_{\vec{k}} = \frac{\hbar^2 k^2}{2m} . \tag{18.7.65}$$

Thus, the free particle propagator is given by

$$G^+(\vec{x}, \vec{y}; t, 0) = -\frac{i}{\hbar}\theta(t)\frac{1}{(2\pi)^3}\int d^3k\, e^{-i\hbar k^2 t/2m}\, e^{i\vec{k}\cdot(\vec{x}-\vec{y})} . \tag{18.7.66}$$

We next evaluate the integral by going to spherical coordinates and aligning the k_z-axis with $\vec{x} - \vec{y}$ to get

$$\int d^3k\, e^{-i\hbar k^2 t/2m}\, e^{i\vec{k}\cdot(\vec{x}-\vec{y})}$$

$$= 2\pi \int_0^\infty k^2\, dk\, e^{-i\hbar k^2 t/2m} \int_{-1}^1 du\, e^{ik|\vec{x}-\vec{y}|u} \tag{18.7.67}$$

$$= \frac{2\pi m}{i\hbar t} \exp\left\{ im\frac{|\vec{x}-\vec{y}|^2}{2\hbar t} \right\} \int_{-\infty}^\infty \exp\left\{ -\left(\frac{i\hbar t + \epsilon}{2m}\right)q^2 \right\} dq \tag{18.7.68}$$

$$= \frac{2\pi m}{i\hbar t} \exp\left\{ im\frac{|\vec{x}-\vec{y}|^2}{2\hbar t} \right\} \sqrt{\frac{2\pi m}{i\hbar t + \epsilon}} . \tag{18.7.69}$$

In the second last line we have added a term $-(\epsilon/2m)q^2$. This was simply to make sure that the integral converges. We can now let $\epsilon \to 0+$ and we see that this serves for us to choose the proper square root so that the integral becomes

$$\left(\frac{2\pi m}{i\hbar t}\right)^{3/2} \exp\left\{ im\frac{|\vec{x}-\vec{y}|^2}{2\hbar t} \right\} .$$

Substituting this back into the expression for the propagator we get the desired result

$$G^+(\vec{x}, \vec{y}; t, 0) = -\frac{i}{\hbar}\theta(t)\left(\frac{m}{2\pi i\hbar t}\right)^{3/2} \exp\left\{ im\frac{|\vec{x}-\vec{y}|^2}{2\hbar t} \right\} . \tag{18.7.70}$$

18.8   Propagator for Simple Harmonic Oscillator

Find the retarded propagator for a one-dimensional particle bound by a harmonic oscillator.

Hint: Use the result of part a) of problem 18.7, as well as the integral representation (5.9.71) of problem 5.11.

Solution

The result of problem 18.7 shows that the retarded propagator is given by

$$G^+(x, y; t, 0) = -\frac{i}{\hbar}\theta(t)\sum_n e^{-iE_n t/\hbar}\phi_n(x)\phi_n^*(y) . \tag{18.8.71}$$

For a simple harmonic oscillator we have that

$$\phi_n(x) = \left(\frac{\alpha^2}{\pi}\right)^{1/4} \frac{1}{2^n n!} H_n(\alpha x) \, e^{-\alpha^2 x^2/2} \tag{18.8.72}$$

with the corresponding energy

$$E_n = (n+1/2)\hbar\omega \tag{18.8.73}$$

where we have also written

$$\alpha^2 = \frac{m\omega}{\hbar} \ .$$

Now, replacing the Hermite polynomials by their integral representation (5.9.71)

$$H_n(x) = \frac{2^n}{\sqrt{\pi}} \int_{-\infty}^{\infty} (x + ip)^n \, e^{-p^2} \, dp \tag{18.8.74}$$

we get

$$\begin{aligned}
G^+(x, y; t, 0) &= -\frac{i}{\hbar}\theta(t) \sum_n \left(\frac{\alpha^2}{\pi}\right)^{1/2} \frac{2^n}{\pi n!} \\
&\times \int_{-\infty}^{\infty} (\alpha x + ip)^n \, e^{-p^2} \, dp \, (\alpha y + iq)^n \, e^{-q^2} \, dq \, e^{-\alpha^2(x^2+y^2)} \, e^{-i(n+1/2)\omega t} \\
&= -\frac{i}{\hbar}\theta(t)\frac{\alpha}{\pi^{3/2}} e^{-\alpha^2(x^2+y^2)} \, e^{-i\omega t/2} \\
&\times \int_{-\infty}^{\infty} \sum_n \frac{1}{n!} \left[2(\alpha x + ip)(\alpha y + iq) \, e^{-i\omega t} \right]^n \, e^{-(p^2+q^2)} \, dp \, dq \\
&= -\frac{i}{\hbar}\theta(t)\frac{\alpha}{\pi^{3/2}} e^{-\alpha^2(x^2+y^2)} \, e^{-i\omega t/2} \\
&\times \int_{-\infty}^{\infty} \exp\left\{ 2(\alpha x + ip)(\alpha y + iq) \, e^{-i\omega t} - p^2 - q^2 \right\} \, dp \, dq \ . \tag{18.8.75}
\end{aligned}$$

If we now change integration variables from p, q to τ, σ where

$$p = \frac{1}{\sqrt{2}}[\tau + \sigma] \qquad q = \frac{1}{\sqrt{2}}[\tau - \sigma] \tag{18.8.76}$$

the argument of the exponential in the integral becomes, after a little algebra,

$$\begin{aligned}
&- (1 + e^{-i\omega t}) \left[\tau - \frac{i}{\sqrt{2}} \frac{\alpha(x+y) \, e^{-i\omega t}}{1 + e^{-i\omega t}} \right]^2 \\
&- (1 - e^{-i\omega t}) \left[\sigma + \frac{i}{\sqrt{2}} \frac{\alpha(x-y) \, e^{-i\omega t}}{1 - e^{-i\omega t}} \right]^2 \\
&- \frac{e^{-i\omega t}}{1 - e^{-2i\omega t}} \left[\alpha^2(x^2 + y^2) \, e^{-i\omega t} - 2\alpha^2 xy \right] \ . \tag{18.8.77}
\end{aligned}$$

This now leads to a pair of standard Gaussian integrals. Thus, we can combine these results to get

$$
G^+(x,y;t,0)
$$

$$
= \frac{-i}{\hbar}\theta(t)\frac{\alpha}{\pi^{3/2}}\exp\left\{-\frac{\alpha^2}{2}\left[(x^2+y^2)\frac{1+e^{-2i\omega t}}{1-e^{-2i\omega t}}-\frac{4xy\,e^{-i\omega t}}{1-e^{-2i\omega t}}\right]\right\}
$$

$$
\times\quad \sqrt{\frac{\pi^2}{1-e^{-2i\omega t}}}\,e^{-i\omega t/2}\quad . \tag{18.8.78}
$$

After replacing α by $\sqrt{m\omega/\hbar}$ and some simplification we have

$$
G^+(x,y;t,0) = \frac{-i}{\hbar}\theta(t)\sqrt{\frac{m\omega}{2\pi i\hbar\sin\omega t}}
$$

$$
\times\quad \exp\left\{\frac{im\omega}{2\hbar\sin\omega t}\left[(x^2+y^2)\sin\omega t-2xy\right]\right\}\quad . \tag{18.8.79}
$$

Bibliography

[17.1] For a totally different way of computing propagators see
R.P. Feynman and A.R.Hibbs, *Quantum Mechanics and Path Integrals*, New York: McGraw-Hill, (1965).

Chapter 19

Scattering - Time Independent

19.1 Equations for Spherical Bessel Functions

Use the integral representations for the spherical Bessel functions [19.1]

$$j_l(x) = \frac{x^l}{i\, 2^{l+1} l!} \int_{-i}^{i} e^{xu} (u^2 + 1)^l \, du$$

$$h_l^{(1)}(x) = \frac{x^l}{i\, 2^{l+1} l!} \int_{\infty\, e^{i\alpha}}^{i} e^{xu} (u^2 + 1)^l \, du$$

$$h_l^{(2)}(x) = \frac{x^l}{i\, 2^{l+1} l!} \int_{-i}^{\infty\, e^{i\alpha}} e^{xu} (u^2 + 1)^l \, du \qquad (19.1.1)$$

to show that

a) $\quad Z_{l-1}(x) + Z_{l+1}(x) = \dfrac{2l+1}{x} Z_l(x) \quad$ for $\quad l \geq 1$

b) $\quad \dfrac{d}{dx} Z_l(x) = Z_{l-1}(x) - \dfrac{l+1}{x} Z_l(x) \quad$ for $\quad l \geq 1$

c) $\quad \dfrac{d}{dx}\left[x^{-l} Z_l(x) \right] = -x^{-l} Z_{l+1}(x) \; .$

Here $Z_l(x)$ may be any one of the four spherical Bessel functions.

Solution

We demonstrate these formulae explicitly for $j_l(x)$. The calculations for the other two Bessel functions is exactly the same.

a)

$$j_l = \frac{x^l}{i\,2^{l+1}l!} \int_{-i}^{i} e^{xu} (u^2 + 1)^l \, du \ . \tag{19.1.2}$$

Now,

$$u\,d(u^2+1)^l = 2lu^2(u^2+1)^{l-1}\,du = \left[2l(u^2+1)^l - 2l(u^2+1)^{l-1}\right]\,du \ . \tag{19.1.3}$$

So,

$$(u^2+1)^l\,du = \frac{1}{2l}u\,d(u^2+1)^l + (u^2+1)^{l-1}\,du \ . \tag{19.1.4}$$

Therefore, writing this out and integrating by parts we get

$$
\begin{aligned}
j_l &= \frac{x^{l-1}}{i\,2^{l+1}(l+1)!} \int_{-i}^{i} u\,e^{xu}\,d(u^2+1)^l + \frac{x}{2l}\,j_{l-1}(x) \\[2mm]
&= -\frac{x^l}{i\,2^{l+1}l!}\frac{1}{2l} \int_{-i}^{i} (u^2+1)^l\,e^{xu}\,(1+xu)\,du + \frac{x}{2l}\,j_{l-1}(x) \\[2mm]
&= -\frac{1}{2l}j_l(x) - \frac{x^{l+1}}{i\,2^{l+1}(l+1)!}\frac{1}{2l} \int_{-i}^{i} (u^2+1)^l\,e^{xu}\,u\,du + \frac{x}{2l}\,j_{l-1}(x) \\[2mm]
&= -\frac{1}{2l}j_l(x) + \frac{x^{l+2}}{i\,2^{l+1}(l+1)!}\frac{1}{2l} \int_{-i}^{i} (u^2+1)^{l+1}\,e^{xu}\,du + \frac{x}{2l}\,j_{l-1}(x) \\[2mm]
&= -\frac{1}{2l}j_l(x) + \frac{x}{2l}\,j_{l+1} + \frac{x}{2l}\,j_{l-1}(x) \ .
\end{aligned}
\tag{19.1.5}
$$

Thus finally,

$$\frac{2l+1}{2l}\,j_l(x) = \frac{x}{2l}\left[j_{l+1}(x) + j_{l-1}(x)\right] \tag{19.1.6}$$

and

$$\frac{2l+1}{x}\,j_l(x) = j_{l+1}(x) + j_{l-1}(x) \ . \tag{19.1.7}$$

b) Again we begin with

$$j_l(x) = \frac{x^l}{i\,2^{l+1}l!} \int_{-i}^{i} e^{xu} (u^2 + 1)^l \, du \ . \tag{19.1.8}$$

Then,

$$
\begin{aligned}
\frac{2l+1}{x}j_l(x) &= \frac{(2l+1)x^{l-1}}{i\,2^{l+1}l!} \int_{-i}^{i} e^{xu} (u^2 + 1)^l \, du \\[2mm]
&= \frac{(2l+1)x^{l-2}}{i\,2^{l+1}l!} \int_{-i}^{i} (u^2 + 1)^l \, de^{xu} \ .
\end{aligned}
\tag{19.1.9}
$$

So,

$$
\begin{aligned}
j_l(x) &= \frac{x^{l-1}}{i\, 2^{l+1}(l-1)!} \int_{-i}^{i} (u^2+1)^l\, de^{xu} \\
&= -\frac{x^{l-1}}{i\, 2^{l+1}l!} \int_{-i}^{i} e^{xu}\, l(u^2+1)^{l-1}\, u\, du \ .
\end{aligned}
\tag{19.1.10}
$$

Thus,

$$
j_l(x) = -\frac{x^{l-1}}{i\, 2^{l+1}(l-1)!} \int_{-i}^{i} e^{xu}\, (u^2+1)^{l-1}\, u\, du
\tag{19.1.11}
$$

and

$$
\begin{aligned}
\frac{dj_l(x)}{dx} &= \frac{l\,x^{l-1}}{i\, 2^{l+1}l!} \int_{-i}^{i} e^{xu}\, (u^2+1)^l\, du \\
&\quad + \frac{x^l}{i\, 2^{l+1}l!} \int_{-i}^{i} e^{xu}\, (u^2+1)^l\, u\, du \\
&= \frac{l}{x}\, j_l(x) - j_{l+1}(x) \ .
\end{aligned}
\tag{19.1.12}
$$

Now use the result of part a) in the form

$$
j_{l+1}(x) = \frac{2l+1}{x}\, j_l(x) - j_{l-1}(x) \ .
\tag{19.1.13}
$$

Then,

$$
\begin{aligned}
\frac{dj_l(x)}{dx} &= j_{l+1}(x) + \left(\frac{l}{x} - \frac{2l+1}{x} \right) j_l(x) \\
&= j_{l-1}(x) - \frac{l+1}{x}\, j_l(x) \ .
\end{aligned}
\tag{19.1.14}
$$

c) Here we also begin with (19.1.1)

$$
j_l(x) = \frac{x^l}{i\, 2^{l+1}l!} \int_{-i}^{i} e^{xu}\, (u^2+1)^l\, du
\tag{19.1.15}
$$

and simply differentiate to get

$$
\begin{aligned}
\frac{d}{dx}[x^{-l}\, j_l(x)] &= \frac{1}{i\, 2^{l+1}l!} \int_{-i}^{i} e^{xu}\, (u^2+1)^l\, u\, du \\
&= \frac{1}{i\, 2^{l+2}l!}\frac{1}{l+1} \int_{-i}^{i} e^{xu}\, d(u^2+1)^{l+1} \\
&= \frac{-x}{i\, 2^{l+2}(l+1)!} \int_{-i}^{i} e^{xu}\, (u^2+1)^{l+1}\, du \\
&= -x^{-l}\, j_{l+1}(x) \ .
\end{aligned}
\tag{19.1.16}
$$

As stated at the beginning, the calculations for $h_l^{(1)}$ and $h_l^{(2)}$ follow exactly the same lines.

19.2 Rodrigues Formula: Spherical Bessel Functions

A useful formula for generating any of the spherical Bessel functions is the Rodrigues formula

$$Z_l(x) = x^l \left(-\frac{1}{x}\frac{d}{dx} \right)^l Z_0 .$$

Use the results of problem 19.1 to get

$$Z_l(x) = -x^{l-1}\frac{d}{dx}\left(\frac{Z_{l-1}}{x^{l-1}} \right)$$

and hence derive the generating formula above.

Solution

To get the desired expression we start with the results of problem 19.1 a)

$$Z_{l-1}(x) = \frac{2l+1}{x} Z_l(x) - Z_{l+1}(x) \tag{19.2.17}$$

and combine it with 19.1 b) to get

$$
\begin{aligned}
\frac{dZ_l(x)}{dx} &= \frac{2l+1}{x} Z_l(x) - Z_{l+1}(x) - \frac{l+1}{x}Z_l(x) \\
&= \frac{l}{x}Z_l(x) - Z_{l+1}(x) .
\end{aligned}
\tag{19.2.18}
$$

Now change l to $l-1$ and rearrange terms to get:

$$Z_l(x) = \frac{l-1}{x} Z_{l-1}(x) - \frac{dZ_{l-1}(x)}{dx} . \tag{19.2.19}$$

Next, we use the identity

$$\frac{d}{dx}\left(\frac{Z_{l-1}}{x^{l-1}} \right) = -\frac{l-1}{x}\frac{Z_{l-1}}{x^{l-1}} + \frac{1}{x^{l-1}}\frac{dZ_{l-1}}{dx} . \tag{19.2.20}$$

Therefore,

$$Z_l(x) = \left(-\frac{d}{dx} + \frac{l-1}{x} \right) Z_{l-1}(x) = -x^{l-1}\frac{d}{dx}\left(\frac{Z_{l-1}}{x^{l-1}} \right) . \tag{19.2.21}$$

Iterating this expression yields the desired result.

19.3   Wronskian for Spherical Bessel Functions

Show that the Wronskian

$$W = j_l(x)n_l'(x) - n_l(x)j_l'(x)$$

satisfies the differential equation

$$\frac{dW}{dx} = -\frac{2}{x}W \ .$$

Solve this equation and use the behaviour of $j_l(x)$, $n_l(x)$ for small x to fix the constant of integration to get

$$W = \frac{1}{x^2} \ .$$

Hint: Start with the differential equations for $j_l(x)$ and $n_l(x)$.

Solution

We start with the differential equation for both j_l and n_l.

$$\left[\frac{d^2}{dx^2} + \frac{2}{x}\frac{d}{dx} + 1 - \frac{l(l+1)}{x^2}\right] j_l(x) \ = \ 0$$

$$\left[\frac{d^2}{dx^2} + \frac{2}{x}\frac{d}{dx} + 1 - \frac{l(l+1)}{x^2}\right] n_l(x) \ = \ 0 \ . \tag{19.3.22}$$

Now we multiply the first by $n_l(x)$ and the second by $j_l(x)$ and subtract to get

$$\frac{dW}{dx} = -\frac{2}{x}W \ . \tag{19.3.23}$$

Therefore,

$$W = \frac{C}{x^2} \ . \tag{19.3.24}$$

We can evaluate the constant C for any value of x. So we choose x near 0 and use the asymptotic form of the spherical Bessel functions as well as the definition of W. Then we find $C = 1$.

19.4   Superposition of Yukawa Potentials

Find an expression, in Born approximation, for the scattering amplitude due to scattering off a potential which is a superposition of Yukawa potentials

$$V(r) = \int_M^\infty d\mu\, \sigma(\mu)\frac{e^{-\mu r}}{r} \ . \tag{19.4.25}$$

Solution

In Born approximation the solution of the Schrödinger equation has the asymptotic form

$$\Psi_{Born}(\vec{r}) \to e^{i\vec{k}\cdot\vec{r}} - \frac{2m}{\hbar^2}\frac{1}{4\pi}\frac{e^{ikr}}{r}\int e^{i\vec{q}\cdot\vec{r}'}V(r')\,d^3r' \qquad (19.4.26)$$

where

$$\vec{q} = \vec{k} - \vec{k}' \qquad (19.4.27)$$

is the momentum transfer. So the scattering amplitude is given by

$$
\begin{aligned}
f(k,\theta) &= -\frac{2m}{\hbar^2}\frac{1}{4\pi}\int e^{i(\vec{k}-\vec{k}')\cdot\vec{r}'}V(r')\,d^3r' \\
&= -\frac{2m}{\hbar^2}\frac{1}{4\pi}\int_M^\infty d\mu\,\sigma(\mu)\int e^{i(\vec{k}-\vec{k}')\cdot\vec{r}'}\frac{e^{-\mu r}}{r}\,d^3r' \\
&= -\frac{2m}{\hbar^2}\int_M^\infty d\mu\,\frac{\sigma(\mu)}{q^2+\mu^2}\ . \qquad (19.4.28)
\end{aligned}
$$

Here,

$$q^2 = 4k^2\sin^2(\theta/2)\ . \qquad (19.4.29)$$

19.5 Born Approximation for Gaussian Potential

a) Calculate the differential cross section in first Born approximation for the potential

$$V(r) = V_0\,e^{-\mu r^2}\ .$$

b) To the same approximation compute the s-wave ($l = 0$) phase shift.

Solution

a) We have

$$V(r) = V_0\,e^{-\mu r^2}\ . \qquad (19.5.30)$$

Then in first Born approximation we have the scattering amplitude

$$f(k,\theta) = -\frac{1}{4\pi}\int e^{i\vec{q}\cdot\vec{r}}\frac{2mV_0}{\hbar^2}e^{-\mu r^2}\,d^3r\ , \qquad (19.5.31)$$

where

$$q = 2k\sin(\theta/2)\ . \qquad (19.5.32)$$

Therefore, we get

$$
\begin{aligned}
f(k,\theta) &= -\frac{2\pi}{4\pi}\frac{2mV_0}{\hbar^2}\int_0^\infty r^2\,e^{-\mu r^2}\,dr\int_{-1}^1 e^{iqru}\,du \\
&= -\frac{2mV_0}{\hbar^2 q}\int_0^\infty r\,e^{-\mu r^2}\sin qr\,dr \\
&= -\frac{2mV_0}{\hbar^2}\frac{\sqrt{\pi}}{4\mu^{3/2}}e^{-q^2/4\mu}\ .
\end{aligned}
\tag{19.5.33}
$$

Hence we find that

$$
\frac{d\sigma}{d\Omega} = |f(k,\theta)|^2 = \frac{\pi}{4}\frac{m^2 V_0^2}{\hbar^4\mu^3}e^{-q^2/2\mu}\ .
\tag{19.5.34}
$$

b) We have in general that

$$
f(k,\theta) = \frac{1}{k}\sum_{l=0}^\infty (2l+1)\,e^{i\delta_l}\,\sin\delta_l\,P_l(\cos\theta)\ .
\tag{19.5.35}
$$

Therefore, we can extract the $l=0$ phase shift from

$$
\frac{1}{k}e^{i\delta_0}\sin\delta_0 = \int_0^\pi f(k,\theta)\sin\theta\,d\theta\ .
\tag{19.5.36}
$$

Applying this to our result above, we find

$$
\frac{1}{k}e^{i\delta_0}\sin\delta_0 = -\frac{mV_0}{2\hbar^2\mu}\sqrt{\frac{\pi}{\mu}}\int_0^\pi \exp\left(-\frac{k^2}{\mu}\sin^2(\theta/2)\right)\sin\theta\,d\theta\ .
\tag{19.5.37}
$$

Therefore we get

$$
\frac{1}{k}e^{i\delta_0}\sin\delta_0 = -\frac{2mV_0}{\hbar^2 k^2}\sqrt{\frac{\pi}{\mu}}\left(1-e^{-k^2/mu}\right)\ .
\tag{19.5.38}
$$

Solving for the phase shift we get

$$
\delta_0 = \frac{1}{2i}\ln\left[1-\frac{2mV_0}{\hbar^2 k^2}\sqrt{\frac{\pi}{\mu}}\left(1-e^{-k^2/mu}\right)\right]\ .
\tag{19.5.39}
$$

19.6 Born Approximation for Square Well

Repeat the previous problem for the potential

$$
V(r) = \begin{cases} -V_0 & \text{for } r<a \\ 0 & \text{for } r\geq a \end{cases}\ .
$$

Solution

This time we have

$$V(r) = \begin{cases} -V_0 & r < a \\ 0 & r \geq a \end{cases} .$$ (19.6.40)

Therefore, as before, with

$$q = 2k \sin(\theta/2)$$ (19.6.41)

we get

$$
\begin{aligned}
f(k, \theta) &= \frac{1}{4\pi} \frac{2mV_0}{\hbar^2} \int_0^a r^2 \, dr \int e^{i\mathbf{q} \cdot \mathbf{r}} \sin \theta d\theta \, d\varphi \\
&= \frac{2mV_0}{q\hbar^2} \int_0^a r \sin qr \, dr \\
&= -\frac{2mV_0 a}{q^2 \hbar^2} \left(\cos qa - \frac{\sin qa}{qa} \right) .
\end{aligned}
$$ (19.6.42)

a) Hence,

$$\frac{d\sigma}{d\Omega} = |f(k, \theta)|^2 = \frac{2mV_0^2 a^2}{q^4 \hbar^4} \left(\cos qa - \frac{\sin qa}{qa} \right)^2 .$$ (19.6.43)

b) Proceeding as for part b) in question 19.6 above we find

$$\frac{1}{k} e^{i\delta_0} \sin \delta_0 = \int_0^\pi f(k, \theta) \sin \theta \, d\theta ,$$ (19.6.44)

so that

$$
\begin{aligned}
&\frac{1}{k} e^{i\delta_0} \sin \delta_0 \\
&= -\frac{2mV_0}{\hbar^2} \int_0^\pi \left[\frac{a \cos(2ak \sin \theta/2)}{(2k \sin \theta/2)^2} - \frac{\sin(2ak \sin \theta/2)}{(2k \sin \theta/2)^3} \right] \sin \theta \, d\theta .
\end{aligned}
$$ (19.6.45)

Thus, letting

$$x = 2ak \sin \theta/2$$ (19.6.46)

we can rewrite the integral as

$$\frac{1}{k} e^{i\delta_0} \sin \delta_0 = -\frac{2mV_0 a^3}{\hbar^2} \int_0^{2ka} \left(\frac{\cos x}{x} - \frac{\sin x}{x^2} \right) dx .$$ (19.6.47)

The integrand is a perfect differential

$$\frac{\cos x}{x} - \frac{\sin x}{x^2} = \frac{d}{dx} \left(\frac{\sin x}{x} \right) .$$ (19.6.48)

So, we immediately get

$$\frac{1}{k} e^{i\delta_0} \sin \delta_0 = \frac{2mV_0 a^3}{\hbar^2} \left(1 - \frac{\sin 2ka}{2ka} \right) .$$ (19.6.49)

Then finally, using the fact that the phase shift is a small angle, we find

$$\delta_0 = \frac{2mV_0 a^2 (ka)}{\hbar^2} \left(1 - \frac{\sin 2ka}{2ka} \right) .$$ (19.6.50)

19.7 Phase Shifts for Delta-function Potential

Compute the phase shifts for scattering by a potential

$$V(r) = V_0 a\, \delta(r - a) \ .$$

Solution

With the potential

$$V(r) = V_0 a\, \delta(r - a) \tag{19.7.51}$$

we have basically free solutions so that we can write the radial solutions as

$$R_l(r) = \begin{cases} B_l\, j_l(kr) & r < a \\ A_l[\cos \delta_l\, j_l(kr) - \sin \delta_l\, n_l(kr)] & r > a \end{cases} \ . \tag{19.7.52}$$

We have already imposed the physical boundary condition that the wavefunction is free of singularities at the origin. We now impose that the wavefunction is continuous at $r = a$ and that the first derivative is discontinuous at $r = a$. The discontinuity is computed by integrating the radial differential equation about $r = a$. Thus, we have

$$B_l\, j_l(ka) = A_l[\cos \delta_l\, j_l(ka) - \sin \delta_l\, n_l(ka)] \tag{19.7.53}$$

and

$$-\frac{\hbar^2}{2m}\left[\left.\frac{dR_l}{dr}\right|_{a+0} - \left.\frac{dR_l}{dr}\right|_{a-0}\right] = -V_0 a\, R_l(a) \ . \tag{19.7.54}$$

Writing out this last equation and substituting for B_l from the first equation we find with

$$Q = \frac{2mV_0 a^2}{\hbar^2 (ka)} \tag{19.7.55}$$

that

$$[\cos \delta_l\, j_l'(ka) - \sin \delta_l\, n_l'] - [\cos \delta_l\, j_l'(ka) - \sin \delta_l\, \frac{n_l(ka)}{j_l(ka)} j_l']$$

$$= Q[\cos \delta_l\, j_l(ka) - \sin \delta_l\, n_l] \ . \tag{19.7.56}$$

Now, using the Wronskian (see problem 9.3)

$$j_l(x)n_l'(x) - j_l'(x)n_l(x) = \frac{1}{x^2} \tag{19.7.57}$$

to simplify, we can solve for $\tan \delta_l$ to get

$$\tan \delta_l = \frac{(2mV_0 a^2)/\hbar^2\, j_l^2(ka)}{(2mV_0 a^2)/\hbar^2\, n_l(ka)j_l(ka) - 1/(ka)} \ . \tag{19.7.58}$$

Another way to solve this problem is to use the partial wave Lippmann-Schwinger equations

$$\psi_l^{(+)}(r) = \psi_l^{(0)}(r) + \int G_l^{(+)}(r, r')U(r')\psi_l^{(+)}(r')\, r'^2 dr' \qquad (19.7.59)$$

where

$$G_l^{(+)}(r, r') = -ik\, j_l(kr_<)\, h_l^{(1)}(kr_>) \ . \qquad (19.7.60)$$

Since we are interested in the asymptotic region we have $r \geq a$ and with

$$U(r) = \frac{2mV_0a}{\hbar^2} \delta(r - a) \qquad (19.7.61)$$

we find that

$$\psi_l^{(+)}(r) = \psi_l^{(0)}(r) - ika^2 \frac{2mV_0a}{\hbar^2} j_l(ka)h_l^{(1)}(kr)\psi_l^{(+)}(a) \ . \qquad (19.7.62)$$

Putting $r = a$ we get

$$\psi_l^{(+)}(a) = \psi_l^{(0)}(a) - ika \frac{2mV_0a^2}{\hbar^2} j_l(ka)h_l^{(1)}(ka)\psi_l^{(+)}(a) \qquad (19.7.63)$$

so that

$$\psi_l^{(+)}(a) = \left[1 + ika \frac{2mV_0a^2}{\hbar^2} j_l(ka)h_l^{(1)}(ka) \right]^{-1} \psi_l^{(0)}(a) \ . \qquad (19.7.64)$$

Thus,

$$\psi_l^{(+)}(r) = \psi_l^{(0)}(r) - \frac{ika \frac{2mV_0a^2}{\hbar^2} j_l(ka)\psi_l^{(0)}(a)}{1 + ika \frac{2mV_0a}{\hbar^2} j_l(ka)h_l^{(1)}(ka)} h_l^{(1)}(kr) \ . \qquad (19.7.65)$$

Now, in general we have

$$\psi_l^{(0)}(r) = i^l(2l + 1)j_l(kr) \qquad (19.7.66)$$

and

$$\psi_l^{(+)}(r) \to \psi_l^{(0)}(r) + i^l(2l + 1)ie^{i\delta_l}\, \sin\delta_l\, h_l^{(1)}(kr) \ . \qquad (19.7.67)$$

Combining these results we find

$$i^l(2l + 1)ie^{i\delta_l}\, \sin\delta_l = -\frac{ika \frac{2mV_0a^2}{\hbar^2} j_l(ka)i^l(2l + 1)j_l(ka)}{1 + ika \frac{2mV_0a^2}{\hbar^2} j_l(ka)h_l^{(1)}(ka)} \ . \qquad (19.7.68)$$

From this it follows that

$$\frac{e^{2i\delta_l} - 1}{2i} = -\frac{ka \frac{2mV_0a^2}{\hbar^2} j_l(ka)^2}{1 + ika \frac{2mV_0a^2}{\hbar^2} j_l(ka)h_l^{(1)}(ka)} \ . \qquad (19.7.69)$$

Therefore, after some simplification, it follows that

$$e^{2i\delta_l} = \frac{1 - ka\frac{2mV_0a^2}{\hbar^2}j_l(ka)n_l(ka)}{1 + ika\frac{2mV_0a^2}{\hbar^2}j_l(ka)h_l^{(1)}(ka)} \ . \tag{19.7.70}$$

If we now write

$$\tan\delta_l = \frac{e^{2i\delta_l} - 1}{2i} \times \frac{2}{e^{2i\delta_l} + 1} \tag{19.7.71}$$

we see that the result coincides with our previous result (19.7.58).

19.8   Phase Shifts for Yukawa Potential

Compute approximate $l = 0$ and $l = 1$ phase shifts for scattering a high energy particle of mass m by a short range potential

$$V(r) = V_0\frac{e^{-\alpha r}}{r} \ .$$

Use whatever seems to be an appropriate approximation.

Solution

For a high energy particle, the Born approximation should be valid. Furthermore, for the Yukawa potential

$$V(r) = V_0\frac{e^{-\alpha r}}{r} \tag{19.8.72}$$

we have the scattering amplitude (in Born approximation) as given, for instance, in [19.2].

$$f(k,\theta) = -\frac{2mV_0}{\hbar^2}\frac{1}{(\vec{k} - \vec{k'})^2 + \alpha^2} = -\frac{2mV_0}{\hbar^2}\frac{1}{4k^2\sin^2(\theta/2) + \alpha^2} \ . \tag{19.8.73}$$

Also the partial wave expansion is given by

$$f(k,\theta) = \frac{i}{2k}\sum_{l=0}^{\infty}(2l + 1)[1 - S_l(k)]P_l(\cos\theta) \tag{19.8.74}$$

where

$$S_l(k) = e^{2i\delta_l(k)} \ . \tag{19.8.75}$$

However, the Born approximation assumes that the scattering amplitude is small. Thus, we must have small phase shifts. In this case we replace $1 - S_l(k)$ by

$$1 - e^{-2i\delta_l(k)} \approx 2i\delta_l(k) \ . \tag{19.8.76}$$

Now projecting out the $l = 0$ partial wave (by integrating with $P_0(\cos \theta) = 1$) we get

$$
\begin{aligned}
\frac{i}{2k} 2i\delta_0(k) &= -\frac{2mV_0}{\hbar^2} \int_0^\pi \frac{\sin \theta \, d\theta}{4k^2 \sin^2(\theta/2) + \alpha^2} \\
&= -\frac{2mV_0}{\hbar^2} \int_0^\pi \frac{2 \sin(\theta/2) \cos(\theta/2) \, d\theta}{4k^2 \sin^2(\theta/2) + \alpha^2} \\
&= -\frac{2mV_0}{\hbar^2 k^2} \ln \left[1 + \frac{4k^2}{\alpha^2} \right].
\end{aligned}
\tag{19.8.77}
$$

Hence we find

$$
\delta_0(k) = \frac{2mV_0}{\hbar^2 k} \ln \left[1 + \frac{4k^2}{\alpha^2} \right].
\tag{19.8.78}
$$

Similarly for the $l = 1$ partial wave we get by integrating with $P_1(\cos \theta) = \cos \theta$.

$$
\begin{aligned}
\frac{i}{2k} 2i\delta_1(k) &= -\frac{2mV_0}{\hbar^2} \int_0^\pi \frac{\sin \theta \cos \theta \, d\theta}{4k^2 \sin^2(\theta/2) + \alpha^2} \\
&= -\frac{2mV_0}{\hbar^2} \int_0^\pi \frac{2 \sin(\theta/2) \cos(\theta/2)[1 - 2\sin^2(\theta/2)] \, d\theta}{4k^2 \sin^2(\theta/2) + \alpha^2} \\
&= -\frac{2mV_0}{\hbar^2 k^2} \left\{ 1 + (1 - \frac{\alpha^2}{k^2}) \ln \left[1 + \frac{4k^2}{\alpha^2} \right] \right\}.
\end{aligned}
\tag{19.8.79}
$$

Therefore,

$$
\delta_1(k) = \frac{2mV_0}{\hbar^2 k} \left\{ 1 + (1 - \frac{\alpha^2}{k^2}) \ln \left[1 + \frac{4k^2}{\alpha^2} \right] \right\}.
\tag{19.8.80}
$$

19.9 Low Energy s-Wave Amplitude

Show that the scattering amplitude for low energy s-waves may be written as

$$
f_0(k) = -\frac{c_0}{1 + ic_0 k - 1/2 \, c_0 k r_0^2}
\tag{19.9.81}
$$

as well as in the form

$$
f_0(k) = \frac{1}{k \cot \delta - ik}
\tag{19.9.82}
$$

where c_0 is the scattering length and r_0 is the effective range [19.3]. Also verify that both versions of the amplitude satisfy the optical theorem.

Solution

In terms of the phase shift the scattering amplitude is

$$
f_0(k) = \frac{i}{2k} \left(1 - e^{2i\delta} \right).
\tag{19.9.83}
$$

This may be rewritten as

$$
\begin{aligned}
f_0(k) &= \frac{i}{2k}\left(1 - e^{2i\delta}\right) \\
&= \frac{e^{i\delta}}{k}\sin\delta \\
&= \frac{\sin\delta}{k\,e^{-i\delta}} = \frac{\sin\delta}{k\cos\delta - ik\sin\delta} \\
&= \frac{1}{k\cot\delta - ik} \ .
\end{aligned}
\tag{19.9.84}
$$

On the other hand for low energies the phase shift is given in terms of the scattering length c_0 and effective range r_0 by

$$
k\cot\delta = -\frac{1}{c_0} + \frac{1}{2}kr_0^2 \ .
\tag{19.9.85}
$$

Substituting this into (19.9.84) we get

$$
\begin{aligned}
f_0(k) &= \frac{1}{-1/c_0 - ik + 1/2\,kr_0^2} \\
&= -\frac{c_0}{1 + ic_0 k - 1/2 c_0 kr_0^2} \ .
\end{aligned}
\tag{19.9.86}
$$

The optical theorem states that the total cross-section σ is given by

$$
\sigma = \frac{4\pi}{k}\Im[f_0(k)] \ .
\tag{19.9.87}
$$

Using the second line of (19.9.84) we see that

$$
\frac{4\pi}{k}\Im[f_0(k)] = \frac{4\pi}{k}\frac{\sin^2\delta}{k} \ .
\tag{19.9.88}
$$

On the other hand the total cross-section is given by

$$
\begin{aligned}
\sigma &= \int \frac{d\sigma}{d\Omega}\,d\Omega \\
&= \int |f(k)|^2\,d\Omega = 4\pi|f(k)|^2 \\
&= \frac{4\pi}{k^2}\sin^2\delta \ .
\end{aligned}
\tag{19.9.89}
$$

So, in this case, the optical theorem is verified.

If we start with (19.9.81) we have

$$
\begin{aligned}
\sigma &= \frac{4\pi}{k}\Im[f_0(k)] \\
&= \frac{4\pi}{k}\frac{-c_0(-c_0 k)}{(1 - 1/2\,c_0 kr_0^2)^2 + c_0^2 k^2} \\
&= \frac{4\pi c_0^2}{(1 - 1/2\,c_0 kr_0^2)^2 + c_0^2 k^2} \ .
\end{aligned}
\tag{19.9.90}
$$

On the other hand by direct computation we see that

$$
\begin{aligned}
\sigma &= \int \frac{d\sigma}{d\Omega} \, d\Omega \\
&= \int |f(k)|^2 \, d\Omega = 4\pi |f(k)|^2 \\
&= \frac{4\pi c_0^2}{(1 - 1/2 \, c_0 k r_0^2)^2 + c_0^2 k^2} \, .
\end{aligned}
\tag{19.9.91}
$$

Thus, in this case we have also verified the optical theorem.

19.10 Spherical Potential Shell

Consider a potential "shell" of value V_0 between $r = a$ and $r = b$ and zero otherwise. Calculate the s-wave phase shift and show that for large V_0 (with respect to what?) resonances occur approximately at energies which would be bound states if V_0 were infinite.

Solution

The general scattering solution for the lth partial wave can be written

$$
R_l(r) = \begin{cases}
j_l(kr) & r < a \\
A_l j_l(Kr) + B_l n_l(Kr) & a < r < b \\
C_l[\cos \delta_l j_l(kr) - \sin \delta_l n_l(kr)] & r > b
\end{cases}
\tag{19.10.92}
$$

where

$$
E = \frac{\hbar^2 k^2}{2m} \quad , \quad E - V_0 = \frac{\hbar^2 K^2}{2m} \, .
\tag{19.10.93}
$$

Imposing continuity of the wavefunction and the first derivative at $r = a$ we get

$$
\begin{aligned}
j_l(ka) &= A_l j_l(Ka) + B_l n_l(Ka) \\
k \, j_l'(ka) &= K[A_l j_l'(Ka) + B_l n_l(Ka)'] \, .
\end{aligned}
\tag{19.10.94}
$$

Then, after some algebra and using the Wronskian for the spherical Bessel functions we find

$$
\begin{aligned}
A_l &= Ka^2[k j_l(ka) n_l'(Ka) - K j_l'(ka) n_l(Ka)] \\
B_l &= -Ka^2[k j_l'(ka) j_l(Ka) - K j_l(ka) j_l'(Ka)] \, .
\end{aligned}
\tag{19.10.95}
$$

Equating logarithmic derivatives at $r = b$ we get

$$
\frac{k[\cos \delta_l j_l'(kb) - \sin \delta_l n_l'(kb)]}{\cos \delta_l j_l(kb) - \sin \delta_l n_l(kb)} = \frac{K[A_l j_l'(Kb) + B_l n_l'(Kb)]}{A_l j_l(Kb) + B_l n_l(Kb)} \, .
\tag{19.10.96}
$$

Dividing the numerator and denominator of the left side by $\cos \delta_l$ and solving for $\tan \delta_l$ we get

$$
\begin{aligned}
&\tan \delta_l \\
&= \frac{K[A_l j_l'(Kb) + B_l n_l'(Kb)] j_l(kb) - k[A_l j_l(Kb) + B_l n_l(Kb)] j_l'(kb)}{k[A_l j_l(Kb) + B_l n_l(Kb)] n_l'(kb) - K[A_l j_l'(Kb) + B_l n_l'(Kb)] n_l(kb)} \, .
\end{aligned}
\tag{19.10.97}
$$

For the special case of $l = 0$ we have

$$j_0(x) = \frac{\sin x}{x} \quad , \quad n_0(x) = -\frac{\cos x}{x} . \tag{19.10.98}$$

Thus, we can either simplify the general case using these expressions or start from scratch, as we now do.

$$R_0(r) = \begin{cases} \frac{\sin kr}{kr} & r < a \\ A\frac{\sin Kr}{Kr} - B\frac{\cos Kr}{Kr} & a < r < b \\ C\frac{\sin(kr+\delta_0)}{kr} & r > b \end{cases} . \tag{19.10.99}$$

Matching the functions and derivatives at $r = a$ we find

$$k\left[\frac{\cos ka}{ka} - \frac{\sin ka}{(ka)^2}\right] = K\left[A\left(\frac{\cos Ka}{Ka} - \frac{\sin Ka}{(Ka)^2}\right) + B\left(\frac{\sin Ka}{Ka} + \frac{\cos Ka}{(Ka)^2}\right)\right]$$

$$\frac{\sin ka}{ka} = A\frac{\sin Ka}{Ka} - B\frac{\cos Ka}{Ka} . \tag{19.10.100}$$

Solving this pair of equations we get

$$A = \frac{K}{k}\sin Ka \sin ka + \cos Ka \cos ka$$

$$B = -\frac{K}{k}\cos Ka \sin ka + \sin Ka \cos ka . \tag{19.10.101}$$

Next we match logarithmic derivatives at $r = b$ and get after a little algebra

$$\cot(kb + \delta_0) = \frac{1}{kb} + \frac{K[(A + B/Kb) - (A/Kb - B)\tan Kb]}{k[A \tan Kb - B]} . \tag{19.10.102}$$

At energies corresponding to a box extending from $r = a$ to $r = b$ we have that

$$R_0(a) = R_0(b) = 0 \tag{19.10.103}$$

so that we get

$$B = A \tan Ka \quad \text{and} \quad B = A \tan Kb . \tag{19.10.104}$$

Equation (19.10.102) now clearly shows that at such energies

$$\cot(kb + \delta_0) = \pm\infty . \tag{19.10.105}$$

Therefore,

$$kb + \delta_0 = n\pi + \pi/2 . \tag{19.10.106}$$

For small values of kb this is the condition for a resonance. So, this approximation means that V_0 is large compared to the energy.

19.11   Expressions for $j_0(x)$ and $n_0(x)$

Verify directly by using the differential equation and their behaviour near $x = 0$ that

$$j_0(x) = \frac{\sin x}{x} \quad \text{and} \quad n_0(x) = -\frac{\cos x}{x} .$$

Solution

We have that near $x = 0$

$$\frac{\sin x}{x} = 1 \quad \text{and} \quad -\frac{\cos x}{x} = -\frac{1}{x} \; . \tag{19.11.107}$$

Thus, they have the correct asymptotic behaviour to be $j_0(x)$ and $n_0(x)$ respectively. Furthermore we find

$$\frac{d}{dx} \left(\frac{\sin x}{x} \right) = \frac{\cos x}{x} - \frac{\sin x}{x^2} \tag{19.11.108}$$

and

$$\frac{d^2}{dx^2} \left(\frac{\sin x}{x} \right) = -\frac{\sin x}{x} - 2\frac{\cos x}{x^2} + 2\frac{\sin x}{x^3} \; . \tag{19.11.109}$$

Therefore,

$$\left(\frac{d^2}{dx^2} + \frac{2}{x}\frac{d}{dx} + 1 \right) \left(\frac{\sin x}{x} \right) = 0 \; . \tag{19.11.110}$$

Similarly

$$\frac{d}{dx} \left(-\frac{\cos x}{x} \right) = \frac{\sin x}{x} + \frac{\cos x}{x^2} \tag{19.11.111}$$

and

$$\frac{d^2}{dx^2} \left(\frac{\cos x}{x} \right) = \frac{\cos x}{x} - 2\frac{\sin x}{x^2} - 2\frac{\cos x}{x^3} \; . \tag{19.11.112}$$

Therefore,

$$\left(\frac{d^2}{dx^2} + \frac{2}{x}\frac{d}{dx} + 1 \right) \left(\frac{\cos x}{x} \right) = 0 \; . \tag{19.11.113}$$

19.12   Effective Range, Scattering Length

Given the potential

$$V(r) = \begin{cases} -V_0 & r < a \\ 0 & r > a \end{cases}$$

find the effective range and the scattering length for the s-wave ($l = 0$).

Solution

As a first step we need to find the s-wave phase shift. To a good approximation this is given by equation [19.2]

$$
\begin{aligned}
\sin \delta_0 &\approx \int_0^a r^2 dr \left(-\frac{2mV_0}{\hbar^2} \right) \frac{\sin^2(kr)}{(kr)^2} \\
&= -\frac{mV_0}{\hbar^2 k^2} \frac{1}{2} \left[ka - \frac{1}{2} \sin 2ka \right] .
\end{aligned}
\tag{19.12.114}
$$

For $k \to 0$ this yields

$$
\begin{aligned}
\sin \delta_0 &\approx -\frac{2mV_0 a^2}{\hbar^2} \left[\frac{1}{3} ka - \frac{1}{15}(ka)^3 \right] \\
&= -\alpha \frac{1}{3} \left[ka - \frac{1}{5}(ka)^3 \right] .
\end{aligned}
\tag{19.12.115}
$$

Here we have introduced the dimensionless parameter

$$
\alpha = \frac{2mV_0 a^2}{\hbar^2} .
\tag{19.12.116}
$$

The scattering length and effective range are defined for $k \to 0$ by

$$
k \cot \delta_0 \approx -\frac{1}{c_0} + \frac{1}{2} k^2 r_0 .
\tag{19.12.117}
$$

Using our result for $\sin \delta_0$ we find

$$
\cos \delta_0 = \sqrt{1 - \sin^2 \delta_0} \approx 1 - \frac{1}{18} \alpha^2 (ka)^2 .
\tag{19.12.118}
$$

Therefore, after a little algebra we get

$$
k \cot \delta_0 \approx -\frac{3}{\alpha a} + \frac{1}{2} k^2 \left(\alpha/3 - 6/(5\alpha) \right) a .
\tag{19.12.119}
$$

So we can read off the answer.

$$
c_0 = \frac{1}{3} \alpha a \quad , \quad r_0 = [\alpha/3 - 6/(5\alpha)] a .
\tag{19.12.120}
$$

19.13   Effective Range, Scattering Length: Yukawa Potential

Repeat the previous problem for the Yukawa potential

$$
V = -V_0 \frac{e^{-\mu r}}{r} .
$$

Solution

Again we start with the approximate equation for $\sin \delta_0$

$$\sin \delta_0 \approx \int_0^a r^2 dr \left(-\frac{2mV}{\hbar^2} \right) \frac{e^{-\mu r}}{\mu r} \frac{\sin^2(kr)}{(kr)^2} \,. \tag{19.13.121}$$

In this case the parameter a is determined by $1/\mu$. However, it is easier, and more exact, to take the upper limit in the integral as ∞. We then get

$$\begin{aligned}
\sin \delta_0 &\approx \int_0^\infty r^2 dr \left(-\frac{2mV}{\hbar^2} \right) \frac{e^{-\mu r}}{\mu r} \frac{\sin^2(kr)}{(kr)^2} \\
&= -\frac{2mV}{\hbar^2 \mu k} \ln \left[1 + \frac{4k^2}{\mu^2} \right] \\
&\approx -\frac{2mV}{\hbar^2 \mu^2} \left[\frac{k}{\mu} - \frac{k^3}{\mu^3} \right] \\
&= -\beta \frac{k}{\mu} \left[1 - \frac{k^2}{\mu^2} \right]
\end{aligned} \tag{19.13.122}$$

where again we have introduced a dimensionless parameter

$$\beta = \frac{2mV}{\hbar^2 \mu^2} \,. \tag{19.13.123}$$

So, after a little algebra we get

$$k \cot \delta_0 \approx -\frac{\mu}{\beta} + \frac{1}{2} k^2 (\beta - 1/\beta) \frac{1}{\mu} \,. \tag{19.13.124}$$

So,

$$c_0 = \frac{\beta}{\mu} \quad , \quad r_0 = (\beta - 1/\beta) \frac{1}{\mu} \,. \tag{19.13.125}$$

19.14   Shape-independent Parameters

Use the results of problems 19.12 and 19.13 to fix the parameters of the Yukawa potential in terms of those of the square well so that both yield the same s-wave scattering length and effective range. The fact that this is possible is what is meant by calling this a "shape-independent" approximation.

Solution

From the previous two problems we get:
For the Yukawa potential

$$c_0 = \frac{\beta}{\mu} \quad , \quad r_0 = (\beta - 1/\beta) \frac{1}{\mu} \tag{19.14.126}$$

where

$$\beta = \frac{2mV}{\hbar^2\mu^2} \ .$$ (19.14.127)

For the square well we get

$$c_0 = \frac{1}{3}\alpha a \quad , \quad r_0 = (\alpha/3 - 6/(5\alpha))\, a$$ (19.14.128)

where

$$\alpha = \frac{2mV_0 a^2}{\hbar^2} \ .$$ (19.14.129)

Equating the two scattering lengths we find

$$\frac{2mV}{\hbar^2\mu^3} = \frac{1}{3}\frac{2mV_0 a^2}{\hbar^2} a \ .$$ (19.14.130)

So that

$$V = \frac{1}{3}(\mu a)^3\, V_0 \ .$$ (19.14.131)

Then, we also find that

$$\beta = \frac{2}{3}\alpha(\mu a) \ .$$ (19.14.132)

So, equating the two effective ranges, we get

$$\frac{1}{3}\alpha(\mu a) - \frac{3}{\alpha(\mu a)} = \left(\frac{\alpha}{3} - \frac{6}{5\alpha}\right)(\mu a) \ .$$ (19.14.133)

This immediately yields that

$$\mu a = \sqrt{\frac{5}{2}} \ .$$ (19.14.134)

So, finally

$$V = \frac{1}{3}\left(\frac{5}{2}\right)^{3/2} V_0 \ .$$ (19.14.135)

19.15   Phase Shifts for Hard Sphere

a) Find the phase shifts for scattering by a hard sphere

$$V(r) = \begin{cases} \infty & r < a \\ 0 & r > a \end{cases} \ .$$

b) Find the total cross-section for an incoming energy

$$E = \frac{\hbar^2 k^2}{2m}$$

in the two limits:

$$k \to 0$$

$$k \to \infty .$$

Give a physical explanation for the factors of 4 and 2.
Hint: For $k \to \infty$ use the asymptotic forms of j_l and n_l to obtain a simple form
for $\sin^2 \delta_l$. Furthermore, replace the sum over l by an integral so that

$$\sigma = \sum_{l=0}^{l=ka} \sigma_l \approx \frac{4\pi}{k^2} \int_0^{ka} (2l+1) \sin^2 \delta_l \, dl .$$

Solution

a) We first have to find the phase shifts for the potential

$$V = \begin{cases} \infty & r \le a \\ 0 & r > a \end{cases} . \tag{19.15.136}$$

If we now define

$$k^2 = \frac{2mE}{\hbar^2} \tag{19.15.137}$$

the radial equation becomes

$$\left[\frac{d^2}{dr^2} + \frac{2}{r}\frac{d}{dr} - \frac{l(l+1)}{r^2} + k^2 \right] R_l(r) = 0 \quad r > a . \tag{19.15.138}$$

The solutions of this equations satisfying the condition that $R_l(r)$ corresponds
to a fixed incoming flux for large values of r is

$$R_l(r) = A_l \, e^{i\delta_l} \left[\cos \delta_l \, j_l(kr) - \sin \delta_l \, n_l(kr) \right] \quad r > a . \tag{19.15.139}$$

The appropriate boundary condition is that $R_l(a) = 0$. Thus, we get

$$\tan \delta_l = \frac{j_l(ka)}{n_l(ka)} . \tag{19.15.140}$$

b) To get the total cross-section we use the equation

$$\sigma = \frac{4\pi}{k^2} \sum_{l=0}^{\infty} (2l+1) \sin^2 \delta_l . \tag{19.15.141}$$

Now, for $k \to 0$

$$\tan \delta_l \approx \sin \delta_l = \frac{j_l(ka)}{n_l(ka)} \tag{19.15.142}$$

and we can use the asymptotic form of the spherical Bessel functions. Thus,

$$\sin \delta_l = \frac{j_l(ka)}{n_l(ka)} \approx -\frac{(ka)^{2l+1}}{(2l+1)!!(2l-1)!!} . \tag{19.15.143}$$

Clearly in the limit only the $l = 0$ term survives and we get

$$\sin^2 \delta_0 = (ka)^2 \ . \tag{19.15.144}$$

Thus,

$$\sigma = 4\pi a^2 \quad \text{for} \quad k \to 0 \ . \tag{19.15.145}$$

Classically, the differential cross-section is related to the impact parameter b (see figure 19.1) by

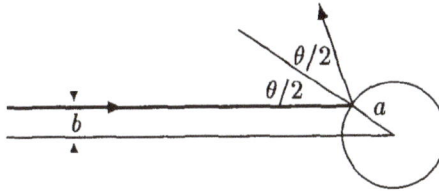

Figure 19.1: Classical scattering off a hard sphere.

$$\frac{d\sigma}{d\Omega} \, d\Omega = b \, db \, d\varphi \tag{19.15.146}$$

or

$$\frac{d\sigma}{d\Omega} = \frac{b}{\sin \theta} \frac{db}{d\theta} \ . \tag{19.15.147}$$

For a hard sphere we have (figure 19.1)

$$b = a \sin \theta/2 \ . \tag{19.15.148}$$

So,

$$\frac{d\sigma}{d\Omega} = \frac{\frac{1}{2} a^2 \sin \theta/2 \cos \theta/2}{\sin \theta} = \frac{1}{4} a^2 \ . \tag{19.15.149}$$

Thus,

$$\sigma = 4\pi \frac{1}{4} a^2 = \pi a^2 \ , \tag{19.15.150}$$

a result that is quite obvious since that is the cross-sectional area that intercepts the incoming beam.

The result for low energy $\sigma = 4\pi a^2$ (19.15.145) is clearly 4 times the classical cross-section of πa^2. The reason for this is that for $k \to 0$ the wavelength λ of the particle gets infinitely large and the particle scatters off the whole surface of area $4\pi a^2$ of the sphere rather than just off the cross-sectional area πa^2 perpendicular to the incident beam.

In the limit as $k \to \infty$ we have that

$$\tan \delta_l = \frac{j_l(ka)}{n_l(ka)} \to -\tan(ka - l\pi/2) \ . \tag{19.15.151}$$

Therefore,

$$\sin^2 \delta_l \to \sin^2(ka - l\pi/2) \ . \tag{19.15.152}$$

Now using the approximation suggested we have that

$$
\begin{aligned}
\sigma &\approx \frac{4\pi}{k^2} \int_0^{ka} (2l + 1) \sin^2 \delta_l \, dl \\
&= \int_0^{ka} (2l + 1) \sin^2(ka - l\pi/2) \, dl \\
&= \frac{8}{k^2} \int_{-ka}^{ka(\pi/2-1)} \left[\frac{4}{\pi}(x + ka) + 1\right] \sin^2 x \, dx \\
&\to \frac{8}{k^2} \left\{ \frac{(ka)^2}{\pi}[\pi^2/4 - \pi] + 2\frac{(ka)^2}{\pi}(\pi/2) \right\} \\
&= 2\pi a^2 \ . \tag{19.15.153}
\end{aligned}
$$

In this case, since we are approaching a classical limit, we would expect the cross-section to be just the classical cross-section of πa^2 and not $2\pi a^2$ that we found. The factor of 2 is due to the way the quantum mechanical cross-section is defined in terms of the *scattered* wave. Thus, one defines the total wavefunction as

$$\psi = \psi_{incident} + \psi_{scattered} \ . \tag{19.15.154}$$

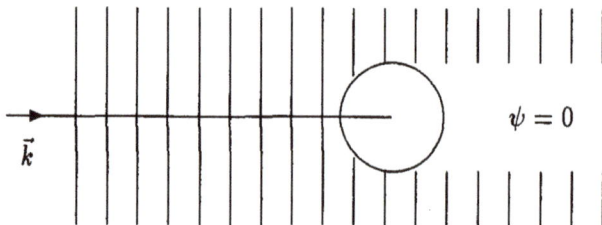

Figure 19.2: Shadow scattering off a hard sphere by high energy particles: The vertical lines depict wavefronts.

In the case of the hard sphere we have for $k \to \infty$ that in the shadow of the sphere the wavefunction vanishes identically (See figure 19.2). This means that, in this region, one defines

$$\psi_{scattered} = -\psi_{incident} \tag{19.15.155}$$

and so one obtains a cross-section that is twice the classical cross-section. This effect is known as *shadow scattering*.

19.16 Resonance for Square Well

Use the exact solution for the square well to find the condition on the potential for the s-wave ($l = 0$) to produce a resonance at an energy $E_0 = (\hbar^2 k_0^2)/2m$. Comment on the result that you obtain.

Solution

The s-wave phase shift is given by

$$\tan \delta_0 = \frac{k\, j_0(Ka) j_0'(ka) - K\, j_0'(Ka) j_0(ka)}{k\, j_0(Ka) n_0'(ka) - K\, j_0'(Ka) n_0(Ka)} . \tag{19.16.156}$$

Using the explicit form in terms of trigonometric functions for j_0 and n_0 we find that

$$\tan \delta_0 = \frac{k\, \tan(Ka) - K\, \tan(ka)}{K + k \tan(ka) \tan(Ka)} . \tag{19.16.157}$$

This can be rewritten as

$$\tan \delta_0 = \tan \left[\tan^{-1} \cdot \left(\frac{k}{K} \tan(Ka) \right) - ka \right] . \tag{19.16.158}$$

For a resonance we require that $\delta_0 = \pi/2$. Therefore we need that

$$Ka = \frac{\pi}{2} . \tag{19.16.159}$$

Writing out the definition of K we get

$$(Ka)^2 = \frac{\pi^2}{4} = \frac{2m(E + V_0)}{\hbar^2} . \tag{19.16.160}$$

So, for a resonance to occur we need

$$E_0 = \frac{\hbar^2 \pi^2}{8ma^2} - V_0 . \tag{19.16.161}$$

This is the energy for a particle in a box whose zero of energy is at $-V_0$.

19.17 Double Slit

a) Find the differential cross-section for scattering of electrons from a double slit. Assume the slits are cut into very thin material and that the incident beam is normal to the plane of the slits. Also assume that the scattering is weak.

Solution

Using the assumptions stated in the problem we may take the scattering potential to be

$$V = \begin{cases} V_0\,\delta(z) & \text{for } \frac{1}{2}(d-a) < |x| < \frac{1}{2}(d+a) \ , \ |y| < \frac{1}{2}h \\ 0 & \text{otherwise} \end{cases} \quad . \quad (19.17.162)$$

Notice that V_0 has the dimensions of energy$\times$length. Also, here we have chosen the slits to lie in the $x-y$ plane and thus we choose the incident beam along the z-axis. Since the scattering is weak we may use the Born approximation. Defining

$$U(\vec{r}) = \frac{2mV(\vec{r})}{\hbar^2} \qquad (19.17.163)$$

we find that

$$\tilde{U}(\vec{q}) = \int e^{i\vec{q}\cdot\vec{r}} U(\vec{r})\, d^3r$$

$$= U_0 \left[\int_{-(d+a)/2}^{-(d-a)/2} dx + \int_{(d-a)/2}^{(d+a)/2} dx \right] \int_{-h/2}^{h/2} dy \int_{-\infty}^{\infty} dz\, \delta(z)\, e^{i\vec{q}\cdot\vec{r}}$$

$$= 4U_0 \frac{\sin(q_y h/2)}{q_x q_y} \left[\sin(q_x(d+a)/2) - \sin(q_x(d-a)/2)\right]$$

$$= 8U_0 \frac{\sin(q_y h/2)}{q_x q_y} \sin(q_x a/2) \cos(q_x d/2) \ . \qquad (19.17.164)$$

Here

$$\vec{q} = \vec{k} - \vec{k}' \ . \qquad (19.17.165)$$

We also have that

$$\vec{k} = (0,0,k) \ , \quad \vec{k}' = (k\sin\theta, 0, k\cos\theta) \qquad (19.17.166)$$

so that

$$\vec{q} = (k\sin\theta, 0, k(1-\cos\theta)) = (k\sin\theta, 0, 2k\sin^2(\theta/2)) \ . \qquad (19.17.167)$$

In this case we find that the differential cross-section is given by

$$\frac{d\sigma}{d\Omega} = \frac{1}{16\pi^2} 64U_0^2 \frac{\sin^2(q_y h/2)}{q_y^2} \frac{\sin^2(q_x a/2)}{q_x} \cos^2(q_x d/2)$$

$$= \frac{4m^2 V_0^2}{\hbar^4} \frac{A^2}{4\pi^2} \frac{\sin^2(ka/2 \sin\theta)}{(ka/2)^2 \sin^2\theta} \cos^2(kd/2 \sin\theta) \ . \qquad (19.17.168)$$

Here, $A = ah$ is the area of the slits.

19.18 Born Approximation: Spherically Symmetric Potential

Show that for a spherically symmetric potential $V(r)$ the <u>total</u> cross-section in first Born approximation is given by

$$\sigma = \frac{m^2}{2\pi\hbar^4 k^2} \int_0^{2k} |\tilde{V}(q)|^2 \, q dq$$

where $\tilde{V}$ is the Fourier transform of V. Use this result together with the properties of the Fourier transform to conclude that for high energies and a potential of finite range a the scattering is appreciable only in the forward direction where the scattering angle θ satisfies

$$\sin(\theta/2) \leq \frac{1}{2ka} \, .$$

Solution

The scattering amplitude is given in Born approximation by

$$f(k,\theta) = -\frac{1}{4\pi} \frac{2m}{\hbar^2} \tilde{V}(\vec{q}) \tag{19.18.169}$$

where

$$q = 2k|\sin\theta/2| \, .$$

The differential cross-section is then

$$\frac{d\sigma}{d\Omega} = |f(k,\theta)|^2 = \frac{m^2}{4\pi^2\hbar^4}|\tilde{V}(\vec{q})|^2 \, . \tag{19.18.170}$$

The total cross-section is now given by

$$\begin{aligned}
\sigma &= \int \frac{d\sigma}{d\Omega} d\Omega \\
&= 2\pi \frac{m^2}{4\pi^2\hbar^4} \int_0^\pi |\tilde{V}(\vec{q})|^2 \sin\theta \, d\theta \\
&= \frac{m^2}{2\pi\hbar^4} \int_0^\pi |\tilde{V}(2k\sin\theta/2)|^2 \, 2\sin\theta/2 \, \cos\theta/2 \, d\theta \, . \tag{19.18.171}
\end{aligned}$$

Therefore, letting $q = 2k\sin\theta/2$ we find that

$$\sigma = \frac{m^2}{2\pi^2\hbar^4 k^2} \int_0^{2k} |\tilde{V}(q)|^2 \, q dq \, . \tag{19.18.172}$$

Now, we are told that the potential V has a finite range a. This means that the Fourier transform $\tilde{V}(q)$ is appreciably different from zero only for $q \leq 1/a$. But, from the definition of q this means that

$$2k\sin\theta/2 < 1/a \, . \tag{19.18.173}$$

Rewriting this we have that for appreciable scattering to occur we need

$$\sin\theta/2 \leq \frac{1}{2ka} \, . \tag{19.18.174}$$

19.19   Scattering from a Separable Potential

Nonlocal potentials occur in situations where reactions are possible as in the case of scattering off a bound state that may break up. *Nonlocal* potentials are such that in configuration space they are of the form

$$\langle \vec{r} | V | \vec{r}' \rangle = V(\vec{r}, \vec{r}') \ .$$

If

$$V(\vec{r}, \vec{r}') = V(\vec{r})\delta(\vec{r} - \vec{r}')$$

then the potential is *local*. Otherwise it is nonlocal. For a nonlocal potential we have that

$$
\begin{aligned}
\langle \vec{r} | V | \psi \rangle &= \int d^3r' \langle \vec{r} | V | \vec{r}' \rangle \langle \vec{r}' | \psi \rangle \\
&= \int d^3r' V(\vec{r}, \vec{r}') \psi(\vec{r}') \ .
\end{aligned}
\tag{19.19.175}
$$

A particularly simple form of nonlocal potential is a *separable* potential. This is of the form

$$V(\vec{r}, \vec{r}') = f(\vec{r}) \, g(\vec{r}') \ .\tag{19.19.176}$$

Solve the scattering problem for a separable potential for an incoming plane wave

$$\psi^{(0)} = \frac{1}{(2\pi)^{3/2}} e^{i\vec{k}\cdot\vec{r}} \ .\tag{19.19.177}$$

Solution

The Schrödinger equation for this case is

$$-\frac{\hbar^2}{2m}\nabla^2\psi + f(\vec{r})\int d^3r' \, g(\vec{r}')\psi(\vec{r}') = E\psi(\vec{r}) \ .\tag{19.19.178}$$

Introducing

$$k^2 = \frac{2mE}{\hbar^2} \quad , \quad F(\vec{r}) = \frac{2m}{\hbar^2} f(\vec{r})\tag{19.19.179}$$

we get

$$(\nabla^2 + k^2)\psi = F(\vec{r})\int d^3r' \, g(\vec{r}')\psi(\vec{r}') = C\, F(\vec{r})\tag{19.19.180}$$

where C is a constant defined by the integral. To solve this equation for incoming boundary conditions we require the incoming Green's function $G^{(+)}$ given by

$$(\nabla^2 + k^2)G^{(+)}(\vec{r}, \vec{r}') = \delta(\vec{r} - \vec{r}') \ .\tag{19.19.181}$$

The solution is

$$G^{(+)}(\vec{r}, \vec{r}') = -\frac{1}{4\pi} \frac{e^{ik|\vec{r}-\vec{r}'|}}{|\vec{r} - \vec{r}'|} \ . \tag{19.19.182}$$

The Schrödinger equation is now solved by

$$\psi^{(+)} = \frac{1}{(2\pi)^{3/2}} e^{i\vec{k}\cdot\vec{r}} - \frac{C}{4\pi} \int d^3r' \frac{e^{ik|\vec{r}-\vec{r}'|}}{|\vec{r} - \vec{r}'|} F(\vec{r}') \ . \tag{19.19.183}$$

To determine the constant C we multiply this equation by $g(\vec{r})$ and integrate over d^3r to get

$$C = \frac{1}{(2\pi)^{3/2}} \tilde{g}(\vec{k}) - \frac{C}{4\pi} \int d^3r' \, d^3r \frac{e^{ik|\vec{r}-\vec{r}'|}}{|\vec{r} - \vec{r}'|} F(\vec{r}') g(\vec{r}) \ . \tag{19.19.184}$$

This is now an algebraic equation for C. Thus, calling

$$A = \frac{1}{4\pi} \int d^3r' \, d^3r \frac{e^{ik|\vec{r}-\vec{r}'|}}{|\vec{r} - \vec{r}'|} F(\vec{r}') g(\vec{r}) \ , \tag{19.19.185}$$

we finally get

$$C = \frac{(2\pi)^{3/2} \tilde{g}(\vec{k})}{1 + A} \tag{19.19.186}$$

where

$$\tilde{g}(\vec{k}) = \int e^{i\vec{k}\cdot\vec{r}} g(\vec{r}) \, d^3r \tag{19.19.187}$$

is the Fourier transform of $g(\vec{r})$. So, we have

$$\psi^{(+)}(\vec{r}) = \frac{1}{(2\pi)^{3/2}} e^{i\vec{k}\cdot\vec{r}} - \sqrt{\frac{\pi}{2}} \frac{\tilde{g}(\vec{k})}{1 + A} \int d^3r' \frac{e^{ik|\vec{r}-\vec{r}'|}}{|\vec{r} - \vec{r}'|} F(\vec{r}') \ . \tag{19.19.188}$$

19.20 Generalized Optical Potential

Suppose two identical spinless particles, mass m_0, can scatter into two identical spinless particles, mass m_i, if their energy is greater than the threshold energy

$$E_i = 2(m_i - m_0)c^2 \ .$$

The outgoing particles, of type i, interact via a potential V_{ii}. The coupling between different types of particles (*channels*) is due to potentials

$$V_{ij} = V_{ji} \ .$$

If there are N coupled channels the resultant Schrödinger equation reads

$$\sum_{j=0}^{N-1} (T_i \delta_{ij} + V_{ij}) |\psi_j\rangle = (E - E_i)|\psi_i\rangle \quad i = 0, 1, 2, \ldots, N-1 \tag{19.20.189}$$

where

$$T_i = \frac{\vec{p}_i^2}{2m_i} \qquad (19.20.190)$$

is the kinetic energy of particle i .

Assume that there are only two channels ($N = 2$) and find an equation of the form

$$(T_0 + V)|\psi_0\rangle = (E - E_0)|\psi_0\rangle$$

for the wavefunction $|\psi_0\rangle$ by eliminating the wavefunction $|\psi_1\rangle$. The resultant "potential" V in this equation is known as a *generalized optical potential*. Note that this potential is complex and allows for absorption or emission of particles of type $i = 0$.

Hint: Use a Green's function to express $|\psi_1\rangle$ in terms of $|\psi_0\rangle$.

Solution

The Schrödinger equation for this case is

$$(T_0 + V_{00} - E + E_0)|\psi_0\rangle = -V_{01}|\psi_1\rangle \qquad (19.20.191)$$
$$(T_1 + V_{11} - E + E_1)|\psi_1\rangle = -V_{10}|\psi_0\rangle \ . \qquad (19.20.192)$$

From the second of these equations we get

$$|\psi_1\rangle = -(T_1 + E - E_1 + i\epsilon + V_{11})^{-1}V_{10}|\psi_0\rangle \ . \qquad (19.20.193)$$

Here we have imposed outgoing wave boundary conditions as indicated by replacing E by $E + i\epsilon$. Substituting this result into (19.20.191) we get

$$[T_0 + V_{00} - V_{01}(T_1 + E - E_1 + i\epsilon + V_{11})^{-1}V_{10}]|\psi_0\rangle = (E - E_0|\psi_0\rangle \qquad (19.20.194)$$

which is the desired equation.

The generalized optical potential is

$$V = V_{00} - V_{01}(T_1 + E - E_1 + i\epsilon + V_{11})^{-1}V_{10} \ . \qquad (19.20.195)$$

A detailed discussion of optical potentials can be found in the article by Feshbach [19.4].

19.21   Free Particle Eigenfunctions

Show that the free particle wavefunctions for states of definite angular momentum (not normalized) may be written

$$\psi_{klm}(r, \theta, \varphi) = (L_x - iL_y)^{l-m}(p_x + ip_y)^l \frac{\sin(kr)}{kr} \ .$$

Solution

For a free particle the states of definite angular momentum satisfy

$$
\begin{aligned}
H\psi_{klm} &= \frac{\hbar^2 \vec{k}^2}{2M}\psi_{klm} \\
L^2\psi_{klm} &= l(l+1)\hbar^2\psi_{klm} \\
L_z\psi_{klm} &= m\hbar\psi_{klm} \ .
\end{aligned}
\tag{19.21.196}
$$

Here,

$$
H = \frac{\vec{p}^2}{2M} \ .
\tag{19.21.197}
$$

The solution for $l = 0$ is

$$
\psi_{k00} = \frac{\sin(kr)}{kr} \ .
\tag{19.21.198}
$$

We now introduce

$$
\begin{aligned}
p_+ &= p_x + ip_y \\
L_\pm &= L_x \pm iL_y \ .
\end{aligned}
\tag{19.21.199}
$$

Then we find that

$$
[p_+, H] = [L_\pm, H] = 0
\tag{19.21.200}
$$

so that these two operators do not change the energy eigenvalue. On the other hand we also find that

$$
\begin{aligned}
{[L_z, p_+]} &= \hbar p_+ \\
{[L^2, p_+]} &= 2\hbar^2 p_+ + 2\hbar(p_+ L_z - p_z L_+) \ .
\end{aligned}
\tag{19.21.201}
$$

If now consider the equation

$$
L^2\psi_{kll} = l(l+1)\hbar^2\psi_{kll}
\tag{19.21.202}
$$

and act on it from the left with p_+ and use the results of (19.21.201) we get

$$
[L^2 p_+ - 2\hbar^2 p_+ - 2\hbar(p_+ L_z - p_z L_+)]\psi_{kll} = l(l+1)\hbar^2 p_+ \psi_{kll} \ .
\tag{19.21.203}
$$

But,

$$
L_+\psi_{kll} = 0 \ .
\tag{19.21.204}
$$

Thus, we find

$$
L^2 p_+ \psi_{kll} = (l+1)(l+2)\hbar^2 p_+ \psi_{kll} \ .
\tag{19.21.205}
$$

Now, starting with

$$
L_z\psi_{klm} = m\hbar\psi_{klm}
\tag{19.21.206}
$$

and acting on it from the left with p_+ we find

$$p_+ L_z \psi_{klm} = (L_z - \hbar) p_+ \psi_{klm} = m\hbar p_+ \psi_{klm} \ . \tag{19.21.207}$$

Therefore,

$$L_z p_+ \psi_{klm} = (m+1)\hbar p_+ \psi_{klm} \ . \tag{19.21.208}$$

Thus, the effect of p_+ is to raise both l and m in ψ_{klm} by one unit. On the other hand we know that L_- lowers the value of m in ψ_{klm} by one unit, but leaves the value of l unchanged. So,

$$
\begin{aligned}
L^2 L_- \psi_{klm} &= l(l+1)\hbar^2 L_- \psi_{klm} \\
L_z L_- \psi_{klm} &= (m-1)\hbar L_- \psi_{klm} \ .
\end{aligned}
\tag{19.21.209}
$$

Finally, we use the fact that, up to normalization,

$$\frac{\sin(kr)}{kr} = \psi_{k00} \tag{19.21.210}$$

is an energy eigenstate corresponding to

$$l = m = 0 \ .$$

Then, we have that

$$
\begin{aligned}
L^2 p_+^l \psi_{k00} &= l(l+1)\hbar^2 p_+^l \psi_{k00} \\
L_z p_+^l \psi_{k00} &= l\hbar p_+^l \psi_{k00}
\end{aligned}
\tag{19.21.211}
$$

so that, up to normalization,

$$p_+^l \psi_{k00} = \psi_{kll} \ . \tag{19.21.212}$$

Now, acting with L_- on ψ_{kll} for $l-m$ times we get, again up to normalization, that

$$L_-^{l-m} \psi_{kll}^l = \psi_{klm} \ . \tag{19.21.213}$$

Therefore, we have proven the desired result.

19.22   Scattering from an Inverse Square Potential

Find the phase shifts for scattering from a potential

$$V(r) = V_0 \frac{a^2}{r^2} \quad V_0 > 0 \ .$$

Solution

The Schrödinger equation for the radial function reads

$$\left[\frac{d^2}{dr^2} + \frac{2}{r}\frac{d}{dr} - \frac{l(l+1)}{r^2} - \frac{b^2}{r^2} + k^2\right] R_l(r) = 0 \ . \tag{19.22.214}$$

Here we have introduced the parameters

$$k^2 = \frac{2mE}{\hbar^2} \quad \text{and} \quad b^2 = \frac{2mV_0a^2}{\hbar^2} \ . \tag{19.22.215}$$

The solution finite at the origin is $j_\lambda(kr)$ where λ is the positive root of

$$\lambda(\lambda+1) = l(l+1) + b^2 \tag{19.22.216}$$

so that,

$$\lambda = -\frac{1}{2} + \sqrt{(l+1/2)^2 + b^2} \ . \tag{19.22.217}$$

The asymptotic form of the solution for large kr is

$$R_l(r) \rightarrow \frac{\sin(kr - \lambda\pi/2)}{kr} \ . \tag{19.22.218}$$

This asymptotic form must coincide with

$$R_l(r) \rightarrow \frac{\sin(kr - l\pi/2 + \delta_l)}{kr} \ . \tag{19.22.219}$$

Therefore,

$$\delta_l = l\pi/2 - \lambda\pi/2 = \left[(l+1/2) - \sqrt{(l+1/2)^2 + b^2}\right]\frac{\pi}{2} \ . \tag{19.22.220}$$

19.23 Neutron-Proton Scattering: Spin Flip

Assume that the scattering amplitudes for the scattering of a slow neutron by a proton are f_3 and f_1 for the triplet and singlet states respectively. Find the probability for the reversal (flip) of the neutron spin if before the collision the neutron spin was up and the proton spin was down.

Solution

The spin wavefunction of the neutron proton system before the collision is

$$|\uparrow, n\rangle|\downarrow, p\rangle \ .$$

This may be rewritten in terms of the triplet and singlet spin wavefunctions as

$$
\begin{aligned}
|\uparrow, n\rangle|\downarrow, p\rangle &= \frac{1}{\sqrt{2}}\left[\frac{1}{\sqrt{2}}(|\uparrow, n\rangle|\downarrow, p\rangle + |\downarrow, n\rangle|\uparrow, p\rangle)\right. \\
&+ \left.\frac{1}{\sqrt{2}}(|\uparrow, n\rangle|\downarrow, p\rangle - |\downarrow, n\rangle|\uparrow, p\rangle)\right] \ .
\end{aligned}
\tag{19.23.221}
$$

Using the given data we have that the scattering amplitude is

$$\frac{1}{\sqrt{2}}\left[\frac{f_3}{\sqrt{2}}(|\uparrow,n\rangle|\downarrow,p\rangle + |\downarrow,n\rangle|\uparrow,p\rangle)\right.$$

$$+ \left.\frac{f_1}{\sqrt{2}}(|\uparrow,n\rangle|\downarrow,p\rangle - |\downarrow,n\rangle|\uparrow,p\rangle)\right]$$

$$= \frac{f_3 + f_1}{2}|\uparrow,n\rangle|\downarrow,p\rangle + \frac{f_3 - f_1}{2}|\downarrow,n\rangle|\uparrow,p\rangle \ . \qquad (19.23.222)$$

The probability for finding the spins flipped is then given by

$$\frac{|(f_3 - f_1)/2|^2}{|(f_3 + f_1)/2|^2 + |(f_3 - f_1)/2|^2} = \frac{|f_3 - f_1|^2}{2(|f_3|^2 + |f_1|^2)} \ . \qquad (19.23.223)$$

19.24   Reflectionless Potential in One Dimension

A particle (mass m) moves in the potential

$$V(x) = -\frac{\hbar^2}{m}\frac{\lambda^2 A e^{-\lambda x}}{\left(A e^{-\lambda x} + 1\right)^2}$$

where λ and $A > 1$ are real constants.
a) Verify that $\phi(k, x)$ are solutions of the Schrödinger equation

$$-\frac{\hbar^2}{2m}\frac{d^2\phi}{dx^2} + [V(x) - E]\phi = 0$$

where

$$\phi(k,\ x) = e^{-ikx}\left[\lambda\frac{A e^{-\lambda x} - 1}{A e^{-\lambda x} + 1} - 2ik\right]$$

and

$$E = \frac{\hbar^2 k^2}{2m} \ .$$

b) Use these solutions to show that there is no reflection from this potential.
c) Find the energy of the bound states, if any.

Solution

a) By direct differentiation we find that

$$-\frac{\hbar^2}{2m}\frac{d^2\phi}{dx^2} = -\frac{\hbar^2}{2m}\left\{-k^2\phi - 2\frac{\lambda^2 A e^{-\lambda x}}{(A e^{-\lambda x} + 1)^2}e^{-ikx}\left[\lambda\frac{A e^{-\lambda x} - 1}{A e^{-\lambda} + 1} - 2ik\right]\right\}$$

$$= (E - V)\phi \ . \qquad (19.24.224)$$

Thus, as required, $\phi(k, x)$ satisfies the Schrödinger equation for all k.

b) To solve the scattering problem we assume an incoming wave from the left. In this case the appropriate solution is

$$\phi(k, x) = e^{ikx}\left[\lambda\frac{A\,e^{-\lambda x} - 1}{A\,e^{-\lambda x} + 1} + 2ik\right]$$

$$+ \; Re^{-ikx}\left[\lambda\frac{A\,e^{-\lambda x} - 1}{A\,e^{-\lambda x} + 1} - 2ik\right] \quad x < 0$$

$$\phi(k, x) = T\,e^{ikx}\left[\lambda\frac{A\,e^{-\lambda x} - 1}{A\,e^{-\lambda x} + 1} + 2ik\right] \quad x > 0 \; . \tag{19.24.225}$$

Matching the solutions and their derivative at $x = 0$ we get

$$\lambda\frac{A-1}{A+1} + 2ik + R\left[\lambda\frac{A-1}{A+1} - 2ik\right] = T\left[\lambda\frac{A-1}{A+1} + 2ik\right]$$

$$ik\left[\lambda\frac{A-1}{A+1} + 2ik\right] - \frac{2\lambda^2}{A} - ikR\left[\lambda\frac{A-1}{A+1} - 2ik\right] - R\frac{2\lambda^2}{A}$$

$$= ikT\left[\lambda\frac{A-1}{A+1} + 2ik\right] - T\frac{2\lambda^2}{A} \; . \tag{19.24.226}$$

This pair of equations may be rewritten

$$R\left[ik\left(\lambda\frac{A-1}{A+1} - 2ik\right) - \frac{2\lambda^2}{A}\right] + (T-1)\left[\left(ik(\lambda\frac{A-1}{A+1} + 2ik\right) - \frac{2\lambda^2}{2A}\right] = 0$$

$$R\left[\lambda\frac{A-1}{A+1} - 2ik\right] - (T-1)\left[\lambda\frac{A-1}{A+1} + 2ik\right] = 0. \tag{19.24.227}$$

The determinant of the coefficients of R and $T-1$ does not vanish for any real k. Therefore, the only allowed solution is

$$T - 1 = 0$$
$$R = 0 \; . \tag{19.24.228}$$

This shows that there is no reflected wave.

c) For the bound state the parameter k must be imaginary to give us square integrable solutions. Thus, we take

$$\phi = C\,e^{\kappa x}\left[\lambda\frac{A\,e^{-\lambda x} - 1}{A\,e^{-\lambda x} + 1} + 2\kappa\right] \quad x < 0$$

$$= C\,e^{-\kappa x}\left[\lambda\frac{A\,e^{-\lambda x} - 1}{A\,e^{-\lambda x} + 1} - 2\kappa\right] \quad x > 0 \; . \tag{19.24.229}$$

Matching the function and derivatives at $x = 0$ yields

$$C\left[\lambda\frac{A-1}{A+1} + 2\kappa\right] - D\left[\lambda\frac{A-1}{A+1} - 2\kappa\right] = 0$$

$$C\left[\kappa\left(\lambda\frac{A-1}{A+1} + 2\kappa\right) - \frac{\lambda A}{(A+1)^2}\right]$$

$$+ \; D\left[\kappa\left(\lambda\frac{A-1}{A+1} - 2\kappa\right) + \frac{\lambda A}{(A+1)^2}\right] = 0 \; . \tag{19.24.230}$$

For a nontrivial solution the determinant of coefficients must vanish. This yields

$$-4\kappa^2 + \lambda^2 \left(\frac{A-1}{A+1}\right)^2 + \frac{2\lambda A}{(A+1)^2} = 0 \ . \tag{19.24.231}$$

The energy is now given by

$$E = -\frac{\hbar^2 \kappa^2}{2m} = -\frac{\hbar^2 \left[\lambda^2 (A-1)^2 + 2\lambda A\right]}{8m(A+1)^2} \ . \tag{19.24.232}$$

19.25   n-p Scattering: Singlet and Triplet States

A neutron and proton scatter off each other in a zero orbital angular momentum state in their centre of mass system. For the triplet spin state the scattering amplitude is a_t, whereas for the singlet spin state the scattering amplitude is a_s.
a) Write the scattering amplitude for this scattering as a matrix in spin space.
b) If we designate the spin up state by $(+)$ and the spin down state by $(-)$ write the differential cross sections for the following reactions.

1) $n(+) + p(+) \rightarrow n(+) + p(+)$
2) $n(+) + p(-) \rightarrow n(+) + p(-)$
3) $n(+) + p(-) \rightarrow n(-) + p(+)$
4) $n(-) + p(-) \rightarrow n(-) + p(-)$
5) $n(-) + p(+) \rightarrow n(-) + p(+)$
6) $n(-) + p(+) \rightarrow n(+) + p(-)$

c) What is the differential cross section for the scattering of unpolarized neutrons by unpolarized protons?

Solution

a) There are two different expressions for the scattering amplitude as a matrix.
1) If we write it for states of total spin namely the states

$$|S, M\rangle = |1, 1\rangle , |1, 0\rangle , |1, -1\rangle , |0, 0\rangle \ .$$

we have a diagonal matrix

$$F(k, \theta)_{S,M;S'M'} = \begin{pmatrix} a_t & 0 & 0 & 0 \\ 0 & a_t & 0 & 0 \\ 0 & 0 & a_t & 0 \\ 0 & 0 & 0 & a_s \end{pmatrix} \ . \tag{19.25.233}$$

On the other hand, if we write the scattering amplitude in terms of the neutron and proton spin states we need the Clebsch-Gordon coefficients

$$\langle 1/2, m_1; 1/2, m_2 | SM \rangle = \begin{pmatrix} 1 & 0 & 0 & 0 \\ 0 & \frac{1}{\sqrt{2}} & 0 & \frac{1}{\sqrt{2}} \\ 0 & \frac{1}{\sqrt{2}} & 0 & -\frac{1}{\sqrt{2}} \\ 0 & 0 & 1 & 0 \end{pmatrix} \ . \tag{19.25.234}$$

Here the rows are labelled by $m_1, m_2 = ++, +-, -+, --$ and the columns are labelled by $(S, M) = (1, 1), (1, 0), (1, -1), (0, 0)$. The scattering amplitude matrix labelled by $m_1, m_2; m'_1, m'_2$ is given by

$$F(k, \theta)_{m_1, m_2; m'_1, m'_2}$$

$$= \sum_{S, M; S', M'} \langle 1/2, m_1; 1/2, m_2 | S M \rangle \langle 1/2, m'_1; 1/2, m'_2 | S' M' \rangle \quad (19.25.235)$$

and the differential cross-section labelled in the same manner is

$$\sigma_{m_1, m_2 \rightarrow m'_1, m'_2} = |F(k, \theta)_{m_1, m_2; m'_1, m'_2}|^2 \; . \qquad\qquad (19.25.236)$$

b) The differential cross sections for the reactions listed may now be written down by inspection from equations (19.25.234) and (19.25.235).

1) $n(+) + p(+) \rightarrow n(+) + p(+)$: $\quad \dfrac{d\sigma}{d\Omega} = |a_t|^2$

2) $n(+) + p(-) \rightarrow n(+) + p(-)$: $\quad \dfrac{d\sigma}{d\Omega} = \dfrac{1}{2}|a_t + a_s|^2$

3) $n(+) + p(-) \rightarrow n(-) + p(+)$: $\quad \dfrac{d\sigma}{d\Omega} = \dfrac{1}{2}|a_t - a_s|^2$

4) $n(-) + p(+) \rightarrow n(-) + p(+)$: $\quad \dfrac{d\sigma}{d\Omega} = \dfrac{1}{2}|a_t + a_s|^2$

5) $n(-) + p(+) \rightarrow n(+) + p(-)$: $\quad \dfrac{d\sigma}{d\Omega} = \dfrac{1}{2}|a_t - a_s|^2$

6) $n(-) + p(-) \rightarrow n(-) + p(-)$: $\quad \dfrac{d\sigma}{d\Omega} = |a_t|^2 \; . \qquad (19.25.237)$

c) For unpolarized neutrons and protons the differential cross section is obtained by averaging (19.25.233) over the initial spin states and summing over the final spin states. So, using the orthogonality of the Clebsch-Gordon coefficients we get

$$\frac{d\sigma}{d\Omega} = \frac{1}{4} \sum_{S, M; S', M'} |F(k, \theta)_{S, M; S' M'}|^2$$

$$= \frac{1}{4} \left[3|a_t|^2 + |a_s|^2 \right] \; . \qquad\qquad (19.25.238)$$

19.26 Phase Shift, Scattering Length, Etc.

The regular s-wave ($l = 0$) function for scattering by some central potential is

$$u(r) = \sin(kr) + \frac{b^2 - a^2}{b^2 + k^2} \frac{k \sinh(br) \cos(kr) - b \cosh(br) \sin(kr)}{b \cosh(br) + a \sinh(br)} \; .$$

a) Calculate the phase shift.
b) Calculate the scattering length and the effective range.

Solution

a) In the asymptotic region $r \to \infty$ the wavefunction must be of the form

$$u(r) \to A \sin(kr + \delta) \ . \tag{19.26.239}$$

Thus, we take the asymptotic form of the given wavefunction. To do this we use that

$$\cosh(br) \to \frac{1}{2} e^{br} \quad \text{and} \quad \sinh(br) \to \frac{1}{2} e^{br} \ . \tag{19.26.240}$$

Then,

$$\begin{aligned}
u(r) \quad \to \quad & \sin(kr) + \frac{b^2 - a^2}{b^2 + k^2} \frac{k \cos(kr) - b \sin(kr)}{b + a} \\
= \quad & \sin(kr) + \frac{b - a}{b^2 + k^2} \left[k \cos(kr) - b \sin(kr) \right] \\
= \quad & \frac{1}{b^2 + k^2} \sqrt{(k^2 + ab)^2 + k^2 (b - a)^2} \times \\
& \left[\frac{k^2 + ab}{\sqrt{(k^2 + ab)^2 + k^2 (b - a)^2}} \sin(kr) \right. \\
& \left. + \frac{k(b - a)}{\sqrt{(k^2 + ab)^2 + k^2 (b - a)^2}} \cos(kr) \right] \ . \tag{19.26.241}
\end{aligned}$$

This is of the desired form (19.26.239) if we define the phase shift δ by

$$\cos \delta \ = \ \frac{k^2 + ab}{\sqrt{(k^2 + ab)^2 + k^2 (b - a)^2}}$$

$$\sin \delta \ = \ \frac{k(b - a)}{\sqrt{(k^2 + ab)^2 + k^2 (b - a)^2}} \ . \tag{19.26.242}$$

b) To get the scattering length c_0 and the effective range r_0 we use the result that for $k \to 0$

$$k \cot \delta \to -\frac{1}{c_0} + \frac{1}{2} k^2 r_0 \ . \tag{19.26.243}$$

But,

$$k \cot \delta = \frac{(k^2 + ab)}{b - a} \ . \tag{19.26.244}$$

Thus, we find that

$$-\frac{1}{c_0} + \frac{1}{2} k^2 r_0 = \frac{k^2 + ab}{b - a} \ . \tag{19.26.245}$$

So,

$$c_0 = -\frac{b - a}{ab} = \frac{1}{b} - \frac{1}{a} \tag{19.26.246}$$

and

$$r_0 = \frac{2}{b - a} \ . \tag{19.26.247}$$

19.27   Scattering off a Diatomic Molecule

A diatomic molecule, such as H_2 or O_2, consisting of two identical atoms is modelled by two identical spherically symmetric scattering centres separated by a distance $\vec{R}$. (See figure 19.3.) Suppose the scattering amplitude for scattering an electron off one of these centres is known to be $f(\theta)$. Find the scattering cross-section for scattering off the molecule.

Hint: Neglect the effect of multiple scattering.

Solution

With reference to figure 19.3 we see that the electron arrives at atom A earlier

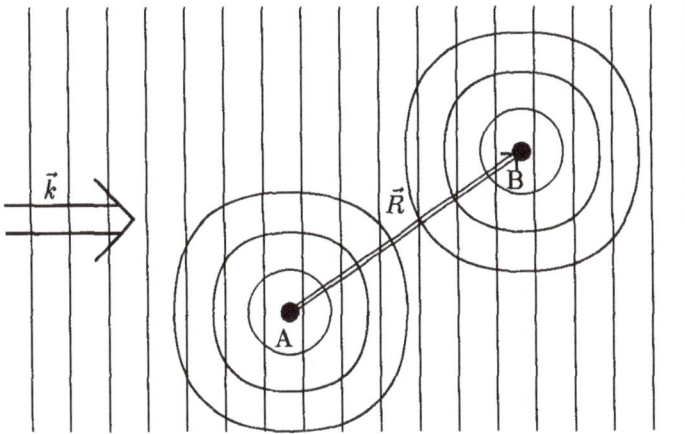

Figure 19.3: Scattering off a diatomic molecule.

than at atom B. So, its phase at atom A relative to the centre of the molecule is

$$e^{i\vec{k}\cdot\vec{R}/2} \quad .$$

Similarly, its phase at B relative to the centre of the molecule is

$$e^{-i\vec{k}\cdot\vec{R}/2} \quad .$$

Thus, asymptotically as $r \to \infty$ the scattered wave must be of the form

$$\psi(r) = e^{i\vec{k}\cdot\vec{R}/2} f(\theta) \frac{e^{ikr_A}}{r_A} + e^{-i\vec{k}\cdot\vec{R}/2} f(\theta) \frac{e^{ikr_B}}{r_B} \quad . \tag{19.27.248}$$

But,

$$kr_A = kr - \frac{1}{2}\vec{k}'\cdot\vec{R} \quad \text{and} \quad kr_B = kr + \frac{1}{2}\vec{k}'\cdot\vec{R} \tag{19.27.249}$$

where $\hbar \vec{k}'$ is the momentum of the scattered electron. We define the "momentum transfer"

$$\vec{q} = \vec{k} - \vec{k}' \tag{19.27.250}$$

and in the denominator of (19.27.248) we replace r_A and r_B by r. Then,

$$
\begin{aligned}
\psi(r) & = f(\theta) \frac{e^{ikr}}{r} \left[e^{i\vec{k}\cdot\vec{R}/2 - i\vec{k}'\cdot\vec{R}/2} + e^{-i\vec{k}\cdot\vec{R}/2 + i\vec{k}'\cdot\vec{R}/2} \right] \\
& = f(\theta) \frac{e^{ikr}}{r} \left[e^{i\vec{q}\cdot\vec{R}/2} + e^{-i\vec{q}\cdot\vec{R}/2} \right] .
\end{aligned}
\tag{19.27.251}
$$

So the scattering amplitude for the molecule in this particular orientation (fixed value of $\vec{R}$) is

$$
\begin{aligned}
F(\vec{k}, \theta)\Big|_{\vec{R}} & = f(\theta) \left[e^{i\vec{q}\cdot\vec{R}/2} + e^{-i\vec{q}\cdot\vec{R}/2} \right] \\
& = 2f(\theta) \cos(\vec{q}\cdot\vec{R}/2) .
\end{aligned}
\tag{19.27.252}
$$

The corresponding differential cross-section is

$$\left(\frac{d\sigma}{d\Omega} \right)_{\vec{R}} = 4|f(\theta)|^2 \cos^2(\vec{q}\cdot\vec{R}/2) . \tag{19.27.253}$$

However, in any given experiment, the direction of the vector $\vec{R}$ is random. So, writing

$$\vec{q}\cdot\vec{R} = qR\cos\alpha \tag{19.27.254}$$

we must average over all values of α. The result is

$$\frac{d\sigma}{d\Omega} = 4|f(\theta)|^2 \frac{1}{\pi} \int_0^\pi \cos^2(\tfrac{1}{2}qR\cos\alpha) \, d\alpha . \tag{19.27.255}$$

If we now change integration variables from α to

$$x = \frac{1}{2}qR\cos\alpha \tag{19.27.256}$$

we get

$$
\begin{aligned}
\frac{d\sigma}{d\Omega} & = \frac{4|f(\theta)|^2}{\pi} \int_{-qR/2}^{qR/2} \frac{\cos^2 x \, dx}{\sqrt{\frac{q^2R^2}{4} - x^2}} \\
& = 2|f(\theta)|^2 \left[1 + J_0(qR) \right] .
\end{aligned}
\tag{19.27.257}
$$

19.28 WKB s-Wave Phase Shift: Attractive Potential

Use the WKB approximation to calculate the s-wave phase shift for scattering off an attractive potential.

Hint: The most general radial solution is asymptotically of the from

$$R_l(r) \rightarrow A_l \left[\cos\delta_l \, j_l(kr) - \sin\delta_l \, n_l(kr) \right] . \tag{19.28.258}$$

Solution

If we replace the radial function $R_l(r)$ by

$$u_l(r) = r R_l(r) \qquad (19.28.259)$$

the Schrödinger equation for $l = 0$ becomes

$$\frac{d^2 u}{dr^2} + q^2(r) u = 0 \qquad (19.28.260)$$

where

$$q^2(r) = \frac{2m[E - V(r)]}{\hbar^2} = k^2 - U(r) . \qquad (19.28.261)$$

Here we also have the additional boundary condition that

$$u(0) = 0 . \qquad (19.28.262)$$

The appropriate WKB solution of this equation is therefore

$$u(r) = \frac{A}{\sqrt{q(r)}} \sin \left[\int_0^r q(r') \, dr' \right] . \qquad (19.28.263)$$

To extract the phase shift we need the asymptotic behaviour of

$$R_0(r) = \frac{1}{r} u(r)$$

to compare it with the general solution given by (19.28.258). This asymptotic behaviour is

$$
\begin{aligned}
R_0(r) \quad &\to \quad \frac{A}{kr} \sin \left[\int_0^r q(r') \, dr' \right] \\
&= \quad \frac{A}{kr} \sin \left[\int_0^r \{q(r') - k\} \, dr' + kr \right] .
\end{aligned}
\qquad (19.28.264)
$$

Comparing this with (19.28.258) and taking the limit $r \to \infty$ we get

$$
\begin{aligned}
\delta_0 \quad &= \quad \int_0^\infty [q(r) - k] \, dr \\
&= \quad \int_0^\infty [\sqrt{k^2 - U(r)} - k] \, dr .
\end{aligned}
\qquad (19.28.265)
$$

19.29 WKB s-Wave Phase Shift: Hulthén Potential

Use the results of problem 19.28 to calculate the s-wave phase shift for the Hulthén potential

$$V(r) = -\frac{V_0}{e^{\alpha r} - 1} . \qquad (19.29.266)$$

Hint: Change the integration variable from r to

$$\sqrt{k^2 - \frac{2mV(r)}{\hbar^2}} .$$

Solution

From problem 19.28 we get that the s-wave phase shift is given by

$$\delta = \int_0^\infty \left[\sqrt{k^2 - \frac{2mV(r)}{\hbar^2}} - k \right] dr \ . \tag{19.29.267}$$

We now change the integration variable from r to

$$z = \sqrt{k^2 - \frac{2mV(r)}{\hbar^2}} \ . \tag{19.29.268}$$

Therefore,

$$z^2 = k^2 + \frac{2mV_0}{\hbar^2} \frac{1}{e^{\alpha r} - 1} \ . \tag{19.29.269}$$

If we call

$$\beta^2 = \frac{2mV_0}{\hbar^2} \tag{19.29.270}$$

we get that

$$e^{\alpha r} = 1 + \frac{\beta^2}{z^2 - k^2} \ . \tag{19.29.271}$$

So, we find that

$$dr = -\frac{2\beta^2}{\alpha} \frac{z \, dz}{(z^2 - k^2)(z^2 - k^2 + \beta^2)} \ . \tag{19.29.272}$$

Hence,

$$\delta = \frac{2\beta^2}{\alpha} \int_k^\infty \frac{z \, dz}{(z + k)(z^2 - k^2 + \beta^2)} \ . \tag{19.29.273}$$

This integral is easily evaluated and yields

$$\delta = \begin{cases} \frac{2k}{\alpha} \ln \frac{2k}{\beta} + \frac{2\sqrt{\beta^2 - k^2}}{\alpha} \arctan \sqrt{\beta^2 - k^2} & \beta \geq k \\[2ex] \frac{2k}{\alpha} \ln \frac{2k}{\beta} + \frac{\sqrt{k^2 - \beta^2}}{\alpha} \ln \left[\frac{k - \sqrt{k^2 - \beta^2}}{k + \sqrt{k^2 - \beta^2}} \right] & \beta \leq k \end{cases} \ . \tag{19.29.274}$$

19.30 WKB Approximation for Phase Shifts

The radial part of the Schrödinger equation for a spherically symmetric potential with

$$rR(r) = u(r)$$

becomes

$$\frac{d^2u}{dr^2} + \left[k^2 - U(r) - \frac{l(l + 1)}{r^2} \right] u = 0 \ .$$

Here,

$$U(r) = \frac{2m}{\hbar^2} V(r) \quad \text{and} \quad k^2 = \frac{2mE}{\hbar^2} \ .$$

By changing the independent variable from r to x where

$$kr = e^x$$

and the dependent variable from $u(r)$ to $v(x)$ where

$$u(r) = e^{x/2} \, v(x)$$

one obtains an equation of the form

$$\frac{d^2 v}{dx^2} + Q^2(x) \, v = 0$$

where

$$Q^2(x) = e^{2x} \left(1 - U/k^2\right) - (l + 1/2)^2 \ .$$

Here, the variable x runs over the region $-\infty < x < \infty$. Assume that $V(r)$ goes to zero rapidly at $r \to \infty$ and is no more singular than $1/r$ near $r = 0$. Also assume that there is only one point x_0 such that

$$Q^2(x_0) = 0 \ .$$

Clearly, for $x \to$ large positive, $Q^2(x) > 0$ and for $x \to$ large negative, $Q^2(x) < 0$. Thus, such a turning point always exists.

Use the WKB approximation to solve the equation for $v(x)$ and then convert back to the variables r and $u(r)$ and look at the asymptotic region $r \to \infty$ to obtain the phase shifts.

Solution

For the equation for $v(x)$ the WKB solution in the asymptotically forbidden region $x < x_0$ must be exponentially damped (since $u(r)$ vanishes at $r = 0$) and is

$$v(x) = \frac{A}{\sqrt{|Q(x)|}} \exp -\left\{ \int_x^{x_0} |Q(x')| \, dx' \right\} \ . \tag{19.30.275}$$

In the classically allowed region $x > x_0$ the solution is

$$\begin{aligned} v(x) &= \frac{1}{\sqrt{Q(x)}} \left[C \exp -i \left\{ \int_{x_0}^x Q(x') \, dx' - \frac{\pi}{4} \right\} \right. \\ &+ \left. D \exp -i \left\{ \int_{x_0}^x Q(x') \, dx' - \frac{\pi}{4} \right\} \right] \ . \end{aligned} \tag{19.30.276}$$

The appropriate matching condition is the one that joins the exponentially damped solution onto the corresponding solution in the classically allowed region. The solution in the classically allowed region corresponding to the solution (19.30.275) is

$$v(x) = \frac{2A}{\sqrt{Q(x)}} \cos \left\{ \int_{x_0}^{x} Q(x')\, dx' - \frac{\pi}{4} \right\} . \tag{19.30.277}$$

Comparing this with the solution (19.30.276) we see that this is accomplished by choosing

$$C = D = 2A . \tag{19.30.278}$$

To find the phase shifts we now revert to the original variables. Thus, we have

$$u(r) = \sqrt{kr}\, v(x) \tag{19.30.279}$$

and

$$Q^2(x) = r^2 \left(k^2 - U(r) - \frac{(l+1/2)^2}{r^2} \right) . \tag{19.30.280}$$

We also define

$$kr_0 = e^{x_0} . \tag{19.30.281}$$

Then,

$$u(r) = 2A \sqrt{\frac{k}{k^2 - U(r) - (l+1/2)^2/r^2}}$$
$$\times \cos \left\{ \int_{r_0}^{r} \sqrt{k^2 - U(r') - (l+1/2)^2/r'^2}\, dr' - \frac{\pi}{4} \right\} . \tag{19.30.282}$$

Now, in general, the asymptotic form of the lth partial wave is

$$u(r) \to \text{Constant} \times \sin(kr - l\pi/2 + \delta_l) \quad \text{for} \quad r \to \infty . \tag{19.30.283}$$

Comparing this with the asymptotic behaviour of our WKB solution we have to rewrite it as

$$u(r) = 2A \sqrt{\frac{k}{k^2 - U(r) - (l+1/2)^2/r^2}}$$
$$\times \sin \left\{ \int_{r_0}^{\infty} \left[\sqrt{k^2 - U(r') - (l+1/2)^2/r'^2} - k \right] dr' \right.$$
$$\left. + k(r - r_0) - \frac{\pi}{4} + \frac{\pi}{2} \right\} . \tag{19.30.284}$$

The phase shift can now be read off

$$\delta_l = (l+1/2)\frac{\pi}{2} - kr_0$$
$$+ \int_{r_0}^{\infty} \left[\sqrt{k^2 - U(r') - (l+1/2)^2/r'^2} - k \right] dr' . \tag{19.30.285}$$

Although this equation is quite different from the one obtained earlier (equation (19.28.265)) they yield similar results.

19.31 Zero-Range Potential

Consider the $l = 0$ free Schrödinger equation and define a solution ψ by the boundary condition that the $r \to 0$ limit of the logarithmic derivative of $r\psi$ is a constant i.e.

$$\lim_{r \to 0} \left[\frac{1}{r\psi} \frac{d}{dr}(r\psi) \right] = \text{a constant independent of energy} . \qquad (19.31.286)$$

The resulting problem mimics a potential and is a solution for a *zero-range* potential. For such a "potential" both the scattering amplitude and the bound state energy (if a bound state exists) are determined by a single parameter namely the constant in $(19.31.286)$.

Now, consider a zero-range potential which is known to have a single $l = 0$ bound state with energy

$$E_0 = -\frac{\hbar^2 \alpha^2}{2m} . \qquad (19.31.287)$$

a) Find the bound state wavefunction $\psi_0(\vec{r})$ as well as the scattering solutions $\psi_{\vec{k}}(\vec{r})$.

b) Show that the solutions are orthogonal and can be normalized to satisfy

$$\int d^3r \, |\psi_0(\vec{r})|^2 = 1$$

$$\int d^3r \, \psi_0(\vec{r}) \psi_{\vec{k}}(\vec{r}) = 0$$

$$\int d^3r \, \psi_{\vec{k}}^*(\vec{r}) \psi_{\vec{q}}(\vec{r}) = \delta(\vec{k} - \vec{q}) . \qquad (19.31.288)$$

c) Show that the solutions form a complete set. That is, show that

$$\psi_0(\vec{r})\psi_0(\vec{r}') + \int d^3k \, \psi_{\vec{k}}(\vec{r})\psi_{\vec{k}}^*(\vec{r}') = \delta(\vec{r} - \vec{r}') . \qquad (19.31.289)$$

Hint: It may be useful to recall that an integral of the form

$$\int d^3k \, e^{ikr}$$

has to be regularized and is defined by

$$\int_0^\infty dk \, e^{ikr} = \lim_{\epsilon \to 0+} \int_0^\infty dk \, e^{ik(r+i\epsilon)}$$

$$= -\lim_{\epsilon \to 0+} \frac{1}{i(r + i\epsilon)}$$

$$= i[P\frac{1}{r} - i\pi\delta(r)] . \qquad (19.31.290)$$

Solution

a) If we set

$$u = r\psi \tag{19.31.291}$$

then the function u, for the bound state, satisfies

$$\frac{d^2u}{dr^2} = -\frac{2mE_0}{\hbar^2} u = \alpha^2 u \ . \tag{19.31.292}$$

So, the bound state solution is of the form

$$u = A e^{-\alpha r} \ . \tag{19.31.293}$$

After imposing the boundary condition (19.31.286) we get

$$\lim_{r \to 0} \frac{1}{u} \frac{du}{dr} = -\alpha \tag{19.31.294}$$

which is indeed a constant. Normalization yields

$$A = \sqrt{\frac{\alpha}{2\pi}} \ . \tag{19.31.295}$$

The scattering solution has to satisfy the free Schrödinger equation and must have only an $l = 0$ outgoing spherical wave. Therefore, it must be of the form

$$\psi_{\vec{k}}(\vec{r}) = \frac{1}{(2\pi)^{3/2}} e^{i\vec{k}\cdot\vec{r}} + f \frac{e^{ikr}}{r} \ . \tag{19.31.296}$$

We now impose the boundary condition (19.31.286) and get

$$\lim_{r \to 0} \left[\frac{1}{r\psi_{\vec{k}}} \frac{d}{dr}(r\psi_{\vec{k}}) \right] = \frac{1}{f} \left[\frac{1}{(2\pi)^{3/2}} + ikf \right] = -\alpha \ . \tag{19.31.297}$$

Therefore,

$$f = -\frac{1}{(2\pi)^{3/2}}(\alpha + ik) \tag{19.31.298}$$

and

$$\psi_{\vec{k}}(\vec{r}) = \frac{1}{(2\pi)^{3/2}} \left[e^{i\vec{k}\cdot\vec{r}} - \frac{1}{\alpha + ik} \frac{e^{ikr}}{r} \right] \ . \tag{19.31.299}$$

b) The normalization of the bound state solutions has already been demonstrated so we next consider the orthogonality between the bound state and scattering solutions.

$$\int d^3r \, \psi_0(\vec{r}) \psi_{\vec{k}}(\vec{r})$$

$$= \frac{1}{(2\pi)^{3/2}} \sqrt{\frac{\alpha}{2\pi}} \int d^3r \left[\frac{e^{-\alpha r}}{r} e^{i\vec{k}\cdot\vec{r}} - \frac{1}{\alpha + ik} \frac{e^{-\alpha r + ikr}}{r^2} \right] \ . \tag{19.31.300}$$

To carry out the integrals we use spherical coordinates and line up the z-axis with the vector $\vec{k}$. In that case the integration over φ yields simply a factor 2π and the integration over θ, after setting $\cos\theta = t$, becomes

$$\int_{-1}^{1} e^{ikrt}\, dt = \frac{1}{ikr}\left(e^{ikr} - e^{-ikr}\right) \ . \tag{19.31.301}$$

Therefore,

$$
\begin{aligned}
&\int d^3r\, \psi_0(\vec{r})\psi_{\vec{k}}(\vec{r}) \\
&= \frac{\sqrt{\alpha}}{2\pi}\left\{\int_0^\infty \frac{1}{ik}\left[e^{-\alpha r}\left(e^{ikr} - e^{-ikr}\right)\right]dr - \frac{1}{\alpha+ik}\frac{2}{\alpha-ik}\right\} \\
&= \frac{\sqrt{\alpha}}{2\pi}\left\{\frac{1}{ik}\left[\frac{1}{\alpha-ik} - \frac{1}{\alpha+ik}\right] - \frac{2}{\alpha^2+k^2}\right\} \\
&= 0 \tag{19.31.302}
\end{aligned}
$$

as required. Next, we consider the (delta-function) normalization of the scattering solutions

$$
\begin{aligned}
&\int d^3r\, \psi_{\vec{k}}^*(\vec{r})\psi_{\vec{q}}(\vec{r}) \\
&= \frac{1}{(2\pi)^3}\left\{\int d^3r\, e^{-i(\vec{k}-\vec{q})\cdot\vec{r}} - \frac{1}{\alpha+iq}\int\frac{d^3r}{r}e^{-ikr}\,e^{i\vec{q}\cdot\vec{r}}\right. \\
&\left. \quad - \frac{1}{\alpha-ik}\int\frac{d^3r}{r}e^{iqr}\,e^{-i\vec{k}\cdot\vec{r}} + \frac{1}{\alpha+iq}\frac{1}{\alpha-ik}\int\frac{d^3r}{r^2}e^{i(q-k)r}\right\} \\
&= \delta(\vec{k}-\vec{q}) + \frac{1}{(2\pi)^3}\{I_1 + I_2 + I_3\} \tag{19.31.303}
\end{aligned}
$$

where

$$
\begin{aligned}
I_1 &= -\frac{2\pi}{iq(\alpha-ik)}\int_0^\infty dr\, e^{-ikr}\left(e^{iqr} - e^{-iqr}\right) \\
&= -\frac{2\pi}{q(k+i\alpha)}\int_0^\infty dr\left(e^{-i(k-q)r} - e^{-i(k+q)r}\right) \tag{19.31.304}
\end{aligned}
$$

$$
I_2 = -\frac{2\pi}{k(q-i\alpha)}\int_0^\infty dr\left(e^{-i(k-q)r} - e^{i(k+q)r}\right) \tag{19.31.305}
$$

and

$$
I_3 = -\frac{4\pi}{(k+i\alpha)(q-i\alpha)}\int_0^\infty dr\, e^{-i(k-q)r} \ . \tag{19.31.306}
$$

Hence we get

$$
\begin{aligned}
&I_1 + I_2 + I_3 \\
&= \frac{2\pi}{(k+i\alpha)(q-i\alpha)}\left\{-\frac{k+i\alpha}{k}\int_0^\infty dr\left(e^{-i(k-q)r} - e^{i(k+q)r}\right)\right. \\
&\left. \quad -\frac{q-i\alpha}{q}\int_0^\infty dr\left(e^{-i(k-q)r} - e^{-i(k+q)r}\right) + 2\int_0^\infty dr\, e^{-i(k-q)r}\right\}. \tag{19.31.307}
\end{aligned}
$$

We now evaluate the terms in the braces. Thus, we have

$$
\begin{aligned}
\{\cdots\} &= i\alpha\left(\frac{1}{q}-\frac{1}{k}\right)\int_0^\infty dr e^{-i(k-q)r} \\
&+ \frac{i\alpha}{k}\int_0^\infty dr e^{i(k+q)r} - \frac{i\alpha}{q}\int_0^\infty dr e^{-i(k+q)r} \\
&+ \int_0^\infty dr e^{i(k+q)r} + \int_0^\infty dr e^{-i(k+q)r} \\
&= i\alpha\left(\frac{1}{q}-\frac{1}{k}\right)\frac{-i}{k-q-i\epsilon} + \frac{i\alpha}{k}\frac{i}{k+q+i\epsilon} - \frac{i\alpha}{q}\frac{-i}{k+q-i\epsilon} \\
&+ \frac{i}{k+q+i\epsilon} + \frac{-i}{k+q-i\epsilon} \\
&= \frac{\alpha}{kq} - \frac{\alpha}{kq} + 2\pi\delta(k+q) \\
&= 0
\end{aligned}
\tag{19.31.308}
$$

since both $k > 0$ and $q > 0$. Thus, we have shown that

$$
\int d^3r\, \psi_{\vec{k}}^*(\vec{r})\psi_{\vec{q}}(\vec{r}) = \delta(\vec{k}-\vec{q}) \ .
\tag{19.31.309}
$$

c) To show completeness we consider

$$
\begin{aligned}
&\int d^3k\, \psi_{\vec{k}}(\vec{r})\psi_{\vec{k}}^*(\vec{r}') \\
&= \frac{1}{(2\pi)^3}\Bigg\{\int d^3k\, e^{i\vec{k}\cdot(\vec{r}-\vec{r}')} - \frac{1}{r}\int\frac{d^3k}{\alpha+ik}e^{ikr}e^{-i\vec{k}\cdot\vec{r}'} \\
&- \frac{1}{r'}\int\frac{d^3k}{\alpha-ik}e^{-ikr'}e^{i\vec{k}\cdot\vec{r}} + \frac{1}{rr'}\int\frac{d^3k}{\alpha^2+k^2}e^{ik(r-r')}\Bigg\} \\
&= \delta(\vec{r}-\vec{r}') + \frac{1}{(2\pi)^3}\{I_1+I_2+I_3\} \ .
\end{aligned}
\tag{19.31.310}
$$

Here we have

$$
\begin{aligned}
&I_1 + I_2 \\
&= -\frac{1}{r}\int\frac{d^3k}{\alpha+ik}e^{ikr}e^{-i\vec{k}\cdot\vec{r}'} - \frac{1}{r'}\int\frac{d^3k}{\alpha-ik}e^{-ikr'}e^{i\vec{k}\cdot\vec{r}} \\
&= -\frac{2\pi}{rr'}\int_0^\infty k\,dk\left\{\frac{e^{ik(r-r')}-e^{ik(r+r')}}{k-i\alpha} + \frac{e^{ik(r-r')}-e^{-ik(r+r')}}{k+i\alpha}\right\} \\
&= -\frac{4\pi}{rr'}\int_0^\infty dk\frac{k^2}{k^2+\alpha^2}e^{ik(r-r')} + \frac{2\pi}{rr'}\int_{-\infty}^\infty k\,dk\frac{e^{ik(r+r')}}{k-i\alpha} \\
&= -\frac{4\pi}{rr'}\int_0^\infty dk\frac{k^2}{k^2+\alpha^2}e^{ik(r-r')} \\
&+ \frac{2\pi}{rr'}\left[\int_{-\infty}^\infty dk\, e^{ik(r+r')} - i\alpha\int_{-\infty}^\infty dk\frac{e^{ik(r+r')}}{k-i\alpha}\right]
\end{aligned}
$$

$$= -\frac{4\pi}{rr'} \int_0^\infty dk \frac{k^2}{k^2+\alpha^2} e^{ik(r-r')} - (2\pi)^2 \alpha \frac{1}{rr'} e^{-\alpha(r+r')} \qquad (19.31.311)$$

At this point we have dropped a term proportional to $\delta(r+r')$ since both $r>0$ and $r'>0$. The first term serves to cancel I_3 and after multiplying by $1/(2\pi)^3$ we see that the last term is just

$$-\psi_0(\vec{r})\psi_0(\vec{r'}) \ .$$

So, combining all of these results we see that

$$\psi_0(\vec{r})\psi_0(\vec{r'}) + \int d^3k \, \psi_{\vec{k}}(\vec{r})\psi^*_{\vec{k}}(\vec{r'}) = \delta(\vec{r}-\vec{r'}) \ . \qquad (19.31.312)$$

19.32   Calogero Equation

The $l=0$ Schrödinger equation for a potential vanishing rapidly for $r \to \infty$ may be written as an integral equation (Lippman-Schwinger equation) that incorporates the boundary condition.

$$\psi(r) = \sin kr + \int_0^\infty dr' \, G(r,r')U(r')\psi(r') \ .$$

The boundary condition that $\psi(0)=0$ is contained in the Green's function

$$G(r,r') = \begin{cases} \frac{1}{k}\sin k(r-r') & r' < r \\ 0 & r' > r \end{cases} \ . \qquad (19.32.313)$$

a) Show that the solution may be written as

$$\psi(r) = C(r)\sin kr + S(r)\cos kr \qquad (19.32.314)$$

and express $C(r)$ and $S(r)$ in terms of $\psi(r)$.
b) Show that the phase shift may be obtained from

$$\tan \delta_0 = \lim_{r\to\infty} t(r) \qquad (19.32.315)$$

where

$$t(r) = \frac{S(r)}{C(r)} \ . \qquad (19.32.316)$$

c) By straightforward differentiation show that $t(r)$ satisfies

$$\frac{dt}{dr} = -\frac{U(r)}{k}[\sin kr + t(r)\cos kr]^2 \ . \qquad (19.32.317)$$

This is called the Calogero equation. [19.5]

d) Assume that $t(r)$ is small and linearize the Calogero equation to get

$$\frac{dt}{dr} + \frac{U(r)}{k}\sin 2kr \, t(r) = -\frac{U(r)}{k}\sin^2 kr \ . \qquad (19.32.318)$$

Use this equation to solve for $t(r)$ and hence find an expression for the s-wave phase shift.

Solution

a) To obtain the desired form we simply write $\sin k(r - r') = \sin kr \cos kr' - \cos kr \sin kr'$ in the integral equation to get

$$
\begin{aligned}
\psi(r) &= \sin kr \left[1 + \frac{1}{k} \int_0^r dr' \cos kr' \, U(r') \psi(r') \right] \\
&\quad + \cos kr \left[-\frac{1}{k} \int_0^r dr' \sin kr' \, U(r') \psi(r') \right] \\
&= C(r) \sin kr + S(r) \cos kr \ .
\end{aligned}
\tag{19.32.319}
$$

So,

$$
\begin{aligned}
C(r) &= 1 + \frac{1}{k} \int_0^r dr' \cos kr' \, U(r') \psi(r') \\
S(r) &= -\frac{1}{k} \int_0^r dr' \sin kr' \, U(r') \psi(r') \ .
\end{aligned}
\tag{19.32.320}
$$

b) For $r \to \infty$ we have the general result that

$$
\psi(r) \to A \sin(kr + \delta_0)
\tag{19.32.321}
$$

where A is a constant. Also for $U(r) \to 0$ sufficiently rapidly for $r \to \infty$ we have that

$$
\begin{aligned}
C(r) &\to C(\infty) \ , \quad \text{a finite constant} \\
S(r) &\to S(\infty) \ , \quad \text{a finite constant} \ .
\end{aligned}
\tag{19.32.322}
$$

This means that

$$
\begin{aligned}
\psi(r) &\to \sqrt{C(\infty)^2 + S(\infty)^2} \ \times \\
&\left[\frac{C(\infty)}{\sqrt{C(\infty)^2 + S(\infty)^2}} \sin kr + \frac{S(\infty)}{\sqrt{C(\infty)^2 + S(\infty)^2}} \cos kr \right] \\
&= A \left[\cos \delta_0 \sin kr + \sin \delta_0 \cos kr \right] \ .
\end{aligned}
\tag{19.32.323}
$$

Therefore,

$$
\tan \delta_0 = \frac{S(\infty)}{C(\infty)} = \lim_{r \to \infty} t(r) \ .
\tag{19.32.324}
$$

c) From equation (19.32.320) for $C(r)$ and $S(r)$ we obtain

$$
\begin{aligned}
\frac{dC}{dr} &= \frac{U}{k} \cos kr \, \psi(r) \\
\frac{dS}{dr} &= -\frac{U}{k} \sin kr \, \psi(r) \ .
\end{aligned}
\tag{19.32.325}
$$

But,

$$
\frac{dt}{dr} = \frac{CS' - C'S}{C^2} = \frac{N}{C^2}
\tag{19.32.326}
$$

where

$$
\begin{aligned}
N &= -\frac{U}{k}[C\sin kr\psi + S\cos kr\psi] \\
&= -\frac{U}{k}\psi^2 \\
&= -\frac{U}{k}[C\sin kr + S\cos kr]^2 \ .
\end{aligned}
$$

(19.32.327)

Therefore,

$$
\begin{aligned}
\frac{dt}{dr} &= -\frac{U}{k}\left[\sin kr + \frac{S}{C}\cos kr\right]^2 \\
&= -\frac{U}{k}[\sin kr + t(r)\cos kr]^2 \ .
\end{aligned}
$$

(19.32.328)

d) If $|t(r)| \ll 1$ we may may expand the right hand side of (19.32.328) and drop the term involving t^2 to get

$$
\begin{aligned}
\frac{dt}{dr} &\approx -\frac{U}{k}\left[\sin^2 kr + 2t(r)\sin kr\cos kr\right] \\
&= -\frac{U}{k}\left[\sin^2 kr + t(r)\sin 2kr\right] \ .
\end{aligned}
$$

(19.32.329)

So,

$$
\frac{dt}{dr} + \frac{U}{k}\sin 2kr\, t = -\frac{U}{k}\sin^2 kr \ .
$$

(19.32.330)

This is a linear equation with integration factor

$$
\exp \int_0^r dr' \frac{U(r')}{k}\sin 2kr' \ .
$$

Therefore, we have that

$$
\begin{aligned}
t(r) &= -\frac{1}{k}\exp\left[-\int_0^r dr' \frac{U(r')}{k}\sin 2kr'\right] \times \\
&\quad \times \int_0^r dr'\, U(r')\sin^2 r' \exp\left[\int_0^{r'} dr'' \frac{U(r'')}{k}\sin 2kr''\right] \ .
\end{aligned}
$$

(19.32.331)

Thus, using (19.32.324) we get

$$
\tan\delta_0 = -\frac{1}{k}\int_0^\infty dr\, U(r)\sin^2 r\, \exp\left[-\int_r^\infty dr' \frac{U(r')}{k}\sin 2kr'\right] \ .
$$

(19.32.332)

Bibliography

[19.1] A.Z. Capri, *Nonrelativistic Quantum Mechanics* 3rd edition, World Scientific Publishing Co. Pte. Ltd., section 19.6, (2002) .

[19.2] ibid, section 19.8.

[19.3] ibid, section 19.11.

[19.4] H. Feshbach, Ann. Phys. (N.Y.), **19**, 287, (1962).

[19.5] F. Calogero, Nuovo Cimento **27**, 261, (1963).

Chapter 20

Systems of Identical Particles

20.1 Periodic Table

Using the Pauli exclusion principle, determine the maximum number of electrons in any energy level n of an atom. Neglect the interactions between the electrons.

Solution

If we neglect the interaction between the electrons then each electron experiences simply a central Coulomb potential and all of them have the same energy levels given by

$$E_n = -\frac{Ze^2}{a}\frac{1}{n^2} . \qquad (20.1.1)$$

If we neglect the spin, all of these energy levels except $n = 1$ are degenerate. Including spin, the degeneracy of the nth energy level is given by

$$g_n = 2\sum_{l=0}^{n-1}(2l+1) = 2n^2 . \qquad (20.1.2)$$

This, according to the Pauli Exclusion Principle, is the maximum number of electrons that may be accommodated in the energy level E_n.

20.2 Identical $s = 1/2$ Particles in $l = 0, 1$ States

Two identical spin 1/2 particles collide.
a) Assume that the wavefunction for the two particles after the collision corresponds to a relative orbital angular momentum with $l = 0$. Write out the possible spinor wavefunctions.
b) Repeat part a) if $l = 1$.

Solution

a) If $l = 0$ the spatial wavefunction is even under an interchange of spatial coordinates. Since spin 1/2 particles are fermions, this requires that the spinor wave function be odd under an interchange of particles. The only possibility is

$$|s, m_s, 1/2, 1/2\rangle = |0, 0, 1/2, 1/2\rangle$$
$$= \frac{1}{\sqrt{2}} [|1/2, 1/2\rangle|1/2, -1/2\rangle - |1/2, -1/2\rangle|1/2, 1/2\rangle] . \tag{20.2.3}$$

b) If $l = 1$ the spatial wavefunction is odd under an interchange of spatial coordinates. Therefore the spinor wavefunction must be even under an interchange of particles. So, we have three possibilities

$$|s, m_s, 1/2, 1/2\rangle = |1, m, 1/2, 1/2\rangle . \tag{20.2.4}$$

When written out these three cases are

$$|1, 1, 1/2, 1/2\rangle = |1/2, 1/2\rangle|1/2, 1/2\rangle$$
$$|1, 0, 1/2, 1/2\rangle = \frac{1}{\sqrt{2}} [|1/2, 1/2\rangle|1/2, -1/2\rangle + |1/2, -1/2\rangle|1/2, 1/2\rangle]$$
$$|1, -1, 1/2, 1/2\rangle = |1/2, -1/2\rangle|1/2, -1/2\rangle . \tag{20.2.5}$$

20.3   Two Identical $s = 0$ Particles

Consider the collision of two identical spin 0 particles. Separate the wave function into centre of mass and relative coordinates. Discuss the symmetry required of the relative wave function. Use this to obtain the modifications required for the scattering amplitude and hence an expression for the differential cross-section.

Solution

In the center of mass system the Hamiltonian is

$$H = \frac{\vec{P}^2}{4m} + \frac{\vec{p}^2}{m} + V(\vec{r}) \tag{20.3.6}$$

where

$$\vec{R} = \frac{1}{2}(\vec{r}_1 + \vec{r}_2) \quad , \quad \vec{r} = \vec{r}_1 - \vec{r}_2 \tag{20.3.7}$$

The wavefunction separates into

$$\Psi(\vec{R}, \vec{r}) = e^{i\vec{R}\cdot\vec{R}} \psi(\vec{r}) . \tag{20.3.8}$$

The first factor (describing the center of mass motion) is clearly symmetric under the interchange of particle labels. Thus, since the spin is 0 and we are dealing with bosons we require that the wavefunction $\psi(\vec{r})$ must also be symmetric under the interchange of particle labels. In other words, we need

$$\psi(\vec{r}) = \psi(-\vec{r}) \ . \tag{20.3.9}$$

In general we would write the asymptotic form of the solution as

$$e^{i\vec{k}\cdot\vec{r}} + f(\theta, \varphi)\, \frac{e^{ikr}}{r} \ . \tag{20.3.10}$$

After symmetrization, this becomes

$$\frac{1}{\sqrt{2}} [\psi(\vec{r}) + \psi(-\vec{r})]$$

$$= \frac{1}{\sqrt{2}} \left[e^{i\vec{k}\cdot\vec{r}} + e^{-i\vec{k}\cdot\vec{r}} + [f(\theta, \varphi) + f(\pi - \theta, \varphi + \pi)]\frac{e^{ikr}}{r} \right] \ . \tag{20.3.11}$$

Now, the detector cannot distinguish between the two particles. Therefore, the differential cross-section is the sum of the cross-sections for particles 1 and 2. However, they have identical scattering amplitudes. Thus,

$$\sigma(\theta, \varphi) = 2 \left| \frac{1}{\sqrt{2}} [f(\theta, \varphi) + f(\pi - \theta, \varphi + \pi)] \right|^2$$

$$= |f(\theta, \varphi) + f(\pi - \theta, \varphi + \pi)|^2 \ . \tag{20.3.12}$$

20.4 Identical $s = 1/2$ Particles in Centre of Mass

Repeat the previous problem for two identical spin 1/2 particles. Again, separate the wavefunction into centre of mass and relative coordinates as well as spin coordinates. Assume the particles scatter from a spin-independent potential.
a) Discuss the symmetry required of the relative wavefunction if the scattering occurs in a singlet ($s = 0$) state. Use this to obtain the modifications required for the scattering amplitude and hence an expression for the differential cross-section.
b) Repeat part a) if the scattering occurs in a triplet ($s = 1$) state.

Solution

We proceed as in the problem above. However, this time the total wavefunction, in the center of mass system, is given by

$$\Psi(\vec{R}, \vec{r}, S, m_s) = e^{i\vec{K}\cdot\vec{R}}\, \psi(\vec{r})\, \chi(S, m_s) \ , \tag{20.4.13}$$

where $\chi(S, m_s)$ is the wavefunction for the total spin of the two particles.
a) If the scattering occurs in a singlet ($s = 0$) state then $\chi(S, m_s)$ is antisymmetric under the interchange of the two particles. In this case we need that the total wavefunction must be antisymmetric under the interchange of the two

particles and hence, that the spatial part of the wavefunction should be symmetric just like in the spin 0 case considered above. Thus, for the singlet case we get that the appropriate scattering amplitude is

$$f_{(s=0)}(\theta, \varphi) = f(\theta, \varphi) + f(\pi - \theta, \varphi + \pi) \qquad (20.4.14)$$

and the differential cross-section is given by

$$\sigma(\theta, \varphi) = \left| f_{(s=0)}(\theta, \varphi) \right|^2 = \left| f(\theta, \varphi) + f(\pi - \theta, \varphi + \pi) \right|^2 . \qquad (20.4.15)$$

Here, we have deliberately dropped the factors of 2 and $1/\sqrt{2}$ that cancel each other.

b) If the total spin is $S = 1$ the spin wave function is symmetric under the interchange of particles. Thus, we need that the scattering amplitude should be antisymmetric. The result is

$$f_{(s=1)}(\theta, \varphi) = f(\theta, \varphi) - f(\pi - \theta, \varphi + \pi) \qquad (20.4.16)$$

and the differential cross-section is given by

$$\sigma(\theta, \varphi) = \left| f_{(s=1)}(\theta, \varphi) \right|^2 = \left| f(\theta, \varphi) - f(\pi - \theta, \varphi + \pi) \right|^2 . \qquad (20.4.17)$$

Again we have deliberately dropped the factors of 2 and $1/\sqrt{2}$ that cancel each other.

20.5   Heisenberg Field Operator for Bosons

Consider a system of non-interacting bosons and write the Hamiltonian in the form

$$H = \sum_{k=0}^{\infty} \hbar \omega_k a_k^\dagger a_k .$$

Find an explicit expression for

$$\exp[iHt/\hbar] \, \psi(\vec{x}, t) \, \exp[-iHt/\hbar]$$

where

$$\psi_s(\vec{x}) = \sum_{k=0}^{\infty} \langle \vec{x} | k \rangle a_k .$$

Hint: Expand the second exponential and commute a_k through showing that

$$a_k \, e^{\lambda a_k^\dagger a_k} = e^{\lambda (a_k^\dagger a_k + 1)} \, a_k .$$

Solution

The field Hamiltonian is

$$H = \sum_{k=0}^{\infty} \hbar\omega_k \, a_k^{\dagger} a_k \; . \tag{20.5.18}$$

So, the field operator in the Heisenberg picture is related to the field operator in the Schrödinger picture by

$$\psi(t, \vec{x}) = \exp\{iHt/\hbar\} \, \psi_S(\vec{x}) \, \exp\{-iHt/\hbar\} \tag{20.5.19}$$

where

$$\psi_S(\vec{x}) = \sum_{k=0}^{\infty} \langle \vec{x}|k\rangle a_k \; . \tag{20.5.20}$$

To evaluate $\exp\{iHt/\hbar\} \, a_k \, \exp\{-iHt/\hbar\}$ we first consider

$$
\begin{aligned}
a_k \left(a_n^{\dagger} a_n \right)^s &= a_k a_n^{\dagger} a_n \left(a_n^{\dagger} a_n \right)^{s-1} \\
&= \left(\delta_{kn} + a_n^{\dagger} a_n \right) a_k \left(a_n^{\dagger} a_n \right)^{s-1} \\
&= \left(\delta_{kn} + a_n^{\dagger} a_n \right) a_k a_n^{\dagger} a_n \left(a_n^{\dagger} a_n \right)^{s-2} \\
&= \left(\delta_{kn} + a_n^{\dagger} a_n \right)^2 a_k \left(a_n^{\dagger} a_n \right)^{s-2} \; .
\end{aligned}
\tag{20.5.21}
$$

So, after n steps, we have

$$a_k \left(a_n^{\dagger} a_n \right)^s = \left(\delta_{kn} + a_n^{\dagger} a_n \right)^s a_k \; . \tag{20.5.22}$$

Hence,

$$
\begin{aligned}
a_k \exp\left(c \, a_n^{\dagger} a_n \right) &= a_k \sum_{s=0}^{\infty} \frac{c^s}{s!} \left(a_n^{\dagger} a_n \right)^s \\
&= \sum_{s=0}^{\infty} \frac{c^s}{s!} \left(\delta_{kn} + a_n^{\dagger} a_n \right)^s a_k \\
&= \exp[c(\delta_{kn} + a_n^{\dagger} a_n)] \, a_k \; .
\end{aligned}
\tag{20.5.23}
$$

So, we get

$$
\begin{aligned}
&\exp(iHt/\hbar) \, a_k \, \exp(-iHt/\hbar) \\
&= \exp\left(it \sum \omega_n a_n^{\dagger} a_n \right) a_k \exp\left(-it \sum \omega_n a_n^{\dagger} a_n \right) \\
&= e^{-i\omega_k t} \, a_k \; .
\end{aligned}
\tag{20.5.24}
$$

Therefore,

$$
\begin{aligned}
\psi(t, \vec{x}) &= \exp\{iHt/\hbar\} \, \psi_S(\vec{x}) \, \exp\{-iHt/\hbar\} \\
&= \sum_{k=0}^{\infty} \langle \vec{x}|k\rangle \, e^{-i\omega_k t} \, a_k \; .
\end{aligned}
\tag{20.5.25}
$$

20.6   Heisenberg Field Operator for Fermions

Repeat the previous problem for fermions.

Solution

The solution for fermions starts as in the problem above. However, now it is even simpler since a_k commutes with $a_n^\dagger a_n$ for $k \neq n$ and for $k = n$ we find

$$a_n a_n^\dagger a_n = -a_n^\dagger a_n a_n + a_n = a_n \; . \tag{20.6.26}$$

This means that

$$
\begin{aligned}
& \exp(iHt/\hbar)\, a_k\, \exp(-iHt/\hbar) \\
= \;& \exp it \left(\sum \omega_n a_n^\dagger a_n \right) a_k \, \exp -it \left(\sum \omega_n a_n^\dagger a_n \right) \\
= \;& e^{-i\omega_k t}\, a_k \; .
\end{aligned} \tag{20.6.27}
$$

So that, as in the problem above, we get

$$
\begin{aligned}
\psi(t,\vec{x}) \; &= \; \exp\{iHt/\hbar\}\, \psi_S(\vec{x})\, \exp\{-iH\}t/\hbar \\
&= \; \sum_{k=0}^{\infty} \langle \vec{x} | k \rangle\, e^{-i\omega_k t}\, a_k \; .
\end{aligned} \tag{20.6.28}
$$

20.7   Two-body Interaction

Obtain the equation

$$V = \frac{1}{2} \int d^3x\, d^3y\, \psi^\dagger(\vec{x}) \psi^\dagger(\vec{y}) V(\vec{x},\vec{y}) \psi(\vec{y}) \psi(\vec{x})$$

for the occupation number space representation of the interaction specified by

$$V(\vec{x}_1, \vec{x}_2, \ldots, \vec{x}_N) = \frac{1}{2} \sum_{j \neq k}^{N} V(\vec{x}_j, \vec{x}_k) \; .$$

Solution

For any configuration space operator F its representation in occupation number space is given by [20.1]

$$
\begin{aligned}
\langle n_0, n_1, \ldots | F | m_0, m_1, \ldots \rangle = \frac{1}{N!} \int d^3x_1 \ldots d^3x_N \langle n_0, n_1, \ldots | \\
\psi^\dagger(\vec{x}_N) \ldots \psi^\dagger(\vec{x}_1) F^N(\vec{x}_1, \vec{x}_2, \ldots, \vec{x}_N) \psi(\vec{x}_1) \ldots \psi(\vec{x}_N) \\
| m_0, m_1, \ldots \rangle \; .
\end{aligned} \tag{20.7.29}
$$

Now we replace F by

$$V(\vec{x}_1, \vec{x}_2, \ldots, \vec{x}_N) = \frac{1}{2} \sum_{j \neq k}^{N} V(\vec{x}_j, \vec{x}_k) = \sum_{j<k}^{N} V(\vec{x}_j, \vec{x}_k) \ . \tag{20.7.30}$$

In this case we easily see that

$$\begin{aligned}
&\langle n_0, n_1, \ldots | V | m_0, m_1, \ldots \rangle \\
=\ &\frac{1}{N(N-1)} \sum_{j<k} \langle n_0, n_1, \ldots | \int d^3 x_j d^3 x_k \psi^\dagger(\vec{x}_j) \\
&\psi^\dagger(\vec{x}_k) V(\vec{x}_j, \vec{x}_k) \psi(\vec{x}_k) \psi(\vec{x}_j) | m_0, m_1, \ldots \rangle \\
=\ &\langle n_0, n_1, \ldots | \frac{1}{2} \int d^3 x \, d^3 y \, \psi^\dagger(\vec{x}) \psi^\dagger(\vec{y}) V(\vec{x}, \vec{y}) \psi(\vec{y}) \psi(\vec{x}) | m_0, m_1, \ldots \rangle \ .
\end{aligned}$$
$$\tag{20.7.31}$$

Therefore,

$$V = \frac{1}{2} \int d^3 x \, d^3 y \, \psi^\dagger(\vec{x}) \psi^\dagger(\vec{y}) V(\vec{x}, \vec{y}) \psi(\vec{y}) \psi(\vec{x}) \tag{20.7.32}$$

as required.

20.8   Diagonalization of Boson Hamiltonian

Show that the Hamiltonian

$$H = \sum_k \left[E_k a_k^\dagger a_k + \lambda_k a_k^\dagger a_k^\dagger + \lambda_k^* a_k a_k \right] \quad , \quad E_k > 2|\lambda_k|$$

can be diagonalized if a_k, $a_k^\dagger$ are bose operators.
Hint: Introduce operators

$$b_k = u_k a_k^\dagger + v_k a_k$$

and choose the constants u_k and v_k appropriately. What happens if a_k, $a_k^\dagger$ are fermi operators? These are simple examples of Bogoliubov transformations.

Solution

We want the transformations

$$\begin{aligned}
b_k &= u_k a_k^\dagger + v_k a_k \\
b_k^\dagger &= v_k^* a_k^\dagger + u_k^* a_k
\end{aligned} \tag{20.8.33}$$

to maintain the bose commutation relations

$$[b_k, b_k^\dagger] = [a_k, a_k^\dagger] = 1 \tag{20.8.34}$$

and that all other commutators vanish. This implies that

$$|v_k|^2 - |u_k|^2 = 1 \ . \tag{20.8.35}$$

So, we can parametrize these coefficients as follows

$$\begin{aligned} v_k &= e^{i\alpha_k}\cosh\theta_k \\ u_k &= e^{i\beta_k}\sinh\theta_k \end{aligned} \tag{20.8.36}$$

Solving for a_k , $a_k^\dagger$ in terms of b_k , $b_k^\dagger$ we get

$$\begin{aligned} a_k &= e^{-i\alpha_k}\cosh\theta_k\, b_k - e^{i\beta_k}\sinh\theta_k b_k^\dagger \\ a_k^\dagger &= e^{i\alpha_k}\cosh\theta_k\, b_k^\dagger - e^{-i\beta_k}\sinh\theta_k b_k \ . \end{aligned} \tag{20.8.37}$$

The Hamiltonian now is a sum of terms of the form

$$\begin{aligned} H_k &= E_k\, a_k^\dagger a_k + \lambda_k a_k^\dagger a_k^\dagger + \lambda_k^* a_k a_k \\ &= E_k\left[\cosh^2\theta_k\, b_k^\dagger b_k - e^{i(\alpha_k+\beta_k)}\sinh\theta_k\cosh\theta_k b_k^\dagger b_k^\dagger \right. \\ &\qquad \left. + \sinh^2\theta_k\, b_k b_k^\dagger - e^{-i(\alpha_k+\beta_k)}\sinh\theta_k\cosh\theta_k b_k b_k\right] \\ &\quad + \lambda_k\left[e^{2i\alpha_k}\cosh^2\theta_k\, b_k^\dagger b_k^\dagger - e^{i(\alpha_k-\beta_k)}\sinh\theta_k\cosh\theta_k b_k^\dagger b_k \right. \\ &\qquad \left. + e^{-2i\beta_k}\sinh^2\theta_k\, b_k b_k - e^{i(\alpha_k-\beta_k)}\sinh\theta_k\cosh\theta_k b_k b_k^\dagger\right] \\ &\quad + \lambda_k^*\left[e^{-2i\alpha_k}\cosh^2\theta_k\, b_k b_k - e^{-i(\alpha_k-\beta_k)}\sinh\theta_k\cosh\theta_k b_k b_k^\dagger \right. \\ &\qquad \left. + e^{2i\beta_k}\sinh^2\theta_k\, b_k^\dagger b_k^\dagger - e^{-i(\alpha_k-\beta_k)}\sinh\theta_k\cosh\theta_k b_k^\dagger b_k\right] \ . \end{aligned} \tag{20.8.38}$$

To get the (off-diagonal) terms involving $b_k^\dagger b_k^\dagger$ as well as those involving $b_k b_k$ to cancel we require that

$$\begin{aligned} &- E_k\, e^{i(\alpha_k+\beta_k)}\sinh\theta_k\cosh\theta_k \\ &+ \lambda_k\, e^{2i\alpha_k}\cosh^2\theta_k + \lambda_k^*\, e^{2i\beta_k}\sinh^2\theta_k = 0 \ . \end{aligned} \tag{20.8.39}$$

Therefore, writing

$$\lambda_k = |\lambda_k|e^{i\sigma_k} \tag{20.8.40}$$

we see that if we choose

$$\alpha_k = -\frac{\sigma_k}{2} \quad , \quad \beta_k = \frac{\sigma_k}{2} \tag{20.8.41}$$

then all we need is

$$E_k\frac{1}{2}\sinh 2\theta_k = |\lambda_k|\cosh 2\theta_k \ . \tag{20.8.42}$$

So, we take

$$\tanh 2\theta_k = \frac{2|\lambda_k|}{E_k} \ . \tag{20.8.43}$$

The Hamiltonian now reads

$$H_k = E_k \left[\cosh 2\theta_k \, b_k^\dagger b_k + \sinh^2 \theta_k \right]$$
$$- (\lambda_k + \lambda_k^*) \left[\sinh 2\theta_k \, b_k^\dagger b_k + \frac{1}{2} \sinh 2\theta_k \right] . \qquad (20.8.44)$$

After substituting for $\cosh \theta_k$ and $\sinh \theta_k$ we get

$$H_k = \frac{1}{\sqrt{E_k^2 - 4|\lambda_k|^2}} \left[E_k^2 - 4|\lambda_k|^2 \cos \sigma \right] \left(b_k^\dagger b_k + \frac{1}{2} \right) - \frac{E_k}{2} . \qquad (20.8.45)$$

If we are dealing with fermion operators the Hamiltonian is already diagonal since

$$a_k^\dagger a_k^\dagger = a_k a_k = 0 . \qquad (20.8.46)$$

20.9   Formula for $e^{-A} B e^A$

Let A and B be arbitrary operators. Derive the following formula

$$e^{-A} B e^A = \sum_{n=0}^{\infty} \frac{1}{n!} [B, A]_n$$

where,

$$[B, A]_0 = B \quad , \quad [B, A]_{n+1} = [[B, A]_n, A] .$$

Solution

We prove the result in two steps. First we prove by induction that

$$B A^n = \sum_{r=0}^{n} \binom{n}{r} A^r [B, A]_{n-r} \qquad (20.9.47)$$

where

$$[B, A]_0 = B \quad , \quad [B, A]_{n+1} = [[B, A]_n, A] . \qquad (20.9.48)$$

Clearly the result is true for $n = 0$. We now assume the result holds for $n - 1$. Thus,

$$B A^{n-1} = \sum_{r=0}^{n-1} \binom{n-1}{r} A^r [B, A]_{n-1-r} . \qquad (20.9.49)$$

Then,

$$B A^n = \sum_{r=0}^{n-1} \binom{n-1}{r} A^r [B, A]_{n-1-r} A$$
$$= \sum_{r=0}^{n-1} \binom{n-1}{r} A^r \{ A[B, A]_{n-1-r} + [B, A]_{n-r} \} . \qquad (20.9.50)$$

The coefficients of the terms $[B, A]_s$ $(1 \le s \le n - 1)$ are

$$\left[\left(\begin{array}{c} n - 1 \\ n - 1 - s \end{array} \right) A^{n-s} + \left(\begin{array}{c} n - 1 \\ n - s \end{array} \right) A^{n-s} \right]$$

$$= A^{n-s} \frac{(n - 1)!}{(n - s)! s!} [(n - s) + s]$$

$$= \left(\begin{array}{c} n \\ s \end{array} \right) A^{n-s} . \tag{20.9.51}$$

For $s = n$ there is only one term, namely

$$\left(\begin{array}{c} n - 1 \\ 0 \end{array} \right) = \left(\begin{array}{c} n \\ n \end{array} \right) . \tag{20.9.52}$$

Also for $s = 0$ there is only one term, namely

$$A^n = A^n \left(\begin{array}{c} n - 1 \\ n - 1 \end{array} \right) = A^n \left(\begin{array}{c} n \\ 0 \end{array} \right) . \tag{20.9.53}$$

Thus, in all cases we have the desired result

$$BA^n = \sum_{r=0}^{n} \left(\begin{array}{c} n \\ r \end{array} \right) A^r [B, A]_{n-r} . \tag{20.9.54}$$

Now consider

$$B\, e^A = \sum_{n=0}^{\infty} \frac{A^n}{n!} = \sum_{n=0}^{\infty} \sum_{s=0}^{n} \frac{1}{(n - s)! s!} A^{n-s} [B, A]_s . \tag{20.9.55}$$

We can interchange the order of summation. Then this equation reads

$$B\, e^A = \sum_{s=0}^{\infty} \sum_{n=s}^{\infty} \frac{A^{n-s}}{(n - s)!} \frac{[B, A]_s}{s!}$$

$$= \sum_{s=0}^{\infty} \sum_{r=0}^{\infty} \frac{A^r}{(r)!} \frac{[B, A]_s}{s!}$$

$$= e^A \sum_{s=0}^{\infty} \frac{[B, A]_s}{s!} . \tag{20.9.56}$$

Multiplying by e^{-A} on the left we obtain the desired result.

20.10   Bogoliubov Transformation: Fermions

Let a_k, $a_k^\dagger$, $(k = 1, 2)$ be fermion operators and define

$$b_1^\dagger = u a_1^\dagger - v a_2 , \quad b_1 = u^* a_1 - v^* a_2^\dagger$$

$$b_2 = v a_1^\dagger + u a_2 , \quad b_2^\dagger = v^* a_1 + u^* a_2^\dagger$$

with

$$|u|^2 + |v|^2 = 1 \quad , \quad uv^* - u^*v = 0 .$$

a) Verify that b_k, $b_k^\dagger$ are fermion operators.

b) Show that for an appropriate choice of the constant c the unitary operator

$$U = \exp\left\{ c\, a_1^\dagger a_2^\dagger - c^*\, a_2 a_1 \right\}$$

has the action

$$b_1^\dagger = U a_1^\dagger U^\dagger \quad , \quad b_1 = U a_1 U^\dagger$$

$$b_2^\dagger = U a_2^\dagger U^\dagger \quad , \quad b_2 = U a_2 U^\dagger .$$

This is known as a Bogoliubov transformation.

Hint: Use the results of the previous problem.

Solution

a) We first verify the anticommutation relations.

$$
\begin{aligned}
b_1 b_1^\dagger + b_1^\dagger b_1 &= [(u^* a_1 - v^* a_2^\dagger), (u a_1^\dagger - v a_2)]_+ \\
&= |u|^2 [a_1^\dagger, a_1]_+ + |v|^2 [a_2^\dagger, a_2]_+ = 1 .
\end{aligned}
\tag{20.10.57}
$$

Similarly

$$
\begin{aligned}
b_2 b_2^\dagger + b_2^\dagger b_2 &= [(v a_1 + u a_2^\dagger), (v^* a_1^\dagger + u^* a_2)]_+ \\
&= |v|^2 [a_1^\dagger, a_1]_+ + |u|^2 [a_2^\dagger, a_2]_+ = 1 .
\end{aligned}
\tag{20.10.58}
$$

It also follows quite trivially that

$$b_1 b_1 = b_2 b_2 = 0 \tag{20.10.59}$$

and hence that

$$b_1^\dagger b_1^\dagger = b_2^\dagger b_2^\dagger = 0 . \tag{20.10.60}$$

This shows that the b_j, $b_j^\dagger$ $\ j = 1, 2$ are indeed Fermi operators.

b) To verify this part we use the result of problem 20.9 and take

$$A = c\, a_1^\dagger a_2^\dagger - c^*\, a_2 a_1 \quad , \quad B = a_j \quad , \quad j = 1, 2 . \tag{20.10.61}$$

Now,

$$[a_j, A]_0 = a_j . \tag{20.10.62}$$

Then,

$$[a_1, A]_1 = [a_1, A] = ca_2^\dagger \qquad , \qquad [a_2, A]_1 = [a_2, A] = ca_1^\dagger$$
$$[a_1, A]_2 = [ca_2^\dagger, A] = -|c|^2 a_1 \qquad , \qquad [a_2, A]_2 = [ca_1^\dagger, A] = -|c|^2 a_2$$
$$[a_1, A]_3 = -[|c|^2 a_1, A] = -c|c|^2 a_2^\dagger \quad , \quad [a_2, A]_3 = [|c|^2 a_2, A] = c|c|^2 a_1^\dagger$$
$$[a_1, A]_4 = -c[|c|^2 a_2^\dagger, A] = |c|^4 a_1 \qquad , \qquad [a_2, A]_4 = c[|c|^2 a_1^\dagger, A] = |c|^4 a_2$$

etc.

Thus,

$$[a_1, A]_{4n} = |c|^{4n} a_1 \qquad , \qquad [a_2, A]_{4n} = |c|^{4n} a_2$$
$$[a_1, A]_{4n+1} = c|c|^{4n} a_2^\dagger \qquad , \qquad [a_2, A]_{4n+1} = -c|c|^{4n} a_1^\dagger$$
$$[a_1, A]_{4n+2} = -|c|^{4n+2} a_1 \qquad , \qquad [a_2, A]_{4n+2} = -|c|^{4n+2} a_2$$
$$[a_1, A]_{4n+3} = -c|c|^{4n+2} a_2^\dagger \quad , \quad [a_2, A]_{4n+3} = c|c|^{4n+2} a_1^\dagger \; .$$

Therefore,

$$e^{-A} a_1 e^A = \sum_{n=0}^{\infty} \frac{1}{n!} [a_1, A]_n$$

$$= a_1 \left(\sum_{n=0}^{\infty} \frac{|c|^{4n}}{(4n)!} - \sum_{n=0}^{\infty} \frac{|c|^{4n+2}}{(4n+2)!} \right)$$

$$+ \frac{c}{|c|} a_2^\dagger \left(\sum_{n=0}^{\infty} \frac{|c|^{4n+1}}{(4n+1)!} - \sum_{n=0}^{\infty} \frac{|c|^{4n+3}}{(4n+3)!} \right)$$

$$= a_1 \cos|c| + \frac{c}{|c|} a_2^\dagger \sin|c| \; . \qquad (20.10.63)$$

Similarly

$$e^{-A} a_2 e^A = a_2 \cos|c| - \frac{c}{|c|} a_1^\dagger \sin|c| \; . \qquad (20.10.64)$$

So if we choose

$$u^* = \cos|c| \quad , \quad v^* = -\frac{c}{|c|}\sin|c| \quad , \quad c^* = c \qquad (20.10.65)$$

then we have exactly the transformation required.

20.11   Diagonalization of Bose Hamiltonian

Diagonalize the Hamiltonian

$$H = \sum_{k=0}^{\infty} \hbar\omega a_k^\dagger a_k + \frac{1}{2}\sum_k V(k)[a_k a_k + a_k^\dagger a_k^\dagger] \; ,$$

where $\omega(k)$ and $V(k)$ are given functions of k and

$$[a_k, a_{k'}^\dagger] = \delta_{k,k'} \quad , \quad [a_k, a_k] = 0 \; .$$

Hint: Use the Bogoliubov transformation

$$b_k = u_k a_k - v_k a_k^\dagger \quad , \quad b_k^\dagger = u_k^* a_k^\dagger - v_k^* a_k$$

with

$$|u_k|^2 - |v_k|^2 = 1 \; .$$

Solution

We have

$$H = \sum_{\vec{k}} \omega(\vec{k}) a_{\vec{k}}^\dagger a_{\vec{k}} + \frac{1}{2} \sum_{\vec{k}} V(\vec{k}) \left(a_{\vec{k}} a_{\vec{k}} + a_{\vec{k}}^\dagger a_{\vec{k}}^\dagger \right) \qquad (20.11.66)$$

where $\omega(\vec{k})$ and $V(\vec{k})$ are given functions of $\vec{k}$ and

$$[a_{\vec{k}}, a_{\vec{k}'}^\dagger] = \delta_{k,k'} \quad , \quad [a_{\vec{k}}, a_{\vec{k}'}] = 0 \; . \qquad (20.11.67)$$

Following the hint we use the Bogoliubov transformation

$$b_{\vec{k}} = u_{\vec{k}} a_{\vec{k}} - v_{\vec{k}} a_{\vec{k}}^\dagger \quad , \quad b_{\vec{k}}^\dagger = u_{\vec{k}}^* a_{\vec{k}}^\dagger - v_{\vec{k}}^* a_{\vec{k}} \qquad (20.11.68)$$

with

$$|u_{\vec{k}}|^2 - |v_{\vec{k}}|^2 = 1 \; . \qquad (20.11.69)$$

This means we can parametrize

$$u_{\vec{k}} = \cosh \chi(\vec{k}) \quad , \quad v_{\vec{k}} = \sinh \chi(\vec{k}) \; . \qquad (20.11.70)$$

Inverting the transformation we find

$$a_{\vec{k}}^\dagger = u_{\vec{k}} b_{\vec{k}}^\dagger + v_{\vec{k}}^* b_{\vec{k}} \quad , \quad a_{\vec{k}} = u_{\vec{k}}^* b_{\vec{k}} + v_{\vec{k}} b_{\vec{k}}^\dagger \; . \qquad (20.11.71)$$

Inserting this into the Hamiltonian we get

$$H = \sum_{\vec{k}} \omega(\vec{k} \left[|u_{\vec{k}}|^2 b_{\vec{k}} b_{\vec{k}}^\dagger + |v_{\vec{k}}|^2 b_{\vec{k}}^\dagger b_{\vec{k}} + u_{\vec{k}} v_{\vec{k}} b_{\vec{k}}^\dagger b_{\vec{k}}^\dagger + u_{\vec{k}}^* v_{\vec{k}}^* b_{\vec{k}} b_{\vec{k}} \right]$$

$$+ \frac{1}{2} \sum_{\vec{k}} V(\vec{k}) \left[(u_{\vec{k}}^* b_{\vec{k}} + v_{\vec{k}} b_{\vec{k}}^\dagger)(u_{\vec{k}}^* b_{\vec{k}} + v_{\vec{k}} b_{\vec{k}}^\dagger) \right.$$

$$\left. + (u_{\vec{k}} b_{\vec{k}}^\dagger + v_{\vec{k}}^* b_{\vec{k}})(u_{\vec{k}} b_{\vec{k}}^\dagger + v_{\vec{k}}^* b_{\vec{k}}) \right] \; . \qquad (20.11.72)$$

After collecting terms, and inserting the parametrization this becomes

$$H = \sum_{\vec{k}} \left[\omega(\vec{k}) \cosh 2\chi(\vec{k}) + V(\vec{k}) \sinh 2\chi(\vec{k}) \right] b_{\vec{k}}^\dagger b_{\vec{k}}$$

$$+ \sum_{\vec{k}} \left[\omega(\vec{k}) \sinh 2\chi(\vec{k}) + \frac{1}{2} V(\vec{k}) \cosh 2\chi(\vec{k}) \right] \left(b_{\vec{k}}^\dagger b_{\vec{k}}^\dagger + b_{\vec{k}} b_{\vec{k}} \right)$$

$$+ \sum_{\vec{k}} \left[\omega(\vec{k}) \frac{1}{2} (\cosh 2\chi(\vec{k}) - 1) + \frac{1}{2} V(\vec{k}) \sinh 2\chi(\vec{k}) \right] \; . \qquad (20.11.73)$$

For the Hamiltonian to be in diagonal form requires that the terms involving $b_{\vec{k}}^{\dagger} b_{\vec{k}}^{\dagger} + b_{\vec{k}} b_{\vec{k}}$ vanish. This means that

$$\frac{1}{2} V(\vec{k}) \cosh 2\chi(\vec{k}) + \omega(\vec{k}) \sinh 2\chi(\vec{k}) = 0 . \qquad (20.11.74)$$

So we find

$$\cosh 2\chi(\vec{k}) \;\; = \;\; \frac{\omega(\vec{k})}{\sqrt{\omega(\vec{k})^2 - V(\vec{k})^2/4}}$$

$$\sinh 2\chi(\vec{k}) \;\; = \;\; \frac{-V(\vec{k})/2}{\sqrt{\omega(\vec{k})^2 - V(\vec{k})^2/4}} . \qquad (20.11.75)$$

Then we have that

$$H = \frac{\omega(\vec{k})^2 - V(\vec{k})^2/2}{\sqrt{\omega(\vec{k})^2 - V(\vec{k})^2/4}} b_{\vec{k}}^{\dagger} b_{\vec{k}} + \frac{\omega(\vec{k})^2/2 + V(\vec{k})^2/4}{\sqrt{\omega(\vec{k})^2 - V(\vec{k})^2/4}} - \frac{\omega(\vec{k})}{2} . \quad (20.11.76)$$

20.12   Diagonalization of Quadratic Hamiltonian

Use a Bogoliubov transformation to diagonalize the Hamiltonian

$$H = \frac{1}{2} \sum_{k>0} [(E_k + |g|)(a_k^{\dagger} a_k + a_{-k}^{\dagger} a_{-k}) + |g|(a_k^{\dagger} a_{-k}^{\dagger} + a_k a_{-k})] .$$

Here, $a_k^{\dagger}$, a_k are Bose creation and annihilation operators respectively.

Solution

The Bogoliubov transformation may be written

$$a_k \;\; = \;\; b_k \cosh \chi_k - b_{-k}^{\dagger} \sinh \chi_k$$

$$a_{-k} \;\; = \;\; b_{-k} \cosh \chi_{-k} - b_k^{\dagger} \sinh \chi_{-k}$$

$$a_k^{\dagger} \;\; = \;\; b_k^{\dagger} \cosh \chi_k - b_{-k} \sinh \chi_k$$

$$a_{-k}^{\dagger} \;\; = \;\; b_{-k}^{\dagger} \cosh \chi_{-k} - b_k \sinh \chi_{-k} \qquad (20.12.77)$$

where $b_k^{\dagger}$, b_k etc. are again Bose creation and annihilation operators. The Hamiltonian, in terms of these new operators, becomes

$$H = \frac{1}{2} \sum_{k>0} b_k^{\dagger} b_k [(E_k + |g|)(\cosh^2 \chi_k + \sinh^2 \chi_{-k})$$

$$- \;\; 2|g| \cosh \chi_k \sinh \chi_{-k}$$

$$+ \;\; b_{-k}^{\dagger} b_{-k} [(E_k + |g|) \cosh^2 \chi_{-k} + \sinh^2 \chi_k) - 2|g| \cosh \chi_{-k} \sinh \chi_k]$$

$$+ \;\; \frac{1}{2} \sum_{k>0} (E_k + |g|)[\sinh^2 \chi_k + \sinh^2 \chi_{-k}$$

$$- \quad \cosh \chi_k \sinh \chi_{-k} - \cosh \chi_{-k} \sinh \chi_k]$$

$$+ \quad \frac{1}{2} \sum_{k>0} b_k^\dagger b_{-k}^\dagger [-(E_k + |g|)(\cosh \chi_k \sinh \chi_k + \cosh \chi_{-k} \sinh \chi_{-k})$$

$$+ \quad |g|(\cosh \chi_k \cosh \chi_{-k} + \sinh \chi_k \sinh \chi_{-k}]$$

$$+ \quad \frac{1}{2} \sum_{k>0} b_k b_{-k} [-(E_k + |g|)(\cosh \chi_k \sinh \chi_k + \cosh \chi_{-k} \sinh \chi_{-k})$$

$$+ \quad |g|(\cosh \chi_k \cosh \chi_{-k} + \sinh \chi_k \sinh \chi_{-k}] \ . \tag{20.12.78}$$

We now choose

$$\chi_{-k} = \chi_k \tag{20.12.79}$$

and

$$(E_k + |g|) \sinh(2\chi_k) = |g| \cosh(2\chi_k) \ . \tag{20.12.80}$$

So,

$$\cosh(2\chi_k) \quad = \quad \sqrt{\frac{E_k + |g|}{E_k}}$$

$$\sinh(2\chi_k) \quad = \quad \sqrt{\frac{|g|}{E_k}} \ . \tag{20.12.81}$$

The Hamiltonian is now diagonal and reads

$$H \quad = \quad \frac{1}{2} \sum_{k>0} (b_k^\dagger b_k + b_{-k}^\dagger b_{-k})[(E_k + |g|) \cosh(2\chi_k) - |g| \sinh(2\chi_k)]$$

$$+ \quad \frac{1}{2} \sum_{k>0} [(E_k + |g|)[\cosh(2\chi_k) - 1] - |g| \sinh(2\chi_k)] \ . \tag{20.12.82}$$

After we replace $\cosh(2\chi_k)$ and $\sinh(2\chi_k)$ by their values we get

$$H \quad = \quad \frac{1}{2} \sum_{k>0} (b_k^\dagger b_k + b_{-k}^\dagger b_{-k}) \left[\frac{(E_k + |g|)^{3/2} - |g|^{3/2}}{\sqrt{E_k}} \right]$$

$$+ \quad \frac{1}{2} \sum_{k>0} \left[\frac{(E_k + |g|)^{3/2} - |g|^{3/2}}{\sqrt{E_k}} - (E_k + |g|) \right] \ . \tag{20.12.83}$$

The energy eigenvalues may now be read off.

20.13   Bogoliubov Transformation: Bose Operators

Consider a finite set N of bose operators a_k such that

$$[a_k, a_p^\dagger] = \delta_{k,p} \quad , \quad [a_k, a_{-p}] = 0 \quad , \quad [a_k^\dagger, a_p^\dagger] = 0 \ .$$

Define

$$b_k = \cosh \lambda \, a_k + \sinh \lambda \, a_k^\dagger$$

$$b_k^\dagger = \cosh \lambda \, a_k^\dagger + \sinh \lambda \, a_k .$$

a) Find the commutation rules for b_k, $b_k^\dagger$.
b) Find a unitary operator

$$V_N = \exp(iT_N)$$

in terms of a_k, $a_k^\dagger$ such that

$$V_N \, a_k \, V_N^\dagger = b_k .$$

c) Show that

$$\lim_{N \to \infty} \langle \Psi | V_N | \Phi \rangle = 0$$

where $|\Psi\rangle$, $|\Phi\rangle$ are any states of the form

$$\prod_k a_k^\dagger |0\rangle .$$

What does this last result mean?

Solution

a) Straightforward computation shows that

$$[b_k, b_k^\dagger] = 1 \quad \text{and} \quad [b_k, b_k] = [b_k^\dagger, b_k^\dagger] = 0 . \tag{20.13.84}$$

b) We now want to find an operator T_N such that

$$
\begin{aligned}
b_k &= \cosh \lambda \, a_k + \sinh \lambda \, a_k^\dagger \\
&= e^{iT_N} \, a_k \, e^{-iT_N} \\
&= \sum_n \frac{1}{n!} [iT_N, a_k]_n
\end{aligned}
\tag{20.13.85}
$$

where (see problem 20.9)

$$[iT_N, a_k]_n = [iT_N, [iT_N, a_k]_{n-1}] \quad , \quad [iT_N, a_k]_0 = a_k . \tag{20.13.86}$$

A short computation shows that we require

$$
\begin{aligned}
[iT_N, a_k]_{2n+1} &= \lambda^{2n+1} a_k^\dagger \\
[iT_N, a_k]_{2n} &= \lambda^{2n} a_k .
\end{aligned}
\tag{20.13.87}
$$

Thus, we try

$$iT_N = \frac{\lambda}{2} \sum_{k=1}^{N} \left(a_k^2 - a_k^{\dagger 2} \right) \tag{20.13.88}$$

and we see that this works.

We can also write this expression as

$$iT_N = \frac{\lambda}{2} \sum_{k=1}^{N} A_k .$$

(20.13.89)

We also have that

$$[A_k, A_q] = 0 \quad \text{for} \quad k \neq q .$$

c) We begin with states of the form

$$|\Phi\rangle = (a_k^\dagger)^n |0\rangle \quad , \quad |\Psi\rangle = (a_k^\dagger)^m |0\rangle .$$

(20.13.90)

Then, for fixed n and m and a sufficiently large N, the expression $\langle\Phi|iT_N|\Psi\rangle$ always contains values of j such that there are terms $\exp A_j$ with no $a_j^\dagger$ to the right or a_j to the left. This leads us to consider terms of the form $\langle 0|e^{A_j}|0\rangle$. To evaluate such a term we temporarily drop the subscript j and expand the exponential. This means we need to evaluate

$$\sum_{n=0}^{\infty} \frac{(\lambda/2)^n}{n!} \langle 0| (a^2 - a^{\dagger 2})^n |0\rangle .$$

(20.13.91)

For the typical mth order term, in this sum, all contributions come from terms having an even value of m or $m = 2n$ and of the form $(-1)^n a^2 \ldots a^{\dagger 2}$ with an equal number of a^2 and $a^{\dagger 2}$ and the $a^{\dagger 2}$ to the right of the a^2 . Using this we find that

$$\langle 0| \left(a^2 - a^{\dagger 2}\right)^2 |0\rangle = \langle 0|a^4 - 2a^{\dagger 2}a^2 - 1 + a^{\dagger 4}|0\rangle = (-1)$$

(20.13.92)

and

$$\langle 0| \left(a^2 - a^{\dagger 2}\right)^4 |0\rangle = \langle 0| \left(a^2 - a^{\dagger 2}\right)^2 \left(a^2 - a^{\dagger 2}\right)^2 |0\rangle = (-1)^2 2! \quad (20.13.93)$$

and generally

$$\langle 0| \left(a^2 - a^{\dagger 2}\right)^{2n} |0\rangle = (-1)^n n! .$$

(20.13.94)

Thus,

$$\sum_{n=0}^{\infty} \frac{(\lambda/2)^n}{n!} \langle 0| \left(a^2 - a^{\dagger 2}\right)^n |0\rangle = \sum_{n=0}^{\infty} \frac{(-1)^n (\lambda/2)^{2n} n!}{2n!}$$

$$< \sum_{n=0}^{\infty} \frac{(-\lambda^2/4)^n}{n!} = e^{-\lambda^2/4} .$$

(20.13.95)

Now, for $\lim_{N\to\infty}$ the number of such terms tends to ∞ and we get a term bounded by

$$\lim_{N\to\infty} [e^{-\lambda^2/4}]^{N-r} \times \text{irrelevant finite factors} .$$

Also r is some finite number. Thus, the limit is 0.

This means that norms are not preserved and the operator $T = \lim T_N$ cannot be unitary. The fancy way of saying this is to say that in this case, "the Bogoliubov transformation is not unitarily implementable".

20.14   Density Matrix for a Subsystem

Consider the normalized wavefunction $\Psi(x_i, q_j)$ $i = i \dots n$, $j = 1 \dots N$ of a system where the coordinates x_i refer to a subsystem. Let $A^{(r)}$ be any operator that acts only on the subsystem. Thus, $H^{(r)}$ represents the total Hamiltonian for the subsystem. Let

$$\langle x|\rho|y\rangle = \int \Psi^*(y_i, q_j)\Psi(x_i, q_j)dq_1 \dots dq_N$$

be the reduced density matrix for the subsystem.

a) Express the expectation value of any operator $A^{(r)}$ in terms of $A^{(r)}$ and ρ.

b) Find the normalization condition satisfied by ρ.

c) Find the equation of motion satisfied by ρ.

Solution

a) The expectation value of $A^{(r)}$ is given by

$$\begin{aligned}
\langle A^{(r)}\rangle &= \int \Psi^*(x_i, q_j)A^{(r)}\Psi(x_i, q_j)dx_1 \dots dx_n dq_1 \dots dq_N \\
&= \int \langle x|A^{(r)}|x\rangle dx_1 \dots dx_n \ .
\end{aligned}$$

(20.14.96)

But,

$$\begin{aligned}
& \langle x|A^{(r)}|x\rangle \\
&= \int \langle x|A^{(r)}|y\rangle dy_1 \dots dy_n dx_1 \dots dx_n dp_1 \dots dp_N \Psi^*(y_i, p_j)\Psi(x_i, p_j) \\
&= \int \langle x|A^{(r)}|y\rangle dy_1 \dots dy_n \langle y|\rho|x\rangle \ .
\end{aligned}$$

(20.14.97)

Therefore,

$$\begin{aligned}
\langle A^{(r)}\rangle &= \int dx_1 \dots dx_n \langle x|A^{(r)}|y\rangle dy_1 \dots dy_n \langle y|\rho|x\rangle \\
&= \mathrm{Tr}[A^{(r)}\rho] \ .
\end{aligned}$$

(20.14.98)

b) The normalization condition follows from the normalization of Ψ by simply putting $A^{(r)} = 1$ to get

$$1 = \int dx_1 \dots dx_n \langle x|\rho|x\rangle = \mathrm{Tr}[\rho] \ .$$

(20.14.99)

c) To get the equation of motion for the reduced density matrix we expand it in terms of the eigenfunctions $\phi_n(x)$ of the Hamiltonian $H^{(r)}$ for the reduced system. Thus, the time-dependent density matrix is

$$\langle x|\rho|y\rangle = \sum_{nm} a_{nm}\phi_n(x)\phi_m^*(y)\, e^{-i(E_n-E_m)t/\hbar} \ . \tag{20.14.100}$$

The coefficients a_{nm} give ρ in the energy representation. If we differentiate this equation with respect to time we find

$$
\begin{aligned}
i\hbar\langle x|\frac{\partial\rho}{\partial t}|y\rangle &= \sum_{nm} a_{nm}[E_n\phi_n(x)\phi_m^*(y)-\phi_n(x)E_m\phi_m^*(y)]\, e^{-i(E_n-E_m)t/\hbar}\\
&= \sum_{nm} a_{nm}\int [\phi_m^*(y)H\phi_n(x)-\phi_n(x)H\phi_m^*(y)]\, e^{-i(E_n-E_m)t/\hbar}\\
&= \sum_{nm} a_{nm}\int [\langle x|H|y'\rangle\phi_n(y')\phi_m^*(y)-\phi_n(x)\phi_m^*(y')\langle y'|H|y\rangle]\\
&\quad \times\ dy_1'\ldots dy_n'\, e^{-i(E_n-E_m)t/\hbar}\\
&= \int [\langle x|H|y'\rangle\rho(y',y)-\rho(x,y')\langle y'|H|y\rangle]dy_1'\ldots dy_n'\\
&= \langle x|H\rho-\rho H|y\rangle\ . \tag{20.14.101}
\end{aligned}
$$

Therefore the equation of motion for the reduced density matrix is

$$i\hbar\frac{\partial\rho}{\partial t}=[H,\rho]\ . \tag{20.14.102}$$

20.15   Density Matrix and S-Matrix

Suppose that the 2×2 S-matrix for scattering of a spin $1/2$ particle off a spin zero target is given by

$$S = f\mathbf{1}+g(\hat{n}\cdot\vec{\sigma})$$

where $\hat{n}$ is a unit vector and

$$|f|^2+|g|^2=1\ .$$

Find the polarization of a beam of particles if the incident beam is totally unpolarized. Notice that this S-matrix is not unitary.

Solution

Since the incident beam is unpolarized, it is described by the density matrix

$$\rho_i = \frac{1}{2}\mathbf{1}\ . \tag{20.15.103}$$

After the scattering the density matrix is

$$
\begin{aligned}
\rho_f &= S\rho_i S^\dagger \\
&= \frac{1}{2}SS^\dagger \\
&= \frac{1}{2}\mathbf{1} + \Re(f^*g)(\hat{n}\cdot\vec{\sigma}) \ .
\end{aligned}
\tag{20.15.104}
$$

Therefore, the polarization after the scattering is given by

$$
\vec{P} = \text{Tr}(\vec{\sigma}\rho_f) = \Re(f^*g)\hat{n} \ .
\tag{20.15.105}
$$

20.16   Zero Energy Bound States of Two Fermions

Two identical spin 1/2 fermions of mass m interact via the potential

$$
V(r) = V_1 + V_2\,\vec{\sigma}_1\cdot\vec{\sigma}_2
$$

where $\vec{\sigma}_1$ and $\vec{\sigma}_2$ are the Pauli matrices associated with the spins of particles 1 and 2 respectively and where

$$
V_1 = \begin{cases} -V & r < a \\ 0 & r > a \end{cases}
\qquad
V_2 = \begin{cases} -U & r < a \\ 0 & r > a \end{cases} \ .
$$

Here, V and U are positive constants. If this system is known to have a single zero energy $l = 0$ bound state and a single zero energy $l = 1$ bound state, find V and U.

Solution

For the $l = 0$ state we have a symmetric spatial wavefunction in the relative coordinates and therefore the spin state must be antisymmetric or $s = 0$. For the $l = 1$ state the spatial wavefunction is antisymmetric and therefore the spin wavefunction corresponds to $s = 1$. Now,

$$
\vec{s}^2 = (\vec{s}_1 + \vec{s}_2)^2
\tag{20.16.106}
$$

so that

$$
\vec{\sigma}_1\cdot\vec{\sigma}_2 = \frac{2}{\hbar^2}[\vec{s}^2 - \vec{s}_1^2 - \vec{s}_2^2] \ .
\tag{20.16.107}
$$

Thus, for the singlet state we get

$$
\vec{\sigma}_1\cdot\vec{\sigma}_2 = \frac{2}{\hbar^2}[0 - 3/4 - 3/4]\hbar^2 = -3 \quad \text{singlet state} \ .
\tag{20.16.108}
$$

For the triplet state we get

$$
\vec{\sigma}_1\cdot\vec{\sigma}_2 = \frac{2}{\hbar^2}[2 - 3/4 - 3/4]\hbar^2 = 1 \quad \text{triplet state} \ .
\tag{20.16.109}
$$

We first consider the singlet state. In this case we have that

$$V(r) = V_1 - 3V_2 = \begin{cases} -V + 3U & r < a \\ 0 & r > a \end{cases} \quad . \tag{20.16.110}$$

The Schrödinger equation for an $l = 0$ bound state with energy $E < 0$ becomes

$$-\frac{\hbar^2}{2m}\frac{1}{r}\frac{d^2(r\psi)}{dr^2} + (-V + 3U - E)\psi = 0 \quad r < a$$

$$-\frac{\hbar^2}{2m}\frac{1}{r}\frac{d^2(r\psi)}{dr^2} - E\psi = 0 \quad r > a . \tag{20.16.111}$$

The solution is

$$r\psi = \begin{cases} A\sin(kr) & r < a & k^2 = \frac{2m(V-3U+E)}{\hbar^2} \\ B\,e^{-\alpha r} & r > a & \alpha^2 = \frac{2m|E|}{\hbar^2} \end{cases} \quad . \tag{20.16.112}$$

Matching the logarithmic derivative at $r = a$ we get

$$k\cot(ka) = -\alpha . \tag{20.16.113}$$

So, if we let $|E| \to 0$ we find that either

$$k = 0 \quad \Rightarrow V - 3U = 0 \tag{20.16.114}$$

or

$$ka = (n + 1/2)\pi \quad \Rightarrow V - 3U = \frac{\hbar^2}{2ma^2}(n+1/2)^2\pi^2 . \tag{20.16.115}$$

The first case $V - 3U = 0$ is not possible since this would mean that for this situation there is no potential at all and hence there is no bound state. So,

$$V - 3U = \frac{\hbar^2\pi^2}{8ma^2} . \tag{20.16.116}$$

We now repeat the solution for the $l = 1$ case. Here the Schrödinger equation reads

$$-\frac{\hbar^2}{2m}\frac{1}{r}\frac{d^2(r\psi)}{dr^2} + \frac{2\hbar^2}{2mr^2}(r\psi) + (-V - U - E)\psi = 0 \quad r < a$$

$$-\frac{\hbar^2}{2m}\frac{1}{r}\frac{d^2(r\psi)}{dr^2} + \frac{2\hbar^2}{2mr^2}(r\psi) - E\psi = 0 \quad r > a . \tag{20.16.117}$$

The solution to this pair of equations is

$$r\psi = \begin{cases} A\frac{\sin(Kr)-Kr\cos(Kr)}{r} & r < a & K^2 = \frac{2m(V+U+E)}{\hbar^2} \\ B\left(\frac{1}{r}+\alpha\right)e^{-\alpha r} & r > a & \alpha^2 = \frac{2m|E|}{\hbar^2} \end{cases} \quad . \tag{20.16.118}$$

Again matching the logarithmic derivatives at $r = a$ we get

$$\frac{(K^2a^2 + 1)\sin(Ka) - Ka\cos(Ka)}{Ka\cos(Ka) - \sin(Ka)} = -\frac{\alpha^2a^2 + \alpha a + 1}{\alpha a + 1} . \tag{20.16.119}$$

Now, letting $|E| \to 0$ we find that the right hand side goes to -1 so that the equation for the energy becomes

$$(K^2 a^2 + 1)\sin(Ka) - Ka\cos(Ka) = -Ka\cos(Ka) + \sin(Ka) \ . (20.16.120)$$

Hence, we find

$$K^2 a^2 \sin(Ka) = 0 \qquad\qquad (20.16.121)$$

so that $Ka = 0$, which is impossible since both U and V are positive, or else

$$Ka = p\pi \quad \Rightarrow U + V = \frac{\hbar^2}{2ma^2}p^2\pi^2 \ . \qquad\qquad (20.16.122)$$

So,

$$U + V = \frac{\hbar^2\pi^2}{2ma^2} \ . \qquad\qquad (20.16.123)$$

Combining this with the result for the singlet case (20.16.115) we get

$$U = \frac{3\hbar^2\pi^2}{32ma^2} \quad , \quad V = \frac{13\hbar^2\pi^2}{32ma^2} \ . \qquad\qquad (20.16.124)$$

20.17 Bose Number Operator: Constant of Motion

Show that for a system of bosons with the Hamiltonian

$$H = \sum_{\vec{k}} E(\vec{k})a_{\vec{k}}^\dagger a_{\vec{k}} + \lambda \sum_{\vec{j}\vec{k}\vec{l}\vec{n}} a_{\vec{j}}^\dagger a_{\vec{k}}^\dagger a_{\vec{l}} a_{\vec{n}} \delta_{\vec{j}+\vec{k}} \delta_{\vec{l}+\vec{n}} V(\vec{j} - \vec{n})$$

the number operator

$$N = \sum_{\vec{p}} a_{\vec{p}}^\dagger a_{\vec{p}}$$

is a constant of the motion.

Solution

We first compute

$$\begin{aligned}
[a_{\vec{p}}^\dagger a_{\vec{p}}, a_{\vec{k}}^\dagger a_{\vec{k}}] &= [a_{\vec{p}}^\dagger a_{\vec{p}}, a_{\vec{k}}^\dagger]a_{\vec{k}} + a_{\vec{k}}^\dagger[a_{\vec{p}}^\dagger a_{\vec{p}}, a_{\vec{k}}] \\
&= a_{\vec{p}}^\dagger a_{\vec{k}}\delta_{\vec{p},\vec{k}} - a_{\vec{k}}^\dagger a_{\vec{p}}\delta_{\vec{p},\vec{k}} \\
&= 0 \ .
\end{aligned} \qquad\qquad (20.17.125)$$

Next we compute

$$\begin{aligned}
[a_{\vec{p}}^\dagger a_{\vec{p}}, a_{\vec{k}}^\dagger a_{-\vec{k}}^\dagger a_{\vec{n}} a_{-\vec{n}}] &= a_{\vec{p}}^\dagger[a_{\vec{p}}, a_{\vec{k}}^\dagger a_{-\vec{k}}^\dagger]a_{\vec{n}} a_{-\vec{n}} + a_{\vec{k}}^\dagger a_{-\vec{k}}^\dagger[a_{\vec{p}}^\dagger, a_{\vec{n}} a_{-\vec{n}}]a_{\vec{p}} \\
&= a_{\vec{p}}^\dagger(a_{-\vec{k}}^\dagger\delta_{\vec{p},\vec{k}} + \delta_{\vec{p},-\vec{k}}a_{\vec{k}}^\dagger)a_{\vec{n}} a_{-\vec{n}} \\
&\quad - a_{\vec{k}}^\dagger a_{-\vec{k}}^\dagger(a_{-\vec{n}}\delta_{\vec{p},\vec{n}} + a_{\vec{n}}\delta_{\vec{p},-\vec{n}})a_{\vec{p}} \ . \qquad (20.17.126)
\end{aligned}$$

Now carrying out the sum over $\vec{p}$, $\vec{k}$, $\vec{n}$ we get

$$[N, \sum_{\vec{j}\vec{k}\vec{l}\vec{n}} a^\dagger_{\vec{j}} a^\dagger_{\vec{k}} a_{\vec{l}} a_{\vec{n}} \delta_{\vec{j}+\vec{k}} \delta_{\vec{l}+\vec{n}} V(\vec{j} - \vec{n})]$$

$$= \sum_{\vec{p}\vec{k}\vec{n}} \left[a^\dagger_{\vec{p}} a^\dagger_{-\vec{p}} (\delta_{\vec{p},\vec{k}} + \delta_{\vec{p},-\vec{k}}) a_{\vec{n}} a_{-\vec{n}} - a^\dagger_{\vec{k}} a^\dagger_{-\vec{k}} (\delta_{\vec{p},\vec{n}} + \delta_{\vec{p},-\vec{n}}) a_{\vec{p}} a_{-\vec{p}} \right] V(-\vec{k} - \vec{n})$$

$$= \sum_{\vec{k}\vec{n}} \left[2 a^\dagger_{\vec{k}} a^\dagger_{-\vec{k}} a_{\vec{n}} a_{-\vec{n}} - 2 a^\dagger_{\vec{k}} a^\dagger_{-\vec{k}} a_{\vec{n}} a_{-\vec{n}} \right] V(-\vec{k} - \vec{n})$$

$$= 0 .$$ (20.17.127)

Combining these results we have that

$$[N, H] = 0$$ (20.17.128)

which was to be shown.

20.18 Bose Operator: More Constants of Motion

Given the Hamiltonian

$$H = a^\dagger_{\vec{k}} a_{\vec{k}} + a^\dagger_{-\vec{k}} a_{-\vec{k}} + g \left((a_0)^2 a^\dagger_{\vec{k}} a^\dagger_{-\vec{k}} + (a^\dagger_0)^2 a_{\vec{k}} a_{-\vec{k}} \right)$$

show that
a) the operator

$$N = a^\dagger_0 a_0 + a^\dagger_{\vec{k}} a_{\vec{k}} + a^\dagger_{-\vec{k}} a_{-\vec{k}}$$

is a constant of the motion as well as that
b) the operator

$$D = a^\dagger_{\vec{k}} a_{\vec{k}} - a^\dagger_{-\vec{k}} a_{-\vec{k}}$$

is a constant of the motion.

Solution

a) If we remember that $\vec{k} \neq 0$ we see that

$$[a_0, a^\dagger_{\pm\vec{k}}] = 0 .$$ (20.18.129)

Also, for Bose operators we have that

$$[a^\dagger_{\vec{j}} a_{\vec{j}}, a_{\vec{j}}] = -a_{\vec{j}}$$

$$[a^\dagger_{\vec{j}} a_{\vec{j}}, a^\dagger_{\vec{j}}] = a^\dagger_{\vec{j}}$$ (20.18.130)

as well as

$$
\begin{aligned}
[a_0^\dagger a_0, (a_0)^2] &= -2(a_0)^2 \\
[a_0^\dagger a_0, (a_0^\dagger)^2] &= 2(a_0^\dagger)^2 \ .
\end{aligned}
\tag{20.18.131}
$$

It then follows that

$$
[a_0^\dagger a_0, H] = -2g\left\{ (a_0^\dagger)^2 a_{\vec{k}} a_{-\vec{k}} - a_{\vec{k}}^\dagger a_{-\vec{k}}^\dagger (a_0)^2 \right\}
\tag{20.18.132}
$$

as well as that

$$
[a_{\vec{k}}^\dagger a_{\vec{k}}, H] = g\left\{ (a_0^\dagger)^2 a_{\vec{k}} a_{-\vec{k}} - a_{\vec{k}}^\dagger a_{-\vec{k}}^\dagger (a_0)^2 \right\}
\tag{20.18.133}
$$

and, by symmetry, that also

$$
[a_{-\vec{k}}^\dagger a_{-\vec{k}}, H] = g\left\{ (a_0^\dagger)^2 a_{\vec{k}} a_{-\vec{k}} - a_{\vec{k}}^\dagger a_{-\vec{k}}^\dagger (a_0)^2 \right\} \ .
\tag{20.18.134}
$$

So, after combining these results we immediately see that

$$
[N, H] = 0 \ .
\tag{20.18.135}
$$

b) Furthermore, it is also immediately clear that

$$
[D, H] = 0 \ .
\tag{20.18.136}
$$

20.19   The Pauli Problem

a) Using the equation

$$
m = \frac{E}{c^2}
$$

calculate the total mass of one gram of electrons confined to a cube 1.0 cm on a side.

b) Repeat the calculation assuming that instead of electrons the particles are bosons with the same mass as the electrons.

Solution

a) This problem shows the dramatic effect of the Pauli exclusion principle. In one gram of electrons there are

$$
N = \frac{1.0}{9.1 \times 10^{-28}} = 1.1 \times 10^{27} \quad \text{electrons} \ .
$$

The energy levels for a particle in a cube of side a are

$$
E_k = \frac{\hbar^2 \pi^2}{2ma^2} k^2
\tag{20.19.137}
$$

where

$$
k^2 = k_x^2 + k_y^2 + k_z^2
\tag{20.19.138}
$$

and each of the k_x, k_y, k_z is an integer.

As a first step we need to evaluate the degeneracy of each level k. This means we need the number of different combinations of k_x, k_y, k_z such that (20.19.138) is satisfied for a given k. In general, with a few exceptions, this degeneracy is 3. Since the number of electrons is very large, we can take the degeneracy to be 3. This means that, according to the Pauli exclusion principle, we can place in each energy level

$$2 \times 3 = 6 \quad \text{electrons} .$$

The factor of 2 is due to the two possible spin values. To get the total energy of the N electrons we have to sum all the energies from $k = 1$ up to $k = N/6$. Thus, the total energy is

$$
\begin{aligned}
E &= \frac{\hbar^2 \pi^2}{2ma^2} \sum_{k=1}^{N/6} k^2 \\
&= \frac{\hbar^2 \pi^2}{2ma^2} \frac{(N/6)[(N/6)+1]}{2} \\
&\approx \frac{\hbar^2 \pi^2 N^2}{144 ma^2} .
\end{aligned}
\tag{20.19.139}
$$

Substituting in the values we get

$$E = \frac{(1.05 \times 10^{-27})^2 \times \pi^2 \times (1.1 \times 10^{-27})^2}{(144) \times (9.1 \times 10^{-28}) \times 1} \approx 10^{26} \quad \text{ergs} . \tag{20.19.140}$$

Now using the equation

$$m = \frac{E}{c^2}$$

we convert this to a mass and find

$$m = \frac{10^{26}}{9 \times 10^{20}} = 10^5 \quad \text{gm} = 100 \quad \text{kg!} . \tag{20.19.141}$$

b) If the particles were bosons they would all be in the ground state and the total energy would be simply

$$E = \frac{\hbar^2 \pi^2}{2ma^2} N \approx 14 \quad \text{ergs} . \tag{20.19.142}$$

In this case the additional mass would be a negligible

$$m = \frac{14}{9 \times 10^{20}} = 1.6 \times 10^{-20} \quad \text{gm} . \tag{20.19.143}$$

20.20   Atomic Isotope Effect

Every nucleus has a finite radius $R = r_0 A^{1/3}$ where

$$r_0 = 1.2 \times 10^{-13} \ \text{cm}$$

and A is the atomic number of the nucleus. Thus, the potential energy experienced by an electron near a nucleus is not simply

$$V(r) = -\frac{Ze^2}{r} \ .$$

If we assume that the charge density in the nucleus is constant then we have instead the potential energy

$$V(r) = \begin{cases} \frac{Ze^2}{R}\left[\frac{r^2}{2R^2} - \frac{3}{2}\right] & r \leq R \\ -\frac{Ze^2}{r} & r \geq R \end{cases} \ . \qquad (20.20.144)$$

a) Use perturbation theory to calculate the isotope shift (that is the dependence on A of the K-electron (1s state) for an atom with Z protons and atomic number A.

b) Use this result to compute the energy splitting for the K-electron between the heaviest lead ($Z = 82$) isotope $A = 214$ and the lightest $A = 195$. Neglect the presence of the other electrons.

Solution

a) The unperturbed Hamiltonian is

$$H_0 = \frac{\vec{p}^2}{2m} - \frac{Ze^2}{r} \ . \qquad (20.20.145)$$

The perturbation is

$$\begin{aligned} H' &= V(r) - \left(-\frac{Ze^2}{r}\right) \\ &= \begin{cases} \frac{Ze^2}{R}\left[\frac{r^2}{2R^2} - \frac{3}{2} + \frac{R}{r}\right] & r \leq R \\ 0 & r \geq R \end{cases} \end{aligned} \qquad (20.20.146)$$

The unperturbed ground state energy of the K-electron is

$$E_0^{(0)} = -\frac{1}{2}\frac{Ze^2}{a/Z} \qquad (20.20.147)$$

where $a = 5.292 \times 10^{-9}$ cm is the Bohr radius. The corresponding wavefunction is

$$\psi_0^{(0)}(r) = \frac{1}{\sqrt{8\pi}}\left(\frac{2Z}{a}\right)^{3/2} e^{-Zr/a} \ . \qquad (20.20.148)$$

The first order correction to $E_0^{(0)}$ is given by

$$E_0^{(1)} = (\psi_0^{(0)}, H' \psi_0^{(0)}) \ . \tag{20.20.149}$$

Thus,

$$E_0^{(1)} = \frac{1}{2} \left(\frac{2Z}{a} \right)^3 \frac{Ze^2}{R} \int_0^R e^{-2Zr/a} \left[\frac{r^2}{2R^2} - \frac{3}{2} + \frac{R}{r} \right] r^2 \, dr \ . \tag{20.20.150}$$

We now let

$$\alpha = \frac{2ZR}{a} \qquad x = \frac{2Zr}{a} \ . \tag{20.20.151}$$

Then,

$$
\begin{aligned}
E_0^{(1)} &= \frac{1}{2} \frac{Ze^2}{R} \int_0^\alpha e^{-x} \left[\frac{x^2}{2\alpha^2} - \frac{3}{2} + \frac{\alpha}{x} \right] x^2 \, dx \\
&= \frac{1}{2} \frac{Ze^2}{a/Z} \frac{2}{\alpha} \left[\frac{12}{\alpha^2} - 3 + \alpha - e^{-\alpha} \left(\frac{12}{\alpha^2} + \frac{12}{\alpha} + 3 \right) \right] \ . \tag{20.20.152}
\end{aligned}
$$

If we now make the dependence on the atomic number A explicit by writing

$$\alpha = \frac{2Zr_0}{a} A^{1/3} = \gamma A^{1/3} \tag{20.20.153}$$

we have the desired dependence on A.

$$
\begin{aligned}
E_0^{(1)} &= \frac{1}{2} \frac{Ze^2}{a/Z} \frac{2}{\gamma} A^{-1/3} \left[\frac{12}{\gamma^2} A^{-2/3} - 3 + \gamma A^{1/3} \right. \\
&\quad - \left. e^{-\gamma A^{1/3}} \left(\frac{12}{\gamma^2} A^{-2/3} + \frac{12}{\gamma} A^{-1/3} + 3 \right) \right] \ . \tag{20.20.154}
\end{aligned}
$$

b) If we take $Z = 82$ and $A = 195$ we get that $\alpha = 0.238$. Substituting these values we find that

$$E_0^{(1)}(A = 195) = \frac{1}{2} \frac{Ze^2}{a/Z} \times 9.91 \times 10^{-3} \ . \tag{20.20.155}$$

Similarly, for $Z = 82$ and $A = 214$ we get that $\alpha = 0.245$. Thus, repeating the calculation we find that in this case

$$E_0^{(1)}(A = 214) = \frac{1}{2} \frac{Ze^2}{a/Z} \times 1.08 \times 10^{-2} \ . \tag{20.20.156}$$

Thus, recalling that

$$\frac{1}{2} \frac{e^2}{a} = 13.6 \ \text{eV}$$

the energy difference between the two isotopes is

$$\Delta E = \frac{1}{2} \frac{Ze^2}{a/Z} \times 9.4 \times 10^{-4} = 86 \ \text{eV} \ . \tag{20.20.157}$$

Bibliography

[20.1] A.Z. Capri, *Nonrelativistic Quantum Mechanics* 3rd edition, World Scientific Publishing Co. Pte. Ltd., section 20.10, (2002) .

Chapter 21

Quantum Statistical Mechanics

21.1   Average Energy of Assembly of SHO's

Compute the average energy of an assembly of identical simple harmonic oscillators using:
a) the microcanonical ensemble
b) the canonical ensemble
c) the grand canonical ensemble .

Solution

a) The microcanonical ensemble is given by

$$\rho = \sum_{E < E_n < E+\Delta} |n\rangle\langle n| . \tag{21.1.1}$$

Let

$$E = (N_1 + \frac{1}{2})\hbar\omega \quad , \quad E + \Delta = (N_2 + \frac{1}{2})\hbar\omega . \tag{21.1.2}$$

Then we have

$$\rho = \sum_{n=N_1}^{N_2} |n\rangle\langle n| . \tag{21.1.3}$$

So,

$$
\begin{aligned}
\langle\langle E \rangle\rangle &= \frac{\sum_{n=N_1}^{N_2} E_n}{\sum_{n=N_1}^{N_2} 1} = \frac{\sum_{n=N_1}^{N_2} (n+1/2)\hbar\omega}{N_2 + 1 - N_1} \\
&= \frac{\hbar\omega}{N_2 + 1 - N_1} \left[\frac{N_2(N_2+1)}{2} - \frac{(N_1-1)N_1}{2} + \frac{N_2 + 1 - N_1}{2} \right]
\end{aligned}
$$

$$= \frac{\hbar\omega}{2}\left[1 + \frac{(N_2 + N_1)(N_2 - N_1 + 1)}{N_2 + 1 - N_1}\right]$$

$$\approx \frac{\hbar\omega}{2}[1 + N_2 + N_1] = E + \frac{\Delta}{2} . \tag{21.1.4}$$

b) For a single harmonic oscillator the partition function is given by $\mathrm{Tr}(e^{-\beta H})$ where

$$H = \hbar\omega(a^\dagger a + 1/2) . \tag{21.1.5}$$

This gives

$$
\begin{aligned}
Z_1 &= \sum_{n=0}^{\infty} e^{-\beta\hbar(n+1/2)} \\
&= \frac{e^{-\beta\hbar\omega/2}}{1 - e^{-\beta\hbar\omega}} \\
&= \frac{1}{2\sinh(\beta\hbar\omega/2)} .
\end{aligned} \tag{21.1.6}
$$

Next we have that

$$Z_N = (Z_1)^N = [2\sinh(\beta\hbar\omega/2)]^{-N} . \tag{21.1.7}$$

Then,

$$
\begin{aligned}
U &= \langle\langle E\rangle\rangle = -\frac{\partial}{\partial\beta}\ln Z_N \\
&= N\frac{\partial}{\partial\beta}\ln[2\sinh(\beta\hbar\omega/2)] \\
&= N\frac{\hbar\omega}{2}\coth(\beta\hbar\omega/2) .
\end{aligned} \tag{21.1.8}
$$

c) In this case we simply compute

$$
\begin{aligned}
Z &= \sum_{N=0}^{\infty} z^N Z_N = \sum_{N=0}^{\infty}\left(\frac{z}{2\sinh(\beta\hbar\omega/2)}\right)^N \\
&= \frac{2\sinh(\beta\hbar\omega/2)}{2\sinh(\beta\hbar\omega/2) - z} .
\end{aligned} \tag{21.1.9}
$$

Then,

$$
\begin{aligned}
U &= \langle\langle E\rangle\rangle = -\frac{\partial}{\partial\beta}\ln Z_G(\beta, z) \\
&= \frac{\hbar\omega}{2}\frac{z\coth(\beta\hbar\omega/2)}{2\sinh(\beta\hbar\omega/2) - z} .
\end{aligned} \tag{21.1.10}
$$

But we also have that

$$N = z\frac{\partial}{\partial z}\ln Z_G(\beta, z) = \frac{z}{2\sinh(\beta\hbar\omega/2) - z} . \tag{21.1.11}$$

Therefore,

$$U = N\frac{\hbar\omega}{2}\coth(\beta\hbar\omega/2) \qquad (21.1.12)$$

which is the same as the result for the canonical ensemble.

21.2 Properties of the Density Matrix

Prove the following properties of a density matrix.

a)

$$\rho^2 \leq \rho .$$

This implies $\rho \geq 0$.

b)

$$\text{Tr}([\rho, A]) = 0 .$$

To see that this is not trivial, consider $\text{Tr}([x, p])$ and discuss a necessary condition on the operator A for this to hold.

Solution

a) To prove

$$\rho^2 \leq \rho \qquad (21.2.13)$$

we begin with ρ in diagonal form. This is always possible since ρ is self-adjoint. In this case we have

$$\rho = \sum_n |n\rangle p_n \langle n| \qquad (21.2.14)$$

with

$$\sum_n p_n = 1 \qquad p_n \geq 0 . \qquad (21.2.15)$$

Now, using completeness we immediately get

$$\rho^2 = \sum_n |n\rangle p_n^2 \langle n| \leq \sum_n |n\rangle p_n \langle n| = \rho . \qquad (21.2.16)$$

b) For this part we simply write out the commutator for a general matrix element and sum over the diagonal elements

$$\sum_m \langle m|[\rho, A]|m\rangle = \sum_{m,n} \{\langle m|n\rangle p_n \langle n|A|m\rangle - \langle m|A|n\rangle p_n \langle n|m\rangle\} = 0 . (21.2.17)$$

Thus, as required,

$$\text{Tr}[\rho, A] = 0 . \qquad (21.2.18)$$

This result relied heavily on the fact that the density matrix is a *bounded* operator, that is, its eigenvalues have a finite bound. Clearly the commutator of the two unbounded operators x and p is $i\hbar$ and its trace blows up. This is the case for all unbounded operators since their eigenvalues are not bounded and therefore neither is the sum of their eigenvalues. So, a necessary condition is that one of the operators be bounded.

21.3   Expectation Values for Spin

In a gas of electrons, a fraction p is known to have their z-component of spin in the up direction. Assume the remainder are random with equal probability for up and down.

a) What is the average value of s_x, s_y, and s_z?

b) If nothing is known about the spins of the remaining fraction $1-p$ of electrons what are the maximum possible values of $\langle\langle s_x \rangle\rangle$, $\langle\langle s_y \rangle\rangle$ and $\langle\langle s_x \rangle\rangle$?

Solution

a) Using the given information we have that

$$\rho = p|+\rangle\langle+| + (1-p)\left[\frac{1}{2}|+\rangle\langle+| + \frac{1}{2}|-\rangle\langle-|\right] = \frac{1}{2}[1 + p\sigma_3] . \qquad (21.3.19)$$

Therefore, we find

$$\langle\langle s_x \rangle\rangle = \mathrm{Tr}(\rho s_x) = \frac{\hbar}{4}\mathrm{Tr}(\sigma_1 + p\sigma_1\sigma_3) = 0$$

$$\langle\langle s_y \rangle\rangle = \mathrm{Tr}(\rho s_y) = \frac{\hbar}{4}\mathrm{Tr}(\sigma_2 + p\sigma_2\sigma_3) = 0$$

$$\langle\langle s_z \rangle\rangle = \mathrm{Tr}(\rho s_z) = \frac{\hbar}{4}\mathrm{Tr}(\sigma_3 + p\sigma_3\sigma_3) = \frac{p}{2}\hbar . \qquad (21.3.20)$$

b) In this case we we can only write

$$\rho = p|+\rangle\langle+| + (1-p)\rho' . \qquad (21.3.21)$$

Then we have that

$$\langle\langle s_x \rangle\rangle = \mathrm{Tr}(\rho s_x) = \frac{\hbar}{2}(1-p)\mathrm{Tr}(\rho'\sigma_1) \qquad (21.3.22)$$

is maximized when

$$\mathrm{Tr}(\rho'\sigma_1) = 1 . \qquad (21.3.23)$$

Then the maximum value is $(\hbar/2)(1-p)$. This requires that

$$\rho' = \begin{pmatrix} \alpha & \beta \\ \beta^* & 1-\alpha \end{pmatrix} \qquad (21.3.24)$$

where

$$\beta + \beta^* = 1 \quad , \quad 0 \le \alpha \le 1.$$
(21.3.25)

Similarly we get that

$$\langle\langle s_y \rangle\rangle = \text{Tr}(\rho s_y) = \frac{\hbar}{2}(1-p)\text{Tr}(\rho' \sigma_2)$$
(21.3.26)

is maximized when

$$\text{Tr}(\rho' \sigma_2) = 1 \quad .$$
(21.3.27)

Then the maximum value is $(\hbar/2)(1-p)$. This requires that

$$\rho' = \begin{pmatrix} \alpha & \beta \\ \beta^* & 1-\alpha \end{pmatrix}$$
(21.3.28)

where

$$\beta - \beta^* = -i \quad , \quad 0 \le \alpha \le 1.$$
(21.3.29)

Finally we have that

$$\langle\langle s_z \rangle\rangle = \text{Tr}(\rho s_z) = p\frac{\hbar}{2} + \frac{\hbar}{2}(1-p)\text{Tr}(\rho' \sigma_3)$$
(21.3.30)

is maximized when

$$\text{Tr}(\rho' \sigma_3) = 1 \quad .$$
(21.3.31)

Then the maximum value is $\hbar/2$. This requires that

$$\rho' = \begin{pmatrix} 1 & \beta \\ \beta^* & 0 \end{pmatrix}$$
(21.3.32)

and

$$\rho = p \begin{pmatrix} 1 & 0 \\ 0 & 0 \end{pmatrix} + (1-p)\rho' = \begin{pmatrix} 1 & (1-p)\beta \\ (1-p)\beta^* & 0 \end{pmatrix} \quad .$$
(21.3.33)

21.4   Expectation Value for Number of Particles

Verify the equation

$$N = \langle\langle \mathsf{N} \rangle\rangle = z\frac{\partial}{\partial z}\ln Z_G$$

where Z_G is the grand canonical partition function and z is the fugacity.

Solution

To verify

$$N = \langle\langle \mathsf{N} \rangle\rangle = z\frac{\partial}{\partial z} \ln Z_G \ , \tag{21.4.34}$$

we begin with

$$N = \frac{\text{Tr}(\mathsf{N}\rho)}{\text{Tr}(\rho)} \tag{21.4.35}$$

where

$$\rho = z^N\, e^{-\beta H} \tag{21.4.36}$$

and

$$Z_G = \text{Tr}\rho \ . \tag{21.4.37}$$

Then,

$$
\begin{aligned}
z\frac{\partial}{\partial z} \ln Z_G &= z\frac{\partial}{\partial z} \ln\left[\text{Tr}\left(z^N e^{-\beta H}\right)\right]\\
&= \frac{\text{Tr}\left(\mathsf{N}z^N e^{-\beta H}\right)}{\text{Tr}(\rho)}\\
&= \frac{\text{Tr}(\mathsf{N}\rho)}{\text{Tr}(\rho)}
\end{aligned}
\tag{21.4.38}
$$

as required.

21.5 Spin 1/2 Polarization

a) Show that any pure state of spin 1/2 is completely polarized.
b) Find the direction of polarization.

Solution

a) The most general pure spin 1/2 state is of the form

$$|\chi\rangle = a|1/2\rangle + b|-1/2\rangle \ , \quad |a|^2 + |b|^2 = 1 \ . \tag{21.5.39}$$

The polarization $\vec{P}$ for such a state is defined by

$$\vec{P} = \langle\chi|\vec{\sigma}|\chi\rangle \ . \tag{21.5.40}$$

In this case we find

$$\vec{P} = (a^*b + ab^*)\hat{e}_x + i(ab^* - a^*b)\hat{e}_y + (|a|^2 - |b|^2)\hat{e}_z \ . \tag{21.5.41}$$

It now follows that

$$\vec{P} \cdot \vec{P} = (a^*b + ab^*)^2 + (ab^* - a^*b)^2 + (|a|^2 - |b|^2)^2$$
$$= (|a|^2 + |b|^2)^2 = 1 . \tag{21.5.42}$$

Thus, the state is completely polarized.

b) If we write the polarization vector in terms of polar angles so that

$$\vec{P} = (\sin\theta\cos\varphi, \sin\theta\sin\varphi, \cos\theta) \tag{21.5.43}$$

we can read off the angles immediately.

$$\begin{aligned}
\sin\theta\cos\varphi &= 2\Re(ab^*) \\
\sin\theta\sin\varphi &= 2\Im(ab^*) \\
\cos\theta &= |a|^2 - |b|^2 .
\end{aligned} \tag{21.5.44}$$

This gives the direction of polarization.

21.6 Density Matrix for Spin s=1

Show that for spin $s = 1$, the density matrix can be completely specified by the unit matrix, the polarization vector $\vec{p}$, and the quadrupole polarization tensor Q_{ik} defined for spin j by

$$\vec{p} = \frac{\langle \vec{J} \rangle}{j\hbar}$$

$$Q_{ik} = \frac{\langle J_i J_k + J_k J_i \rangle}{j(j+1)\hbar^2} - \frac{2}{3}\delta_{ik} .$$

Hint: Show that the unit matrix $\mathbf{1}$, the polarization vector $\vec{p}$, and the quadrupole polarization tensor Q_{ik} form a basis for expanding any 3×3 matrix.

Solution

We first need to show that the unit matrix $\mathbf{1}$, the polarization vector $\vec{p}$, and the quadrupole polarization tensor Q_{ik} form a basis for the expansion of the density matrix. To do this we need 9 linearly independent matrices. We start by considering the 9 matrices $(1/\hbar^2)J_i J_j$ $i, j = 1, 2, 3$ where

$$\frac{J_1}{\hbar} = \frac{1}{\sqrt{2}}\begin{pmatrix} 0 & 1 & 0 \\ 1 & 0 & 1 \\ 0 & 1 & 0 \end{pmatrix} \quad \frac{J_2}{\hbar} = \frac{1}{\sqrt{2}}\begin{pmatrix} 0 & -i & 0 \\ i & 0 & -i \\ 0 & i & 0 \end{pmatrix}$$

$$\frac{J_3}{\hbar} = \begin{pmatrix} 1 & 0 & 0 \\ 0 & 0 & 0 \\ 0 & 0 & -1 \end{pmatrix} . \tag{21.6.45}$$

We break these nine matrices up as follows.

$$\frac{1}{\hbar^2}J_i J_j = \frac{1}{2\hbar^2}(J_i J_j + J_j J_i) + \frac{1}{2\hbar^2}(J_i J_j - J_j J_i)$$

$$= \frac{1}{2}\left[\frac{1}{\hbar^2}(J_i J_j + J_j J_i) - \frac{4}{3}\delta_{ij}\right] + \frac{2}{3}\delta_{ij} + \frac{i}{2\hbar^2}\epsilon_{ijk}J_k$$

$$= \frac{2}{3}\delta_{ij} + \frac{i}{2\hbar^2}\epsilon_{ijk}J_k + Q_{ij} \ . \tag{21.6.46}$$

Notice that all of these matrices, except the unit matrix, are traceless. Also, only five of the Q_{ij} can be independent since

$$Q_{11} + Q_{22} + Q_{33} = 0 \quad \text{and} \quad Q_{ij} = Q_{ji} \ . \tag{21.6.47}$$

To show that these 8 traceless matrices plus the unit matrix form a basis for 3×3 matrices we need to show that if

$$S = a_0 1 + \vec{a} \cdot \vec{J} + \sum_{ij} a_{ij}Q_{ij} = 0 \tag{21.6.48}$$

then,

$$a_0 = 0, \ \vec{a} = 0, \ a_{ij} = 0 \ .$$

To complete this proof we simply evaluate (by multiplying out the various matrices) the following traces.

$$\begin{aligned}
\text{Tr}(J_i^2) &= 2\hbar^2 \\
\text{Tr}(Q_{ii}^2) &= 2/3 \\
\text{Tr}(Q_{ij}^2) &= 1/2 \quad i \neq j \\
\text{Tr}(J_i J_j) &= 0 \quad i \neq j \\
\text{Tr}(J_k Q_{ij}) &= 0 \\
\text{Tr}(Q_{kl}Q_{ij}) &= 0 \quad i \neq k \ \text{or} \ j \neq l \ .
\end{aligned} \tag{21.6.49}$$

We now take the trace of (21.6.48) and find that this yields

$$3a_0 = 0 \ .$$

Next we multiply (21.6.48) from the left by J_k and take the trace to get

$$2a_k = 0 \ .$$

Finally we mutliply (21.6.48) from the left by Q_{kl} and take the trace to get

$$a_{ij} = 0 \ .$$

This shows that these 9 matrices are linearly independent. Thus, we can write

$$\rho = a_0 1 + a_1 \frac{J_1}{\hbar} + a_2 \frac{J_2}{\hbar} + a_3 \frac{J_3}{\hbar} +$$
$$+ \quad a_{11}Q_{11} + a_{12}Q_{12} + a_{13}Q_{13} + a_{22}Q_{22} + a_{23}Q_{23} \ . \tag{21.6.50}$$

Normalization requires that $a_0 = 1/3$. We now find that if we define

$$\vec{p} = \frac{\langle \vec{J} \rangle}{2\hbar} \tag{21.6.51}$$

then

$$p_x = a_1 \quad , \quad p_y = a_2 \quad , \quad p_z = a_3 . \tag{21.6.52}$$

Also the remaining five constants $a_{11}, \ldots, a_{23}$ are expressed in terms of the five linearly independent quantities

$$\langle Q_{ij} \rangle . \tag{21.6.53}$$

21.7   Polarization Vector for Spin j

Show that for the case of general spin j, if we again define the polarization vector by

$$\vec{p} = \frac{\langle \vec{J} \rangle}{j\hbar}$$

and are given a "hamiltonian"

$$H = -\gamma \vec{J} \cdot \vec{B}$$

where $\vec{B}$ is a magnetic field, then we have

$$\frac{d\vec{p}}{dt} = \gamma \vec{p} \times \vec{B}$$

$$\frac{d^2\vec{p}}{dt^2} = 0 .$$

Solution

This is a straightforward computation.

$$
\begin{aligned}
i\hbar \frac{d\vec{p}}{dt} &= \frac{1}{j\hbar} i\hbar \frac{d}{dt} \text{Tr} \left\{ \vec{J} \rho \right\} \\
&= \frac{1}{j\hbar} \text{Tr} \left\{ \vec{J} [H, \rho] \right\} \\
&= -\frac{\gamma}{j\hbar} \text{Tr} \left\{ \vec{J} (\vec{J} \cdot \vec{B}) \rho - \vec{J} \rho (\vec{J} \cdot \vec{J}) \right\} \\
&= -\frac{\gamma}{j\hbar} \text{Tr} \left\{ [\vec{J} (\vec{J} \cdot \vec{B}) - \vec{J} (\vec{J} \cdot \vec{J})] \rho \right\} \\
&= \frac{\gamma}{j\hbar} i\hbar \text{Tr} \left\{ (\vec{J} \times \vec{B}) \rho \right\} \\
&= i\hbar \gamma \vec{p} \times \vec{B} .
\end{aligned}
\tag{21.7.54}
$$

Therefore,

$$\frac{d\vec{p}}{dt} = \gamma \vec{p} \times \vec{B} .$$

(21.7.55)

It now follows that

$$
\begin{aligned}
\frac{d^2\vec{p}}{dt^2} &= \gamma \frac{d}{dt}\left(\vec{p} \times \vec{B}\right) \\
&= \gamma^2 \left(\vec{p} \times \vec{B}\right) \times \vec{B} = 0 .
\end{aligned}
$$

(21.7.56)

21.8   Composite Density Matrix

Show that if $\rho^{(1)}$ and $\rho^{(2)}$ are two density matrices and $\tilde{\rho}$ is defined by

$$\tilde{\rho}_{m,n;m',n'} = \rho^{(1)}_{m,m'}\rho^{(2)}_{n,n'}$$

(21.8.57)

then $\tilde{\rho}$ satisfies the general properties of a density matrix
a)

$$\tilde{\rho} = \tilde{\rho}^\dagger$$

b)

$$\mathrm{Tr}\tilde{\rho} = 1$$

c)

$$\tilde{\rho}^2 \le \tilde{\rho}$$

as well as the equations
d)

$$\rho^{(1)}_{m,m'} = \sum_n \tilde{\rho}_{m,n;m',n}$$

(21.8.58)

and

$$\rho^{(2)}_{n,n'} = \sum_m \tilde{\rho}_{m,n;m,n'} .$$

(21.8.59)

Solution

The density matrix $\tilde{\rho}$ is defined by

$$\tilde{\rho}_{m,n;m',n'} = \rho^{(1)}_{m,m'}\rho^{(2)}_{n,n'} .$$

(21.8.60)

a) It then follows that

$$
\begin{aligned}
(\tilde{\rho}^\dagger)_{m,n;m',n'} &= \tilde{\rho}^*_{m',n';m,n} \\
&= \rho^{(1)*}_{m',m}\rho^{(2)*}_{n',n} \\
&= (\rho^{(1)\dagger})_{m,m'}(\rho^{(2)\dagger})_{n,n'} \\
&= \rho^{(1)}_{m,m'}\rho^{(2)}_{n,n'} .
\end{aligned}
$$

(21.8.61)

b)

$$\mathrm{Tr}\tilde{\rho} = \sum_{mn} \tilde{\rho}_{m,n;m,n}$$

$$= \sum_{mn} \rho^{(1)}_{m,m}\rho^{(2)}_{n,n} = 1 .$$ (21.8.62)

c)

$$(\tilde{\rho})^2_{m,n;k,l} = \sum_{m'n'} \tilde{\rho}_{m,n;m',n'}\tilde{\rho}_{m',n';k,l}$$

$$= \sum_{m'n'} \rho^{(1)}_{m,m'}\rho^{(2)}_{n,n'}\rho^{(1)}_{m',k}\rho^{(2)}_{n',l}$$

$$= (\rho^{(1)})^2_{m,k}(\rho^{(2)})^2_{n,l} .$$ (21.8.63)

Therefore,

$$(\tilde{\rho})^2 = (\rho^{(1)})^2(\rho^{(2)})^2 \le \rho^{(1)}\rho^{(2)} = \tilde{\rho} .$$ (21.8.64)

d)

$$\sum_n \tilde{\rho}_{m,n;m',n} = \rho^{(1)}_{m,m'}\sum_n \rho^{(2)}_{n,n}$$

$$= \rho^{(1)}_{m,m'} .$$ (21.8.65)

Similarly,

$$\sum_m \tilde{\rho}_{m,n;m,n'} = \left(\sum_m \rho^{(1)}_{m,m}\right)\rho^{(2)}_{n,n'}$$

$$= \rho^{(2)}_{n,n'} .$$ (21.8.66)

21.9 Arbitrariness of Composite Density Matrix

Show that if we have

$$\rho^{(1)} = \alpha_1\sigma^{(1)} + \beta_1\tau^{(1)} , \quad \alpha_1 + \beta_1 = 1 , \quad \alpha_1 , \beta_1 > 0$$

$$\rho^{(2)} = \alpha_2\sigma^{(2)} + \beta_2\tau^{(2)} , \quad \alpha_2 + \beta_2 = 1 , \quad \alpha_2 , \beta_2 > 0$$

and

$$\mathrm{Tr}\sigma^{(i)} = \mathrm{Tr}\tau^{(i)} = 1 .$$

Then any combination

$$\tilde{\rho} = \alpha\sigma^{(1)}\otimes\sigma^{(2)} + \beta\sigma^{(1)}\otimes\tau^{(2)} + \gamma\tau^{(1)}\otimes\sigma^{(2)} + \delta\tau^{(1)}\otimes\tau^{(2)}$$

with

$$\alpha + \beta = \alpha_1 \quad \alpha + \gamma = \alpha_2$$

$$\gamma + \delta = \beta_1 \qquad \beta + \delta = \beta_2$$

satisfies

$$\rho^{(1)}_{m,m'} = \sum_n \tilde{\rho}_{m,n;m',n} \tag{21.9.67}$$

and

$$\rho^{(2)}_{n,n'} = \sum_m \tilde{\rho}_{m,n;m,n'} \tag{21.9.68}$$

and is a possible density matrix for the composite system. This establishes the necessity of the condition that $\rho^{(1)}$ and $\rho^{(2)}$ correspond to pure states in order that

$$\tilde{\rho}_{m,n;m',n'} = \rho^{(1)}_{m,m'} \rho^{(2)}_{n,n'} \tag{21.9.69}$$

give a unique solution for a density matrix for the composite system.

Solution

We have that

$$
\begin{aligned}
\tilde{\rho}_{m,n;m',n'} &= \alpha \sigma^{(1)}_{m,m'} \sigma^{(2)}_{n,n'} + \beta \sigma^{(1)}_{m,m'} \tau^{(2)}_{n,n'} \\
&+ \gamma \tau^{(1)}_{m,m'} \sigma^{(2)}_{n,n'} + \delta \tau^{(1)}_{m,m'} \tau^{(2)}_{n,n'} .
\end{aligned}
\tag{21.9.70}
$$

It now follows that

$$
\begin{aligned}
\sum_n \tilde{\rho}_{m,n;m',n} &= \alpha \sigma^{(1)}_{m,m'} + \beta \sigma^{(1)}_{m,m'} + \gamma \tau^{(1)}_{m,m'} + \delta \tau^{(1)}_{m,m'} \\
&= (\alpha + \beta) \sigma^{(1)}_{m,m'} + (\gamma + \delta) \tau^{(1)}_{m,m'} \\
&= \alpha_1 \sigma^{(1)}_{m,m'} + \beta_1 \tau^{(1)}_{m,m'} \\
&= \rho^{(1)}_{m,m'} .
\end{aligned}
\tag{21.9.71}
$$

In exactly the same fashion we find that

$$
\begin{aligned}
\sum_m \tilde{\rho}_{m,n;m,n'} &= \alpha \sigma^{(2)}_{n,n'} + \beta \tau^{(2)}_{n,n'} + \gamma \sigma^{(2)}_{n,n'} + \delta \tau^{(2)}_{n,n'} \\
&= (\alpha + \gamma) \sigma^{(2)}_{n,n'} + (\beta + \delta) \tau^{(2)}_{n,n'} \\
&= \alpha_2 \sigma^{(2)}_{n,n'} + \beta_2 \tau^{(2)}_{n,n'} \\
&= \rho^{(2)}_{n,n'} .
\end{aligned}
\tag{21.9.72}
$$

21.10   Two Energy Levels Bose Gas

Consider a "gas" of Bose particles with energy either $+E$ or $-E$. The Hamiltonian for this system is

$$H = E(a_2^\dagger a_2 - a_1^\dagger a_1)$$

where a_i , $a_i^\dagger$ ($i = 1, 2$) are the usual annihilation and creation operators for bosons.

a) Show that the canonical partition function is given by

$$Z_N = \frac{\sinh[\beta E(N + 1)]}{\sinh \beta E}$$

and that the grand canonical partition function is given by

$$Z_G = \left[1 - 2z \cosh \beta E + z^2\right]^{-1}$$

b) Compute the internal energy U, and the average number of particles $\langle\langle N \rangle\rangle$ and express U as a function of β and $\langle\langle N \rangle\rangle$ rather than as a function of β and z.

Solution

The Hamiltonian is

$$H = E(a_2^\dagger a_2 - a_1^\dagger a_1) = E(N_2 - N_1) \ . \tag{21.10.73}$$

a) To compute the canonical partition function we use the fact that the total number of particles is fixed. So,

$$N = N_1 + N_2 \quad \text{or} \quad N_1 = N - N_2 \tag{21.10.74}$$

and

$$H = E(2N_2 - N) \ . \tag{21.10.75}$$

So,

$$
\begin{aligned}
Z_N &= \text{Tr} e^{-\beta H} = \sum_{N_2=0}^{N} e^{-\beta(2N_2 - N)} \\
&= e^{\beta EN} \frac{1 - e^{-\beta 2E(N+1)}}{1 - e^{-\beta 2E}} \\
&= \frac{\sinh[\beta E(N + 1)]}{\sinh \beta E} \ .
\end{aligned}
\tag{21.10.76}
$$

For bosons we have

$$
\begin{aligned}
Z_G &= \prod_{k=1}^{2} \left(1 - z e^{-\beta E_k}\right)^{-1} \\
&= \left(1 - z e^{-\beta E}\right)^{-1} \left(1 - z e^{\beta E}\right)^{-1} \\
&= \left(1 - 2z \cosh \beta E + z^2\right)^{-1} \ .
\end{aligned}
\tag{21.10.77}
$$

An alternate way to compute Z_G is to use

$$
\begin{aligned}
Z_G &= \sum_{N=0}^{\infty} z^N Z_N \\
&= \frac{1}{2\sinh(\beta E)} \sum_{N=0}^{\infty} z^N \left(e^{\beta(N+1)} - e^{-\beta(N+1)} \right) \\
&= \frac{1}{2\sinh(\beta E)} \left(\frac{e^{\beta E}}{1 - z e^{\beta E}} - \frac{e^{-\beta E}}{1 - z e^{-\beta E}} \right) \\
&= \frac{1}{2\sinh(\beta E)} \left(\frac{1}{e^{-\beta E} - z} - \frac{1}{e^{\beta E} - z} \right) \\
&= \frac{1}{2\sinh(\beta E)} \frac{e^{\beta E} - e^{-\beta E}}{1 - z(e^{\beta E} + e^{-\beta E}) + z^2} \\
&= \left(1 - 2z\cosh(\beta E) + z^2 \right)^{-1} .
\end{aligned}
\tag{21.10.78}
$$

b) The internal energy is given by

$$
\begin{aligned}
U &= -\frac{\partial}{\partial \beta} Z_G(\beta, z) = \frac{\partial}{\partial \beta} \ln \left[1 - 2z\cosh\beta E + z^2 \right] \\
&= \frac{-2z E \sinh\beta E}{1 - 2z\cosh\beta E + z^2} .
\end{aligned}
\tag{21.10.79}
$$

The average number of particles is given by

$$
\begin{aligned}
N &= \langle\langle \mathsf{N} \rangle\rangle = z\frac{\partial}{\partial z} \ln Z_G(\beta, z) \\
&= \frac{2z\cosh\beta E - 2z^2}{1 - 2z\cosh\beta E + z^2} .
\end{aligned}
\tag{21.10.80}
$$

Therefore,

$$
U = -N \frac{-2E\sinh\beta E}{2\cosh\beta E - z} .
\tag{21.10.81}
$$

We can further solve for z in terms of N using the equation for N. This is simply a quadratic and yields

$$
z = \frac{1}{N+2} \left\{ (N+1)\cosh\beta E \pm \sqrt{(N+1)^2 \cosh^2\beta E - N(N+2)} \right\}.
\tag{21.10.82}
$$

Now

$$
z = e^{\beta N} \geq 1 .
\tag{21.10.83}
$$

Therefore we see that for $E \to 0$ we have to choose the $+$ sign in the above expression.

21.11   Density Matrix: Particles Coupled by Spring

Consider two point masses connected by a spring so that the Hamiltonian is

$$H = \frac{p_1^2}{2m_1} + \frac{p_2^2}{2m_2} + \frac{1}{2}k\,(x_1 - x_2)^2 \ . \tag{21.11.84}$$

Find the density matrix for the nth excited state

$$\rho_n\,(x_1, x_1'; x_2, x_2') = \psi_{K,n}(x_1, x_2)\psi_{K,n}(x_1', x_2')^* \tag{21.11.85}$$

where

$$H\psi_{K,n}(x_1, x_2) = \left[\frac{\hbar^2 K^2}{2(m_1 + m_2)} + (n+1/2)\hbar\omega\right]\psi_{K,n}(x_1, x_2') \tag{21.11.86}$$

and

$$\omega = \sqrt{k/m} \ , \quad m = \frac{m_1 m_2}{m_1 + m_2} \ . \tag{21.11.87}$$

Finally compute the reduced density matrix for particle 1 and show that it is not idempotent; i.e. that correlations between the two particles persist.

Solution

In the centre of mass system defined by

$$\begin{aligned} X &= \frac{m_1 x_1 + m_2 x_2}{M} \\ x &= x_1 - x_2 \end{aligned} \tag{21.11.88}$$

with

$$\begin{aligned} M &= m_1 + m_2 \\ m &= \frac{m_1 m_2}{m_1 + m_2} \end{aligned} \tag{21.11.89}$$

the Hamiltonian becomes

$$H\phi = -\frac{\hbar^2}{2M}\frac{\partial^2 \phi}{\partial X^2} - \frac{\hbar^2}{2m}\frac{\partial^2 \phi}{\partial x^2} + \frac{1}{2}m\omega^2 x^2 \phi \tag{21.11.90}$$

where

$$\omega = \sqrt{k/m} \ . \tag{21.11.91}$$

The eigenfunctions for the nth excited state written in the centre of mass system are

$$\phi(x, X) = \frac{1}{2\pi}e^{iKX}\,u_n(x) \tag{21.11.92}$$

where

$$-\frac{\hbar^2}{2m}\frac{d^2 u_n(x)}{dx^2} + \frac{1}{2}m\omega^2 x^2 u_n(x) = (n+1/2)\hbar\omega \, u_n(x) \; . \qquad (21.11.93)$$

In the laboratory frame the wavefunction is

$$\psi(x_1, x_2) = \frac{1}{2\pi}\exp\left[iK\left(\frac{m_1}{M}x_1 + \frac{m_2}{M}x_2\right)\right] u_n(x_1 - x_2) \; . \qquad (21.11.94)$$

The density matrix is therefore given by

$$\rho(x_1, x_1'; x_2, x_2') = \frac{1}{(2\pi)^2}\exp\left[iK\left(\frac{m_1}{M}(x_1 - x_1') + \frac{m_2}{M}(x_2 - x_2')\right)\right]$$
$$\times u_n(x_1 - x_2)u_n(x_1' - x_2') \; . \qquad (21.11.95)$$

The reduced density matrix is

$$\rho_1(x_1, x_1') = \int \rho(x_1, x_1'; x_2, x_2) \, dx_2 = \frac{1}{(2\pi)^2}\exp\left[iK\left(\frac{m_1}{M}(x_1 - x_1')\right)\right]$$
$$\times \int u_n(x_1 - x_2), u_n(x_1' - x_2) \, dx_2 \; . \qquad (21.11.96)$$

If we set $x_2 = x_1' - z$ this can be rewritten as

$$\rho_1(x_1, x_1')$$
$$= -\frac{1}{(2\pi)^2}\exp\left[iK\left(\frac{m_1}{M}(x_1 - x_1')\right)\right] \int dz \, u_n(z)u_n(z + (x_1 - x_1')) \; . \qquad (21.11.97)$$

Since this is not just of the form

$$\frac{1}{(2\pi)^2}\exp\left[iK\left(\frac{m_1}{M}(x_1 - x_1')\right)\right] u_n(x_1)u_n(x_1') \qquad (21.11.98)$$

the density matrix ρ_1 is not idempotent and we see that correlations persist. To convince ourself of the fact that correlations persist we check that ρ_1 is not idempotent.

$$\int \rho_1(x_1, x_1')\rho_1(x_1', x_1'') \, dx_1' = \frac{1}{(2\pi)^4}\exp\left[iK\left(\frac{m_1}{M}(x_1 - x_1'')\right)\right]$$
$$\times \int dz \, u_n(z)u_n(z + (x_1 - x_1'))dz' \, u_n(z')u_n(z' + (x_1' - x_1'')) \, dx_1'$$
$$\neq -\frac{1}{(2\pi)^2}\exp\left[iK\left(\frac{m_1}{M}(x_1 - x_1'')\right)\right]$$
$$\times \int dz \, u_n(z)u_n(z + (x_1 - x_1''))$$
$$= \rho_1(x_1, x_1'') \; . \qquad (21.11.99)$$

21.12   Particles: Dissimilar, Bose, and Fermi

Two particles of mass m, confined to a cubical box of sides L, interact with each other via an interaction

$$V = V_0 \, \delta(\vec{r}_1 - \vec{r}_2)$$

where the particle positions are given by $\vec{r}_1$ and $\vec{r}_2$. Use first order perturbation theory to discuss the effect of this interaction on the energy levels of this system for the following cases.
a) The particles are distinguishable.
b) The particles are identical bosons with spin 0.
c) The particles are identical fermions with spin 1/2.

Solution

For the non-interacting particles the energies are just the sums of the individual energies for a particle in a box and so are given by

$$E_{\vec{n}_1, \vec{n}_2} = \frac{\hbar^2 \pi^2}{2mL^2}(\vec{n}_1^2 + \vec{n}_2^2) \tag{21.12.100}$$

where

$$\vec{n}_i = (n_{ix}, n_{iy}, n_{iz}) \quad i = 1, 2 \ .$$

The eigenfunctions for the two states are

$$u_{\vec{n}_i}(\vec{r}_i) = \left(\frac{2}{L}\right)^{3/2} \sin(n_{ix}\pi x/L) \, \sin(n_{iy}\pi y/L) \, \sin(n_{iz}\pi z/L) \ . \tag{21.12.101}$$

a) Distinguishable Particles
In this case the unperturbed wavefunctions are

$$\psi_{\vec{n}_1, \vec{n}_2}(\vec{r}_1, \vec{r}_2) = u_{\vec{n}_1}(\vec{r}_1)u_{\vec{n}_2}(\vec{r}_2) \ . \tag{21.12.102}$$

To simplify the computation we work with one coordinate at a time. The change in energy in first order perturbation theory is given by

$$E^{(1)}_{\vec{n}_1, \vec{n}_2} = (\psi_{\vec{n}_1, \vec{n}_2}, V\psi_{\vec{n}_1, \vec{n}_2}) \ . \tag{21.12.103}$$

Thus, we first compute

$$
\begin{aligned}
I_x &= \frac{2}{L} \int_0^L dx_1 \, dx_2 \, \sin^2(n_{1x}\pi x_1/L) \, \sin^2(n_{2x}\pi x_2/L)\delta(x_1 - x_2) \\
&= \frac{2}{L} \int_0^L dx \, \sin^2(n_{1x}\pi x/L) \sin^2(n_{2x}\pi x/L) \\
&= \frac{2}{L}\frac{1}{4} \int_0^L dx \, [1 - \cos(2n_{1x}\pi x/L)][1 - \cos(2n_{2x}\pi x/L)] \\
&= \frac{1}{4}(2 + \delta_{n_{1x}n_{2x}}) \ . \tag{21.12.104}
\end{aligned}
$$

So, for distinguishable particles the energy correction is given by

$$E^{(1)}_{\vec{n}_1,\vec{n}_2} = \frac{V_0}{64}(2 + \delta_{n_{1x}n_{2x}})(2 + \delta_{n_{1y}n_{2y}})(2 + \delta_{n_{1y}n_{2y}}) \ . \qquad (21.12.105)$$

b) If we have Bose particles of zero spin, the wavefunction is given by

$$\psi_{\vec{n}_1,\vec{n}_2}(\vec{r}_1,\vec{r}_2) = \frac{1}{\sqrt{2}}[u_{\vec{n}_1}(\vec{r}_1)u_{\vec{n}_2}(\vec{r}_2) + u_{\vec{n}_1}(\vec{r}_2)u_{\vec{n}_2}(\vec{r}_1)] \qquad (21.12.106)$$

The change in energy in first order perturbation theory is again given by

$$E^{(1)}_{\vec{n}_1,\vec{n}_2} = (\psi_{\vec{n}_1,\vec{n}_2}, V\psi_{\vec{n}_1,\vec{n}_2}) \ . \qquad (21.12.107)$$

After carrying out the integrations over the delta functions we get exactly double the integral in the previous case for each direction. Thus,

$$E^{(1)}_{\vec{n}_1,\vec{n}_2} = \frac{V_0}{8}(2 + \delta_{n_{1x}n_{2x}})(2 + \delta_{n_{1y}n_{2y}})(2 + \delta_{n_{1y}n_{2y}}) \ . \qquad (21.12.108)$$

c) In the case of fermions of spin 1/2 we have two possibilities.
1) If the particles are in a singlet state ($s = 0$) then the spatial wavefunction is as for the bose case above and we get

$$E^{(1)}_{\vec{n}_1,\vec{n}_2} = \frac{V_0}{8}(2 + \delta_{n_{1x}n_{2x}})(2 + \delta_{n_{1y}n_{2y}})(2 + \delta_{n_{1y}n_{2y}}) \ . \qquad (21.12.109)$$

2) If the particles are in a triplet state ($s = 1$) the spatial wavefunction is antisymmetric. In this case carrying out the delta function integrations causes each integrand to vanish and we get

$$E^{(1)}_{\vec{n}_1,\vec{n}_2} = 0 \ . \qquad (21.12.110)$$

21.13   A Three-Level Laser

Consider a three-level system with energies $E_3 > E_2 > E_1$ with populations N_3, N_2, N_1 respectively. If this system is irradiated with monochromatic radiation of frequency

$$\nu = \frac{E_3 - E_1}{h}$$

it is possible, at equilibrium, to populate level 3 higher than level 2 so that $N_3 > N_2$. This is called a population inversion and permits the possibility of a laser with the laser frequency

$$\nu_{laser} = \frac{E_3 - E_2}{h} \ .$$

Assume that the only radiation present is the pumping frequency

$$\nu = \frac{E_3 - E_1}{h}$$

and use the Einstein Coefficients A_{ij}, B_{ij} (see problem 1.8) to determine the ratio of N_3/N_2 as well as the ratio N_2/N_1.
Hint: Write out the equation for the rate of change of the three populations and impose the equilibrium conditions.

Solution

The rate of change of the population in level 3 is entirely due to emission of radiation. Thus,

$$\frac{dN_3}{dt} = -N_3 B_{13}\rho(\nu) - N_3(A_{23} + A_{13}) \ . \tag{21.13.111}$$

The first term on the right describes the *induced* transitions from level 3 to level 1, while the second term describes the *spontaneous* transitions from level 3 to level 2 as well as from level 3 to level 1. For level 2 we have only spontaneous transitions from level 3 to level 2 and from level 2 to level 1. Thus,

$$\frac{dN_2}{dt} = N_3 A_{32} - N_2 A_{21} \ . \tag{21.13.112}$$

Level 1 is only able to absorb radiation. Therefore,

$$\frac{dN_1}{dt} = N_1 B_{31}\rho(\nu) \ . \tag{21.13.113}$$

At equilibrium, we have that

$$\frac{dN_2}{dt} = 0 = N_3 A_{32} - N_2 A_{21} \tag{21.13.114}$$

so that

$$\frac{N_3}{N_2} = \frac{A_{21}}{A_{32}} \ . \tag{21.13.115}$$

Thus, if $A_{21} > A_{32}$ we have population inversion. Also, at equilibrium, the rate of change of population 3 equals the rate of change of population 1. Therefore,

$$N_3 B_{13}\rho(\nu) + N_3(A_{23} + A_{13}) = N_1 B_{31}\rho(\nu) \ . \tag{21.13.116}$$

From this we get that

$$\frac{N_3}{N_1} = \frac{B_{31}\rho(\nu)}{B_{13}\rho(\nu) + (A_{23} + A_{13})} \ . \tag{21.13.117}$$

Now, writing

$$\frac{N_2}{N_1} = \frac{N_3}{N_1}\frac{N_2}{N_3} \tag{21.13.118}$$

we find that

$$\frac{N_2}{N_1} = \frac{B_{31}\rho(\nu)A_{32}}{A_{21}[B_{13}\rho(\nu) + (A_{23} + A_{13})]} \ . \tag{21.13.119}$$

21.14   Integrals from Quantum Statistical Mechanics

In quantum statistical mechanics one encounters integrals of the form

$$I_{\pm}(k) = \int_0^{\infty} \frac{x^k \, dx}{e^x \pm 1} \; . \tag{21.14.120}$$

a) Show that $I_{+}(k)$ can be expressed in terms of $I_{-}(k)$.
b) Evaluate $I_{\pm}(k)$ in terms of the Riemann zeta function [21.1]

$$\zeta(k) = \sum_{n=1}^{\infty} \frac{1}{n^k} \; . \tag{21.14.121}$$

Solution

a) To express $I_{+}(k)$ in terms of $I_{-}(k)$ we use the identity

$$\frac{1}{e^x + 1} = \frac{1}{e^x - 1} - \frac{2}{e^{2x} - 1} \; . \tag{21.14.122}$$

Then,

$$I_{+}(k) = \int_0^{\infty} \frac{x^k \, dx}{e^x - 1} - 2 \int_0^{\infty} \frac{x^k \, dx}{e^{2x} - 1} \; . \tag{21.14.123}$$

The first integral is simply $I_{-}(k)$ and in the second integral we replace the integration variable x by $t = 2x$. Then this integral becomes $2^{-k} I_{-}(k)$. Thus, we have that

$$I_{+}(k) = (1 - 2^{-k}) I_{-}(k) \; . \tag{21.14.124}$$

b) To express $I_{-}(k)$ in terms of the Riemann zeta function we expand the denominator in the integrand in a binomial series.

$$
\begin{aligned}
I_{-}(k) &= \int_0^{\infty} \frac{x^k \, dx}{e^x - 1} \\
&= \sum_{n=1}^{n=\infty} \int_0^{\infty} x^k \, e^{-nx} \, dx \\
&= \Gamma(k+1) \sum_{n=1}^{\infty} \frac{1}{n^{k+1}} \\
&= \Gamma(k+1) \zeta(k+1) \; .
\end{aligned}
\tag{21.14.125}
$$

Bibliography

[21.1] E.T.Whittaker and G.N.Watson, *A Course of Modern Analysis - 4th Edition*, Chapter 13, Cambridge, (1963).

[21.2] Kerson Huang, *Statistical Mechanics*, John Wiley and Sons, Inc., (1963).

References

There are several books that deal with problems and solutions in quantum mechanics. The following is a fairly exhaustive list of such books presently available in English.

1. F. Constantinescu and E. Magyari, *Problems in Quantum Mechanics*, Pergamon Press (1971).

2. S. Flügge, *Practical Quantum Mechanics* I, Springer-Verlag, (1971).
   S. Flügge, *Practical Quantum Mechanics* II, Springer-Verlag, (1971).

3. I.I. Gol'dman and V.D. Krivchenkov, *Problems in Quantum Mechanics*, Addison-Wesley Publishing Co., (1961).

4. D. ter Haar, *Selected Problems in Quantum Mechanics*, Academic Press, (1964).

5. C.S. Johnson and L.G. Peedersen, *Problems and Solutions in Quantum Chemistry and Physics*, Addison-Wesley Publishing Co., (1974).

6. Lim, Yung-kuo, *Problems and Solutions on Quantum Mechanics: Major American Universities Ph.D. Qualifying Questions and Solutions*, World Scientific, (1998).

7. L.P. Kok and J. Visser, *Quantum Mechanics: Problems and their Solutions*, Coulomb Press Leyden, (1987).

8. H.A. Mavromatis, *Exercises in Quantum Mechanics, 2nd reveised edition*, Kluwer Academic Publishers, (1992).

9. G.L. Squires, *Problems in Quantum Mechanics with Solutions*, Cambridge University Press, (1995).

On quantum mechanics itself there are very many excellent texts on the market. A small selection of representative texts is listed below.

10. I. Bialnycki-Birula, M. Cieplak, J. Kaminsky, *Theory of Quanta*, Oxford University Press, (1992).

11. D. Bohm, *Quantum Theory*, Prentice-Hall, (1951).

12. A.Z. Capri, *Nonrelativistic Quantum Mechanics 3rd Edition*, World Scientific, (2002).

13. C. Cohen-Tannoudji, B. Diu & F. Laloe, *Quantum Mechanics*, John-Wiley and Hermann, (1977).

14. A.S. Davydov, *Quantum Mechanics*, Pergamon Press, (1965).

15. R.H. Dicke and J.P. Wittke, *Introduction to Quantum Mechanics*, Addison-Wesley Publishing Co., (1960).

16. P.A.M. Dirac, *The Principles of Quantum Mechanics*, Oxford University Press, (1958).

17. H. Friedrich, *Theoretical Atomic Physics - Second Edition*, Springer-Verlag, (1998).

18. S. Gasiorowicz, *Quantum Physics*, John Wiley and Sons, Inc., (1974).

19. L.D. Landau and E.M. Lifshitz, *Quantum Mechanics*, Pergamon Press, (1985).

20. P.M. Mathews and K. Venkatesan, *A Textbook of Quantum Mechanics*, Tata McGraw-Hill Publishing Co. Ltd., (1977).

21. E. Merzbacher, *Quantum Mechanics*, John Wiley and Sons, Inc., (1970).

22. A. Messiah, *Quantum Mechanics Vol.1 & Vol.2*, North-Holland Publishing Co., (1961).

23. M.M. Morrison, *Understanding Quantum Physics*, Prentice-Hall, (1990).

24. J.J. Sakurai, *Modern Quantum Mechanics*, Benjamin/Cummings, (1985).

25. L.I. Schiff, *Quantum Mechanics*, McGraw-Hill Book Co. Inc., (1968).

26. R. Shankar, *Principles of Quantum Mechanics*, Plenum Press, (1981).

Index

www.ingramcontent.com/pod-product-compliance
Ingram Content Group UK Ltd.
Pitfield, Milton Keynes, MK11 3LW, UK
UKHW030107110125
453466UK00021B/300

9 789810 246501